PROBABILITY, RANDOM SIGNALS, AND STATISTICS

A *Textgraph* with Integrated Software for Electrical and Computer Engineers

X. Rong Li

CRC Press
Boca Raton London New York Washington, D.C.

Library of Congress Cataloging-in-Publication Data

Li, X. Rong
Probability, random signals, and statistics : a textgraph with integrated software for electrical & computer engineers / X. Rong Li.
p. cm.
Includes bibliographical references and index.
ISBN 0-8493-0433-4 (alk. paper)
Signal processing—Statistical methods. 2. Stochastic processes. 3. Statistical communication theory. I. Title.
TK5102.9.L49 1999
621.382′2—dc21 99-25417
CIP

Direct all inquiries to CRC Press LLC, 2000 N.W. Corporate Blvd., Boca Raton, Florida 33431.

International Standard Book Number 0-8493-0433-4
Library of Congress Card Number 99-25417
Printed in the United States of America 2 3 4 5 6 7 8 9 0
Printed on acid-free paper

To Peizhu, Helen, and Linda

Contents

PREFACE

This book is a natural outgrowth of my notes for a senior undergraduate course on probability and random signals that I have been teaching twice a year over the past several years at the University of New Orleans and the University of Hartford. It is intended as an elementary introduction to the principles and applications of probability, random variables, random processes and signals, and statistics. The prerequisite is a basic knowledge of calculus, signals, and systems.

In short, the distinctive features of this book include

- Unique "textgraph" format
- Succinct and lucid presentation
- Well-integrated with companion software and computers
- Good balance between basic and more advanced topics
- 200 worked-out realistic examples from a myriad of engineering applications
- 50 computer exercises
- 40 self-test problems with complete solutions
- Chapter summary and requirements
- Valuable instructor's manual with 40 suggested exam problems, solutions to all homework problems and computer exercises, and all larger-type pages as transparencies.

The book is intended to be both instructor-friendly and student-friendly. It is written in an innovative format of "***textgraph***"[1] consisting of separate pages in fonts of different sizes, respectively:

- Larger-typeface pages form a relatively concise text for basic topics, written in a way suitable for use as viewgraphs/transparencies for classroom teaching directly.
- Pages with smaller type provide more detailed explanations, examples, and/or more advanced materials.

This format has several advantages. First, it provides a good balance between basic and more advanced topics such that the latter is covered without losing the emphasis on the former. This should be ideal for both average and good students. Secondly, the main task of the instructor of a course is converting the text into his or her own course notes, which requires a clear and concise presentation of carefully selected topics of the subject. The "textgraph" format makes

[1]The word "textgraph" first appeared in the graduate text *Estimation and Tracking: Principles, Techniques, and Software* by Y. Bar-Shalom and myself, which, however, differs in format from this one.

such a conversion unnecessary. In fact, it gives the instructor time to cover more material, to explain important concepts and to review examples in class. Thirdly, it helps the student see the forest rather than just the trees by eliminating the need to copy down large numbers of equations necessary in such a course. For probability and random signals a clear understanding of basic concepts and principles is far more important than mathematical skills and tricks, unlike some other subjects in mathematics, science, and engineering. The "textgraph" format frees the student from spending too much valuable classroom time on the nitty gritties of the mathematics involved.

As part of the requirement of a textgraph, the book was written in such a way that the material presented in the larger typeface is self-contained.

Another distinctive feature of the book is that it is well-integrated with a companion software package named P&R, found in the enclosed CD-ROM. Its updated version can be downloaded from the Downloads Section of the web site: **http://www.crcpress.com/**. Written in MATLAB[2] and running under a Windows environment, P&R is entirely menu-driven and extremely user-friendly. It does not require the knowledge of any programming language (including MATLAB). It is demonstrated via examples throughout the book that many practical problems can be solved easily using this package.

You have not really learned a theory unless you know how to use it. There is no doubt that the student can gain tremendously, particularly in terms of how to put the theory to use, with the help of carefully designed computer exercises. That is why some of the more recent texts on the same subject include the use of computers. The use of computers suggested in these texts requires some familiarity with a computer language, most often MATLAB. Most students in my class, however, are not skillful enough in MATLAB and as a result these computer exercises are too much of a burden. The companion software for this textgraph solves this problem well. It has been extremely well received by my students. For example, many of the more than 50 computer exercises included in this textgraph, especially the more practical ones, would require too much computer programming effort without the companion software. All the practical examples solved by P&R in the book can be repeated by the student *easily* and thus it helps the student get a real feel of how random problems are solved. In other words, the software makes it possible to include many practical examples, which are easily repeatable by the student, and many computer exercises which give the student practice in solving "real world" problems.

Engineering students love examples. As one of my students put it, "we learn the material through examples only." This is true even more so for a course on probability and random signals. Engineering probability is a subject in which the mathematics involved appears deceptively simple to students. The major difficulty is knowing which equations or formulas to apply for a particular problem. In addition to conveying a clear and good understanding of the basic concepts and principles as much as possible, such a know-how ("know-which," more precisely) is hard to teach other than through good examples. Many of my students over the years have complained that not enough *good* examples were included in the half-dozen texts I have tried. Of course, examples are such that for students "the more the better." In addition to the 40 self-test problems, I have integrated about 200 examples into this textgraph, each with a complete and detailed solution.

---

[2]MATLAB is the registrated trademark of The MathWorks, Inc., Natwick, MA, USA.

These examples were carefully made and selected to demonstrate not only the basic concepts and principles, but also how to put the theory to use, as well as how to stimulate the student's interest. They cover a wide variety of application areas, including communications, signal processing, radar technology, control systems, quality control, electric power, reliability, statistical inference, stochastic simulation, computer engineering, operations research, and everyday life problems.

Still another distinctive feature of the book is its succinctness. This is partly due to my writing style and partly due to the requirements of the viewgraph/transparency part of the "textgraph" format. Most students and myself do not like excess verbosity. A number of compact and unambiguous notations have been adopted, which helps to achieve the goal.

Guide to the Instructor

This book is intended to be extremely instructor-friendly. It includes many features that an instructor loves. Writing course notes, selecting and grading homework problems and computer exercises, and creating exam problems are three of the most time-consuming tasks for the instructor. An instructor can save a lot of time if he or she adopts this book. The "textgraph" format makes it unnecessary to prepare his or her own notes. Complete and detailed solutions of all problems and computer exercises at the end of each chapter are provided in the *Instructor's Manual* (ISBN 0-8493-0092-4), which makes the second task easier. What is more innovative is that the Instructor's Manual also includes 40 suggested exam problems, with complete and detailed solutions. The following two additional items are also included in the Instructor's Manual: (1) a floppy disk that updates the companion software to include menu-driven solutions of all computer exercises, including solutions to the exercises with parameters modified arbitrarily by the instructor; and (2) a copy of all the pages in the larger typeface of the book, which is provided to help the instructor make viewgraphs/transparencies directly.

The companion software is capable of the following tasks:

- Generate random numbers
- Plot probability density (or mass) functions
- Look up tables of probability distributions and percentiles
- Plot histograms and scatter diagrams of data
- Identify distributions of data
- Generate Gaussian processes, random sinusoids, and random pulse trains
- Estimate correlation functions and power spectra of random processes
- Generate system responses
- Some other miscellaneous tasks
- Solutions to all computer exercises (instructor's version only)

A more detailed description of the software can be found in Appendix A.

Each chapter is written in such a fashion that whenever possible more important and basic topics are covered first and the least important topics are put at the end.

Chapter 1 provides a general introduction to and a global picture of probability, random signals, and statistics. It explains what they are about and why they are important. It also tries to stimulate student interest.

Basic concepts of probability, along with the total probability theorem and Bayes' rule, are studied in Chapter 2. The material covered in this chapter forms the first hurdle the student must overcome.

Chapter 3 introduces the concept of a random variable, its distribution and expectation. It also includes the topic of random number generation, which allows the student to play with random numbers early on using the companion software and/or computers to gain better understanding of the material covered and to arm the student with a powerful tool for solving practical problems.

Chapter 4 extends the topics of Chapter 3 to two or more random variables. The emphasis is on the relation between random variables, which was unfortunately not treated well in most previous texts. I have tried hard to present a clear picture of such relations. The companion software should prove to be particularly useful for this chapter.

An introduction to elements of statistics is given in Chapter 5. This chapter was written in such a way that provides three possible levels of coverage that are self-contained, depending on how much statistics the instructor wants to cover in his or her course. I am aware of the fact that different instructors in different or the same schools may choose quite different levels of coverage of elements of statistics in such a course.

Chapter 6 presents the basic concepts of random processes and its time-domain analysis. This chapter is probably the most difficult one for the student. I have tried hard to reduce the difficulty as much as possible. For example, correlation functions are explained using common sense and sample correlation functions are studied through examples and computer exercises to provide the student with a real feel for what correlation functions are. The companion software should prove to be particularly useful for this chapter.

The power spectrum of a random process is covered in Chapter 7. The emphasis is on the interpretations of various components of a power spectrum and on white noise.

The response of a linear system to random input is the topic of Chapter 8. It is an application of the material covered in the previous chapters, in particular Chapters 6 and 7. It is also the heart of random signal processing.

A variety of optimal linear systems are treated in Chapter 9. It deals with the synthesis of linear systems that are optimal in some sense, as opposed to the analysis of a linear system with random input in Chapter 8.

Approximate teaching hours for each chapter

Chapter	1	2	3	4	5	6	7	8	9	1–9
Hour(s)	1	7	10	7	4	5	4	5	3	46

I have been able to cover most of the material presented in this book in a three-credit course in one semester. The following table lists the approximate hours per chapter for classroom teaching, assuming all materials presented using the larger-type pages are covered. For convenience, one hour stands for the actual 50 minutes of classroom teaching time such that a three-credit course has three hours each week.

It is clear from the above table that the material might be slightly more than what can be covered in a three-credit course in one semester of 14 weeks. Some topics may have to be skipped. For this purpose, the following table gives a breakdown of sections in terms of importance in my opinion.

Importance of each section

Most Important	*Important*	*Less Important*	*Least Important*
1.1, 1.2, 1.3, 1.4	1.6	1.5	
2.1, 2.2, 2.3, 2.4, 2.5, 2.6			
3.1, 3.2, 3.3, 3.4, 3.5, 3.7	3.8, 3.10	3.11, 3.9	3.6, 3.12
4.1, 4.2, 4.3, 4.4, 4.5, 4.6, 4.7	4.8, 4.9		4.10
5.1, 5.2	5.3, 5.4	5.5, 5.6, 5.7	
6.1, 6.2, 6.3, 6.4, 6.5, 6.8	6.6, 6.7	6.9	
7.1, 7.2, 7.3, 7.5	7.6	7.4	
8.2, 8.3, 8.5	8.1, 8.6	8.4	8.7
9.1, 9.2	9.3, 9.4		

A couple of times I have also used a draft of the book, along with supplementary material, for a graduate-level introductory course on engineering probability and random processes.

I look forward to receiving comments and/or complaints on this book. I can be reached at **xli@uno.edu** or at the Department of Electrical Engineering, University of New Orleans, Lakefront, New Orleans, LA 70148, USA.

Guide to the Student

This book is indended to be very friendly to you no matter if you are a good, average, or below average student. The "textgraph" format makes it crystal clear what is important and what is less important. If you are a below average student, you do not need to worry too much about the material covered by the smaller typeface, especially the more advanced topics. You should be able to get the most out of this book by studying the material presented in the larger type only given your disadvantageous background knowledge or limited study time. If, however, you are a good student, you should study the material written in the smaller face for better mastering of the subject matter. You should take a middle ground if you are an average student. For example, you should usually study the explanations and additional examples written in the smaller type but you may be better off skipping the more advanced topics.

The book includes many worked-out examples. Almost every example attempts to illustrate something important. You should study these examples well unless the corresponding section is not covered by your instructor.

The brief summary and basic requirements presented at the end of each chapter are intended to highlight the important concepts, topics, and/or results as simply as possible so that you can be more focused and you can check how well you understand the material covered in the chapter. I advise you to study the corresponding summary and requirements each time your instructor has finished a chapter or at least when you are preparing for an exam.

A number of well-designed and balanced self-test problems are provided at the end of each chapter, along with complete solutions. I have actually used most of these problems to test my students. You should try to solve these problems yourself in a limited time frame without referring to the solutions first. Then compare your answers with the solutions and see how well you did. Some of these problems might be slightly harder than your actual exam problems, obviously depending on your instructor.

The companion software is extremely easy to use and quite powerful in solving many problems of this course. I guarantee that you will be well rewarded if you spend time to play with it. The best starting point is to repeat the examples solved by the software presented in the book.

More challenging problems and examples are identified by a star $*$.

Acknowledgments

Many people have in different ways helped me in the process of writing this book, in particular my graduate students. I am greatly indebted to Chen He and Xiaorong (Sean) Zhi who helped me tremendously on the development of the companion software, so much so that it is as much theirs as it is mine. They also generated many of the high-quality figures. Chen He also helped me generate the data files and solutions of computer exercises. Sean Zhi also typed a few chapters of the first draft. Two more graduate students of mine also merit special thanks: Ning Li and Lei Lu for their help in typing some of the solutions to the homework problems and computer exercises and entering index items in the LaTeX format. I am also grateful to the many students who took my class and made various comments and complaints on the many versions of my notes, which grew into this book.

Dr. Guolin Pan at Analog Devices, Dr. Jean Dezert at French National Establishment for Aerospace Research (ONERA), and my colleague Dr. Wayne Barcelo at the University of New Orleans have reviewed most or all chapters of the book and made many valuable comments and suggestions. Dr. Louis Godbout at the University of Hartford lent his notes to me, which had an impact on the presentation of this book, in particular the selection of engineering examples. They all deserve my deep appreciation.

National Science Foundation has supported me over the years via several grants, including a Career award and a Research Initiation award. I take this opportunity to acknowledge my great gratitude to NSF and in particular to my program director, Dr. Kishan Baheti. I am also very grateful to Dr. Rabinder Madan at the Office of Naval Research, for his strong support of my research over the years. Without the above support I would not have had as many graduate students to help me on this book and the book might never have become a reality.

I also appreciate Nora Konopka, Sara Seltzer, and Suzanne Lassandro of CRC Press for their help during the preparation of the final draft and the production of the book.

I have mixed feelings in saying that I had no secretary to thank for help on this book. I myself typed the numerous drafts of the manuscript.

Finally, I am very happy that I have this opportunity to express in formal writing my love and gratitude to my wife, Peizhu, for her understanding, encouragement, and support throughout the years during my pursuit of academic excellence. My love goes also to our daughters, Helen and Linda. With them, my life is so much more cheerful and meaningful.

X. Rong Li
xli@uno.edu

1

INTRODUCTION

The most important questions of life are, for the most part, really only problems of probability.

Laplace

Life is a school of probability.

Walter Bagehot

Probability is the very guide of life.

Cicero

This chapter provides a general introduction to the topics covered in the book.

Main Topics

- What Is Randomness?
- What Are Probability and Random Processes?
- Why Study Probability and Random Processes?
- Key Features of the Book
- Rules for the Presentation

1.1 Randomness, Random Signals and Systems

The primary goal of this book is to study probabilistic tools and methods for solving various types of problems encountered in engineering that contain a degree of *randomness*, in particular, random signals and random systems. It is then natural to ask "What is randomness?" Loosely speaking, ***randomness*** simply means *not predictable with certainty*. For example, an event is random if and only if its outcome cannot be predicted with certainty. Otherwise, the event is ***deterministic***. Here are some examples:

- Your final grade of this course is random at this moment.
- Your body weight at a certain time next year is random now.
- The age of a friend whose birth date you do not know is random.

A ***signal*** is a *time function* that carries useful information for the problem under consideration. A ***random signal*** is a signal whose values cannot be predicted for sure. Examples include:

- Dow Jones Industrial Average Index, which jumps up and down with uncertainty although with a growing trend over time.
- Your grade point average at this university, whose future value cannot be predicted with certainty.
- The output voltage from a solar detector, which depends on weather conditions that are impossible to be exactly predicted.
- The bit stream through a digital communication channel, which fluctuates with time between 0 and 1 due to e.g., unpredictable interference.
- The exact position of a gas molecule in a box.

A random time function containing no useful information about the problem is known as ***noise***. Examples include:

- Thermal noise in an electronic circuit.
- Background hiss from a radio receiver.
- Picture interference ("snow") on a TV screen.
- Sea sounds in a sonar system.

A noise process may also be said to be a random signal (in its wide sense).

The state of a ***random system*** as a time function is random.

1.2 Probability and Random Processes

There are primarily two reasons for studying randomness.

- Many practical problems "in fact" involve random effects.
- Many practical problems are too complex to be described accurately in a deterministic manner and thus may be (artificially but) more effectively modeled as problems with random effects.

There is always a gap between mathematical models and reality. Real-world problems are usually neither completely deterministic nor truly random. What is important is that the model used is tractable and leads to satisfactory results. With this in mind, if all the primary driving forces of a phenomenon are known accurately, a deterministic model is usually good. If they are not known accurately or no primary driving forces exist, then a random model may be more appropriate.

There are two sciences that study random problems: probability and statistics. They are essential for solving problems involving randomness, in particular, problems with random signals and systems.

Probability and statistics can be viewed as an analysis-synthesis pair:

- ***Probability*** deals with topics of analyzing a random problem.
- ***Statistics*** studies various aspects of how to establish probabilistic models from observations of a random phenomenon.

In a narrower sense, probability deals with either a time-invariant problem involving randomness or a snapshot of a time-varying problem with randomness. Time-varying random problems are the topics of ***random processes***, which are more useful in engineering practice in a direct sense.

Probability is essential for the study of random processes because the theory of random processes is a generalization/extension of (and thus more complex than) probability. Thus, probability has to be studied prior to the more interesting topics of random processes. More relevant and realistic examples in engineering practice are given in a later part of the book although every effort is made to include good engineering examples and problems in the first part.

1.3 Typical Engineering Applications

- Communications
- Computer networks
- Decision theory and decision making
- Estimation and filtering
- Information processing
- Power engineering
- Quality control
- Reliability
- Signal detection
- Signal and data processing
- Stochastic systems

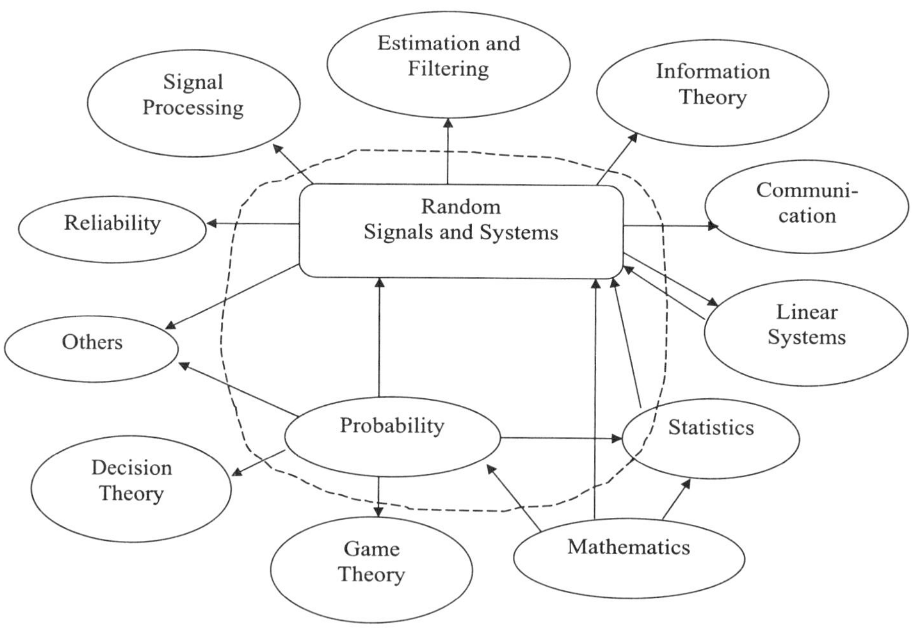

Figure 1.1: Coverage of this book and relation to some other subjects.

1.4 Why Study Probability and Random Processes?

Whereas a random phenomenon is not predictable with *certainty*, it usually contains some elements of predictability. In other words, it can usually be predicted to some degree, albeit not perfectly, as the examples in Section 1.1 suggested. The reason for this condition is that there are a variety of regularities, known as ***statistical regularities***, associated with a random phenomenon which exhibit themselves only over a large number of occurrences of such phenomena. Were this not true, the study of probability and random processes would gain nothing. In fact, the aim of the entire probability and random process theory is to study these regularities so as to make appropriate predictions of various aspects of a random phenomenon.

The theory of probability and random processes is abstract. It is abstract so that it can be applied to a wide spectrum of problems, making it extremely useful. This abstractness does not prevent us from learning the theory by taking either a top-down approach through numerous examples, as provided in this book, or a bottom-up approach by deriving various laws from the most fundamental ones. Because the theory is abstract, it is also easier and more economical to learn since no exact knowledge of any particular practical background field and no expensive or sophisticated facilities are needed. The prerequisites for this book are calculus and a basic knowledge of signals and systems.

The **importance** of the subject matter covered by this book:

- What you will learn will have many applications, not just for random signals and systems.
- This is the only course that is dedicated to random (uncertain) effects, while there are many courses dealing with nonrandom (deterministic) effects. You will be handicapped in engineering practice without mastering the material covered in this book.

Specifically, this book will

- Provide you with the appropriate concepts to help you understand random phenomena, in particular random signals and random systems.
- Provide you with the appropriate tools (theory, models, and techniques) and training to help you analyze problems and/or design systems with uncertainty.

1.5 Key Features of the Book

- *"Textgraph" Format.* This book is in an innovative form of "textgraph":
  - Larger typeface pages form a concise *text* for basic topics suitable for use as view*graphs*/transparencies for classroom teaching directly.
  - Smaller typeface pages provide more detailed discussions, examples, and more advanced materials.

  This format gives the instructor more time to explain concepts and review examples in class and frees the student from copying down large numbers of equations necessary in such a course. It also provides a good balance between basic and more advanced topics such that the latter is covered without losing the emphasis on the former.
- *Integration with Software.* The book is integrated closely with the companion software P&R. Written in MATLAB and running under Windows 95, 98, and NT, P&R is menu-driven and user-friendly. It does not require the knowledge of any programming language. It provides a useful tool for solving many problems. Many examples in the book are solved by using the software. The computer exercises are integrated particularly well with the software.
- *Many Realistic Examples.* A large number of realistic examples are provided with complete and detailed solutions. These examples were drawn from a wide variety of application areas, including communications, signal processing, radars, control systems, electric power, quality control, reliability, statistical inference, stochastic simulation, computer engineering, and operations research, as well as everyday life problems.
- *Self-Test Problems.* Self-test problems, along with complete and detailed solutions, are provided at the end of each chapter to help the reader test how well he or she understands the material and studies for exams.
- *Hands-On Emphasis.* Not only is the understanding of important concepts emphasized, special attention is also given early on to help and encourage the reader to apply what has been learned.
- *Clear Distinction of Importance.* Important concepts and results are written in bold typeface or boxed equations.
- *Summary and Requirements.* Each chapter includes a brief summary and specific requirements to help the reader focus on the more important material and to make it clear how thoroughly the material should be mastered.

1.6 Rules for the Presentation

- Text in the larger typeface presents more important and basic topics (or concepts, results, etc.), which are essential and should be mastered well.
- Text in the smaller typeface presents one or more of the following:
  - more detailed discussions
  - additional examples
  - more advanced topics
  - less important results

  They should be read (with less attention though), especially if the reader wants to have a better understanding of the subject matter.
- The most important terms (or results, concepts, etc.) are emphasized with words in ***italic bold*** face.
- Boxed equations are the most important ones.
- Numbered equations are either referenced elsewhere in the book or are more important than the ones without a number.
- (2.3)–(2.6) stands for Equation (2.3) *through* Equation (2.6).
- The following innovative shorthand notations are adopted:
  - (2.3) $-$ (2.6) stands for

    $$\text{LHS of (2.3)} - \text{LHS of (2.6)} \;=\; \text{RHS of (2.3)} - \text{RHS of (2.6)}$$

    where LHS = left-hand side, RHS $=$ right-hand side.
    Similar notations are used for other operations, such as $+, \times, /$, and expectation. For example, $\frac{d}{dx}$(2.3) stands for

    $$\frac{d}{dx}[\text{LHS of (2.3)}] \;=\; \frac{d}{dx}[\text{RHS of (2.3)}]$$
  - Others:

    $\overset{P1}{=}$ stands for "equal due to $P1$"

    $\overset{(2.3)}{=}$ stands for "equal due to Equation (2.3)"

    $\overset{?}{=}$ stands for "equal, but why?"

    $\overset{\text{shorthand}}{=}$ stands for "denoted in shorthand as"

    $\overset{\Delta}{=}$ stands for "equal by definition"

2

BASIC CONCEPTS IN PROBABILITY

We see that the theory of probability is at bottom only common sense reduced to calculation; it makes us appreciate with exactitude what reasonable minds feel by a sort of instinct, often without being able to account for it. ... It is remarkable that this science, which originated in the consideration of games of chance, should become the most important object of human knowledge.

Laplace

This chapter covers fundamental topics on probabilities of events.

Main Topics

- Basics of Set Theory
- Fundamental Concepts in Probability
- Conditional Probability
- Independent Events
- Total Probability Theorem and Bayes' Rule
- Combined Experiments and Bernoulli Trials

The materials covered in this chapter are essential for the study of the remaining chapters. Emphasis should be put on the understanding of concepts and how they can be applied.

2.1 Basics of Set Theory

2.1.1 Basic Definitions

- ***set*** = a collection of objects, denoted by an upper case Latin letter
  Example: $A = \{1, 4\}, \quad D = \{d_1, d_2, d_3\}$.
- ***element*** = an object in a set, denoted by a lower case Latin letter
  We say "a is an element of A," "a is in A," or "a belongs to A," denoted as $a \in A$.
- ***empty set*** = ***null set*** = a set with no elements, denoted by $\emptyset$
- ***space*** = the set with all the elements for the problem *under consideration* (sometimes called ***universal set***), denoted by S

Convention:

$$\text{Upper case Latin letter} = \text{set}$$
$$\text{Lower case Latin letter} = \text{element}$$

If *every* element of set A is also an element of set B, then A is said to be a ***subset*** of B, denoted as $A \subset B$ or $B \supset A$. Set A is said to be ***equal*** to set B if $A \subset B$ *and* $A \supset B$, denoted as $A = B$. In this case, A and B have exactly the same elements. Two sets are said to be ***disjoint*** if they do not have any element in common.

Example 2.1: Consider the set of all positive integers smaller than 7:

$$\begin{aligned} \text{rule method}: \quad & A = \{x : 0 < x < 7, x \text{ an integer}\} \\ \text{tabular method}: \quad & A = \{1, 2, 3, 4, 5, 6\} \\ & \quad = \text{the space (universal set) of a 6-face die} \end{aligned}$$

Tabular form is not universally applicable.

Example 2.2: Consider the set of all positive numbers smaller than 6:

$$\text{rule method}: \quad B = \{x : 0 < x < 6, x \text{ a real number}\}$$

There is no tabular form for this set because it is *uncountable*.

Example 2.3: Consider the set of all positive integers:

$$C = \{x : x > 0, x \text{ an integer}\} = \{1, 2, 3, \ldots\}$$

Example 2.4: The set of human genders G = {female, male}

2.1 Basics of Set Theory

Set theory is the foundation of the entire modern mathematics. This is even more true for probability theory. Manipulation (i.e., calculation) of probability is based on set operation. The well-known Russian mathematician A. N. Kolmogorov recognized this and developed the prevailing axiomatic system of probability. It is one of the most important achievements in probability theory.

Two sets A and B are said to be ***equivalent*** if there is a ***one-to-one correspondence*** between (the elements of) A and B. For example, set $G = \{\text{male, female}\}$ is equivalent to set $\{0, 1\}$, which is equivalent to $\{1, 2\}$.

A set is ***finite*** if it has finitely many elements. It is equivalent to the set $\{1, 2, \ldots, n\}$ for some natural number (i.e., positive integer) n. For example, set A of Example 2.1 is finite; set G of Example 2.4 is also equivalent to set $\{1, 2\}$ and is finite.

A set is ***infinite*** if it has infinitely many elements. An infinite set that is equivalent to the set of all natural numbers is said to be *denumerable.* A finite, denumerable, or empty set is said to be ***countable***; otherwise, it is an ***uncountable*** set. A denumerable set is said to be "countable" because there is a *definite and explicit* way by which all the elements of the set can be counted (without missing) provided that the counting process goes on (for ever if necessary). For an uncountable set, no such way exists. An uncountable set has to have infinitely many elements. However, not every set that has infinitely many elements is uncountable. As a matter of fact, a denumerable set is the smallest set that is infinite and all denumerable sets are equivalent to the set of natural numbers and thus are equivalent to each other. All other infinite sets are uncountable.

Set C of Example 2.3 is denumerable and thus countable; set B of Example 2.2 is uncountable because there is no definite and explicit manner to count all of its elements.

In fact, the set of all rational numbers[1] is countable because there is a one-to-one correspondence between it and the set of all natural numbers. However, the set of all irrational numbers is uncountable — there are much more irrational numbers than rational numbers! This is quite a surprise to people who are new to this theory.

If two sets A and B are equivalent, they must have the same number of elements; otherwise there cannot be a one-to-one correspondence between them. Conversely, two sets are said to have the same number of elements if they are equivalent. This is obviously true for finite sets but not so clear to beginners for infinite sets. For example, let $A = \{1, 2, 3, 4, \ldots\}$ be the set of all positive integers and $B = \{2, 4, 6, 8, \ldots\}$ be the set of all positive even integers. Then they are equivalent since every element of B has a corresponding element in A (e.g., the one with exactly half of the value) and vice versa. Note that the above sets A and B are equivalent even though every element of B belongs to A and some elements of A do not belong to B (i.e., B is a subset of A but A is not a subset of B, hence $A \neq B$). Consequently, we say that A and B have the same number of elements and thus A is not larger than B although A appears to be "larger" than B! This is interesting and surprising to beginners. In fact, sets A and B and the set of all rational numbers all have the same number of elements, which is equal to the smallest infinity.

[1] A rational number is one that can be expressed in the form of n/m for some integers n and m. A real number that is not rational is called irrational. For example, $\sqrt{2}$, e, and π are irrational numbers.

2.1.2 Basic Set Operations

Definitions:

- The set of all elements of A *or* B is called the ***union*** (or ***sum***) of A and B, denoted as $A \cup B$ or $A + B$. Union of *disjoint* sets A and B may be denoted as $A \uplus B$.
  Convention: "A or B" = "either A or B or both."
- The set of all elements *common* to A *and* B is called the ***intersection*** (or ***product***) of A and B, denoted as $A \cap B$ or AB.
- The set of all elements of A that are not in B is called the ***difference*** of A and B, denoted as $A - B$.
- The set of all elements in the space S but not in A is called the ***complement*** of A, denoted as $\overline{A}$. It is equal to $S - A$.

A simple and instructive way of illustrating the relationships among sets is the so-called ***Venn diagram***, as illustrated below.

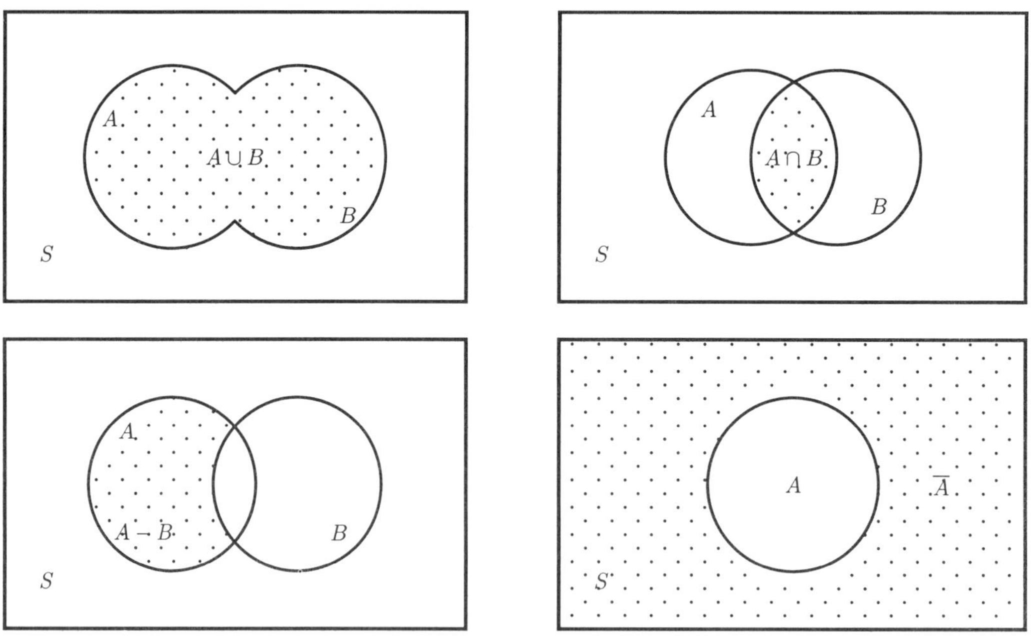

Figure 2.1: Basic set operations.

Example 2.5: Set Operations

For A, B, and C considered in Examples 2.1, 2.2, and 2.3:

$A \subset C$
$A \cup B = \{x : 0 < x \leq 6, x \text{ a real number}\}$
$A \cup C = C$
$B \cup C = \{x : x \text{ a positive integer or a real number satisfying } 0 < x < 6\}$
This set has a mixed type.
$A \cap B = \{1, 2, 3, 4, 5\}$
$A \cap C = A$
$B \cap C = \{1, 2, 3, 4, 5\}$
$A - B = \{6\}$
$A - C = \emptyset$
$B - A = \{x : 0 < x < 6, x \text{ a noninteger real number}\}$
$B - C = \{x : 0 < x < 6, x \text{ a noninteger real number}\}$
$C - A = \{x : x \geq 7, x \text{ an integer}\} = \{7, 8, 9, \ldots\}$
$C - B = \{x : x \geq 6, x \text{ an integer}\} = \{6, 7, 8, \ldots\}$

Space S depends on what we are considering. If we are considering only positive real numbers, then $S = \{x : x > 0, x \text{ real}\}$. Thus,

$\overline{A} = \{x : x \text{ a positive real number other than } 1, 2, 3, 4, 5, 6\}$
$\overline{B} = \{x : x \geq 6, x \text{ a real number}\}$
$\overline{C} = \{x : x \text{ a noninteger positive real number}\}$

If, however, we are considering all real numbers, then $S = \{x : x \text{ real}\}$. Thus

$\overline{A} = \{x : x \text{ a real number other than } 1, 2, 3, 4, 5, 6\}$
$\overline{B} = \{x : x \leq 0 \text{ or } x \geq 6, x \text{ a real number}\}$
$\overline{C} = \{x : x \leq 0 \text{ or } x \text{ a noninteger positive real number}\}$

2.1.3 Basic Algebra of Sets

	Algebra of sets	*Algebra of numbers*
	Union $\cup$	sum "+"
	Intersection $\cap$	product "·"
1	$A \cup B = B \cup A$	$a + b = b + a$
2	$A \cap B = B \cap A$	$a \cdot b = b \cdot a$
3	$A \cup (B \cup C) = A \cup B \cup C$	$a + (b + c) = a + b + c$
4	$A \cap (B \cap C) = A \cap B \cap C$	$a \cdot (b \cdot c) = a \cdot b \cdot c$
5	$A \cap (B \cup C) = (A \cap B) \cup (A \cap C)$	$a \cdot (b + c) = a \cdot b + a \cdot c$
6	$A \cup (B \cap C) = (A \cup B) \cap (A \cup C)$	see below

Since $A \cap A = A$, $(A \cap B) \subset A$, $(A \cap C) \subset A$, and

$$(a + b)(a + c) = a \cdot a + a \cdot b + a \cdot c + b \cdot c$$

Line 6 in the table above follows from

$$(A \cup B) \cap (A \cup C) = \underbrace{(A \cap A) \cup (A \cap B) \cup (A \cap C)}_{=A} \cup (B \cap C) = A \cup (B \cap C)$$

This illustrates that set algebra has its own rules.

De Morgan's laws:

$$\overline{A \cup B} = \overline{A} \cap \overline{B} \tag{2.1}$$

$$\overline{A \cap B} = \overline{A} \cup \overline{B} \tag{2.2}$$

Similarly,

$$\overline{A \cup B \cup C} = \overline{A \cup D} = \overline{A} \cap \overline{D} = \overline{A} \cap \overline{B \cup C} = \overline{A} \cap (\overline{B} \cap \overline{C}) = \overline{A} \cap \overline{B} \cap \overline{C}$$

$$\overline{A_1 \cup \cdots \cup A_n} = \overline{A_1} \cap \cdots \cap \overline{A_n}$$

$$\overline{A_1 \cap \cdots \cap A_n} = \overline{A_1} \cup \cdots \cup \overline{A_n}$$

$$\overline{(A_1 \cup A_2) \cap (A_3 \cup A_4)} = (\overline{A_1} \cap \overline{A_2}) \cup (\overline{A_3} \cap \overline{A_4})$$

Rules: (1) interchange $\cup$ and $\cap$; (2) interchange $(*)$ and $\overline{(*)}$. However, care should be taken when dealing with multiple nests, as demonstrated below.

Example 2.6:

$$\overline{\overline{\underbrace{(\overline{A} \cap B)}_{D}} \cup \overline{C}} = \overline{\overline{D} \cup \overline{C}} = D \cap C = (\overline{A} \cap B) \cap C = \overline{A \cup \overline{B}} \cap C \tag{2.3}$$

The set union $A \cup B$ is sometimes denoted by $A + B$, and is called the "***sum*** of A and B," or simply "A plus B." The set intersection $A \cap B$ is sometimes denoted by AB, especially when dealing with probability, and is called the "***product*** of A and B," or simply "A times B." A reason for such notations is that set union and intersection obeys similar laws as that of the sum and product of numbers, respectively, as illustrated in the previous table. In some publications, $A + B$ is used to mean $A \uplus B$, i.e., $A \cup B$ when A and B are disjoint. We will maintain the following convention: $A + B \equiv A \cup B \not\equiv A \uplus B$, where $A \uplus B$ is the ***disjoint union*** of A and B.

(2.1)–(2.3) can be verified easily using ***Venn diagram***. The Venn diagram for (2.1) is shown in Fig. 2.2, where in (b) the total shaded area and the densely shaded area stand for $\overline{A} \cup \overline{B}$ and $\overline{A} \cap \overline{B}$, respectively. See also Problems 2.2 and 2.3. However, a Venn diagram is not effective for complex set operations.

If a set identity holds, then it is guaranteed by the so-called *duality principle* that another identity also holds by: (1) interchanging union $\cup$ and intersection $\cap$; and (2) interchanging S and $\emptyset$ in the first identity. For example, the table on the last page consists of three dual pairs: interchanging $\cup$ and $\cap$ in Line 5 yields Line 6 and vice versa; interchanging $\cup$ and $\cap$ in Line 3 yields Line 4 and vice versa; interchanging $\cup$ and $\cap$ in Line 1 yields Line 2 and vice versa. Another example is the following pair of identities connected by the duality principle is

$$S \cup A = S, \qquad \emptyset \cap A = \emptyset$$

The analogy of set algebra and algebra of numbers and the Venn diagrams are introduced here primarily to facilitate the understanding of set algebra. They are not necessarily good tools for actual algebraic manipulation of sets.

The good news is that the algebra of sets used in this book will not be too involved.

While the concept of a set has nothing to do with the relationship among its elements, a countable set is often called a ***discrete set*** in engineering practice and the term "continuous set" is sometimes used nonrigorously to mean a continuous region of a space, such as an interval of the real line or a continuous region of the x-y plane. For Examples 2.1 through 2.4, B is "continuous" while the others are discrete. We will try to avoid this latter terminology.

We could have "a set of sets," that is, a set whose elements are sets. In such cases, we usually say "a collection of sets" or "a family of sets." Note that $\{\emptyset\} \neq \emptyset$ since the number of elements in $\emptyset$ is zero while the number of elements in $\{\emptyset\}$ is 1 (i.e., the empty set is included in $\{\emptyset\}$ as the unique element).

$\overline{A \cup B}$ shaded $= \overline{A} \cup \overline{B}$, densely shaded $= \overline{A} \cap \overline{B}$

Figure 2.2: Proof of (2.1) by Venn diagram.

2.2 Fundamental Concepts in Probability

2.2.1 Definitions

- ***random experiment*** = experiment (action) whose result is uncertain (cannot be predicted with certainty) before it is performed
- ***trial*** = single performance of the random experiment
- ***outcome*** = result of a trial
- ***sample space*** S = the set of all possible outcomes of a random experiment
- ***event*** = subset of the sample space S (to which a probability can be assigned)
  = a collection of possible outcomes
- ***sure event*** = sample space S (an event for sure to occur)
- ***impossible event*** = empty set $\emptyset$ (an event impossible to occur)

We say *an event has* ***occurred*** *if and only if the outcome observed belongs to the set of the event*, as explained below.

Example 2.7: Die-Rolling Events

Rolling a die is a random experiment. An outcome can be any number from 1 to 6. Sample space = $\{1, 2, 3, 4, 5, 6\}$. Some possible events are

$$
\begin{aligned}
A &= \{\text{an even number shows up}\} = \{2, 4, 6\} \qquad \text{(3 outcomes)} \\
B &= \{\text{a number greater than 5 shows up}\} \\
&= \{6\} \qquad \text{(single outcome)} \\
C &= \{\text{2 shows up}\} = \{2\} \qquad \text{(single outcome)} \\
D &= \{\text{a number greater than 6 shows up}\} = \emptyset \qquad \text{(no outcome)} \\
E &= \{\text{2 and 4 show up}\} = \emptyset \qquad \text{(no outcome)} \\
F &= \{\text{2 or 4 shows up}\} = \{2, 4\} \qquad \text{(2 outcomes)} \\
G &= \{\text{a number from 1 to 6 shows up}\} \\
&= \{1, 2, 3, 4, 5, 6\} = S \qquad \text{(all outcomes)}
\end{aligned}
$$

Thus, if "2" showed up, then we say that events A, C, F, and G have all *occurred*.

2.2.2 Probability of an Event

Traditional definitions of the probability of an event A:

$$\textbf{\textit{Classical:}}\ P\{A\} = \frac{\#\text{ of possible outcomes for event } A}{\#\text{ of possible outcomes for space } S}$$

$$\textbf{\textit{Relative frequency:}}\ P\{A\} = \lim_{N\to\infty} \frac{\#\text{ of occurrences of event } A}{N\ (\text{total \# of trials})}$$

$$\textbf{\textit{Geometric:}}\ P\{A\} = \frac{\text{geometric measure of set } A}{\text{geometric measure of space } S}$$

These definitions are very natural but limited:

- The classical definition is virtually applicable only to events with finitely (or countably) many outcomes that are equally probable.
- The geometric definition is an extension for events with uncountably many outcomes that are uniformly probable.
- The relative-frequency definition is more general than the other two definitions but is still limited. It is difficult to be applied to problems in which outcomes are not equally probable.

Example 2.8: Classical Probability: Die Rolling

Consider Example 2.7. The probabilities of events are

$$P\{A\} = P\{\text{an even number shows up}\} = \frac{1+1+1}{6} = \frac{1}{2}$$

$$P\{E\} = P\{2 \text{ and } 4 \text{ show up}\} = 0/6 = 0$$

$$\Longrightarrow \textit{Impossible event has zero probability}$$

$$P\{F\} = P\{2 \text{ or } 4 \text{ shows up}\} = 2/6 = 1/3$$

$$P\{G\} = P\{\text{a number from 1 to 6 shows up}\} = 6/6 = 1$$

$$\Longrightarrow \textit{Sure event has unity probability}$$

Note, however, as illustrated in the next example,

- An event of zero probability is not necessarily an impossible event.
- An event of unity probability is not necessarily a sure event.

These counter-intuitive results are possible only when the sample space has infinitely many elements.

Example 2.9: Geometric Probability: Waveform Sampling

The following voltage waveform is to be sampled at a random time τ over the period $-1 \leq t \leq 5$.

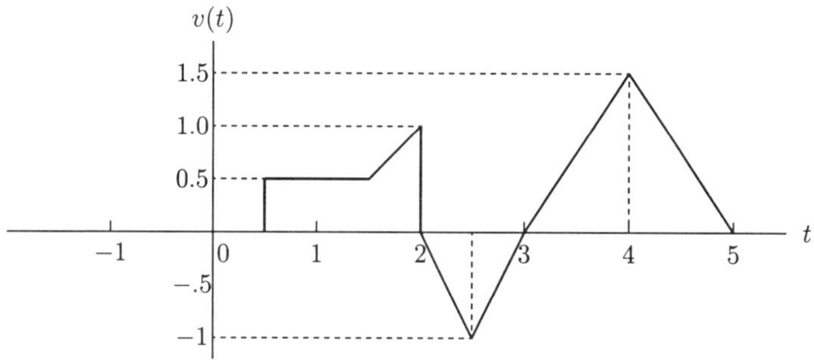

(a) Determine the probability that the sampled value $v(\tau) < -0.25$:

$$A = \{\text{sampled value } v(\tau) < -0.25\}$$

$$P\{A\} = \frac{\text{time in which sampled value } v(\tau) < -0.25}{\text{total time}} = \frac{3/4}{6} = \frac{1}{8}$$

(b) Determine the probability that the sampled value $v(\tau) \geq 1.0$:

$$B = \{\text{sampled value } v(\tau) \geq 1.0\}, \quad P\{B\} = \frac{(1/3)(5-3)}{6} = \frac{1}{9}$$

(c) Determine the probability that the sampled value $v(\tau) = 0.5$:

$$C = \{\text{sampled value } v(\tau) = 0.5\}, \quad P\{C\} = (1.5 - 0.5)/6 = 1/6$$

(d) Determine the probability that the sampled value $v(\tau) = 1.2$:

$$D = \{\text{sampled value } v(\tau) = 1.2\}, \quad P\{D\} = 0/6 = 0$$

(e) Determine the probability that the sampled value satisfies $-1 \leq v(\tau) < 1.5$ but not equal to -0.5:

$$E = \{-1 \leq v(\tau) < 1.5, v(\tau) \neq -0.5\}, \quad P\{E\} = 6/6 = 1$$

Note:

- D is not an impossible event but $P\{D\} = 0$.
- E is not a sure event but $P\{E\} = 1$.

To be more rigorous, an experiment is a random one only if all the following three conditions are satisfied: (1) it can be repeated under the same conditions; (2) it has more than one outcome and they are all known before the experiment; and (3) which outcome will show up cannot be determined prior to the experiment.

Everyone has the experience that the relative frequency of the occurrence of a random event under "identical" conditions is quite stable. This is the physical foundation for the relative-frequency definition of probability. Of course, were this not the case, we most likely (with an extremely high "probability"!) would not have a probability theory. The stability of relative frequency reflects the regularity about the occurrence of an event, which is usually imbedded in the apparent irregularities (uncertainties) involved in the event occurrence. The regularity will overcome the irregularities only through a large number of trials since the irregularities tend to cancel out each other. As a result, probability laws exhibit themselves only through a large number of trials.

In general, a probabilistic solution to a problem with random effects includes the following steps:

S1. Determine (prior) probabilities of certain elementary events.
S2. Obtain probabilities of other events according to certain rules.
S3. Make certain inference based on the obtained probabilities.

The probability theory calculates the probabilities of other (composite) events based on these elementary event probabilities (Step 2). Both Steps 3 and 1 belong to the science of statistics.

It should be understood that the probability of an event based on the relative-frequency concept is defined to be the limit as the number of trials increases. In other words, probability is the limit of relative frequency at infinitely many trials. Otherwise, probability would be random, depending on how many trials are performed. With this interpretation, both classical and geometric definitions of probability can be thought of as special cases of the relative-frequency definition. This is the case for Examples 2.8 and 2.9.

The relative-frequency interpretation of probability is important in application of probability theory, especially for Step 1 above. Specifically, the probabilities of certain events can be assigned according to the relative frequency based on experimental results (or based on intuition or common sense). In history, the theory of probability (or chance) started from the study of gambling, which supports this interpretation.

The underlying assumption for the relative-frequency interpretation of probability (or the classical theory of probability completed by Laplace) is that every elementary event is equally probable (likely), which is sometimes called the *principle of indifference* or the *principle of sufficient reason.* Even though justifiable in some cases, this assumption may be questionable in other cases in which the elementary events are not equally likely. In some other cases involving infinitely many possible outcomes, how to define equiprobable elementary events cannot be done unambiguously and thus may lead to confusing paradoxes. That is part of the reasons why the axiomatic approach to probability is prevailing. This is presented next.

2.2.3 Axioms of Probability Theory

A set of events $A_1, A_2, \ldots, A_n$ is said to be ***mutually exclusive*** or ***disjoint*** if

$$A_i \cap A_j = \emptyset \qquad \forall i \neq j$$

That is, at most one event can occur (if one occurs, any other cannot occur).

In the contemporary theory of probability, the following properties have been identified as fundamental (necessary and sufficient) for probability as a measure, which are taken as axioms:

- *Axiom* 1 (nonnegativity): Probability of any event A is bounded by 0 and 1:

$$\boxed{0 \leq P\{A\} \leq 1} \tag{2.4}$$

- *Axiom* 2 (unity): Any sure event (the sample space) has unity probability:

$$\boxed{P\{S\} = 1} \tag{2.5}$$

- *Axiom* 3 (finite additivity): If $A_1, A_2, \ldots, A_n$ are disjoint events, then

$$P\{A_1 \uplus A_2 \uplus \cdots \uplus A_n\} \triangleq P\Big\{\biguplus_{i=1}^{n} A_i\Big\} = \sum_{i=1}^{n} P\{A_i\} \tag{2.6}$$

- *Axiom* 3' (countable additivity): If $A_1, A_2, \ldots$ are disjoint events, then

$$\boxed{P\Big\{\biguplus_{i=1}^{\infty} A_i\Big\} = \sum_{i=1}^{\infty} P\{A_i\}} \tag{2.7}$$

All other probability laws can be derived from these axioms. Keep in mind that (2.6) and (2.7) are valid only for mutually exclusive events.

These axioms imply that probability can be interpreted as *mass* associated with various events. They are clearly reasonable from relative-frequency perspective:

$$\begin{aligned} 0 \leq P\{A\} &= \frac{N_A}{N} \leq 1 \\ P\{S\} &= \frac{N}{N} = 1 \\ P\Big\{\biguplus_{i=1}^{n} A_i\Big\} &= \frac{N_{A_1} + \cdots + N_{A_n}}{N} = \frac{N_{A_1}}{N} + \cdots + \frac{N_{A_n}}{N} = \sum_{i=1}^{n} P\{A_i\} \end{aligned}$$

Many people know more than one person named John Smith. So ambiguity may arise if we use a person's name to define a person, although this is natural, convenient, and common practice. The traditional definitions of probability suffer from a similar ambiguity in some cases. However, science and engineering cannot tolerate any ambiguity. This was partly the reason why the axiomatic method of probability was developed. It is similar in some sense to defining a person by the exact date, time, and place of his or her birth, which is precise and avoids ambiguity but is less convenient.

The axiomatic method is one of the most important approaches of science, especially modern mathematics. It is the most rigorous approach, omnipresent in the entire domain of mathematics. It consists in setting forth certain basic statements about the concepts to be studied that are self-evident, known as axioms or postulates, and then using classical logic to deduce all other results. These "taken-for-granted" basic statements must be consistent, independent, and complete in the sense that they do not lead to any contradiction, they do not imply each other, and they suffice to deduce all other results of the subject. A typical example of this method is that of defining a mathematical entity by a consistent, independent, and complete set of its fundamental properties.

The axiomatic method has several important advantages over other methods. For example, it makes it clear what can be used as the root or most fundamental for any further study of the subject. If an axiomatic system exists for a theory, it is usually believed that the theory is quite mature. The axiomatic method can be traced back to the ancient Greece around the Aristotle's time. It was popularized by Euclid's celebrated work *Elements*, which was written about 300 B.C. In his two books that provided a foundation for the classical mechanics, Archimedes (287–212 B.C.) employed the method featured in Euclid's work. Newton's famous work *Principia*, published in 1686, is organized in a deductive way that can be considered an early form of the axiomatic system. The treatise on analytic mechanics published by Lagrange in 1788 is a masterpiece of logical perfection, containing many elements of an axiomatic system. Hilbert's classic work *Grundlagen der Geometrie* on the foundation of geometry, published in 1899, has been generally regarded as the first that displays the axiomatic method in its modern form.

The axiomatic system of probability was developed by the Russian mathematician A. N. Kolmogorov in 1933 based on set theory and measure theory. It was one of the early success of the modern axiomatic method in a branch of mathematics other than geometry. It overcomes many limitations of the traditional definitions of probability.

The axioms of probability clearly make perfect sense. Axioms 1 and 2 follow from our custom of treating probability as a positive percentage in everyday life. The condition in Axiom 3 that the events are disjoint cannot be removed. Take die rolling as an example. Let $A = \{2,4,6\}, B = \{1,3,5\}, C = \{1,2,3\}$. Then clearly

$$P\{A \cup B\} = P\{1,2,3,4,5,6\} = 1 = P\{1,3,5\} + P\{2,4,6\} = P\{A\} + P\{B\}$$

but

$$P\{A \cup C\} = P\{1,2,3,4,6\} = 5/6 \neq 1/2 + 1/2 = P\{A\} + P\{C\}$$

because A and C are not mutually exclusive since both A and C occur if face 2 shows up.

Note that Axiom 3' implies Axiom 3 but not the other way round. In other words, a property that is valid for an arbitrary but finite number does not necessarily hold for infinity.

2.2.4 Probability of the Union of Two Events

The union of events A and B in space S is the set of all outcomes of A or B (or both). In other words, if any outcome of either A or B occurs, then we say the ***union of events*** A and B, denoted by $A \cup B$, occurs.

By Axiom 3, if $A \cap B = \emptyset$ (A and B cannot both occur), then

$$P\{\text{either } A \text{ or } B \text{ occurs}\} = P\{A \uplus B\} = P\{A\} + P\{B\}$$

What if $A \cap B \neq \emptyset$ (A and B may both occur)? Note that

$$A \cup B = A \uplus (\overline{A} \cap B)$$

which can be shown easily by Venn diagram. Clearly, $A \cap (\overline{A} \cap B) = \emptyset$. Hence,

$$P\{A \cup B\} = P\{A \uplus (\overline{A} \cap B)\} \overset{\text{Axiom 3}}{=} P\{A\} + P\{\overline{A} \cap B\} \tag{2.8}$$

Similarly, $B = (A \cap B) \uplus (\overline{A} \cap B)$ and $(A \cap B) \cap (\overline{A} \cap B) = \emptyset$, and thus

$$P\{B\} = P\{(A \cap B) \uplus (\overline{A} \cap B)\} \overset{?}{=} P\{A \cap B\} + P\{\overline{A} \cap B\} \tag{2.9}$$

Thus, (2.8) – (2.9) yields the ***probability of union of two events*** or ***addition rule of probability***:

$$\boxed{P\{A \cup B\} = P\{A\} + P\{B\} - P\{A \cap B\}} \tag{2.10}$$

This is intuitively correct:

$$P\{\text{either } A \text{ or } B \text{ or both occur}\} = P\{A \text{ occurs}\} + P\{B \text{ occurs}\} - \underbrace{P\{A \text{ and } B \text{ both occur}\}}_{\text{double counted}}$$

since $P\{A \text{ and } B \text{ both occur}\}$ is double counted in $P\{A \text{ occurs}\} + P\{B \text{ occurs}\}$.

2.2.5 Probability of the Complement of an Event

If A is an event in space S, then $S = A \uplus \overline{A}$, and

$$1 \overset{\text{Axiom 2}}{=} P\{S\} = P\{A \uplus \overline{A}\} \overset{\text{Axiom 3}}{=} P\{A\} + P\{\overline{A}\}$$

Thus, the ***probability of the complement of an event*** is

$$\boxed{P\{\overline{A}\} = 1 - P\{A\}} \tag{2.11}$$

2.2.6 Joint Probability

The ***joint events*** A and B is the intersection of sets A and B, which is the set of outcomes common to both A and B. As such, the ***joint probability*** of events A and B is the probability that they both occur. This probability is denoted by

$$P\{A \cap B\} \triangleq P\{A, B\} \triangleq P\{AB\} \tag{2.12}$$

which is given by, from (2.10),

$$\boxed{P\{A \cap B\} = P\{A\} + P\{B\} - P\{A \cup B\}} \tag{2.13}$$

This can be interpreted as follows:

$$\begin{aligned} P\{A \text{ and } B \text{ both occur}\} = P\{&A \text{ occurs regardless } B \text{ occurs or not}\} \\ &+ P\{B \text{ occurs regardless } A \text{ occurs or not}\} \\ &- P\{A \text{ or } B \text{ or both occur}\} \end{aligned}$$

If $A \cap B = \emptyset$, then

$$P\{A \cap B\} = P\{A\} + P\{B\} - P\{A \uplus B\} \overset{\text{Axiom 3}}{=} 0$$

or

$$P\{A \cap B\} = P\{\emptyset\} = P\{\overline{S}\} \overset{(2.11)}{=} 1 - P\{S\} = 0$$

which makes sense: "*mutually exclusive events have zero joint probability*" and "*impossible events have zero probability.*"

Example 2.10: Axiomatic Probabilities of Die-Rolling Events

Rolling a die is a random experiment whose outcomes can be any face from 1 to 6. Let F_i be the event that face i shows up. Assume the die is a fair one — each outcome is equiprobable. Let us use the probability axioms to (determine how to) assign a probability to every outcome.

$$1 \overset{\text{Axiom 2}}{=} P\{S\} = P\Big\{\biguplus_{i=1}^{6} F_i\Big\} \overset{\text{Axiom 3}}{=} \sum_{i=1}^{6} P\{F_i\} \overset{\text{fair die}}{=} 6P\{F_i\}$$

which yields $P\{F_i\} = 1/6, i = 1, \ldots, 6$ and hence:

$$P\{\text{an even numbered face shows up}\} \overset{\text{Axiom 3}}{=} P\{F_2\} + P\{F_4\} + P\{F_6\} = 1/2$$

2.3 Conditional Probability

The probability of an event A *under the condition* that event B has occurred is called the ***conditional probability*** of A given B (or probability of A conditioned on B), defined as

$$\boxed{P\{A|B\} \triangleq \frac{P\{A \cap B\}}{P\{B\}}} \qquad \text{if } P\{B\} \neq 0 \tag{2.14}$$

The relative-frequency interpretation of this is

$$\begin{aligned} P\{A|B\} &= \frac{\#\text{ of outcomes for event } (A \cap B)}{\#\text{ of outcomes for event } B} = \frac{N_{A\cap B}}{N_B} \\ &= \frac{N_{A\cap B}/N}{N_B/N} = \frac{P\{A \cap B\}}{P\{B\}} \end{aligned}$$

Since $A \cap S = A$ and $P\{S\} = 1$, we have $P\{A\} = \frac{P\{A\cap S\}}{P\{S\}} = P\{A|S\}$. A comparison of this with (2.14) indicates that *conditional probability of A given B is simply the probability of A assuming the entire sample space is B*, which is reasonable since event B has occurred (and the occurrence of any outcome outside B is impossible), as illustrated below.

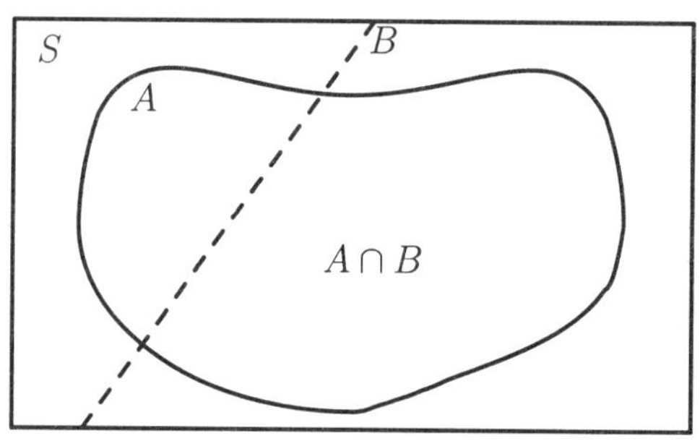

Figure 2.3: Illustration of conditional probability.

(2.14) implies that the following ***product rule*** or ***multiplication rule*** holds

$$\boxed{P\{AB\} \triangleq P\{A \cap B\} = P\{A|B\}P\{B\}} \qquad \text{if } P\{B\} \neq 0 \tag{2.15}$$

Whenever we define certain probability, theoretically we have to check if it satisfies all the axioms of probability. The conditional probability as defined above does satisfies the axioms. Specifically,

- It satisfies Axiom 1 since $P\{A \cap B\}$ and $P\{B\}$ are both nonnegative and thus $P\{A|B\}$ is also nonnegative.
- It satisfies Axiom 2 since by letting $A = S$, we have $P\{S|B\} = \frac{P\{S \cap B\}}{P\{B\}} = \frac{P\{B\}}{P\{B\}} = 1$.
- It satisfies Axiom 3 since for two disjoint events A_1 and A_2, events $A_1 \cap B$ and $A_2 \cap B$ are also disjoint and thus

$$P\{(A_1 \cup A_2) \cap B\} = P\{(A_1 \cap B) \uplus (A_2 \cap B)\} \overset{\text{Axiom 3}}{=} P\{A_1 \cap B\} + P\{A_2 \cap B\}$$

which implies that Axiom 3 holds for the conditional probability:

$$\begin{aligned} P\{(A_1 \cup A_2)|B\} &= \frac{P\{(A_1 \cup A_2) \cap B\}}{P\{B\}} = \frac{P\{A_1 \cap B\}}{P\{B\}} + \frac{P\{A_2 \cap B\}}{P\{B\}} \\ &= P\{A_1|B\} + P\{A_2|B\} \end{aligned}$$

There are three ways to calculate the conditional probability $P\{A|B\}$:

- Calculate $P\{A \cap B\}$ and $P\{B\}$ in the sample space S and then use (2.14).
- Calculate $P\{A\}$ in the reduced sample space $S_B = B$ consisting of all outcomes of event B.
- Calculate $P\{A\}$ assuming B has occurred.

By the definition of $P\{B|A\}$, we have $P\{A \cap B\} = P\{B|A\}P\{A\}$, which can also be obtained by interchanging A and B in (2.15). Similarly, we have the following ***product rule*** or ***multiplication rule***:

$$P\{ABC\} = P\{C|A, B\}P\{B|A\}P\{A\}$$

In general, we have the following ***chain rule***, ***multiplication rule*** or ***product rule***:

$$P\{A_1 A_2 \cdots A_n\} = P\{A_n|A_{n-1} \cdots A_1\} \cdots P\{A_2|A_1\}P\{A_1\} \tag{2.16}$$

Example 2.11: Consider rolling a die. Let

$$\begin{aligned} A &= \{\text{number 1 shows up}\} \\ B &= \{\text{an odd number shows up}\} \\ C &= \{\text{number 1 or 2 shows up}\} \end{aligned}$$

Then,

$$\begin{aligned} P\{A|B\} &= \frac{P\{A \cap B\}}{P\{B\}} = \frac{P\{A\}}{P\{B\}} = \frac{1/6}{3/6} = \frac{1}{3} \\ P\{C|B\} &= \frac{P\{C \cap B\}}{P\{B\}} = \frac{P\{A\}}{P\{B\}} = \frac{1/6}{3/6} = \frac{1}{3} \\ P\{B|C\} &= \frac{P\{C \cap B\}}{P\{C\}} = \frac{P\{A\}}{P\{C\}} = \frac{1/6}{2/6} = \frac{1}{2} \end{aligned}$$

Example 2.12: Double Luck with Lottery

The probability that a person will win the lottery twice can be calculated as follows. Let $A_i = \{\text{win the } i\text{th time}\}$. Then

$$P\{\text{win twice}\} = P\{A_1, A_2\} = P\{A_2|A_1\}P\{A_1\} \stackrel{?}{=} P\{A_2\}P\{A_1\}$$

where $P\{A_2|A_1\}$ is the probability of winning the second time after winning the first time, which is equal to $P\{A_2\}$ in reality. When this is the case, we say that the two wins are statistically independent, to be studied later.

Example 2.13: A pair of resistor and capacitor was chosen at random from a box of nine pairs of resistors and capacitors with the following resistance and capacitance, respectively.

Pair	1	2	3	4	5	6	7	8	9
R ($k\Omega$)	1	1	1	2.5	2.5	2.5	5	5	5
C (μF)	20	40	50	20	40	50	20	40	50
$\tau = RC$ (ms)	20	40	50	50	100	125	100	200	250

(a) What is the probability that the time constant $\tau < 50$ or $\tau > 125$?
Let $T = \{50 \le \tau \le 125\}$. Then

$$P\{\tau < 50 \text{ or } \tau > 125\} = P\{\overline{T}\} = 1 - P\{T\} \stackrel{\text{table}}{=} 1 - \frac{5}{9} = \frac{4}{9}$$

(b) Determine $P\{\overline{T}|C \neq 20\}$:

$$P\{\overline{T}|C \neq 20\} = \frac{P\{\overline{T} \cap (C \neq 20)\}}{P\{C \neq 20\}} \stackrel{\text{table}}{=} \frac{3/9}{6/9} = \frac{1}{2}$$

(c) Determine $P\{C \neq 20|\overline{T}\}$:

$$P\{C \neq 20|\overline{T}\} = \frac{P\{\overline{T} \cap (C \neq 20)\}}{P\{\overline{T}\}} \stackrel{\text{(a),(b)}}{=} \frac{3/9}{4/9} = \frac{3}{4}$$

(d) Determine $P\{C \neq 20, R \neq 5|\overline{T}\}$:

$$P\{C \neq 20, R \neq 5|\overline{T}\} = \frac{P\{\overline{T} \cap (C \neq 20) \cap (R \neq 5)\}}{P\{\overline{T}\}} = \frac{1/9}{4/9} = \frac{1}{4}$$

Example 2.14: Life Expectancy

Let t be the age of a person when he/she dies. The probability that he/she dies at an age not older than t_0 is given by

$$P\{t \le t_0\} = \int_0^{t_0} \alpha(t)dt$$

where $\alpha(t)$ is a function determined from mortality records by

$$\alpha(t) = \begin{cases} 3 \times 10^{-9} t^2 (100-t)^2 & 0 \le t \le 100 \text{ years} \\ 0 & \text{otherwise} \end{cases}$$

(This model is approximate — it implies a zero probability to have a life expectancy longer than 100 years.)

(a) The probability that a person will die between the ages of 60 and 70 is

$$P\{60 \le t \le 70\} = \frac{\text{\# of people who die between 60 and 70}}{\text{total population}}$$
$$= \int_0^{70} \alpha(t)dt - \int_0^{60} \alpha(t)dt = \int_{60}^{70} \alpha(t)dt = 0.154$$

(b) The probability that a person will die between the ages of 60 and 70 assuming that his/her current age is 60 is

$$P\{60 \le t \le 70 | t \ge 60\} = \frac{\text{\# of people who die between 60 and 70}}{\text{total population of age older than 60}}$$
$$= \frac{P\{60 \le t \le 70\}}{P\{t \ge 60\}} = \frac{\int_{60}^{70} \alpha(t)dt}{\int_{60}^{100} \alpha(t)dt} = 0.486$$

(c) The probability that a person will die between the ages of 20 and 50 assuming that his/her current age is 60 is

$$P\{20 \le t \le 50 | t \ge 60\} = \frac{P\{(20 \le t \le 50) \cap (t \ge 60)\}}{P\{t \ge 60\}} = \frac{P\{\emptyset\}}{P\{t \ge 60\}} = 0$$

The results of this example are useful for e.g., determination of the premiums of a life insurance policy.

Example 2.15: Winning Strategy for a TV Game

A TV game that was popular in some European countries is as follow. A car is behind one of three doors. A player chooses one door. The player wins the car if it turns out to be behind the selected door. After the player chooses, the TV host will open another door which does not have the car because the host knows where the car is. After the host opens that door, the player is allowed to switch to choose the third door or stick to the original choice.

Question: Is it better to switch to the third door?

Assume for notational simplicity that the player has chosen door B and the host has opened door A. Then, since by now we know that door A does not have the car, switching to choose door C will win if and only if door B does not have the car, denoted by $\overline{B}$; that is,

$$P\{\text{Winning by switching}\} = P\{C|\overline{A}, \overline{B}\}P\{\overline{B}\} = P\{\overline{B}\} = 1 - P\{B\} = 2/3$$

Clearly, switching and not switching are mutually exclusive because they cannot both win. Thus, we have

$$P\{\text{Winning by not switching}\} \leq 1 - P\{\text{Winning by switching}\} = 1/3$$

In fact, the probability of winning by choosing door B in the first place is $1/3$. By not switching, the probability of winning stays unchanged because the new information that door A does not have the car is not utilized since it would be the same if instead door C was opened by the host. Thus,

$$P\{\text{Winning by not switching}\} = P\{B\} = 1/3$$

Consequently, the chance of winning is doubled by switching! This answer would be hard to come by without a good understanding of probability concepts.

The above analysis can be extended to the general n-door problem as follows. Assume for simplicity that the original choice of the player was door B, door A was opened by the host, and door C is chosen if the player changes his or her choice. Then, switching will win if and only if door B does not have the car and door C turns out to have the car among the $n-2$ remaining doors; that is,

$$P\{\text{Winning by switching}\} = P\{C|\overline{A}, \overline{B}\}P\{\overline{B}\} = \frac{1 - P\{B\}}{n-2} = \frac{1}{n-2}\frac{n-1}{n}$$

which is greater than $P\{\text{Winning by not switching}\} = P\{C\} = \frac{1}{n}$.

2.4 Independent Events

Two events A and B are said to be ***independent*** if the probability of occurrence of one event is not affected by the occurrence of the other event, that is,

$$P\{A|B\} = P\{A\} \qquad \text{and} \qquad P\{B|A\} = P\{B\} \tag{2.17}$$

where $P\{A\}$ and $P\{B\}$ are assumed nonzero. An equivalent but more compact form of (2.17) is

$$P\{A \cap B\} = P\{A\}P\{B\} \tag{2.18}$$

Thus, formally two events A and B are said to be statistically (or probabilistically) ***independent*** if (2.18) holds. The equivalence of (2.17) (with $P\{A\} \neq 0 \neq P\{B\}$) and (2.18) can be seen as follows:

$$(2.17) \Longrightarrow P\{A \cap B\} \overset{(2.14)}{=} P\{A|B\}P\{B\} \overset{(2.17)}{=} P\{A\}P\{B\} \Longrightarrow (2.18)$$

$$(2.18) \Longrightarrow \left\{\begin{matrix} P\{A\}P\{B\} \overset{(2.18)}{=} P\{A \cap B\} \overset{(2.14)}{=} P\{A|B\}P\{B\} \\ P\{A\}P\{B\} \overset{(2.18)}{=} P\{B \cap A\} \overset{(2.14)}{=} P\{B|A\}P\{A\} \end{matrix}\right\} \Longrightarrow (2.17)$$

Events are said to be statistically ***dependent*** if they are not independent. Independence simplifies the calculation of joint probability greatly:

$$\text{joint probability} \overset{\text{if independent}}{=} \text{product of probabilities}$$

Independence of n Events

For n events $A_1, A_2, \ldots, A_n$, if

$$\begin{aligned} P\{A_i \cap A_j\} &= P\{A_i\}P\{A_j\} \qquad \forall i \neq j \\ P\{A_i \cap A_j \cap A_k\} &= P\{A_i\}P\{A_j\}P\{A_k\} \qquad \forall i \neq j \neq k \\ &\vdots \\ P\{A_1 \cap A_2 \cap \cdots \cap A_n\} &= P\{A_1\}P\{A_2\} \cdots P\{A_n\} \end{aligned}$$

then events $A_1, A_2, \ldots, A_n$ are said to be statistically ***independent***. Otherwise, they are dependent.

Statistical independence is a mathematical abstraction of the concept of (practical or intuitive) independence as used in everyday life. They are, however, not identical. Practical independence of events A and B means that they have no mutual effect (or relation, information, constraints, etc.). However, statistical independence of events A and B *merely* implies that the knowledge of one event gives no information, *in the sense of changing the odds*, about the occurrence of the other. For example, if $A \subset S$ (sample space), then the occurrence of A implies that of S and thus A and S are not independent in the practical sense. They are, however, always statistically independent since $P\{A \cap S\} = P\{A\} = P\{A\}P\{S\}$. This is because although the occurrence of A does imply that of S, it does not change the probability of S. Also, if $A \cap B = \emptyset$, then A and B cannot be independent in the practical sense. They are, however, statistically independent if and only if $P\{A\} = 0$ or $P\{B\} = 0$. Thus, not all statistical independent events are independent in the practical sense. However, practical independence implies statistical independence: If events are practically independent, then they are statistically independent. In practice, whether events are statistically independent or not is usually determined by whether they are practically independent or not; that is, by physical meaning of the events via common sense or intuition rather than the mathematical definition.

In fact, (2.18) is more general than (2.17) in that the former, but not the latter, is still meaningful if $P\{A\} = P\{B\} = 0$. Furthermore, it can be shown easily (see problem 2.15) that $P\{A|B\} = P\{A\}$ and $P\{B|A\} = P\{B\}$ are equivalent (i.e., one implies the other) if $P\{A\} \neq 0$ and $P\{B\} \neq 0$.

The symbol $\forall$ is simply the mathematical shorthand for "for all" or better "for every."

Every subset of a set of independent events is independent.

Independence of a set of events requires that the probability of intersections of *any* group of the events in the set be equal to the product of the probabilities of the events in the group. For example, the following equation $P\left\{\bigcap_{i=1}^{k} A_i\right\} = \prod_{i=1}^{k} P\{A_i\}, k < n$, has to be true in order for the events $A_1, \ldots, A_n$ to be independent. This equation alone, however, does not guarantee the independence of any group of events $A_1, \ldots, A_n$.

A set of events $A_1, \ldots, A_n$ is sometimes said to be ***pairwise independent*** (or ***independent by pairs***) if

$$P\{A_i \cap A_j\} = P\{A_i\}P\{A_j\} \qquad \forall i \neq j$$

In general, if n events are independent then they are pairwise independent. But the fact that they are pairwise independent does not imply they are independent. This is a surprise to many people. For a set of events, if none of the events *individually* will affect the occurrence of any other event in the set, then these events are pairwise independent. They are independent if none of the events *individually or collectively* will affect the occurrence of any other event in the set. For example, $P\{ABC\} = P\{A\}P\{B\}P\{C\}$ neither implies nor is implied by the following

$$P\{AB\} = P\{A\}P\{B\} \qquad P\{BC\} = P\{B\}P\{C\} \qquad P\{CA\} = P\{C\}P\{A\}$$

See problem 2.26.

Example 2.16: Independent vs. Disjoint Events

Can two events be both independent and ***disjoint*** (i.e., ***mutually exclusive***)? Note that

$$P\{A\}P\{B\} \overset{\text{independent}}{=} P\{A \cap B\} \overset{\text{disjoint}}{=} 0$$

indicates that if two events are both independent and disjoint, then at least one of them has zero probability — *nonzero-probability events cannot be both independent and disjoint.* Intuitively, if two events are disjoint, the occurrence of one precludes the other and thus they cannot be independent. Note the difference:

$$\text{joint probability} \overset{\text{if independent}}{=} \text{product of probabilities} \tag{2.19}$$

$$\text{union probability} \overset{\text{if disjoint}}{=} \text{sum of probabilities} \tag{2.20}$$

Example 2.17: Reliability of Communication Channel

Consider the following communication network. Assume the links are independent and the probability that a link is operational is 0.95.

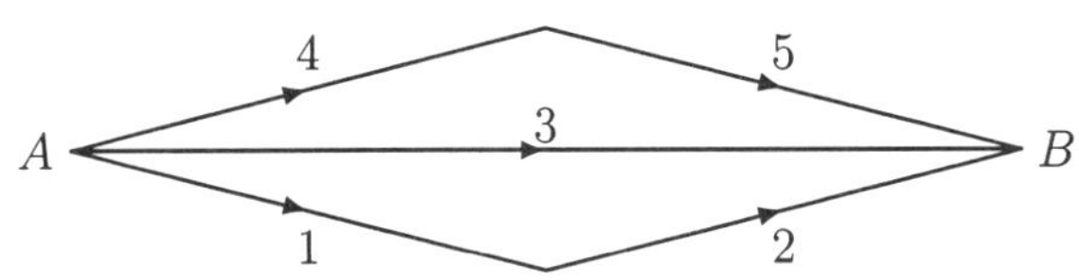

Since independence of links implies that paths are independent, the probability of being able to transmit from A to B can be calculated as follows:

$$\begin{aligned}
P\{\text{path 1-2 OK}\} &\overset{?}{=} P\{\text{link 1 OK}\}P\{\text{link 2 OK}\} = 0.95 \times 0.95 = 0.9025 \\
P\{\text{path 1-2 fails}\} &= 1 - P\{\text{path 1-2 OK}\} = 1 - 0.9025 = 0.0975 \\
P\{\text{path 3 fails}\} &= 1 - P\{\text{link 3 OK}\} = 1 - 0.95 = 0.05 \\
P\{\text{all paths fail}\} &\overset{?}{=} P\{\text{path 1-2 fails}\}P\{\text{path 4-5 fails}\}P\{\text{path 3 fails}\} \\
&= 0.0975 \times 0.0975 \times 0.05 = 0.000475
\end{aligned}$$

Finally,

$$\begin{aligned}
P\{\text{able to transmit from } A \text{ to } B\} &= 1 - P\{\text{all paths fail}\} \\
&= 1 - 0.000475 = 0.999525 \quad \text{(very high)}
\end{aligned}$$

Example 2.16 indicates that independent events are quite different from ***disjoint*** (or ***mutually exclusive***) events. Recall that two events A and B are mutually exclusive (disjoint) if $A \cap B = \emptyset$. In other words, disjoint events are those which have no common outcome — the occurrence of one event precludes the occurrence of the others. As a result, the occurrence of one of the disjoint events gives definite information about the probabilities of the others — they cannot occur — and thus either these events are not independent or at most one of them has nonzero probability. In short, if events with zero probabilities are not considered (i.e., for nontrivial events), disjoint events cannot be independent and independent events cannot be disjoint. In practice, if the occurrence of one event does not affect the chance of the others, then they are independent; if the occurrence of one event does preclude the occurrence of the others, then they are mutually exclusive.

It is interesting to know that a confusion of independent and disjoint events actually led to the first serious investigation into probability and is one of the things that gave birth to probability theory. In the 17th century a Frenchman named De Meré was perplexed about certain gambling schemes he was using. At the beginning, he bet he would receive at least one 6 in 4 rolls of a die because he thought his chance of winning was 2/3, based on the following reasoning: Since 6 shows up with probability (chance) 1/6 and it is independent and thus disjoint from roll to roll, his chance of winning was

$$\frac{1}{6} + \frac{1}{6} + \frac{1}{6} + \frac{1}{6} = \frac{2}{3}$$

Although the calculation was wrong, he was lucky to make money because the correct probability of his winning is actually $0.5177 > 0.5$ (see problem 2.42). de Meré later switched to betting that he would receive at least one 12 in 24 rolls of two dice and falsely reasoned that his chance of winning was

$$\frac{1}{36} + \cdots + \frac{1}{36} = 24 \cdot \frac{1}{36} = \frac{2}{3}$$

due to again a confusion of independence and disjoint events. He was not lucky this time because the probability of his winning for this scheme is actually $0.4914 < 0.5$ (see problem 2.43). Perplexed by his loss of money, de Meré wrote to Pascal, who contacted Fermat and the two mathematicians began the first investigation into probability theory.[2]

The first " $\stackrel{?}{=}$ " in Example 2.17 follows from the independence of links since

$$\{\text{path 1-2 OK}\} = \{(\text{link 1 OK}) \cap (\text{link 2 OK})\}$$

The last " $\stackrel{?}{=}$ " follows from the independence of paths.

In Example 2.17, the probability of being able to transmit from A to B can be calculated alternatively as,

$$P\{\text{able to transmit from } A \text{ to } B\} = P\{(\text{path 1-2 OK}) \cup (\text{path 3 OK}) \cup (\text{path 4-5 OK})\}$$

which requires, however, the use of the result of problem 2.10 since the independence of different paths implies that the events that these paths are OK are not mutually exclusive.

[2] A somewhat different account of this story can be found in Warren Weaver, *Lady Luck — The Theory of Probability*, Dover, New York, 1982.

Example 2.18: Security of Nuclear Power Plant

A nuclear power plant will shut down if systems A and B or A and C fail simultaneously. A, B, and C are independent systems and their probabilities of failure are 0.01, 0.03, and 0.02, respectively.

(a) What is the probability that the plant will stay on line?
Let Fail = 0 and OK = 1. Then

Case	*A*	*B*	*C*	*Plant P*	*Probabilities*
0	0	0	0	0	$(0.01)(0.03)(0.02)$
1	0	0	1	0	$(0.01)(0.03)(1-0.02)$
2	0	1	0	0	$(0.01)(1-0.03)(0.02)$
3	0	1	1	1	$(0.01)(1-0.03)(1-0.02)$
4	1	0	0	1	$(1-0.01)(0.03)(0.02)$
5	1	0	1	1	$(1-0.01)(0.03)(1-0.02)$
6	1	1	0	1	$(1-0.01)(1-0.03)(0.02)$
7	1	1	1	1	$(1-0.01)(1-0.03)(1-0.02)$

$$\begin{aligned}
P\{\text{plant shut down}\} &\stackrel{?}{=} (0.01)(0.03)(0.02)+(0.01)(0.03)(1-0.02) \\
&\quad +(0.01)(1-0.03)(0.02) \\
&= 0.000494 \\
P\{\text{plant on line}\} &= 1-P\{P=\text{“0”}\}=1-0.000494 \\
\text{or } &= P\{\text{Cases 3 through 7}\} \\
&= 0.999506 \quad \text{(very high)}
\end{aligned}$$

(b) What is the probability that the plant stays on line given that A failed?

$$\begin{aligned}
P\{\text{plant on line}|A\text{ failed}\} &= P\{P=\text{“1”}|A=\text{“0”}\} \\
&= \frac{P\{(P=\text{“1”})\cap(A=\text{“0”})\}}{P\{A=\text{“0”}\}} \\
&= \frac{(0.01)(1-0.03)(1-0.02)}{0.01} \\
&= 0.9506 \quad \text{(still not low)}
\end{aligned}$$

2.5 Total Probability Theorem and Bayes' Rule

2.5.1 Total Probability Theorem

A set of events $A_1, A_2, \cdots, A_n$ is said to be

- ***mutually exclusive*** (or ***disjoint***) if $A_i \cap A_j = \emptyset, \forall i \neq j$, meaning that *at most one* event can occur (if one occurs then any other cannot occur).
- (collectively) ***exhaustive*** if $A_1 \cup A_2 \cup \cdots \cup A_n = S$, meaning that *at least one* of the events will occur.
- a ***partition*** of sample space S if *one and only one* of the events will occur. Symbolically,

$$\text{partition} = \text{mutually exclusive} + \text{exhaustive}$$

Clearly, the probabilities of the member events of a partition $A_1, A_2, \cdots, A_n$ sum up to unity:

$$P\{A_1\} + \cdots + P\{A_n\} = P\{A_1 \uplus A_2 \uplus \cdots \uplus A_n\} = P\{S\} = 1 \quad (2.21)$$

Consider an event B in S and a partition $A_1, A_2, \cdots, A_n$ of S. Clearly,

$$B = B \cap S = B \cap (A_1 \uplus \cdots \uplus A_n) = (B \cap A_1) \uplus \cdots \uplus (B \cap A_n) \quad (2.22)$$

Since $(B \cap A_1), \ldots, (B \cap A_n)$ are mutually exclusive, we have

$$\begin{aligned} P\{B\} &= P\{(B \cap A_1) \uplus (B \cap A_2) \uplus \cdots \uplus (B \cap A_n)\} \\ &\overset{\text{Axiom 3}}{=} P\{B \cap A_1\} + P\{B \cap A_2\} + \cdots + P\{B \cap A_n\} \end{aligned}$$

But, for $P\{A_i\} \neq 0$,

$$P\{B \cap A_i\} \overset{(2.15)}{=} P\{B|A_i\}P\{A_i\}$$

Hence, we have the following result, known as ***total probability theorem***:

$$\boxed{P\{B\} = P\{B|A_1\}P\{A_1\} + P\{B|A_2\}P\{A_2\} + \cdots + P\{B|A_n\}P\{A_n\}} \quad (2.23)$$

This theorem is valid for any event B and any partition $A_1, A_2, \cdots, A_n$ of the sample space S. It facilitates greatly the calculation of $P\{B\}$ in many situations because both $P\{B|A_i\}$ and $P\{A_i\}$ may be much easier to calculate than a direct calculation of $P\{B\}$.

Total probability theorem is often useful for the calculation of the unconditional probability of an event $P\{B\}$ knowing various conditional probabilities of the events $P\{B|A_i\}$ and the probabilities of the conditioning events $P\{A_i\}$. Intuitively, it provides a way to find an "effect" from its "causes": It calculates the probability of an "effect" (event B) from the probabilities of all its possible "causes" (events A_i's) and the relationships between these possible "causes" and "effect" ($P\{B|A_i\}$).

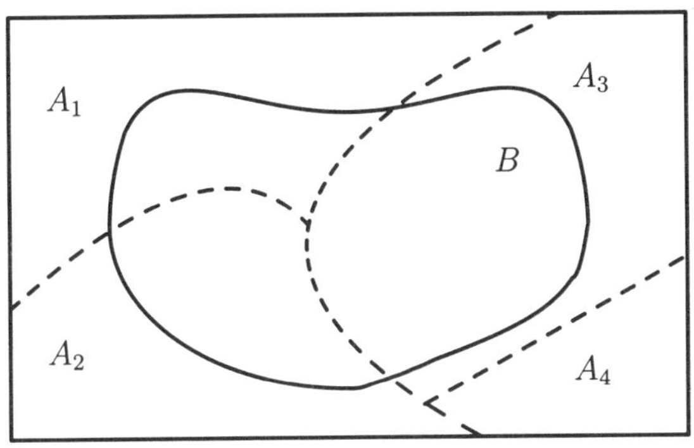

(a) sample space partitioning and event B

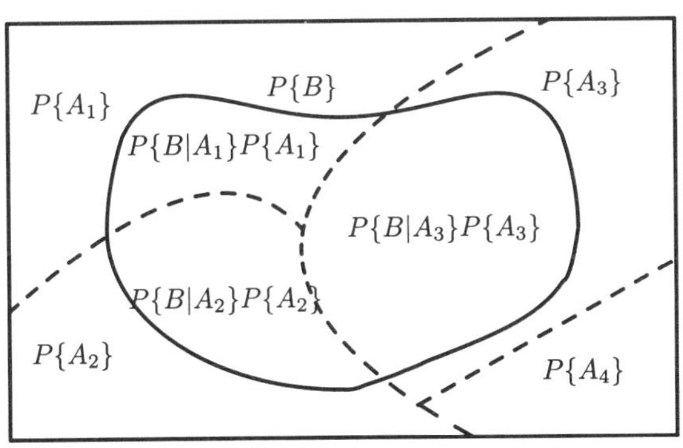

(b) probability decomposition

Figure 2.4: Illustration of total probability theorem.

2.5.2 Bayes' Rule

Since

$$P\{A|B\}P\{B\} \overset{(2.15)}{=} P\{A \cap B\} = P\{B \cap A\} \overset{(2.15)}{=} P\{B|A\}P\{A\}$$

we have the ***Bayes' rule***, ***Bayes' formula***, or ***Bayes' theorem***:

$$\boxed{P\{A|B\} = \frac{P\{B|A\}P\{A\}}{P\{B\}}} \tag{2.24}$$

In particular, for any partition $A_1, A_2, \cdots, A_n$ of the sample space S, we have

$$\begin{aligned} P\{A_i|B\} &= \frac{P\{B|A_i\}P\{A_i\}}{P\{B\}} \\ &\overset{(2.23)}{=} \frac{P\{B|A_i\}P\{A_i\}}{P\{B|A_1\}P\{A_1\} + \cdots + P\{B|A_n\}P\{A_n\}} \end{aligned} \tag{2.25}$$

How Bayes' rule should be interpreted divides statistics into two schools: Bayesian and non-Bayesian. The Bayesian school interprets the various probabilities involved in the Bayes' rule as follows:

$P\{A_i\}$ = ***a priori probability*** of event A_i
= probability of event A_i without knowing event B has occurred
$P\{A_i|B\}$ = ***a posteriori probability*** of event A_i
= probability of event A_i knowing event B has occurred

In this sense, Bayes' rule provides a way of calculating the a posteriori probability by combining the a priori probability with the evidence from the current experiment in which the occurrence of event B has been observed.

Bayes' rule is usually used to find the conditional probability $P\{A|B\}$ of event A given another event B knowing the reversely conditional probability $P\{B|A\}$. Intuitively, it is often used to find a "cause" from the "effect": given an "effect" (event B), it calculates the probability of a particular possible "cause" (event A_i) among all its possible "causes" (events $A_1, \ldots, A_n$).

Total probability theorem and Bayes' rule are two of the most important and useful probability laws.

Example 2.19: Probability of Correct Communication

A binary (with element 0 or 1) digital communication channel has the following error probabilities

$$P\{R_1|S_0\} = 0.1$$
$$P\{R_0|S_1\} = 0.05$$

where

$$S_0 = \{\text{"0" sent}\} \qquad R_0 = \{\text{"0" received}\}$$
$$S_1 = \{\text{"1" sent}\} \qquad R_1 = \{\text{"1" received}\}$$

Since only "0" or "1" can be received, we have

$$P\{R_0|S_0\} = P\{\overline{R_1}|S_0\} = 1 - P\{R_1|S_0\} = 0.9$$
$$P\{R_1|S_1\} = P\{\overline{R_0}|S_1\} = 1 - P\{R_0|S_1\} = 0.95$$

Suppose that it is discovered that "0" is received with probability 0.8 (i.e., $P\{R_0\} = 0.8$).

(a) Determine the probability that "1" is sent: Let $x = P\{S_1\}$. Since $S_0 \uplus S_1 = S$ (sample space), $P\{S_0\} = 1 - P\{S_1\} = 1 - x$. Then by total probability theorem,

$$0.8 = P\{R_0\} \overset{(2.23)}{=} P\{R_0|S_1\}P\{S_1\} + P\{R_0|S_0\}P\{S_0\} = 0.05x + 0.9(1-x)$$

Solving the above equation yields

$$P\{\text{"1" sent}\} = P\{S_1\} = x = \frac{0.1}{0.85} = 0.1176$$

(b) Determine the probability that "1" was sent given that "1" is received:

$$P\{S_1|R_1\} \overset{(2.24)}{=} \frac{P\{R_1|S_1\}P\{S_1\}}{P\{R_1\}} = \frac{(0.95)(0.1176)}{0.2} = 0.5588$$

(c) Determine the probability that "0" was sent given that "0" is received:

$$P\{S_0|R_0\} = \frac{P\{R_0|S_0\}P\{S_0\}}{P\{R_0\}} = \frac{(0.9)(1-1176)}{0.8} = 0.9926$$

Note that it is much more reliable to transmit "0" for this channel.

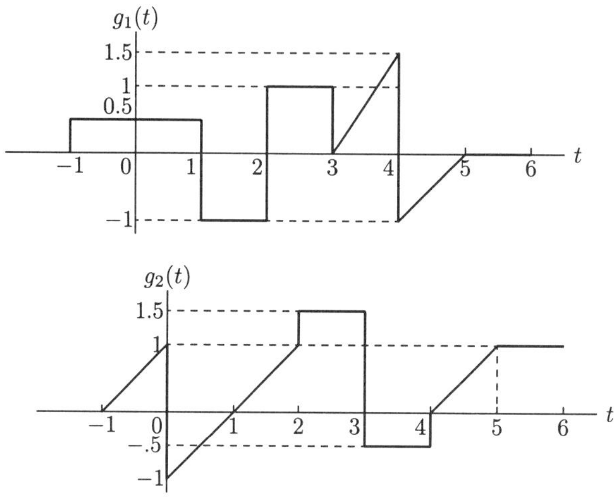

Example 2.20: Random Selection and Sampling of Waveform

Consider the waveforms and the following random experiment:

S1. First, select one waveform at random.
S2. Then, sample the selected waveform at a random time τ, $-1 \leq \tau < 6$.

If the sampled value $g(\tau) \geq 0.5$, what is the probability that it was sampled from $g_1(t)$? Let

$$A_i = \{\text{waveform } g_i(t) \text{ is sampled}\}, \qquad i = 1, 2 \Longrightarrow \text{ a partition}$$
$$B = \{\text{sampled value } g(\tau) \geq 0.5\}$$

By total probability theorem,

$$\begin{aligned} P\{B\} &= P\{B|A_1\}P\{A_1\} + P\{B|A_2\}P\{A_2\} \\ &= \frac{2+1+2/3}{7} \times \frac{1}{2} + \frac{1/2+1/2+1+1/2+1}{7} \times \frac{1}{2} = 0.5119 \end{aligned}$$

Thus, by Bayes' rule,

$$\begin{aligned} P\{g_1(t) \text{ sampled}|g(\tau) \geq 0.5\} = P\{A_1|B\} &= \frac{P\{B|A_1\}P\{A_1\}}{P\{B\}} \\ &= \frac{(11/21)(1/2)}{0.5119} = 0.5116 \end{aligned}$$

Example 2.21: Random Selection of Capacitor

Given the table below, consider the following random experiment: (1) first select a box; and (2) then choose a capacitor from the box.

	Box #			
Capacitance (μF)	1	2	3	*Total*
0.1	35	25	40	100
0.5	75	95	70	240
1.0	60	10	65	135
Total	170	130	175	475

Assume that the box selection and the capacitor selection are both with equal probability. If a $0.1\mu F$ capacitor is selected, what is the probability that it came from box 3? Let

$$A_i = \{\text{capacitors in box } i\} \quad \Longrightarrow \quad P\{A_i\} = \frac{1}{3}, \quad i = 1, 2, 3$$
$$B = \{0.1\mu F \text{ chosen}\}$$

Then, from the table,

$$P\{B|A_1\} = \frac{35}{170}, \qquad P\{B|A_2\} = \frac{25}{130}, \qquad P\{B|A_3\} = \frac{40}{175}$$

A_1, A_2, A_3 form a partition:

- A capacitor cannot be in both A_i and A_j — mutually exclusive
- A capacitor must be in one of A_i — exhaustive.

Thus, by total probability theorem and Bayes' rule,

$$P\{B\} = P\{B|A_1\}P\{A_1\} + P\{B|A_2\}P\{A_2\} + P\{B|A_3\}P\{A_3\}$$
$$= \frac{35}{170}\frac{1}{3} + \frac{25}{130}\frac{1}{3} + \frac{40}{175}\frac{1}{3} = 0.2089 \neq \frac{100}{475} \quad \text{(why?)}$$
$$P\{\text{box } 3|0.1\mu F\} = P\{A_3|B\} = \frac{P\{B|A_3\}P\{A_3\}}{P\{B\}} = \frac{(40/175)(1/3)}{0.2089} = 0.3647$$

Note that $P\{\text{box } 3|0.1\mu F\} \neq 40/100$. What if we assume (unrealistically for this problem) the following?

$$P\{A_1\} = \frac{170}{475}, \qquad P\{A_2\} = \frac{130}{475}, \qquad P\{A_3\} = \frac{175}{475}$$

See Example 2.31.

2.6 Combined Experiments and Bernoulli Trials

2.6.1 Combined Experiments

Consider the following random experiment:

S1. First roll a die.
S2. Then toss a coin (independent from Step 1).

The sample space of these two individual experiments are, respectively,

$$S_1 = \{1, 2, 3, 4, 5, 6\}$$
$$S_2 = \{H, T\}$$

This is a ***combined experiment*** with sample space

$$\begin{aligned} S &= S_1 \times S_2 \qquad \text{(Cartesian product of } S_1 \text{ and } S_2) \\ &= \{(1,H),(1,T),(2,H),(2,T),(3,H),(3,T), \\ &\quad (4,H),(4,T),(5,H),(5,T),(6,H),(6,T)\} \end{aligned}$$

What is the probability that an even number and head H will show up?

$$A = \{(2,H),(4,H),(6,H)\}$$
$$P\{A\} = \frac{3}{12}$$

Alternatively, this can be obtained as follows:

$$\begin{aligned} A_1 &= \{\text{an even number shows up in die rolling}\} \\ A_2 &= \{\text{head shows up in coin tossing}\} \\ P\{A\} &= P\{A_1 \times A_2\} \\ &\stackrel{?}{=} P\{A_1\}P\{A_2\} = \frac{3}{6} \times \frac{1}{2} \\ &= \frac{3}{12} \end{aligned}$$

The combined experiment of more than two experiments can be handled similarly.

2.6.2 Bernoulli Trials

Bernoulli trials are a special random combined experiment for which

- Only two outcomes (A and $\overline{A}$) are possible on any single trial.
- Repeated trials are independent (from trial to trial).

A typical problem is: What is the probability that A occurs exactly k times out of N (independent) trials? Let

$$\begin{aligned} p &= P\{A\} \text{ (on a single trial)} \\ q &= P\{\overline{A}\} = 1 - p \text{ (on a single trial)} \\ B &= \{A \text{ occurs exactly } k \text{ times out of } N \text{ trials}\} \end{aligned}$$

Consider events

$$B_1 = \{\underbrace{A, A, \ldots, A}_{k \text{ times}}, \underbrace{\overline{A}, \overline{A}, \ldots, \overline{A}}_{(N-k) \text{ times}}\} = \{A \text{ occurs on and only on each of first } k \text{ trials}\}$$

$$B_2 = \{\underbrace{A, A, \ldots, A}_{k-1 \text{ times}}, \overline{A}, A, \underbrace{\overline{A}, \overline{A}, \ldots, \overline{A}}_{(N-k-1) \text{ times}}\}$$

$$\vdots$$

$$B_M = \{\underbrace{\overline{A}, \overline{A}, \ldots, \overline{A}}_{(N-k) \text{ times}}, \underbrace{A, A, \ldots, A}_{k \text{ times}}\}$$

where M is the number of distinct orders for choosing k out of N trials:

$$M = \binom{N}{k} = \frac{N!}{k!(N-k)!} = \textbf{\textit{binomial coefficient}} \text{ of } k \text{ out of } N \qquad (2.26)$$

All these events have identical probability, given by, for $i = 1, \ldots, M$,

$$P\{B_i\} = P\{A\} \cdots P\{A\} P\{\overline{A}\} \cdots P\{\overline{A}\} = p^k (1-p)^{N-k} = p^k q^{N-k}$$

Since such sequences consist of mutually exclusive events, we have finally

$$\begin{aligned} P\{B\} &= P\{A \text{ occurs exactly } k \text{ times in } \textit{any order} \text{ out of } N \text{ trials}\} \\ &= P\{B_1 \uplus B_2 \uplus \cdots \uplus B_M\} = \sum_{i=1}^{M} P\{B_i\} = M P\{B_i\} = \binom{N}{k} p^k q^{N-k} \end{aligned}$$

or

$$\boxed{P\{A \text{ occurs exactly } k \text{ times in } N \text{ trials}\} = \tbinom{N}{k} [P\{A\}]^k [P\{\overline{A}\}]^{N-k}} \qquad (2.27)$$

The n factorial $n!$ is defined as $n! = n(n-1)(n-2)\cdots(2)(1)$, where n is a positive integer. Note that $0! \triangleq 1$. Factorial $n!$ increases *extremely* fast with n. For example, $70! > 10^{100}$.

To find the probability of an event using the classical definition, knowledge of combinatorics is essential.

There are the following four ways to choose a sample of m elements from a set of n distinguishable objects:

Table 2.1: Sample selection combinatorics.

Order matters?	*Repetition allowed?*	*The sample is known as*	*# of ways to select the sample*
Yes	Yes	m-permutation with replacement	n^m
Yes	No	m-permutation	$n!/(n-m)!$
No	Yes	m-combination with replacement	$\frac{(n+m-1)!}{m!(n-1)!}$
No	No	m-combination	$\frac{n!}{m!(n-m)!}$

The meaning of combination with or without replacement and permutation with or without replacement is explained below. The binomial coefficient $\binom{n}{m}$ is equal to the ***number of combination*** $\frac{n!}{m!(n-m)!}$.

Example 2.22: Sample Selection Combinatorics

Selection of a 2-element sample from the set $\{a, b\}$ has the following combinatorics:

Sample type	*# of ways to select the sample*	*set of samples*
permutation with replacement	$2^2 = 4$	$\{aa, ab, ba, bb\}$
permutation	$2!/(2-2)! = 2$	$\{ab, ba\}$
combination with replacement	$\frac{(2+2-1)!}{2!(2-1)!} = 3$	$\{ab, aa, bb\}$
combination	$\frac{2!}{2!(2-2)!} = 1$	$\{ab\}$

With replacement, the same element may be used to form a sample *repeated.* This is reflected in the samples aa and bb. The difference between ***permutation*** and ***combination*** is that ab and ba are considered to be different in permutation but the same in combination since the latter does not take order into account.

Note that $\binom{N}{m} = \binom{N}{N-m}$, $\binom{N}{0} = 1$, and $(a+b)^n = \sum_{m=1}^{n} \binom{n}{m} a^m b^{n-m}$.

Two individual experiments E_1 and E_2 with sample spaces S_1 and S_2, respectively, are ***independent*** if and only if every event A_1 of E_1 is independent of every event A_2 of E_2:

$$P\{A_1 \times A_2\} = P\{A_1 \times S_2\}P\{S_1 \times A_2\} = P\{A_1\}P\{A_2\}$$

Example 2.23: Repeated Die Rolling

A die will be rolled 5 times.

(a) What is the probability that "3" will show up exactly twice?
First identify for (2.27) that $k = 2$, $N = 5$, and

$$p = P\{A\} = \frac{1}{6} \qquad q = P\{\overline{A}\} = \frac{5}{6}$$

Then

$$P\{\text{"3" shows up exactly twice in 5 trials}\} = \binom{5}{2} p^2 q^{5-2} = \frac{5!}{2!3!}\left(\frac{1}{6}\right)^2\left(\frac{5}{6}\right)^3$$
$$= 0.16075$$

(b) What is the probability that "4" will show up at least twice?

$$\begin{aligned} P\{\text{"4" shows up at least twice}\} &= 1 - P\{\text{"4" does not show up}\} \\ &\quad - P\{\text{"4" shows up once}\} \\ &= 1 - \binom{5}{0} p^0 q^{5-0} - \binom{5}{1} p^1 q^{5-1} \\ &= 1 - \frac{5!}{0!5!}\left(\frac{1}{6}\right)^0\left(\frac{5}{6}\right)^5 - \frac{5!}{1!4!}\left(\frac{1}{6}\right)\left(\frac{5}{6}\right)^4 \\ &= 0.1962 \end{aligned}$$

Alternatively,

$$\begin{aligned} P\{\text{"4" shows up at least twice}\} &= P\{\text{"4" shows up twice}\} \\ &\quad + P\{\text{"4" shows up 3 times}\} \\ &\quad + P\{\text{"4" shows up 4 times}\} \\ &\quad + P\{\text{"4" shows up 5 times}\} \\ &= \binom{5}{2} p^2 q^3 + \binom{5}{3} p^3 q^2 + \binom{5}{4} p^4 q + \binom{5}{5} p^5 \\ &= 0.1962 \end{aligned}$$

(c) What is the probability that "4" shows up at least 5 times?

$$P\{\text{"4" shows up at least 5 times}\} = P\{\text{"4" shows up 5 times}\} = \binom{5}{5} p^5$$
$$= 0.0001286$$

Example 2.24: Probability of Typos

A typist makes an error while typing a letter 0.3% of the time. There are two types of errors. Type A and B errors occur 80% and 20% of the time, respectively, whenever an error occurs.

(a) What is the probability of no error in 10 letters? Let

$$E_i^N = \{i \text{ errors in } N \text{ letters}\} \quad \Longrightarrow \quad P\{E_1^1\} = 0.003$$
$$A_i^N = \{i \text{ type A errors in } N \text{ letters}\} \quad \Longrightarrow \quad P\{A_1^1|E_1^1\} = 0.80$$

Then, from (2.27),

$$\begin{aligned} P\{E_0^{10}\} = P\{\text{no error in 10 letters}\} &= \binom{10}{0}(P\{E_1^1\})^0(1 - P\{E_1^1\})^{10} \\ &= \binom{10}{0}(0.003)^0(1 - 0.003)^{10} = 0.9704 \end{aligned}$$

(b) What is the probability of no type A error in 10 letters? By (2.23),

$$P\{A_1^1\} = P\{A_1^1|E_1^1\}P\{E_1^1\} + \underbrace{P\{A_1^1|\overline{E_1^1}\}}_{=0 \text{ (why?)}} P\{\overline{E_1^1}\} = (0.8)(0.003) + 0 = 0.0024$$

Then,

$$\begin{aligned} P\{A_0^{10}\} = P\{\text{no type A error in 10 letters}\} &= \binom{10}{0}(P\{A_1^1\})^0(1 - P\{A_1^1\})^{10} \\ &= \binom{10}{0}(0.0024)^0(1 - 0.0024)^{10} = 0.9763 > P\{E_0^{10}\} \quad \text{(why?)} \end{aligned}$$

(c) Given that exactly one error has occurred in 10 letters, what is the probability that it is a type A error? By Bayes' rule,

$$\begin{aligned} P\{A_1^{10}|E_1^{10}\} &= \frac{P\{E_1^{10}|A_1^{10}\}P\{A_1^{10}\}}{P\{E_1^{10}\}} = \frac{P\{A_1^{10}\}}{P\{E_1^{10}\}} \qquad (\text{since } P\{E_1^{10}|A_1^{10}\} = 1) \\ &= \frac{\binom{10}{1}(P\{A_1^1\})^1(1 - P\{A_1^1\})^9}{\binom{10}{1}(P\{E_1^1\})^1(1 - P\{E_1^1\})^9} = \frac{(0.0024)(1 - 0.0024)^9}{(0.003)(1 - 0.003)^9} = 0.8043 \end{aligned}$$

Note that

$$P\{A_1^{10}|E_1^{10}\} = 0.8043 \neq 0.8 = P\{A_1^1|E_1^1\} = P\{\text{a type A error given an error}\}$$

You are invited to provide an explanation (see problems 2.38 and 2.39).

Example 2.25: Bernoulli Trials by P&R

Bernoulli trials can be performed easily using the companion software P&R as follows:[3]

S1. Click "Miscellaneous" in the main window of P&R.
S2. Click "Bernoulli Trials" and choose "Trial Generator." The "**Bernoulli Trial Generator**" window will appear.
S3. Enter the parameters (i.e., 20 trials and $P\{A\} = 0.36$) as shown in Fig. 2.5. Click "Ok." The total times event A occurs in this experiment is then given and the outcome of the trials plotted as a sequence of "0" and "1," as shown in Fig. 2.5.

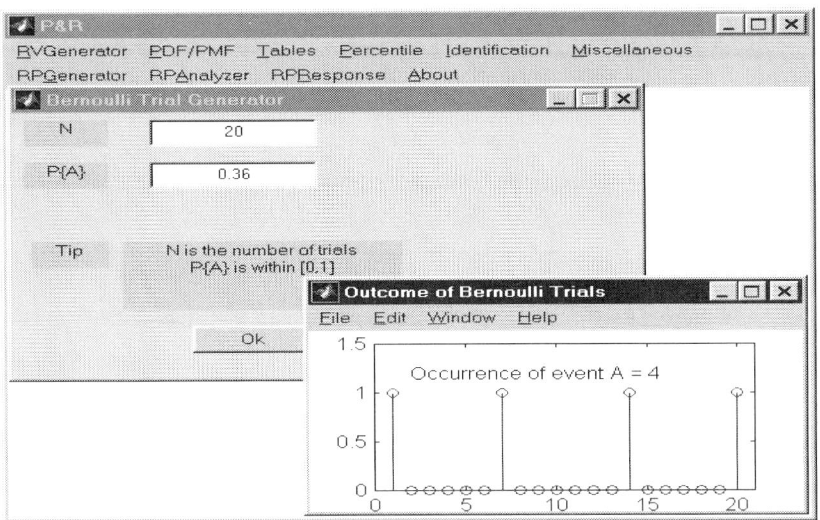

Figure 2.5: Bernoulli trials by P&R.

Example 2.26: Probability Calculation for Bernoulli Trials by P&R

The probability of an event in Bernoulli trials can also be computed easily using the companion software P&R as follows:

S1. Click "Miscellaneous" in the main window of P&R.
S2. Click "Bernoulli Trials" and choose "Probability Calculator." The "**Probability Calculator for Bernoulli Trials**" window will appear.
S3. Enter the parameters (i.e., 35 trials, event A occurs *at least* 7 times and $P\{A\} = 0.29$) as shown in Fig. 2.6. Click "Ok." The probability of the event A occurs at least 7 times is then given, as shown in Fig. 2.6.

[3] You are strongly encouraged to repeat all P&R examples. You will definitely learn faster and more this way. These examples are fairly easy to follow and repeat.

The probability of an event occurs exactly, at least, or at most k times can be calculated this way.

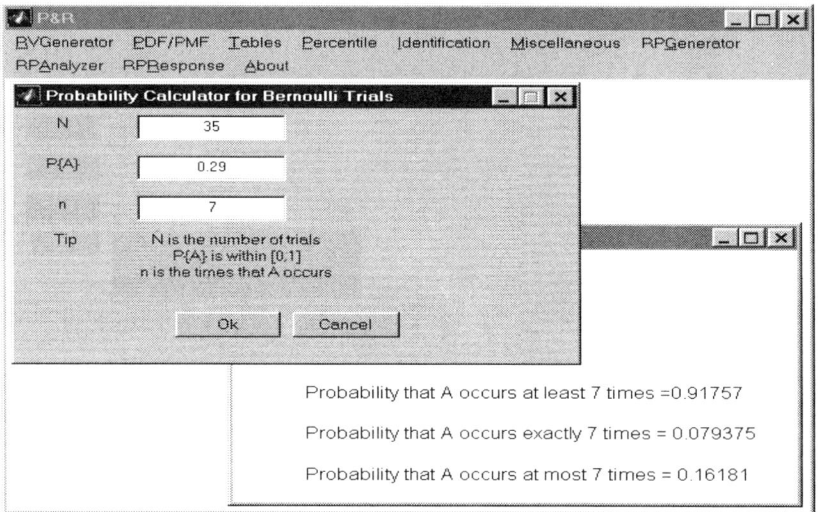

Figure 2.6: Probability Calculation for Bernoulli trials by P&R.

2.7 Summary and Requirements

Probability is a number assigned to an event that satisfies the following three axioms: (1) It is nonnegative and not larger than unity; (2) a sure event has unity probability; and (3) probability of the union of *mutually exclusive* events is equal to the sum of probabilities of the events. Special attention should be paid to the condition of the last axiom above. Probability of an event can be interpreted as (the limit of) the relative frequency of the occurrence of the event.

Random events are subsets of the sample space and thus manipulation of events is based on set operation.

The *probability of the union of two events* A and B is

$$P\{A+B\} \triangleq P\{A\cup B\} = P\{A\} + P\{B\} - P\{A\cap B\}$$

where $P\{A\cap B\}$ is the *joint probability* of events A and B, given by

$$P\{AB\} \triangleq P\{A\cap B\} = P\{A\} + P\{B\} - P\{A\cup B\} = P\{A|B\}P\{B\}$$

where $P\{A|B\} \triangleq \frac{P\{A\cap B\}}{P\{B\}}$ is the *conditional probability* of A given B. The probability of the complement of an event A is $P\{\overline{A}\} = 1 - P\{A\}$.

Two events are *independent* if the occurrence of one does not affect the occurrence of the other. Mathematically, that is, $P\{A \cap B\} = P\{A\}P\{B\}$. They are disjoint (or mutually exclusive) if the occurrence of one excludes the occurrence of the other; that is, $A \cap B = \emptyset$.

$$P\{AB\} \overset{\text{if } A,B \text{ independent}}{=} P\{A\}P\{B\}$$

$$P\{A+B\} \overset{\text{if } A,B \text{ disjoint}}{=} P\{A\}+P\{B\}$$

Total probability theorem is given by

$$P\{B\} = P\{B|A_1\}P\{A_1\} + P\{B|A_2\}P\{A_2\} + \cdots + P\{B|A_n\}P\{A_n\}$$

where $A_1, \ldots, A_n$ form a partition of the sample space. It calculates the probability of an "effect" (event B) from the probabilities of all its possible "causes" (events A_i's) and the relationships between the possible "causes" and the "effect" ($P\{B|A_i\}$).

Bayes' rule

$$P\{A_i|B\} = \frac{P\{B|A_i\}P\{A_i\}}{P\{B\}} = \frac{P\{B|A_i\}P\{A_i\}}{P\{B|A_1\}P\{A_1\} + \cdots + P\{B|A_n\}P\{A_n\}}$$

is often used to find a "cause" from the "effect": given an "effect" (event B), it calculates the probability of a particular possible "cause" (event A_i) among all its possible "causes" (events $A_1, \ldots, A_n$).

Independent trials where each trial is identical and has only two possible outcomes (A and $\overline{A}$) are known as *Bernoulli trials*. The probability of the number of the occurrence of A is given by

$$P\{A \text{ occurs exactly } k \text{ times in } N \text{ trials}\} = \binom{N}{k}[P\{A\}]^k[P\{\overline{A}\}]^{N-k}$$

Basic Requirements

- Understand the concept of probability and its relative-frequency interpretation. Be familiar with the properties of probability.
- Know how to find probability of an event for simple problems using the classical, geometric, and/or relative-frequency definitions.
- Be familiar with the formulas of probabilities of the union, intersection, and complement of events.
- Know how to judge whether two events are independent from both common sense and definitions. Have a good understanding of the difference between independent and mutual exclusive events.
- Comprehend conditional probability and know how to calculate it.
- Know when and how to apply the total probability theorem and Bayes' rule.
- Learn what type of problem can be formulated as Bernoulli trials and know how to apply (2.27).

In addition, be clear about the limitations of the traditional definitions of probability.

The major difficulties with the classical problems of probability are the manipulation of events (i.e., how to express an event in terms of some known events) and the correct recognition of the problem type. The most difficult topics of the chapter are the total probability theorem and the Bayes' rule, especially how they can be applied to solve problems.

2.8 Additional Examples

2.27 *Jury verdict.* DNA evidence is introduced on a murder trial. Suppose it is correct that if the accused is not guilty then the chance that the two DNA samples (one from the accused and the other found at the murder scene) have an exact match is one out of fifty million. A match is found. Do you think the accused is guilty or not? On what scientific ground is your judgment based?

Solution: If an event with an extremely small probability, based on a probabilistic model of the event, has occurred on a *single* trial, then in practice we have strong reasons (beyond a reasonable doubt) to believe that the underlying probabilistic model is incorrect. Since one out of fifty million is extremely small, we should think the underlying assumption that the accused is not guilty is incorrect and thus a reasonable verdict should be that the accused is guilty. Abandoning the underlying probabilistic model due to the occurrence of an extremely small probability event on a single trial is common practice in science and engineering and should also be acceptable by most people concerning court cases. Of course, such a judgment is based on the assumption that the statistics (one out fifty million in this case) is correct (or at least with an extremely high confidence of small errors) as well as some other assumptions, such as that the samples and the test are reliable and the probability that there is a conspiracy to frame the accused is extremely small.

2.28 *A/D conversion.* The following voltage waveform is to be sampled at a random time τ over the period $-1 \leq t \leq 5$.

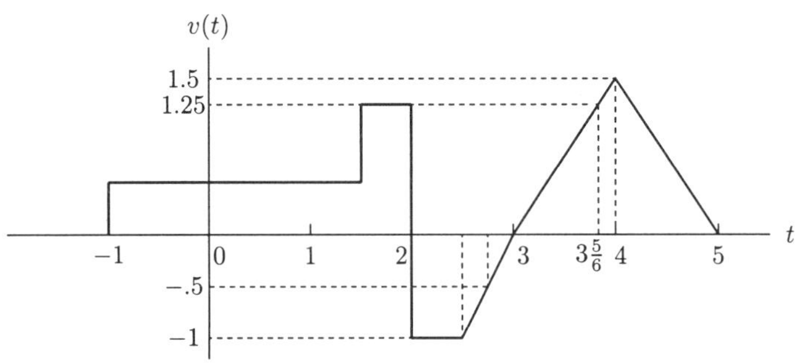

(a) What is the probability that the sampled value $v(\tau) < -0.5$?
(b) What is the probability that the sampled value $v(\tau) \geq 1.25$?
(c) What is the probability that the sampled value $v(\tau) = 1.25$?

(d) What is the probability that the sampled value $v(\tau) = -0.5$?
(e) What is the probability that the sampled value satisfies $-1 \le v(\tau) < 1.5$ but not equal to -0.5?

Solution:

(a) $P\{v(\tau) < -0.5\} = \frac{3/4}{6} = \frac{1}{8}$.
(b) $P\{v(\tau) \ge 1.25\} = \frac{1/2+2/6}{6} = \frac{5}{36}$.
(c) $P\{v(\tau) = 1.25\} = \frac{0.5}{6} = 1/12$.
(d) $P\{v(\tau) = -0.5\} = \frac{0}{6} = 0$.
(e) $P\{-1 \le v(\tau) < 1.5, v(\tau) \ne -0.5\} = 6/6 = 1$.

2.29 *Life insurance premium.* The probabilities that a husband and a wife will be alive 15 years from now are assigned as 0.9 and 0.95, respectively, by an insurance company to determine premiums. Find the probability that in 15 years (a) both will be alive; (b) neither will be alive; (c) at least one will be alive; (d) only one will be alive; (e) the husband will be alive alone. Assume that the husband and the wife will be alive independently.

Solution: Let H = husband alive, W = wife alive.

(a) $P\{\text{both alive}\} = P\{H \cap W\} \overset{\text{indenpendence}}{=} P\{H\}P\{W\} = (0.9)(0.95) = 0.855$.
(b) $P\{\text{neither alive}\} = P\{\overline{H} \cap \overline{W}\} \overset{\text{indenpendence}}{=} P\{\overline{H}\}P\{\overline{W}\} = (1-0.9)(1-0.95) = 0.05$.
(c) $P\{\text{at least one alive}\} = P\{H \cup W\} = P\{H\} + P\{W\} - P\{H \cap W\} = 0.9 + 0.95 - 0.855 = 0.995$.
(d) $P\{\text{only one alive}\} = P\{(H \cap \overline{W}) \uplus (\overline{H} \cap W)\} = P\{H \cap \overline{W}\} + P\{\overline{H} \cap W\} = P\{H\}P\{\overline{W}\} + P\{\overline{H}\}P\{W\} = (0.9)(1-0.95) + (1-0.9)(0.95) = 0.14$.
(e) $P\{\text{husband alive alone}\} = P\{H \cap \overline{W}\} = P\{H\}P\{\overline{W}\} = (0.9)(1-0.95) = 0.045$.

2.30 *Quality control.* A box of 100 resistors with certain resistance and tolerance is given below

Resistance (Ω)	*2% tolerance*	*5% tolerance*	*Total #*
20	10	16	26
50	25	14	39
100	29	6	35
Total	64	36	100

(a) What is the probability of drawing a resistor of 50Ω with 2% tolerance?
(b) What is the probability of drawing a resistor of 50Ω or having 2% tolerance?
(c) What is the probability of drawing a resistor of 50Ω given that the resistor has 2% tolerance?
(d) What is the probability of *not* drawing a resistor of 100Ω with 5% tolerance?

(e) What is the probability of drawing a resistor with 2% tolerance given that it is of 20Ω?

Solution:

(a) $P\{(50\Omega) \cap (2\%)\} = \frac{25}{100} = 0.25$.

(b) Since 50Ω and 2% are not mutually exclusive, we have

$$P\{(50\Omega) \cup (2\%)\} = P\{50\Omega\} + P\{2\%\} - P\{(50\Omega) \cap (2\%)\} = \frac{39}{100} + \frac{64}{100} - \frac{25}{100} = 0.78$$

(c) $P\{50\Omega|2\%\} = \frac{P\{(50\Omega)\cap(2\%)\}}{P\{2\%\}} = \frac{25/100}{64/100} = \frac{25}{64} = 0.3906$.

(d) $P\{\overline{(100\Omega) \cap (5\%)}\} = 1 - P\{(100\Omega) \cap (5\%)\} = 1 - \frac{6}{100} = \frac{47}{50} = 0.94$.

(e) $P\{2\%|20\Omega\} = \frac{P\{(20\Omega)\cap(2\%)\}}{P\{20\Omega\}} = \frac{10/100}{26/100} = \frac{5}{13} = 0.3846$.

2.31 *Quality control.* Given the table below, which lists the number of capacitors of a certain capacitance and tolerance in a box, assume a capacitor is chosen at random (with equal probability) out of all 475 capacitors.

	Tolerance			
Capacitance (μF)	1%	2%	5%	*Total*
0.1	35	25	40	100
0.5	75	95	70	240
1.0	60	10	65	135
Total	170	130	175	475

(a) What is the probability that a $0.1\mu F$ capacitor is chosen?

(b) What is the probability that a capacitor of 5% tolerance is chosen?

(c) If a $0.1\mu F$ capacitor is selected, what is the probability that it has tolerance 5%?

(d) Comparing with Example 2.21, why $P\{\text{box } 3|0.1\mu F\} = 0.3647 \neq \frac{40}{100}$ there?

Solution:

(a) $P\{0.1\mu F\} = \frac{100}{475}$.

(b) $P\{5\%\} = \frac{175}{475}$.

(c) $P\{5\%|0.1\mu F\} = \frac{P\{(5\%)\cap(0.1\mu F)\}}{P\{0.1\mu F\}} = \frac{40/475}{100/475} = \frac{40}{100}$.

(d) In Example 2.21, we assume selection of a box has equal probability $\frac{1}{3}$, which is not proportional to the number of capacitors in them: $170/130/175$. As a result, Example 2.21 and the current problem are based on two different random experiments. If we assume (which is realistic for the current example but not for Example 2.21)

$$P\{A_1\} = 170/475, \qquad P\{A_2\} = 130/475, \qquad P\{A_3\} = 175/475$$

then, it can be obtained easily that $P\{\text{box } 3|0.1\mu F\} = \frac{40}{100}$. The key is that by the above assumption of $P\{A_i\}$, each capacitor has equal probability of being chosen and thus it is equivalent to having all capacitors in one box.

2.32 *Random selection.* A technician has two electronic parts cabinets with drawer arrangements as below.

4-drawer cabinet	*3-drawer cabinet*
PNP	*PNP*
PNP	*PNP*
NPN	*NPN*
NPN	

The technician selects one cabinet at random and withdraws a transistor from one of the drawers which is also chosen at random. Assume that each cabinet and each drawer within the selected cabinet are equally likely to be selected.

(a) What is the probability that a PNP transistor is chosen?
(b) Given that an NPN transistor is chosen, what is the probability that it came from the 3-drawer cabinet?

Solution:

(a) By total probability theorem,

$$P\{PNP\} = P\{PNP|4\}P\{4\} + P\{PNP|3\}P\{3\} = \frac{1}{2}\cdot\frac{1}{2} + \frac{2}{3}\cdot\frac{1}{2} = \frac{7}{12} = 0.5833$$

(b) By Bayes' rule,

$$P\{3|PNP\} = \frac{P\{PNP|3\}P\{3\}}{P\{PNP\}} = \frac{(1/3)(1/2)}{1-7/12} = \frac{2}{5} = 0.4$$

2.33 *All roads lead to Rome.* Toss a coin three times. Find the probability that head shows up at least once.

Solution: Let $A = \{\text{head shows up at least once}\}$, H = head, T = tail.

Method 1: Since each toss has two possible outcomes H and T with equal probability, there are $2^3 = 8$ possible outcomes when toss three times. Then

$$P\{A\} = 1 - P\{\text{all tails}\} = 1 - \frac{1}{8} = \frac{7}{8}$$

Method 2: In all 8 possible outcomes, 7 are in favor of A. Thus

$$P\{A\} = \frac{7}{8}$$

Method 3: Let $A_n = \{\text{head shows up exactly } n \text{ times}\}$, $n = 1, 2, 3$. Clearly,

$$P\{A_1\} = P\{HTT\} + P\{THT\} + P\{TTH\} = \frac{3}{8}$$

$$P\{A_2\} = P\{HHT\} + P\{HTH\} + P\{THH\} = \frac{3}{8}$$

$$P\{A_3\} = P\{HHH\} = \frac{1}{8}$$

Since A_1, A_2, A_3 are mutually exclusive,

$$P\{A\} = P\{A_1 \uplus A_2 \uplus A_3\} = P\{A_1\} + P\{A_2\} + P\{A_3\} = \frac{3}{8} + \frac{3}{8} + \frac{1}{8} = \frac{7}{8}$$

Method 4: Tossing three times is an experiment with three independent trials, each with two possible (equally probable) outcomes. By Bernoulli trials, $p = 1/2$,

$$P\{A\} = 1 - P\{\text{exactly 0 times out of 3 trails}\} = 1 - \binom{3}{0}(1/2)^0(1/2)^3 = \frac{7}{8}$$

Method 5: Tossing three times is an experiment with three independent trials, each with two possible (equally probable) outcomes. By Bernoulli trials, $p = 1/2$,

$$\begin{aligned} P\{A\} &= P\{\text{once out of 3 trails}\} + P\{\text{twice out of 3 trails}\} + P\{\text{3 times out of 3 trails}\} \\ &= \binom{3}{1}(1/2)^1(1/2)^2 + \binom{3}{2}(1/2)^2(1/2)^1 + \binom{3}{3}(1/2)^3(1/2)^0 = \frac{7}{8} \end{aligned}$$

2.34 *Preference of multiple-choice problems.* A test consists of 5 multiple-choice problems, each with 4 choices and only one is correct and only one is allowed to choose. A poor student does not know the correct answer to any of the problems. He/she selects the choices at random. What are the probabilities that he/she has 1, 2, 3, 4, or 5 correct choices?

Solution: Let $P\{n\} = \{n \text{ correct choices}\}$. Then since the probability of a correct choice for each problem is 1/4, by Bernoulli trials,

$$P\{n\} = \binom{5}{n}(1/4)^n(1 - 1/4)^{5-n}$$

Thus,

$$\begin{aligned} P\{\text{no correct choice}\} = P\{0\} &= \binom{5}{0}(1/4)^0(1 - 1/4)^{5-0} = 243/1024 = 0.2373 \\ P\{\text{1 correct choice}\} = P\{1\} &= \binom{5}{1}(1/4)^1(1 - 1/4)^{5-1} = 405/1024 = 0.3955 \\ P\{\text{2 correct choice}\} = P\{2\} &= \binom{5}{2}(1/4)^2(1 - 1/4)^{5-2} = 270/1024 = 0.2637 \\ P\{\text{3 correct choice}\} = P\{3\} &= \binom{5}{3}(1/4)^3(1 - 1/4)^{5-3} = 90/1024 = 0.08789 \\ P\{\text{4 correct choice}\} = P\{4\} &= \binom{5}{4}(1/4)^4(1 - 1/4)^{5-4} = 15/1024 = 0.01465 \\ P\{\text{5 correct choice}\} = P\{5\} &= \binom{5}{5}(1/4)^5(1 - 1/4)^{5-5} = 1/1024 = 0.000977 \end{aligned}$$

Note that these probabilities do sum up to unity. The student would have scored 0 point were the test not in the form of multiple choices. He/she, however, will have only 23.73% chance of getting 0 point now. Clearly, multiple-choice problems are in favor of poor students and the average test score will be higher than what it should be.

2.35* *Probability of modem error.*[4] A digital modem makes an error in transmitting a bit of a message with 0.01% probability. 60% of the errors are type A errors: the actual bit is 0 but the modem sends/receives it as 1. The other type (type B) of errors are: the actual bit is 1 but the modem sends/receives it as 0.

(a) What is the probability of no error when sending a 100 bit message?
(b) What is the probability of no type B error when sending a 100-bit message?
(c) Given that an error has occurred when sending a 100-bit message, what is the probability that it is not a type B error?
(d) Given that two errors have occurred when sending a 50-bit message, what is the probability that exactly one type B error occurs?

Solution: Let

$$\begin{aligned} E_i^N &= \{\text{exactly } i \text{ errors in } N \text{ bits}\} \\ A_i^N &= \{\text{exactly } i \text{ type A errors in } N \text{ bits}\} \\ B_i^N &= \{\text{exactly } i \text{ type B errors in } N \text{ bits}\} \end{aligned}$$

The problems then can be treated as Bernoulli independent trials. The single event probabilities are

$$P\{E_1^1\} = 0.01\%$$

and, by total probability theorem,

$$\begin{aligned} P\{A_1^1\} &= P\{A_1^1|E_1^1\}P\{E_1^1\} + \underbrace{P\{A_1^1|\overline{E_1^1}\}}_{=0}P\{\overline{E_1^1}\} = P\{A_1^1|E_1^1\}P\{E_1^1\} \\ &= (60\%)(0.01\%) = 0.006\% \\ P\{B_1^1\} &= P\{B_1^1|E_1^1\}P\{E_1^1\} + \underbrace{P\{B_1^1|\overline{E_1^1}\}}_{=0}P\{\overline{E_1^1}\} = P\{B_1^1|E_1^1\}P\{E_1^1\} \\ &= (40\%)(0.01\%) = 0.004\% \end{aligned}$$

where $P\{A_1^1|\overline{E_1^1}\} = 0$ and $P\{B_1^1|\overline{E_1^1}\} = 0$ because $\overline{E_1^1}$ stands for no error in sending a single bit.

(a) Here $p = P\{E_1^1\}$,

$$P\{E_0^{100}\} = \binom{100}{0}(0.01\%)^0(1 - 0.01\%)^{100} = 99\%$$

[4]A number with a star means that the example or problem is more challenging.

(b) Here $p = P\{B_1^1\}$,

$$P\{B_0^{100}\} = \binom{100}{0}(0.004\%)^0(1 - 0.004\%)^{100} = 99.6\%$$

(c)

$$\begin{aligned}
P\{\overline{B_1^{100}}|E_1^{100}\} = P\{A_1^{100}|E_1^{100}\} &= \frac{P\{(A_1^{100}) \cap (E_1^{100})\}}{P\{E_1^{100}\}} = \frac{P\{A_1^{100}\}}{P\{E_1^{100}\}} \\
&= \frac{\binom{100}{1}(P\{A_1^1\})^1(1 - P\{A_1^1\})^{99}}{\binom{100}{1}(P\{E_1^1\})^1(1 - P\{E_1^1\})^{99}} = \frac{(0.006\%)^1(1 - 0.006\%)^{99}}{(0.01\%)^1(1 - 0.01\%)^{99}} \\
&= (0.6)^1\left(\frac{0.99994}{0.9999}\right)^{99} = 0.6024
\end{aligned}$$

(d)

$$\begin{aligned}
P\{B_1^{50}|E_2^{50}\} &= 1 - P\{B_0^{50}|E_2^{50}\} - P\{B_2^{50}|E_2^{50}\} \\
&= 1 - \frac{P\{(B_0^{50}) \cap (E_2^{50})\} + P\{(B_2^{50}) \cap (E_2^{50})\}}{P\{E_2^{50}\}} \\
&= 1 - \frac{P\{A_2^{50}\} + P\{B_2^{50}\}}{P\{E_2^{50}\}} \\
&= 1 - \frac{\binom{50}{2}\left[(P\{A_1^1\})^2(1 - P\{A_1^1\})^{48} + (P\{B_1^1\})^2(1 - P\{B_1^1\})^{48}\right]}{\binom{50}{2}(P\{E_1^1\})^2(1 - P\{E_1^1\})^{48}} \\
&= 1 - \frac{(0.00006)^2(0.99994)^{48} + (0.00004)^2(0.99996)^{48}}{(0.0001)^2(0.9999)^{48}} = 0.4788
\end{aligned}$$

where

$$\begin{aligned}
\{(B_0^{50}) \cap (E_2^{50})\} &= \{\text{two errors (but not type B) in sending 50-bit message}\} \\
&= \{\text{two type A errors in sending 50-bit message}\} = A_2^{50}
\end{aligned}$$

2.9 Problems

Words we use create states on us.
So you want to change "problems" to "challenges."

— Antony Robbins, *Personal Power*

2.1 *Venn diagram.* Draw Venn diagrams to verify

$$\begin{aligned}
A \cap (B \cup C) &= (A \cap B) \cup (A \cap C) \\
A \cup (B \cap C) &= (A \cup B) \cap (A \cup C)
\end{aligned}$$

2.2 *Venn diagram.* Draw Venn diagrams to show (2.2).

2.3 *Venn diagram.* Draw Venn diagram to verify

$$\overline{\overline{(\overline{A} \cap B)} \cup \overline{C}} = (\overline{A} \cap B) \cap C$$

2.4 *Set operation.* Determine whether each of the following is true or false: (a) $A \cup B = A\overline{B} \cup B$; (b) $(AB)(A\overline{B}) = \emptyset$; (c) $A \cup B = \overline{A}B$; (d) $\overline{A \cup B} \cap C = \overline{A}\,\overline{B}\,\overline{C}$; (e) If $A \subset B$, then $A = AB$, $A \cup B = B$, $\overline{B} \subset \overline{A}$; (f) If $AB = \emptyset$ and $C \subset B$, then $AC = \emptyset$.

2.5 *Set operation.* Let sample space $S = \{0, 1, \ldots, 9\}$ and two sets $B = \{4, 5, 9\}$, $C = \{2, 3, 5\}$. Find $\overline{C} \cup B$, $\overline{B}C$, and $\overline{C\overline{BC}}$.

2.6 *Probability of set difference.* Find $P\{A - B\}$ in terms of $P\{A\}$ and $P\{B\}$ if (a) A and B are mutually exclusive or (b) $A \subset B$.

2.7 *Properties of probability.* Assume in this problem that events $A \subset B$. (a) Is it always true that $P\{A\} \leq P\{B\}$? Justify your answer. (b) Find $P\{A \cup B\}$ in terms of $P\{A\}$ and $P\{B\}$. (c) Use the axiomatic approach to show that $P\{B - A\} = P\{B\} - P\{A\}$.

2.8 *Disjoint probability.* Find $P\{\overline{AB}\}$ if $P\{A\} = 0.8$, $P\{A - B\} = 0.5$.

2.9 *Summation law.* Given $P\{A\} = 0.4$, $P\{B\} = 0.25$, and $P\{A \cup B\} = 0.55$, find $P\{A\overline{B}\}$.

2.10 *Probability of union of three events.* Derive an expression for $P\{A \cup B \cup C\}$ in terms of $P\{A\}$, $P\{B\}$, $P\{C\}$, $P\{AB\}$, $P\{BC\}$, $P\{CA\}$, and $P\{ABC\}$.

2.11 *Probability of three events.* Given $P\{A\} = 0.2$, $P\{B\} = P\{C\} = 0.25$, $P\{BC\} = 0$, and $P\{AB\} = P\{CA\} = 0.1$, find

(a) the probability that at least one of the events A, B, and C occurs
(b) the probability that none of the events A, B, and C occurs

2.12 *Minimum number of hits to kill.* Assume that an aircraft carrier will be sunk with a probability of 0.3 if hit by a torpedo. At least how many torpedoes are required to hit the aircraft carrier so that it is sunk with a probability of 0.9?

2.13 *Elementary probability of sampling with replacement.* In a lottery, 10 different numbers are in a box. Each of 10 persons gets a number from the box at random (without putting it back). Which one in the sequence to get the number has the maximum winning probability? How large is the probability?

2.14 *Birthday problem.* Assume that the birthday of a student is random over the 365 days of a year. A class has 20 students. Find the probability that at least two students have the same birthday.

2.15 *Equivalent formulas for independence.* Show that if $P\{A\} \neq 0$ and $P\{B\} \neq 0$, then $P\{A|B\} = P\{A\}$ and $P\{B|A\} = P\{B\}$ are equivalent; that is, one implies the other.

2.16 *Die-rolling problem.* For the die rolling experiment, let $F_i = \{\text{face } i \text{ shows up}\}$. Find

(a) $P\{F_i,\, i \text{ even}|F_1 \cup F_2 \cup F_3 \cup F_4\}$
(b) $P\{F_1 \cup F_2 \cup F_3 \cup F_4|F_i, i \text{ even}\}$
(c) $P\{F_1 \cap F_2 \cap F_3 \cap F_4|F_i, i \text{ even}\}$
(d) $P\{F_2 \cup F_6|F_1 \cup F_2 \cup F_3 \cup F_4\}$

2.17 *Relation of various probabilities.* Consider two events A and B. (a) Under what condition(s) is $P\{A\} = P\{A \cup B\}$? (b) Under what condition(s) is $P\{A\} = P\{A \cap B\}$? (c) Under what condition(s) is $P\{A\} = P\{A|B\}$?

2.18 *Relation of various probabilities.* Given $P\{A\} = 0.4$, $P\{B\} = 0.15$, and $P\{A|B\} = 0.65$, find $P\{A \cup B\}$.

2.19 *Probabilities of two events.* Two cities A and B have an identical probability 0.24 of raining in an arbitrary day, and the probability 0.15 of raining in the same day. Find

(a) the probability that it will rain in either A or B (or both)
(b) the probability that it is raining in A knowing that it is raining in B
(c) the probability that it is raining in B knowing that it is raining in A

2.20 *Lower bound on event probability.* Give a lower bound for $P\{A\}$ in terms of $P\{B\}$ and $P\{C\}$ if A always occurs whenever B and C occur simultaneously.

2.21 *Hit probability.* A and B are shooting at the same target independently with probabilities 0.6 and 0.15 of hitting it. Find

(a) the probability that the target is hit by either A or B
(b) the probability that the target is hit by A knowing that it is hit

2.22 *Quality control.* In a drawer of 12 capacitors, there are 3 defected ones. Two capacitors are taken out from the drawer, one at a time without replacement. Find

(a) the probability that the *first* capacitor is defective
(b) the probability that the *second* capacitor is defective knowing that the first one is defective
(c) the probability that both capacitors are defective
(d) the probability that the *second* capacitor is defective

2.23* *Fraud probability.* Students A and B may make an identical and uncommon mistake in solving an exam problem. If the students should work independently, A and B would have 0.05 and 0.02 probabilities of making this mistake. Assume that the probability that A and B make the same mistake (independently or dependent) is 0.008. Find

(a) the probability that they make the mistake independently
(b) the probability that A copies B's mistake
(c) the probability that B copies A's mistake

2.24 *Reliability.* A hospital has three independent power supplies with reliability of 0.99, 0.995, and 0.95, respectively. Find the probability of the hospital's power outage.

2.25 *Reliability.* An alarm system consists of N independent subsystems connected in parallel, each with reliability 0.97. What is the smallest N that guarantees 0.9995 reliability of the overall system?

2.26 *Independence vs. pairwise independence.* A regular tetrahedron has three of its four sides in red, green, and yellow, respectively. Its fourth side is in a mixed color with red, green, and yellow simultaneously. Consider a random experiment of tossing this tetrahedron (the side facing down is chosen). Let

$$R = \{\text{a side with red color shows up}\}$$
$$G = \{\text{a side with green color shows up}\}$$
$$Y = \{\text{a side with yellow color shows up}\}$$

Note that in the above "a side with red color" does not mean "a side with red color alone."

(a) Find probabilities $P\{R\}$, $P\{G\}$, and $P\{Y\}$.
(b) Find probabilities $P\{RG\}$, $P\{GY\}$, $P\{YR\}$, and $P\{RGY\}$.
(c) Are events R, G, and Y independent?
(d) Are events R, G, and Y pairwise independent?

2.27 *Condition for independence.* Show that if $P\{A|B\} + P\{\overline{A}|\overline{B}\} = 1$ and $0 < P\{A\} < 1$, $0 < P\{B\} < 1$, then A and B are independent.

2.28 *Condition for independence.* If $P\{A|B\} = P\{A|\overline{B}\}$ and $0 < P\{B\} < 1$, show that A and B are independent.

2.29 *Probability of independent events.* Events A, B, and C are independent. Given $P\{A\} = 0.3$, $P\{B\} = 0.4$, $P\{C\} = 0.5$, find (a) $P\{A \cup B\}$; (b) $P\{A \cup B \cup C\}$; (c) $P\{A - B\}$.

2.30 *Probability of independent events.* Given two independent events A and B with $P\{A\} = 0.3$, $P\{B\} = 0.4$, find (a) $P\{A \cup B\}$; (b) $P\{\overline{A}\,\overline{B}\}$; (c) $P\{B|A\}$; (d) $P\{\overline{B}|\overline{A}\}$.

2.31 *Cause finding.* Suppose that independent missiles A, B, and C have probabilities 0.4, 0.5, 0.6 of hitting the enemy's headquarters, respectively. Assume that the enemy's headquarters will be destroyed with a probability 0.3, if hit by a single missile, 0.6 if hit by two missiles, and 1 if hit by all three missiles. Find

(a) the probability that the headquarters is destroyed

(b) the probability that the headquarters is destroyed by a single missile knowing that it is destroyed
(c) the probability that the headquarters is destroyed by missile A knowing that it is destroyed

2.32 *Power system security.* A power system will collapse if a ground fault occurs on any two of lines A, B, and C simultaneously. Assume that ground faults on lines A, B, and C are independent and have identical probability of 0.01.

(a) What is the probability that the system will collapse?
(b) What is the probability that the system will collapse given that a ground fault on line A has occurred?
(c) What is the probability that a ground fault on line A occurred given that the system has collapsed?

2.33 *Winning probability.* At 14-14 tie in a volleyball game, the game will be over if a team scores additional 2 points more than the other team. Suppose that teams A and B have the probabilities 0.55 and 0.45 of wining a single point, respectively. Find the probability that team A will win.

2.34* *Spot-check.* To control the quality of 200 oscilloscopes, 4 of them are taken out at random and tested independently. If at least one is found defective, then all 200 oscilloscopes will be rejected. Assume that a defective oscilloscope will be found defective with 0.98 probability and a defect-free one will be mistakenly declared defective with 0.01 probability. Find the probability of accepting these 200 oscilloscopes if in fact 5 of them are defective.

2.35 *Probability of success.* Suppose that event A has the probability p of at least one occurrence in N independent trials. Find

(a) the probability of event A on a single trial
(b) the probability that A occurs at most once
(c) the probability of event A on a single trial if $p = 79/81$ and $N = 4$

2.36 *Probability of winning.* Consider the following combined experiment

- Roll a die.
- Toss a coin.
- Play a fair game (with no tie and equal probability for win and loss).

Assume that these experiments are independent of each other.

(a) What is the probability that an odd number and head show up and win the game?
(b) What is the probability that an odd number shows up and win the game given that the head shows up?

2.37* *Probability of a code.* The code of a safe consists of four distinct digits. Find

(a) the probability that the code includes neither digit 3 nor digit 4
(b) the probability that the code includes digit 3 but not digit 4

2.38* *Effect of nonoccurrence.* For Example 2.24, explain why $P\{A_1^{10}|E_1^{10}\} \neq P\{A_1^1|E_1^1\}$.

2.39 *Effect of nonoccurrence.* For Example 2.24, show that $P\{A_1^N|E_1^N\} > P\{A_1^{N-1}|E_1^{N-1}\} > P\{A_1^1|E_1^1\}, \forall N > 2$.

2.40* *Probability of communication error.* Consider a communication link which consists of two independent and identical channels of self-test problem 2.5 in series. Find

(a) the probability that a "1" is received correctly
(b) the probability that a single digit is received correctly
(c) the probability that a 10-digit number is received correctly
(d) the probability that a 10-digit number is received with exactly one incorrect digit
(e) the probability that a 10-digit number is received with some error
(f) the probability that a "1" is incorrectly received as a "0" in a 10-digit number given that the number is received with exactly one incorrect digit

2.41* *Probability of communication error.* Consider a communication link which consists of two independent and identical channels of self-test problem 2.5 in parallel, meaning that a bit is sent through both channels at the same time. Suppose that the receiver accepts the bit if they match, otherwise, it requests that the bit be sent again through both channels. Find

(a) the probability that a "1" is received correctly
(b) the probability that a single digit is received correctly
(c) the probability that a 10-digit number is received incorrectly
(d) the probability that a 10-digit number is received with exactly one incorrect digit
(e) the probability that a "1" is incorrectly received as a "0" in a 10-digit number given that the number is received with exactly one incorrect digit

2.42 *De Meré's perplexity.* Consider rolling a fair die 4 times. Let $A_i = \{6 \text{ shows up on roll } i\}$. Find the probability of de Meré's first scheme (p. 32); that is, the probability of at least one 6 out of 4 rolls.

2.43 *De Meré's perplexity.* Consider rolling a pair of fair dice 24 times. Let $A_i = \{12 \text{ shows up on roll } i\}$. Find the probability of de Meré's second scheme (p. 32); that is, the probability of at least one 12 out of 24 rolls.

2.44 *Set number preference.* In a tennis match, a player has the probability 0.6 of winning each single set. Find the probability that he will win (a) two out of three sets; (b) three out of five sets; and (c) four out of seven sets. If you were the player, would you prefer to have more sets in a match? Why? The more the better?

2.45 *Puzzling questions.* Raise two technical questions of your own that you are puzzled by and that are closely related to the material covered in this chapter. These questions should be important, not too simple, and not too involved.

2.10 Computer Exercises

2.1 *Probability calculation for Bernoulli trials.*

(a) Write a MATLAB function subroutine `bn_exact.m`, with the top line $P =$ `bn_exact` (p, N, k), implementing (2.27) to calculate $P = P\{A$ occurs exactly k times in N trials$\}$, where $p = P\{A\}$ is the probability of event A on any single trial; N is the total number of trials; and k is the number of occurrences of event A.

(b) Use `bn_exact.m` to solve part (a) of Example 2.23. Make sure the results are correct. Compare the results with those obtained using the companion software P&R.

(c) Find P for $p = 0.36643, N = 120, k = 47$.

2.2 *Probability calculation for Bernoulli trials.*

(a) Write a MATLAB function subroutine `bn_atleast.m`, with the top line $P =$ `bn_atleast`(p, N, k), calculating $P = P\{A$ occurs at least k times in N trials$\}$ based on subroutine `bn_exact.m`, where $p = P\{A\}$ is the probability of event A on any single trial; N is the total number of trials; and k is the number of occurrences of event A.

(b) Use `bn_atleast.m` to solve parts (b) and (c) of Example 2.23, and problems 2.42 and 2.43. Make sure the results are correct.

(c) Find P for $p = 0.36643, N = 120, k = 47$.

(d) Use the companion software P&R to find P and compare the result with the above.

2.3 *Probability calculation for Bernoulli trials.*

(a) Write a MATLAB function subroutine `bn_atmost.m`, with the top line $P =$ `bn_atmost`(p, N, k), calculating $P = P\{A$ occurs at most k times in N trials$\}$ based on subroutine `bn_exact.m`, where $p = P\{A\}$ is the probability of the occurrence of event A on any single trial; N is the total number of trials; and k is the number of occurrences of event A.

(b) Find P for $p = 0.36643, N = 120, k = 46$.

(c) Use the companion software P&R to find P and compare the result with the above.

2.4 *Bernoulli trials.*

(a) Use the companion software P&R to conduct 60 Bernoulli trials in which event A has a probability 0.23 of occurrence on each trial. Record the number of occurrences of event A.

(b) Repeat (a) 10 times. How many times did event A have at least 45 occurrences out of 60 trials in your experiment? What is the corresponding theoretical probability?
(c) Is the relative frequency of occurrence of event A approximately equal to 0.23? Why or why not?

2.11 Self-Test Problems

If you can't solve a problem,
you can always look up the answer.
But please, try first to solve it by yourself;
then you'll learn more and you'll learn faster.

Donald E. Knuth, *The TEXbook* (1983)

2.1 Answer the following questions briefly.

(a) Are events A and $\overline{A}$ mutually exclusive? Given two mutually exclusive events A and B, is it always true that $B = \overline{A}$?
(b) Is it reasonable to use traditional definition to find the probability that a particular face of an unfair die will show up?
(c) Is it always true that $P\{A + B\} = P\{A \cup B\} = P\{A\} + P\{B\}$?
(d) Can we use the three axioms of probability alone to find the probability of an event?
(e) Are $P\{A \cap B\}$, $P\{AB\}$, and $P\{A, B\}$ different notations of the same thing?
(f) Is it true that $P\{A \cap B\} = P\{A\}P\{B\}$ if A and B are mutually exclusive?
(g) Is it true that $P\{A \cup B\} = P\{A\} + P\{B\}$ if A and B are independent?
(h) Under what condition for the set of events $A_1, \ldots, A_n$, can the total probability theorem be applied to event B?

2.2 Let A, B, C be three random events. Express the following in terms of A, B, and C.

(a) B occurs, A, C do not occur.
(b) A and B occur, but C does not occur.
(c) A, B and C all occur.
(d) One and only one event occurs.
(e) At most one event occurs.
(f) At least one event occurs.
(g) None of the events occurs.
(h) At least two events occur.
(i) At most two events occur.

2.3 Given $P\{AB\} = P\{\overline{A}\,\overline{B}\}$ and $P\{A\} = 0.3$, find $P\{B\}$.

2.4 An Internet user A would like to send an important message to user B. Assume that there are two links from A to B, each with 6 relay nodes connected in series. A message may reach B through a link only if every relay node in that link is operational. Assume that a node is operational with probability 0.99, which is independent of any other node.

(a) Find the probability that one link is operational regardless of the other link.
(b) Find the probability that at least one link is operational.
(c) Find the probability that both links are operational.
(d) What is the probability that a message will reach B if it is sent through a link chosen at random?
(e) What is the probability that a message will reach B if it is sent through both links?

2.5 A binary (with element 0 or 1) computer communication channel has the following error probabilities

$$P\{R_1|S_0\} = 0.1 \qquad P\{R_0|S_1\} = 0.05$$

where

$$S_0 = \{\text{“0” sent}\} \qquad R_0 = \{\text{“0” received}\}$$
$$S_1 = \{\text{“1” sent}\} \qquad R_1 = \{\text{“1” received}\}$$

Suppose that 0 is *sent* with probability of 0.8. Find

(a) the probability that “1” is received
(b) the probability that “1” was sent given that “1” is received
(c) the probability that “0” was sent given that “0” is received

2.6 A professor answers 4 times questions raised by students in a class. Let W denote “wrong answer” and R “right answer.” Suppose that the professor’s answer is incorrect with probability 1/10 each time, that is, $P\{W\} = 0.1$. Find

(a) the probability of $WWRW$ in that order
(b) the probability that the professor answers correctly exactly 3 times
(c) the probability that the professor answers correctly at least 3 times

2.12 Solutions to Self-Test Problems

2.1 (a) Two events A and B are mutually exclusive if $A \cap B = \emptyset$. Since events A and $\overline{A}$ satisfy $A \cap \overline{A} = \emptyset$ as well as $A \cup \overline{A} = S$, they are mutually exclusive. However, not every pair of mutually exclusive events A and B are complement events of each other (i.e., it is not always true $B = \overline{A}$) because it could be the case $A \cup B \neq S$.
(b) The traditional definitions of probability assume that each outcome has equal probability. Since an unfair die will have unequal probability for different faces, these definitions should not be used.

(c) Since $A + B$ and $A \cup B$ are different notations of the same thing, $P\{A + B\} = P\{A \cup B\}$. However, $P\{A + B\} = P\{A\} + P\{B\}$ if and only if $P\{A \cap B\} = 0$ since $P\{A+B\} \overset{(2.10)}{=} P\{A\}+P\{B\}-P\{A\cap B\}$. Note that $P\{A\cap B\} = 0$ if and only if A and B are mutually exclusive, $P\{A\} = 0$ or $P\{B\} = 0$.

(d) Any probability has to satisfy these three axioms. Conversely, if an assignment of numbers satisfies these three axioms, then it can be used as probability. In other words, the three axioms are actually the fundamental properties of probability and all the other properties and laws of probability can be derived from these three properties. As properties, these axioms alone cannot provide probability of any event. They just specify the necessary and sufficient conditions for a quantity to be probability.

(e) As stated in (2.12), $P\{A \cap B\}$, $P\{AB\}$, and $P\{A, B\}$ are different notations of the same thing.

(f) No, it is true only if A and B are independent.

(g) No, it is true only if A and B are mutually exclusive.

(h) The events $A_1, \dots, A_n$ should be a partition of the sample space. More specifically and more precisely, they should satisfy the following equation

$$P\{A_1\} + P\{A_2\} + \cdots + P\{A_n\} = 1$$

Even more precisely, $A_1, \dots, A_n$ need only to satisfy (2.22):

$$B = A_1B \uplus A_2B \uplus \cdots \uplus A_nB$$

where $A_1B, A_2B, \dots, A_nB$ are mutually exclusive (i.e., $A_iB \cap A_jB = \emptyset, \forall i \neq j$); that is, B must be the *disjoint union* of $A_1B, A_2B, \dots, A_nB$.

2.2 (a) $B\overline{A}\,\overline{C}$; (b) $BA\overline{C}$; (c) ABC; (d) $(A\overline{B}\,\overline{C}) \cup (\overline{A}B\overline{C}) \cup (\overline{A}\,\overline{B}C)$; (e) $\overline{AB \cup BC \cup CA}$, or equivalently, $(\overline{A}\,\overline{B}\,\overline{C}) \cup (A\overline{B}\,\overline{C}) \cup (\overline{A}B\overline{C}) \cup (\overline{A}\,\overline{B}C)$; (f) $A \cup B \cup C$; (g) $\overline{A}\,\overline{B}\,\overline{C}$; (h) $AB \cup BC \cup CA$; (i) $\overline{ABC}$ or equivalently, $(\overline{A}\,\overline{B}\,\overline{C}) \cup (A\overline{B}\,\overline{C}) \cup (\overline{A}B\overline{C}) \cup (\overline{A}\,\overline{B}C) \cup (AB\overline{C}) \cup (\overline{A}BC) \cup (A\overline{B}C)$.

2.3 Note that

$$\begin{aligned} P\{A\} + P\{B\} - P\{A \cup B\} &\overset{(2.13)}{=} P\{A \cap B\} \overset{\text{given}}{=} P\{\overline{A}\,\overline{B}\} \overset{(2.1)}{=} P\{\overline{A \cup B}\} \\ &\overset{(2.11)}{=} 1 - P\{A \cup B\} \end{aligned}$$

This leads to

$$P\{A\} + P\{B\} = 1 \quad \Longrightarrow \quad P\{B\} = 1 - P\{A\} = 1 - 0.3 = 0.7$$

2.4 (a) $P\{\text{a link OK}\} = P\{\text{each node in the link OK}\} = p^6 = (0.99)^6 = 0.9415$

(b)

$$\begin{aligned} P\{\text{both links down}\} &= P\{\text{neither link OK}\} = [1 - P\{\text{a link OK}\}]^2 \\ &= (1 - 0.9415)^2 = 0.00342 \\ P\{\text{at least one link OK}\} &= 1 - 0.00342 = 0.9966 \end{aligned}$$

(c) $P\{\text{both links OK}\} = (0.9415)^2 = 0.8864$
(d) $P\{\text{message reaches } B\}|_{\text{use one link}} = P\{\text{a link OK}\} = 0.9415$
(e) $P\{\text{message reaches } B\}|_{\text{use both links}} = P\{\text{at least one link OK}\} = 0.9966$

2.5

$$P\{R_0|S_0\} = P\{\overline{R_1}|S_0\} = 1 - P\{R_1|S_0\} = 0.9$$
$$P\{R_1|S_1\} = P\{\overline{R_0}|S_1\} = 1 - P\{R_0|S_1\} = 0.95$$

(a) By total probability theorem,

$$P\{R_1\} = P\{R_1|S_0\}P\{S_0\}+P\{R_1|S_1\}P\{S_1\} = (0.1)(0.8)+(0.95)(1-0.8) = 0.198$$

(b) By Bayes' rule,

$$P\{S_1|R_1\} = \frac{P\{R_1|S_1\}P\{S_1\}}{P\{R_1\}} = \frac{(0.95)(1-0.8)}{0.198} = 0.9596$$

(c) By total probability theorem and Bayes' rule,

$$\begin{aligned} P\{S_0|R_0\} &= \frac{P\{R_0|S_0\}P\{S_0\}}{P\{R_0\}} = \frac{P\{R_0|S_0\}P\{S_0\}}{P\{R_0|S_0\}P\{S_0\} + P\{R_0|S_1\}P\{S_1\}} \\ &= \frac{(0.9)(0.8)}{(0.9)(0.8) + (0.05)(0.2)} = 0.9863 \end{aligned}$$

2.6 (a) $P\{WWRW\} = (0.1)(0.1)(1-0.1)(0.1) = 0.0009.$
(b)

$$\begin{aligned} P\{\text{exactly 3 correct answers}\} &= P\{\text{exactly 1 wrong answer}\} \\ &= \binom{4}{1}(0.1)^1(0.1)^3 = 0.2916 \end{aligned}$$

(c)

$$\begin{aligned} P\{\text{at least 3 correct answers}\} &= P\{\text{at most 1 wrong answer}\} \\ &= \binom{4}{0}(0.1)^0(0.1)^4 + \binom{4}{1}(0.1)^1(0.1)^3 = 0.9477 \end{aligned}$$

3

THE RANDOM VARIABLE

Moral: There is no safety in numbers, or in anything else.

James Thurber, *The Fairly Intelligent Fly*

Chapter 2 covers probabilistic tools effective primarily for simple events with countably many or equally probable outcomes. They are ineffective or complicated for problems with more involved events.

This chapter deals with the random variable approach, a powerful and effective tool for many random problems. It is based on describing every outcome of a random experiment as a unique *number* and thus makes many powerful tools in calculus applicable.

Main Topics

- Concept of Random Variable
- Cumulative Distribution Function
- Probability Density Function
- Uniform Distribution
- Gaussian Distribution
- Expectation and Moments
- Functions of a Random Variable
- Generation of Random Numbers

3.1 Concept of Random Variable

Recall that a (dependent) variable $u = f(v)$ is a function $f(\cdot)$ of an independent variable v such that to every value of v there corresponds a value of u. That is, $f(\cdot)$ is obtained by assigning a (unique) number to every value of v.

A random variable is simply a dependent variable as a function of an independent variable — the outcomes of a random experiment. It is a *numerical description* of the outcomes. Specifically, to *every* outcome ω of a random experiment, we assign a *unique* number $X(\omega)$. The *function* $X(\cdot)$ thus defined is then a ***random variable* (*RV*)** (since its value is uncertain prior to performing the experiment). The value $x = X(\omega')$ for a *given* outcome ω' is called its ***realization*** or *the* ***value*** *on which the RV X takes.*

Convention:

Upper case Latin letter	=	RV
Lower case Latin letter	=	the value

RVs may be grouped into three types, depending on the *range* of the function:

- A ***discrete RV*** is one that may take on only discrete values.
- A ***continuous RV*** is one having a continuous range of values.
- A ***mixed RV*** is one with both discrete and continuous values.

Only discrete RVs may be defined on a discrete sample space S. However, a discrete RV may have a sample space consisting of a continuous range of points or of a mixture of such regions and isolated points.

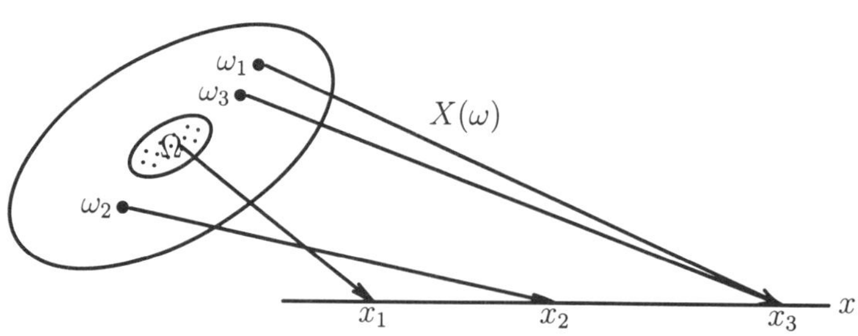

Figure 3.1: Mapping of a random variable.

The Moral cited above was drawn out from the following story about a *Fairly Intelligent Fly* told by James Thurber:

> A large spider in an old house built a beautiful web in which to catch flies. Every time a fly landed on the web and was entangled in it the spider devoured him, so that when another fly came along he would think the web was a safe and quiet place in which to rest. One day a fairly intelligent fly buzzed around above the web so long without lighting that the spider appeared and said, "Come on down." But the fly was too clever for him and said, "I never light where I don't see other flies and I don't see any other flies in your house." So he flew away until he came to a place where there were a great many other flies. He was about to settle down among them when a bee buzzed up and said, "Hold it, stupid, that's flypaper. All those flies are trapped." "Don't be silly," said the fly, "they're dancing." So he settled down and became stuck to the flypaper with all the other flies.

Probability studies various statistical regularities of random phenomena *mathematically and thus numerically*. It is thus natural and convenient to *quantify* various random events. Since a random event has different outcomes due to random effects, the quantity describing the event will also take different values. This leads to the concept of the random variable. The introduction of the random variable makes it possible to apply many powerful tools in calculus.

In this chapter, the approach based on the concept of random variable is studied, which is an extremely powerful tool for many probabilistic problems. It is based on describing every outcome of a random experiment as a unique *number* and thus converts various events into their equivalent $\{x_1 < X \leq x_2\}$ form, where x_1 and x_2 are two real numbers. As a result, many powerful tools in calculus can be applied. The study in this chapter is limited to problems that can be handled with a single random variable. Multiple random variables will be studied later.

A random variable as a single-value function differs from a conventional function in the following: Prior to an experiment, while the possible values of a RV are known, the exact value on which a RV takes is random depending on the outcome of the experiment, but with definite statistical regularity in the sense that each value has a certain probability of being taken on. Also, although the domain of a conventional function is the set (or a subset) of all real numbers, a RV is defined over a sample space whose elements are not necessarily (actually often not) real numbers.

Example 3.1: For the random experiment of tossing a coin, the following assignments defines three RVs X, Y, and Z:

$$X(\text{head}) = 0, \quad X(\text{tail}) = 1$$
$$Y(\text{head}) = 1, \quad Y(\text{tail}) = 0.5$$
$$Z(\text{head}) = 0, \quad X(\text{tail}) = -1$$

Example 3.2: Some Random Variables Defined on Die Rolling

Consider rolling a die. Let F_i be the event that face i shows up. Three RVs X, Y, and Z are defined by the following assignments:

$$\begin{array}{lll} X(F_i) = i & & \\ Y(F_i) = -i & & \\ Z(F_1) = 3.3, & Z(F_2) = -1.2, & Z(F_3) = 3.3 \\ Z(F_4) = -2.1, & Z(F_5) = 0.2, & Z(F_6) = -3.7 \end{array}$$

After defining RVs, the calculation of event probabilities reduces to that of probabilities of the RVs taking on certain values. For example,

$$\begin{aligned} P\{\text{an even number shows up}\} &= P\{X = 2, 4, \text{ or } 6\} = P\{Z \le 0\} \\ &= P\{Y = -2, -4, \text{ or } -6\} \end{aligned}$$

Example 3.3: Events in the Form of $\{X \le x\}$

This example demonstrates that $\{X \le x\}$ is an event. Consider the RV X of Example 3.2 of rolling a die, that is, $X(F_i) = i$. Since X can only take on integers $1, 2, \ldots, 6$, we have

$$\{X \le 3.4\} = \{X \le 3.8\} = \{F_1, F_2, F_3\}, \quad \{X \le 0.5\} = \emptyset, \quad \{X \le 8\} = S$$

Thus, $P\{X \le 3.4\} = P\{X \le 3.8\} = 3/6 = 0.5$.

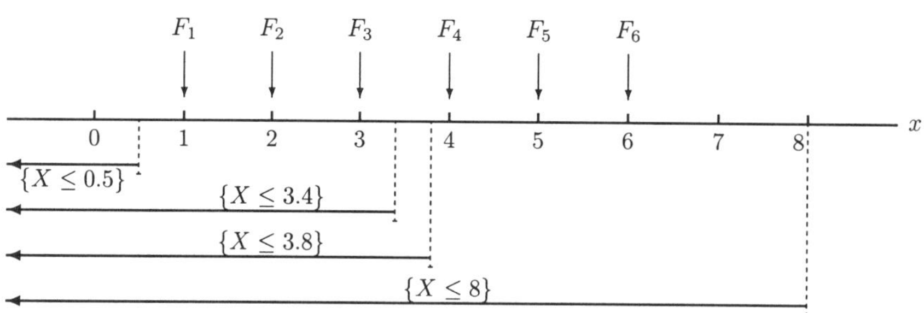

Figure 3.2: Illustration of $\{X \le x\}$ as an event for Example 3.3.

Not every function can be used to define a RV. For example, a multi-valued function is illegal here: There may have more than one number to describe some individual outcomes. This is so stipulated to avoid unnecessary ambiguities in the numerical description of the outcomes. However, the same number may be assigned to (used to describe) different outcomes, such as the RV Z in Example 3.2 and the RV of Example 3.4 below. In addition, we assign ∞ and $-\infty$ only to either an empty set or a set of outcomes with zero probability; that is,

$$P\{X(\omega) = \infty\} = P\{X(\omega) = -\infty\} = 0 \tag{3.1}$$

Otherwise, we could have $P\{X \geq \infty\} \neq 0$ and $P\{X \leq -\infty\} \neq 0$, which is not convenient.

Since the sample spaces of Examples 3.1 and 3.2 are discrete, only discrete RVs can be defined.

Example 3.4: A $10\mu F$ capacitor with 10% tolerance is taken off the shelf. Its capacitance C is unknown but is known in the continuous range of $9\mu F < C \leq 11\mu F$. C is quantized into 4 levels of continuous sections of equal size such that a discrete RV X is generated, where $X = 1, 2, 3, 4$.

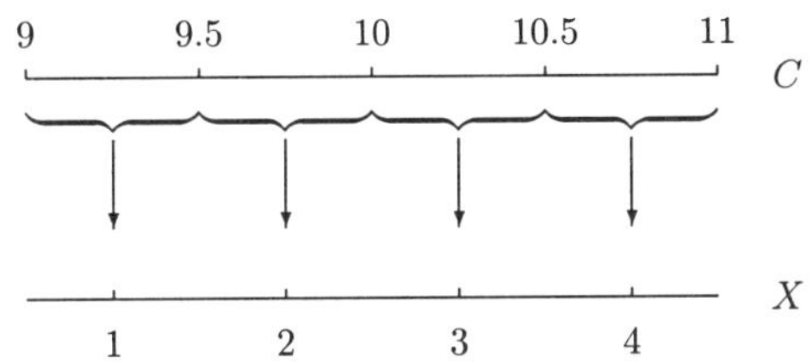

Note that X is a RV since its value is random before C is determined. For example,

$$P\{X = 1\} = P\{9\mu F < C \leq 9.5\mu F\}$$

Example 3.5: Consider the random experiment of testing the life span of a bulb. Let X be the life span (in hours) of the bulb. Then X is a random variable because its value for a particular bulb is uncertain before the bulb is tested. Note that X is a nonnegative continuous RV because its range of possible values is $x \geq 0$.

Example 3.6: The teaching performance of a professor will be evaluated by all 20 students in his class. Each student can give him either "bad," "fair," "good," "very good," or "excellent." Let X_i and Y_i be the numbers assigned to the evaluation given by student i determined as in the following table:

Evaluation	*bad*	*fair*	*good*	*very good*	*excellent*
X_i	1	2	3	4	5
Y_i	0	1	2	3	4

Then both X and Y are RVs defined for the same random experiment. Let $U = \frac{1}{20}\sum_{i=1}^{20} X_i$ and $V = \frac{1}{20}\sum_{i=1}^{20} Y_i$. Then both U and V are also RVs, which can be used as the final performance grades of the professor.

3.2 Cumulative Distribution Function

3.2.1 Motivation and Definition

$\{X \leq x\}$ is an event that corresponds to the set of all outcomes ω_i's for which $X(\omega_i) \leq x$. Its probability is of interest in many cases. This leads to defining the ***cumulative distribution function*** (***CDF***) of a RV X as

$$\boxed{F(x) = F_X(x) = P\{X \leq x\}} \tag{3.2}$$

i.e., $F_X(x)$ is the probability that RV X takes on a value not greater than x.

It can be shown that

- A discrete RV is one having a stairway-type (and thus discontinuous) CDF (see Figure 3.3).
- A continuous RV is one having an absolutely continuous CDF (see Figure 3.4).

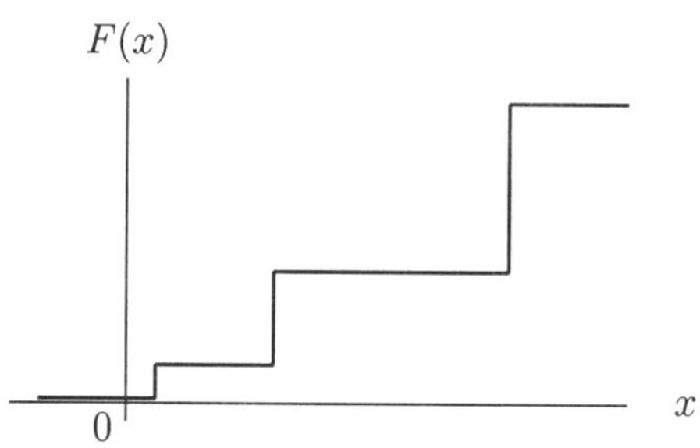

Figure 3.3: The general pattern of the CDF of a discrete RV.

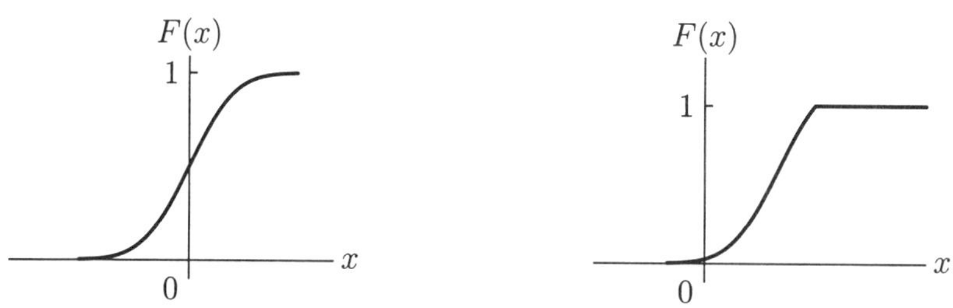

Figure 3.4: Some general patterns of the CDF of a continuous RV.

More rigorously, in order for X to be a RV, $\{X \leq x\}$ must be an event for every real x. In fact, $\{X \leq x\}$ is the set of all the outcomes ω_i's (i.e., a subset of the sample space) such that $X(\omega_i) \leq x$, where $X(\omega_i)$ is a *given* number depending on ω_i. For instance, in Example 3.2,

$$\begin{aligned}\{X \leq 3\} &= \{F_1, F_2, F_3\} \\ \{Y \leq -3\} &= \{F_4, F_5, F_6\} \\ \{Z \leq 0\} &= \{F_2, F_4, F_6\}\end{aligned}$$

because for instance, $Z(F_2) = -1.2 \leq 0, Z(F_4) = -2.1 \leq 0, Z(F_6) = -3.7 \leq 0$. In Example 3.1, $\{X < 0\} = \{\text{head}\}, \{Y \leq 0\} = \emptyset$ and in Example 3.4, $\{X \leq 3\} = \{9\mu F < C \leq 10.5\mu F\}$.

Clearly, events in the form of $\{X > x\}$ or $\{x_1 < X \leq x_2\}$ can be represented in terms of $\{X \leq x\}$ as

$$\begin{aligned}\{X > x\} &= \{-\infty \leq X \leq \infty\} - \{X \leq x\} = S - \{X \leq x\} = \overline{\{X \leq x\}} \\ \{x_1 < X \leq x_2\} &= \{X \leq x_2\} - \{X \leq x_1\}\end{aligned}$$

Events in the form of $\{X = x\}$ will be studied in Section 3.3.

The ***cumulative distribution function*** (CDF) is a function (of x) distributed over possible values of the RV X, hence the name. To completely describe a RV, not only the set of all its possible values must be specified, but also a description of the probability of each value on which it may take is essential. More generally, it is better to have a description of the probability that the RV takes on a value within an arbitrary interval $(a, b]$. The CDF provides such a description and thus contains all probabilistic information of the RV. It will be shown in the next section that a description (called probability density or mass function) of the probability of each value on which a RV may take can be obtained easily from the CDF.

The CDF is a conventional real-valued function with some nice properties, to be studied next. Such functions are studied extensively in calculus. As such, introduction of the CDF builds a bridge between probability and calculus and thus random phenomena can be studied using the familiar methods and tools developed in calculus.

If all possible values of a RV can be listed or counted, then the RV is discrete. Otherwise, it is either continuous or mixed. Since a continuous RV has an (absolutely) continuous CDF and a discrete RV has a stairway-type CDF, if the CDF of a RV is neither (absolutely) continuous nor of a stairway type, it has to be a mixed RV.

A CDF $F(x)$ is absolutely continuous if it is the integral of a nonnegative and integrable function $f(x)$ ($f(x)$ cannot include any delta function); that is, $F(x) = \int_{-\infty}^{x} f(v)dv$. This means that the probability mass is distributed smoothly (i.e., without concentration of probability mass at any point). An absolutely continuous CDF must be continuous but a continuous CDF is not necessarily absolutely continuous, just like not every continuous function $G(x)$ has a derivative function $g(x)$ such that $G(x) = \int_{-\infty}^{x} g(v)dv$. If the CDF is continuous but not absolutely continuous then the RV is not continuous. However, we will not deal with such cases and thus from now on we will not distinguish between continuous and absolutely continuous CDFs.

3.2.2 Properties of CDF

1. $0 \leq F(x) \leq 1,\ \forall x$, since $F(x)$ is probability.
2. $F(-\infty) = P\{X \leq -\infty\} = P\{X = -\infty\} = 0$, which follows from (3.1).
3. $F(\infty) = P\{X \leq \infty\} = 1$, since $\{X \leq \infty\}$ is a sure event.
4. $F(x)$ is nondecreasing as x increases:

$$F(x_2) \geq F(x_1) \qquad \forall x_1 < x_2 \tag{3.3}$$

This follows from the fact that $\{X \leq x_1\} \subset \{X \leq x_2\}$. Specifically,

$$\begin{aligned} F(x_2) = P\{X \leq x_2\} &\stackrel{?}{=} P\{(X \leq x_1) \uplus (x_1 < X \leq x_2)\} \\ &\stackrel{?}{=} P\{X \leq x_1\} + P\{x_1 < X \leq x_2\} \\ &= F(x_1) + P\{x_1 < X \leq x_2\} \geq F(x_1) \end{aligned} \tag{3.4}$$

5. By (3.4),

$$\boxed{P\{x_1 < X \leq x_2\} = F(x_2) - F(x_1)} \tag{3.5}$$

A special case is:

$$P\{X > x\} = P\{x < X \leq \infty\} = F(\infty) - F(x) = 1 - F(x) \tag{3.6}$$

6. The CDF of a discrete RV X taking on values $x_1, x_2, \ldots, x_n$ is given by

$$F(x) = P\{X \leq x\} = \underbrace{P\{X = x_1\} + P\{X = x_2\} + \cdots + P\{X = x_j\}}_{\text{if } x_i < x_{i+1},\ \forall i \text{ and } x_j \leq x < x_{j+1}}$$

$$F(x) = \sum_{i=1}^{n} P\{X = x_i\} u(x - x_i) = \sum_{x_i \leq x} P\{X = x_i\} = \sum_{x_i \leq x} p_i \tag{3.7}$$

where $p_i = P\{X = x_i\},\ i = 1, \ldots, n$, are known as ***point masses*** and $u(x - x_i)$ is the ***unit step function*** with a jump at $x = x_i$, defined by

$$u(x - a) = \begin{cases} 0 & x < a \\ 1 & x \geq a \end{cases} \tag{3.8}$$

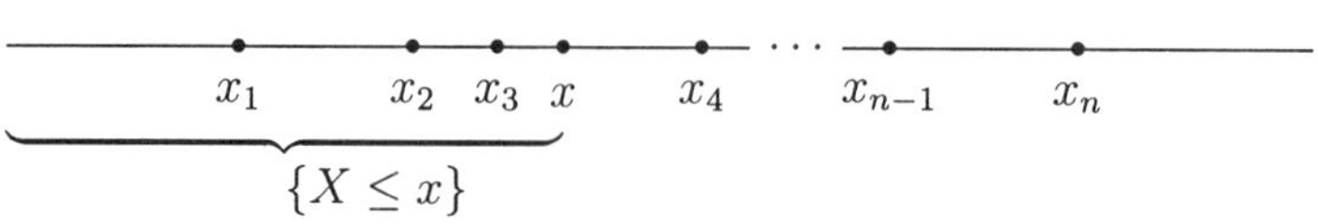

$\forall x$ is a common mathematical shorthand for "for every x."

The third equation in (3.4) follows from Axiom 3 and the fact that $(-\infty, x_1]$ and $(x_1, x_2]$ are mutually exclusive. The second equation follows from the fact:

$$(-\infty, x_2] = (-\infty, x_1] \cup (x_1, x_2]$$

(3.5) is extremely important: It relates the probability of the event $\{x_1 < X \le x_2\}$ to the CDF and here lies the primary use of the CDF. Thus the probability of the events in the form of $\{x_1 < X \le x_2\}$ or $\{X > x\}$ can be found easily from the probabilities of events in the form of $\{X \le x\}$.

A point mass $p_i = P\{X = x_i\}$ is the *height of the jump* at x_i in the CDF.

(3.7) can be viewed as the definition of the CDF for a discrete RV. It basically states that

CDF of a discrete RV = sum of point masses located to the left of (and including) x

Note that the definition (3.8) of the unit step function differs slightly from the following commonly used one:

$$u(x-a) = \begin{cases} 0 & x < a \\ 1 & x > a \end{cases} \tag{3.9}$$

where the value of $u(x-a)$ is not defined at $x = a$. For (3.7) to be correct, however, $u(x-a)$ has to be defined to be equal to 1 at $x = a$, as in (3.8), which will make a difference if $x = x_i$ for some x_i.

Two additional properties of a CDF that are important are

7. A CDF defined by (3.2) is always "continuous from right"; that is,

$$F(x^+) \triangleq \lim_{\epsilon \to 0} F(x + \epsilon) = F(x)$$

See Fig. 3.3. It is not necessarily "continuous from left." However, see the remarks below.

8. The CDF can have at most countably many discontinuous points.

Proofs and discussions of these properties can be found in many more advanced textbooks, e.g., A. Papoulis, *Probability, Random Variables, and Stochastic Processes*, McGraw-Hill, New York, 1991.

In fact, if a function satisfies properties 1, 2, 3, 4, and 7, then it must be the CDF of some RV. These properties can be used to test if a function can be a valid CDF of a RV.

While we define CDF as $P\{X \le x\}$, it is sometimes defined as $P\{X < x\}$. The former definition is commonly used in North America and Western Europe and the latter in the former Soviet Union countries. The only difference resulted is that the latter leads to "continuous from left" rather than "continuous from right." As a result, for a continuous RV there is no difference but for a discrete RV, the difference lies in whether the point x is counted in the interval or not.

Note that CDF is not unique to a RV; that is, two distinct RVs may have the same CDF. This can be understood easily by considering Example 3.2. Let RV U be defined by $U(F_i) = i+1, i = 1,2,3,4,5$ and $U(F_6) = 1$. Then U and X are different but have identical CDF.

Example 3.7: Determination of CDF: Power Consumption

Consider 3 independent machines, each having 75% of the time in operation and a power consumption of 1kW when in use. We wish to predict the average total power consumption of the 3 machines.

Let $X = \{\text{total \# of machines in operation}\}$. Note that X is discrete since it can only take on 0, 1, 2, or 3. Given X, the total power consumption is known and thus the problem becomes how to find the CDF of X.

Note that this is a Bernoulli trial problem with a total of 3 trials. Then, from (2.27) with $p = 0.75$,

$$
\begin{aligned}
F_X(0) &= P\{X \le 0\} = P\{X = 0\} = \binom{3}{0}(0.75)^0(1-0.75)^3 = 0.0156 \\
F(1) - F(0) &= P\{X \le 1\} - P\{X \le 0\} \\
&= P\{(X=1) \uplus (X=0)\} - P\{X=0\} \\
&= P\{X=1\} = \binom{3}{1}(0.75)^1(1-0.75)^2 = 0.1406 \\
F(2) - F(1) &= P\{X=2\} = \binom{3}{2}(0.75)^2(1-0.75)^1 = 0.4219 \\
F(3) - F(2) &= P\{X=3\} = \binom{3}{3}(0.75)^3(1-0.75)^0 = 0.4219
\end{aligned}
$$

Check: $\quad 0.0156 + 0.1406 + 0.4219 + 0.4219 = 1$

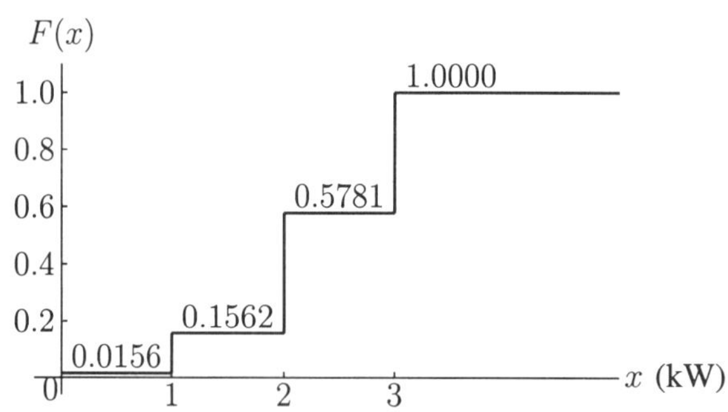

Note that the CDF is of a stairway type since the RV is discrete.

Example 3.8: From CDF to Probability — Machine Failure

Let X be the service time (i.e., time before failure) of a machine. Suppose that it has the following (approximate) CDF

$$F(x) = \begin{cases} 0 & x \le 1 \\ 0.75(x-1) & 1 < x \le 2 \\ 0.75 & 2 < x \le 6 \\ (0.25/3)(x-6) + 0.75 & 6 < x \le 9 \\ 1 & x > 9 \end{cases}$$

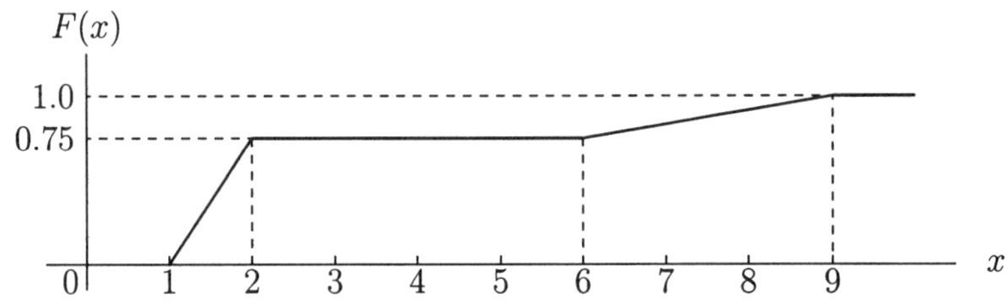

This is a continuous RV since its CDF is continuous, though not smooth. Find the following probabilities:

$$\begin{aligned}
P\{X > 7.5\} &\overset{(3.6)}{=} 1 - F(7.5) = 1 - [(0.25/3)(7.5-6) + 0.75] = 0.125 \\
P\{1.5 < X \le 7.5\} &\overset{(3.5)}{=} F(7.5) - F(1.5) \\
&= (0.25/3)(7.5-6) + 0.75 - 0.75(1.5-1) = 0.5 \\
P\{X > 7.5 | X > 2\} &= \frac{P\{(X > 7.5) \cap (X > 2)\}}{P\{X > 2\}} = \frac{P\{X > 7.5\}}{P\{X > 2\}} \\
&= \frac{P\{X > 7.5\}}{1 - F(2)} = \frac{0.125}{1 - 0.75} = 0.5 \\
P\{X > 1.5 | X > 5.5\} &= \frac{P\{(X > 1.5) \cap (X > 5.5)\}}{P\{X > 5.5\}} = \frac{P\{X > 5.5\}}{P\{X > 5.5\}} = 1
\end{aligned}$$

Note that $P\{X > 1.5 | X > 5.5\} = 1$ makes sense since $X > 1.5$ will always be true given that $X > 5.5$ is true.

3.3 Probability Density Function

The ***probability density function*** (***PDF***) of a continuous RV X is defined as

$$\boxed{f(x) = f_X(x) = \frac{dF(x)}{dx}} \tag{3.10}$$

which is a *density* function indicating where the RV values are more (or less) consolidated.

For a discrete RV X taking on values $x_1, \ldots, x_n$, its ***probability mass function*** (***PMF***) is the sequence of its point masses, defined by, for $x_j = x_1, x_2, \ldots, x_n$,

$$p_j = p(x_j) = P\{X = x_j\} = \sum_{i=1}^{n} P\{X = x_i\}\delta_{i-j} = \sum_{i=1}^{n} p_i \delta_{i-j} \tag{3.11}$$

where p_i's are ***point masses*** and δ is the ***Kronecker delta function***

$$\delta_{i-j} = \begin{cases} 1 & i = j \\ 0 & i \neq j \end{cases} \tag{3.12}$$

By introducing the ***delta function***

$$\delta(x - x_i) = \frac{d}{dx}u(x - x_i) = \begin{cases} \infty & x = x_i \\ 0 & \text{elsewhere} \end{cases} \tag{3.13}$$

the PDF of a RV X taking on values $x_1, \ldots, x_n$ with point masses p_i's is

$$f(x) = \frac{d}{dx}F(x) \overset{\frac{d}{dx}(3.7)}{=} \sum_{i=1}^{n} P\{X = x_i\}\delta(x - x_i) = \sum_{i=1}^{n} p_i \delta(x - x_i) \tag{3.14}$$

which is a probabilistically weighted sum of delta functions.

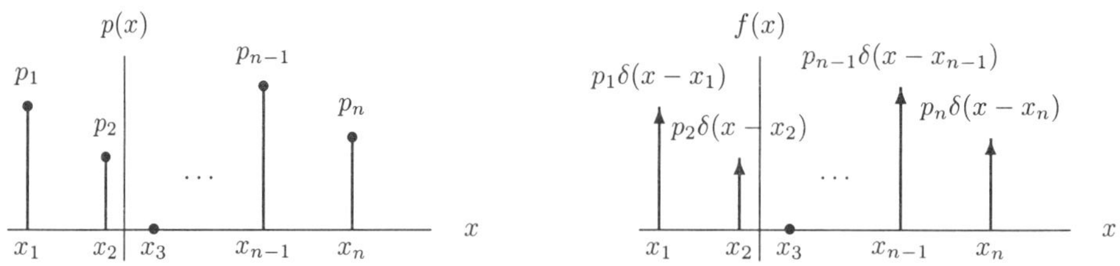

Figure 3.5: PMF and PDF of a discrete RV.

The introduction of the CDF and PDF makes it possible to study random events systematically by well-established, deterministic mathematical tools.

PDFs are usually more convenient than CDFs to describe a RV. Most RVs are known by their PDFs. The PDF has a clear physical interpretation, conceptually in analogy with the mass density function, which indicates where the mass is concentrated:

$$P\{x < X \leq x + \Delta x\} = f(x)\Delta x \qquad \text{for small } \Delta x$$

However, although every RV has a CDF, not every RV has a PDF well-defined over the entire region because the required derivative may not exist at places where the CDF as a curve is not smooth, such as at a sharp corner or at places where the CDF has a jump.

The Kronecker delta function is also known as the ***discrete delta function***.

To broaden the scope of application of the PDF, the delta function is introduced here to handle possible jumps in the CDF. The ***delta function*** is also known as the ***Dirac function*** or unit ***impulse function***. It is the most well-known representative of the so-called *generalized functions*. It has the following unique property, in addition to (3.13),

$$\int_{-\infty}^{\infty} g(x)\delta(x-a)dx = g(a) \tag{3.15}$$

or more precisely,

$$\int_{a-c}^{a+b} g(x)\delta(x-a)dx = g(a), \qquad b, c > 0 \tag{3.16}$$

$$\int_{a+c}^{a+b} g(x)\delta(x-a)dx = \int_{a-c}^{a-b} g(x)\delta(x-a)dx = 0, \qquad b, c > 0 \tag{3.17}$$

The rule is that if the integral is over an interval that includes the point of the delta function that is infinite then it is equal to the function evaluated at the location of the delta function; otherwise it is equal to zero. This unique property is the most important fact concerning the delta function to be used.

The following two interpretations of the delta function are appropriate for an engineering student, as depicted in Fig. 3.6:

- It is the derivative of the unit step function, given by (3.13).
- Consider a single rectangular impulse centered at $x = a$. A delta function $\delta(x-a)$ is simply the limit of this impulse as its width goes to zero while holding its area equal to unity. Simply put, it is an impulse at $x = a$ with zero width, unit area and thus infinite height. The name "impulse function" came from this interpretation.

The delta function can also be interpreted as a density function with all its mass (which is unity) concentrated entirely on point $x = a$.

Note that delta (impulse) function and Kronecker delta function are related but different. The former is defined over a continuous region, while the latter is over discrete points only. A delta function is equal to infinity where its argument is zero, and zero elsewhere; while a Kronecker delta function is equal to unity where its argument (subscript) is zero, zero at other discrete points and undefined elsewhere.

Although $y_1(a) \triangleq p_1\delta(x-a)|_{x=a}$ and $y_2(a) \triangleq p_2\delta(x-a)|_{x=a}$ are both infinity, it is fine to compare them:

$$\frac{y_1(a)}{y_2(a)} = \frac{p_1\delta(x-a)}{p_2\delta(x-a)} = \frac{p_1}{p_2}$$

which can be interpreted as the ratio of the heights of the impulses. This is similar to the comparison of infinities in calculus.

The introduction of the delta function makes it possible and convenient to use a single system to represent both continuous and discrete RVs.

The type of a RV can be determined by its PDF or CDF easily:

- The PDF of a discrete RV must be a weighted sum of delta functions.
- The PDF of a continuous RV cannot involve any delta function.
- The PDF of a mixed RV consists of both delta functions and other functions.

These are equivalent to the following, respectively:

- The CDF of a discrete RV consists of discontinuous points and constant portions (i.e., stairway type) only.
- The CDF of a continuous RV cannot have any discontinuous point.
- The CDF of a mixed RV must have both discontinuous points and smooth (but not constant) portions.

It can be shown that the PDF of a continuous RV can have at most finitely many discontinuous points.

The PMF $p_1, \ldots, p_n$ satisfies

$$\sum_{j=1}^{n} p_j = 1 \qquad p_j \geq 0, \qquad j = 1, 2, \ldots, n$$

Conversely, any set of numbers that satisfies this equation may be the set of the PMF of some discrete RV.

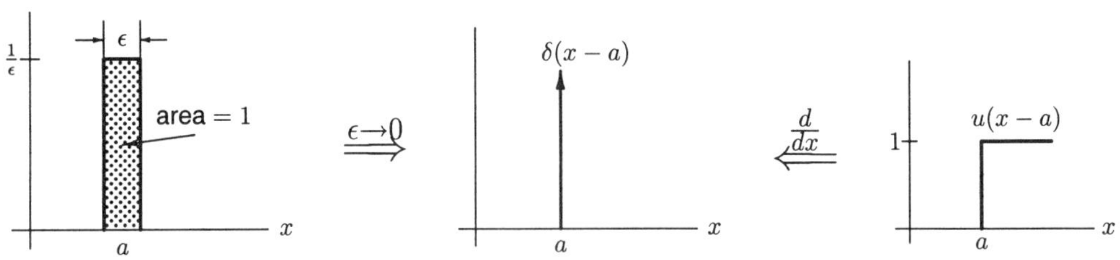

Figure 3.6: Interpretations of the delta function.

Properties of PDF

1. Nonnegativity: $f(x) \geq 0,\ \forall x$, since $F(x)$ is nondecreasing (i.e., $\frac{dF(x)}{dx} \geq 0,\ \forall x$).
2. Relation with CDF: Integrating PDF over $(-\infty, x]$ yields CDF:
$$F(x) \overset{(3.10)}{=} \int_{-\infty}^{x} f(v)dv = \text{area under } f(x) \text{ over interval } (-\infty, x] \tag{3.18}$$
3. Normalization property: By (3.18) and property 3 of CDF,
$$\int_{-\infty}^{\infty} f(x)dx = F(\infty) = 1 \tag{3.19}$$
4. Area under $f(x)$ over interval $(x_1, x_2]$ is equal to $P\{x_1 < X \leq x_2\}$:
$$\boxed{P\{x_1 < X \leq x_2\} = \int_{x_1}^{x_2} f(x)dx} \tag{3.20}$$
which follows from (3.18) and (3.5):
$$\int_{x_1}^{x_2} f(x)dx = \int_{-\infty}^{x_2} f(x)dx - \int_{-\infty}^{x_1} f(x)dx = F(x_2) - F(x_1) = P\{x_1 < X \leq x_2\}$$
5. For a *continuous* RV X and any *given* value c, although $\{X = c\}$ is not an impossible event, its probability is zero: $P\{X = c\} = 0,\ \forall c$, since
$$P\{X = c\} = \lim_{\epsilon \to 0} P\{c - \epsilon < X \leq c\} = \lim_{\epsilon \to 0} \int_{c-\epsilon}^{c} f(v)dv \overset{?}{=} 0 \tag{3.21}$$
where $\overset{?}{=}$ follows from the fact that the PDF of a continuous RV does not involve a delta function. This is also clear from the fact that the CDF of a continuous RV is continuous and thus $P\{X = c\} = F(c+0) - F(c-0) = 0$.

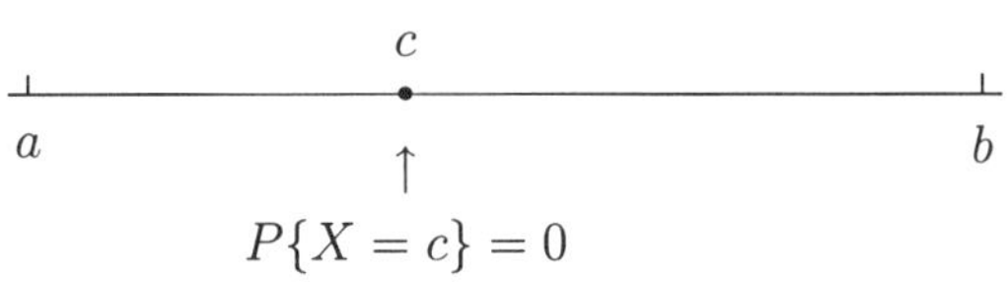

Property 5 can be understood via the following example: Throwing a point of zero size over an interval $(a, b]$, the probability that the point is exactly on $x = c$ for any given c in between a and b is zero. This is not true for a discrete or mixed RV because its PDF involves a delta function.

Example 3.9: CDF, PDF and PMF of Waveform Sampling

The following waveform $g(t)$ of a period $T = 8$ is sampled at a random time τ.

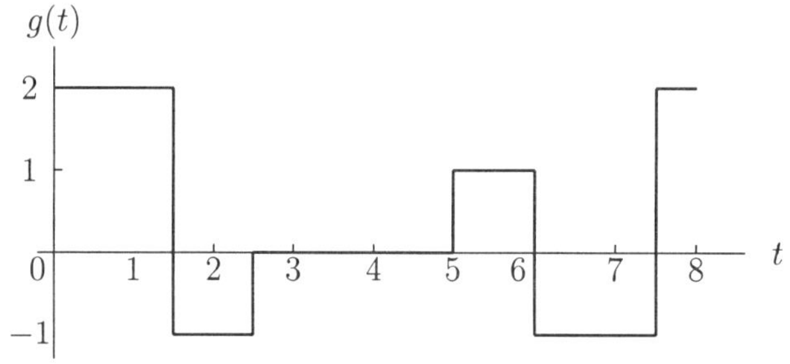

Let $X = \{\text{sampled value of } g(\tau)\}$. Note that X is a discrete RV taking on four possible values: $-1, 0, 1, 2$. The corresponding point masses are

$$P\{X = -1\} = 2.5/8 = 0.3125 \qquad P\{X = 0\} = 2.5/8 = 0.3125$$
$$P\{X = 1\} = 1/8 = 0.125 \qquad P\{X = 2\} = 2/8 = 0.25$$

Hence, its PDF, PMF and CDF are, respectively,

$$f(x) \overset{(3.14)}{=} 0.3125\delta(x+1) + 0.3125\delta(x) + 0.125\delta(x-1) + 0.25\delta(x-2)$$
$$p(x_i) \overset{(3.11)}{=} 0.3125\delta_{x_i+1} + 0.3125\delta_{x_i} + 0.125\delta_{x_i-1} + 0.25\delta_{x_i-2}$$
$$F(x) \overset{(3.7)}{=} 0.3125u(x+1) + 0.3125u(x) + 0.125u(x-1) + 0.25u(x-2)$$

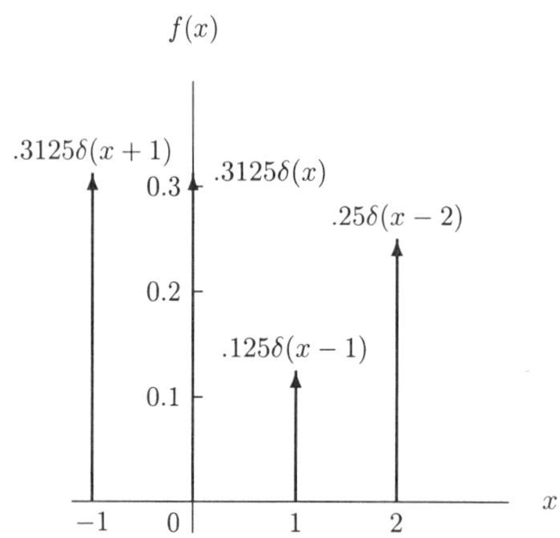

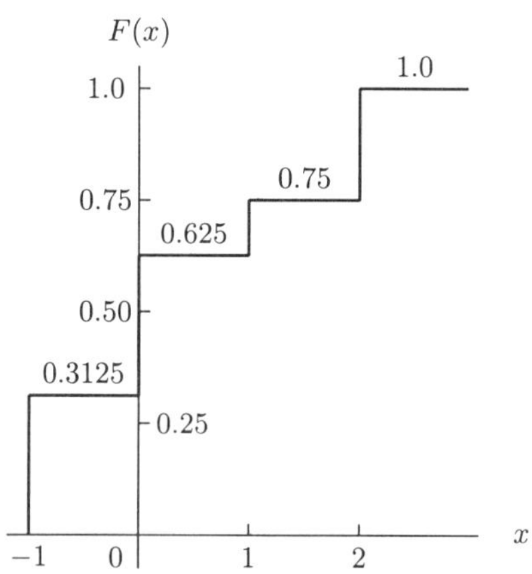

Example 3.10: The following waveform $v(t)$ of a period $T = 1$ is sampled at a random time τ.

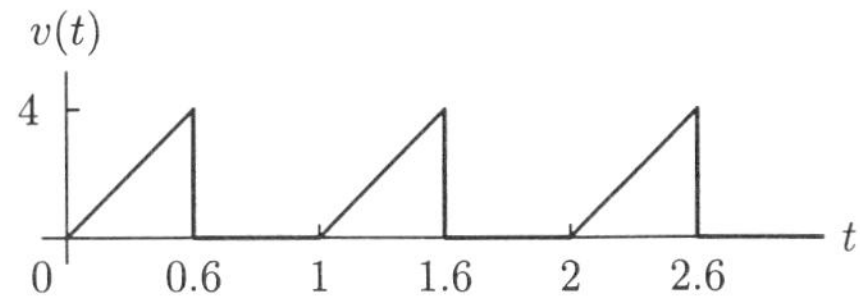

Let $X = \{\text{sampled value of } v(\tau)\}$. It will be clear that X is a mixed RV.

(a) Find PDF and CDF of X: By inspection,

$$\begin{aligned} v(\tau) = 0 \text{ is true 40\% of the time} \quad &\Longrightarrow \quad P\{X = 0\} = 0.4 \\ v(\tau) \text{ is always in between 0 and 4} \quad &\Longrightarrow \quad P\{0 \le X \le 4\} = 1 \\ \text{points in } (0,4) \text{ are equally probable} \quad &\Longrightarrow \quad f(x) = \frac{1-0.4}{4}, 0 < x < 4 \end{aligned}$$

where $(1 - 0.4)$ is the probability mass uniformly distributed over $(0, 4)$. Hence

$$f(x) = 0.4\delta(x) + \frac{0.6}{4}[u(x) - u(x-4)]$$

$$F(x) = \int_{-\infty}^{x} f(x)dx = \begin{cases} 0 & x < 0 \\ 0.4 + \frac{0.6}{4}x & 0 \le x < 4 \\ 1 & x \ge 4 \end{cases}$$

(b) Find $P\{X = 3\}$ and $P\{\frac{1}{2} < X \le 3\}$:

$$P\{X = 3\} = 0 \quad \text{(since no delta function at } x = 3)$$

$$P\{\tfrac{1}{2} < X \le 3\} = F(3) - F(\tfrac{1}{2}) = 0.4 + \frac{0.6}{4} \cdot 3 - \left(0.4 + \frac{0.6}{4} \cdot \frac{1}{2}\right) = 0.365$$

$$\text{or} \quad = \int_{\frac{1}{2}}^{3} f(x)dx = 0.4 \int_{\frac{1}{2}}^{3} \delta(x)dx + \frac{0.6}{4} \int_{\frac{1}{2}}^{3} [u(x) - u(x-4)]dx = 0.365$$

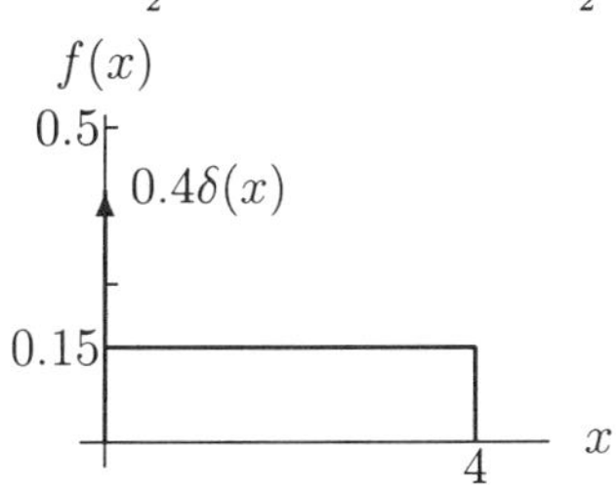

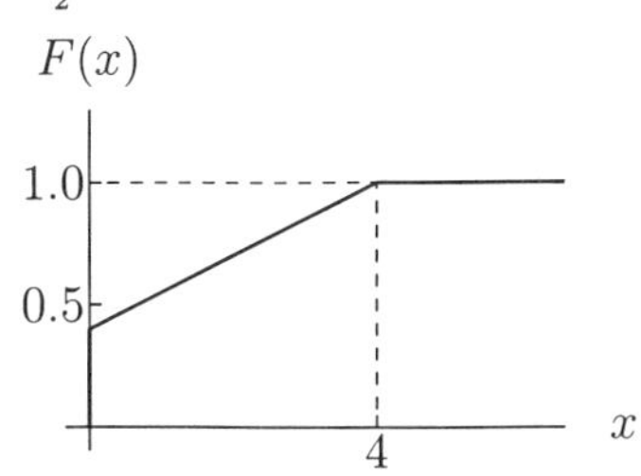

A typical mistake by many students when obtaining the CDF from a PDF is the following:

$$F(x) = \int_{-\infty}^{\infty} f(u)du$$

For a discrete RV, such as the one in Example 3.9, its PDF, CDF, and PMF can be determined in general by the following procedure:

1. Find the locations $x_1, \ldots, x_n$ of all point masses p_i's.
2. Use (3.14), (3.7), and (3.11) to write out its PDF, CDF, and PMF, respectively.

The determination of the PDF and CDF of a continuous or mixed RV is usually more involved. It could be quite difficult sometimes.

For a continuous RV, although its CDF has to be continuous, its PDF can be discontinuous as well as continuous. The PDF is in general piecewise continuous; that is, the PDF of a continuous RV can have at most finitely many discontinuous points.

The PDF and CDF of a continuous RV can be determined in general by the following procedure:

1. Find the intervals outside of which the PDF is identically zero (it is possible that no such interval exists).
2. Find the places where two adjacent continuous pieces of the PDF connect.
3. Determine each individual continuous piece of the PDF.
4. Obtain the CDF from the PDF.

For a mixed RV, its PDF and CDF can be determined in general by the following procedure:

1. Find the locations $x_1, \ldots, x_n$ of all point masses p_i's.
2. Use (3.14) and (3.7) to write out the discrete part of its PDF and CDF, respectively.
3. Proceed as for a continuous RV for the remaining part.

To summarize, note the following:

$$\begin{aligned} \text{a delta function at } x_i \text{ in the PDF} &\iff \text{a jump at } x = x_i \text{ in the CDF} \\ \text{weight (or quantity) of the point mass} &= \text{height of the jump} \end{aligned}$$

When obtaining a PDF from the corresponding CDF, do not forget to include a delta function at each point where the CDF is discontinuous and determine the associated scaling factor of the delta function (i.e., the point mass), such as at $x = 0$ in Example 3.10.

After the PDF or CDF of a RV is obtained, its validity should be checked by verifying the appropriate properties. For a PDF, its nonnegative property (i.e., property 1) and the normalization property (3.19) (i.e., property 3) and for a CDF, its first 4 properties are usually checked, which can be done easily. If both PDF and CDF are obtained, then (3.10) should be checked.

The PDF of a RV is not unique: If its values at finitely many points are changed (provided not delta functions), the CDF of the RV remains unchanged because integration over a single point is zero.

3.4 Uniform Distribution

A ***uniform random variable*** X over (a,b) is one having a nonzero constant PDF over the single interval (a,b):

$$f(x) = \mathcal{U}(x;a,b) \triangleq \begin{cases} \dfrac{1}{b-a} & a < x < b \\ 0 & \text{elsewhere} \end{cases} \tag{3.22}$$

for some a and b. In this case, we use notation

$$\boxed{X \sim \mathcal{U}(a,b)}$$

Note that

$$\int_{-\infty}^{+\infty} f(x)dx = \int_a^b \frac{1}{b-a}dx = 1 \qquad \text{(total area = 1)}$$

$$F(x) = \int_{-\infty}^{x} f(v)dv = \begin{cases} 0 & x < a \\ \displaystyle\int_a^x \frac{1}{b-a}dv = \frac{x-a}{b-a} & a \le x \le b \\ 1 & x > b \end{cases}$$

Clearly, a uniform RV is a continuous one. Thus, $P\{X = x_0\} = 0$ for any given x_0 and an evenly distributed *discrete* RV (e.g., X or Y of Example 3.2 for die rolling) cannot be a uniform RV.

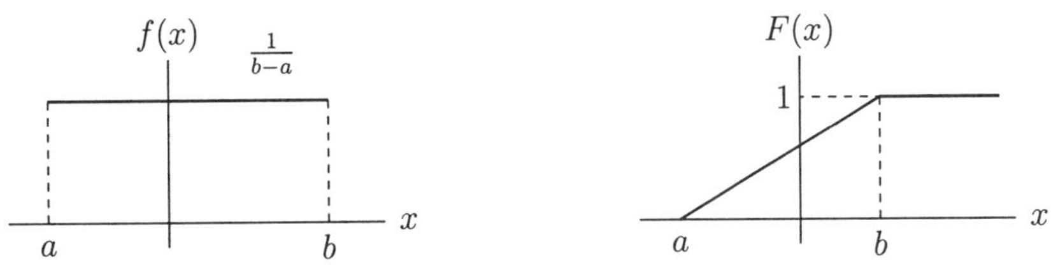

Figure 3.7: The PDF and CDF of a uniform random variable.

Example 3.11: A capacitor has capacitance $X \sim \mathcal{U}(90\mu F, 110\mu F)$. What is the probability that X is between $95\mu F$ and $102\mu F$?

$$\begin{aligned} P\{95 < X < 102\} &= P\{95 \le X < 102\} = P\{95 \le X \le 102\} \\ &= P\{95 < X \le 102\} = \int_{95}^{102} \frac{1}{110-90}dx = 0.35 \end{aligned}$$

A uniform RV is usually a good model for random effects that (are assumed to) occur equally probable over *continuous* sample points. This arises in many practical situations, such as

- Quantization errors: The error introduced by quantizing (i.e., rounding off the sampled values of) a continuous curve (or waveform, signal, etc.) into discrete levels is (approximately) uniformly distributed over the interval between the two adjacent quantization levels.
- Random phase angle: The unknown phase angle of a sinusoid in many cases may be fairly well assumed to be uniformly distributed over a range of 2π radians.
- Random pulse location: The time position of pulses in a periodic sequence of pulses may be reasonably assumed to be uniformly distributed over one period.

An arbitrary random variable may be obtained by a procedure from uniform random variables. This is particularly important for the generation of random numbers, which is required for computer simulation of problems with random effects. That is why only uniform random number generators are usually provided in many software packages.

Another nice property of a uniform RV is that it is easy to handle mathematically. It is among the most important and simplest.

The uniform distribution is only defined for continuous RVs and thus a discrete RV cannot be uniformly distributed. This is the case even if the RV is *evenly* distributed over a number of discrete points (such as X of Example 3.2), or in other words the RV has a constant PMF, rather than a constant PDF over a finite interval. Such a random variable may be said to have an ***even distribution*** or ***discrete uniform distribution***.

Since a uniform RV is a continuous one, whether its PDF is zero or not at the two end points of the interval a and/or b is immaterial. That is why we shall use $\mathcal{U}(a,b)$ to denote also $\mathcal{U}[a,b)$, $\mathcal{U}(a,b]$, and $\mathcal{U}[a,b]$, where $[a,b)$, $(a,b]$, and $[a,b]$ denote, respectively,

$$(a,b] = \{a < x \leq b\}, \qquad [a,b) = \{a \leq x < b\}, \qquad [a,b] = \{a \leq x \leq b\}$$

A continuous RV with a PDF having more than one piece of nonzero constant portion is not a uniform RV. For example, the following PDF

$$f(x) = \begin{cases} \dfrac{0.25}{b-a} & a < x < b \\ \dfrac{0.75}{d-c} & b < c < x < d \\ 0 & \text{elsewhere} \end{cases}$$

is not the PDF of a uniform RV.

3.5 Gaussian Distribution and Central Limit Theorem

A ***Gaussian*** (or ***normal***) ***random variable*** X is one having the PDF:

$$\boxed{f(x) = \mathcal{N}(x; \bar{x}, \sigma^2) \triangleq \frac{1}{\sigma\sqrt{2\pi}} e^{-\frac{(x-\bar{x})^2}{2\sigma^2}}} \tag{3.23}$$

for $-\infty < x < \infty$. Its CDF is

$$F(x) = \frac{1}{\sigma\sqrt{2\pi}} \int_{-\infty}^{x} e^{-\frac{(x-\bar{x})^2}{2\sigma^2}} dx \tag{3.24}$$

A shorthand for such a RV is (read as "X is Gaussian distributed with $\bar{x}$ and σ^2")

$$\boxed{X \sim \mathcal{N}(\bar{x}, \sigma^2)} \tag{3.25}$$

where $\bar{x}$ and σ^2 are two parameters, called ***mean*** and ***variance***, which will be studied later. Note that this CDF has no closed-form expression.

Although the expression of the Gaussian PDF is weird, the Gaussian distribution is extremely popular and useful. It is the most important distribution and will be treated extensively later. It arises in so many practical situations that it is honored as the "***normal***" distribution above all other distributions. It deserves this honor well in view of the *central limit theorem*, to be studied later.

Some typical examples of Gaussian RVs are: thermal noise, measurement error, test score of a large class, and height or weight of a large group of people.

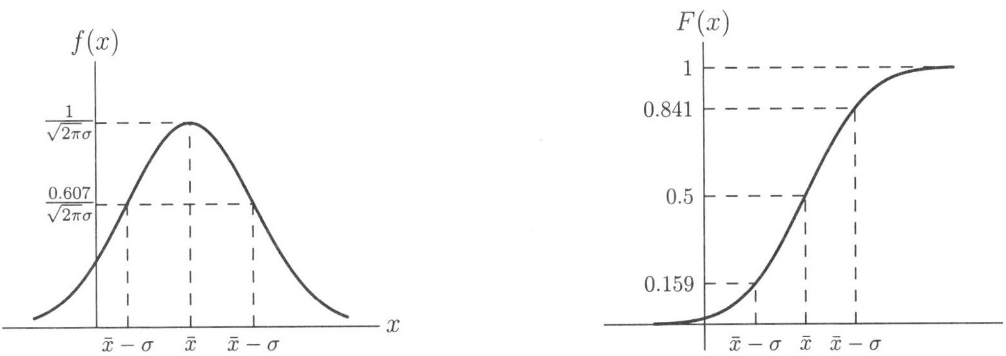

Figure 3.8: The PDF and CDF of a Gaussian random variable.

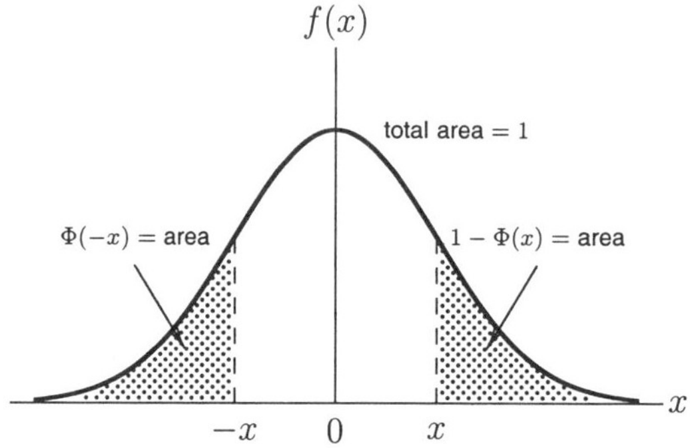

Figure 3.9: The probability areas of a standard Gaussian random variable.

A $\mathcal{N}(0,1)$ RV is called ***standard Gaussian*** (or ***standard normal***). Its CDF is tabulated with notation

$$\Phi(x) \triangleq F(x)\Big|_{x\sim\mathcal{N}(0,1)} = \frac{1}{\sqrt{2\pi}}\int_{-\infty}^{x} e^{-\frac{v^2}{2}}dv \tag{3.26}$$

The table of $\Phi(x)$ is sometimes given only for $x \geq 0$. For $x < 0$, by even symmetry of $\mathcal{N}(x;0,1)$, we may use

$$\Phi(-x) = 1 - \Phi(x) \tag{3.27}$$

How to use the table of $\Phi(x)$ for a nonstandard Gaussian RV $\mathcal{N}(\bar{x},\sigma^2)$? The answer is simple:

$$\begin{aligned} F(x) &= \frac{1}{\sigma\sqrt{2\pi}}\int_{-\infty}^{x} e^{-\frac{(v-\bar{x})^2}{2\sigma^2}}dv \\ &= \frac{1}{\sigma\sqrt{2\pi}}\int_{-\infty}^{\frac{x-\bar{x}}{\sigma}} e^{-\frac{u^2}{2}}d(\sigma u) \quad \left(\text{let } u = \frac{v-\bar{x}}{\sigma} \text{ then } v = \sigma u + \bar{x}\right) \\ &= \Phi\left(\frac{x-\bar{x}}{\sigma}\right) \end{aligned} \tag{3.28}$$

$\Phi(x)$ and in general the CDF $F(x)$ of a general Gaussian RV $\mathcal{N}(\bar{x},\sigma^2)$ are tabulated in the companion software P&R: You can get the value of $F(x)$ by specify x, $\bar{x}$ and σ^2 in P&R, as shown in Example 3.12.

A function closely related with $\Phi(\cdot)$ is the so-called ***error function***:

$$\text{erf}(x) = \frac{2}{\sqrt{\pi}}\int_0^x e^{-t^2}dt$$

The Gaussian PDF function was first introduced in 1733 by the French mathematician, A. De Moivre, to approximate probabilities of events associated with coin tossing. (De Moivre did it because he made his living at a coffee shop by calculating the probabilities of gambling bets in various games of chance for gamblers.) The usefulness of the distribution, however, was not really revealed until the great German mathematician, K. F. Gauss, used it in 1809 to describe the distribution of measurement errors for predicting the locations of astronomical bodies. During the second half of the nineteenth century, however, statisticians found that almost all well-behaved (i.e., normal) data sets have a distribution that fits the bell-shaped Gaussian curve. As a consequence, following the lead of the British statistician, K. Pearson, most statisticians started to refer to this curve as the ***normal distribution***, especially outside North America where many people still call it the Gaussian distribution.

The bell-shaped Gaussian PDF curve is centered at $\bar{x}$. Its only two knee points (i.e., the point at which the derivative changes its sign) are $\bar{x} \pm \sigma$. How flat it is depends on σ^2.

Gaussian RVs have many nice mathematical properties, which make them among the most easily tractable models for random phenomena. For example, a weighted sum of (independent) Gaussian RVs is also Gaussian. These nice properties are the actual (or at least primary) reason why Gaussian models are used in so many practical situations. Of course, having these nice properties is the primary reason for it to be chosen by Nature as the honorable "*normal*" distribution.

The Gaussian distribution is also the limit distribution of several other popular distributions, such as the binomial and Poisson distributions (see Section 3.6). Some other (e.g., chi-square) distributions can also be derived from the Gaussian distribution, as described in Section 3.6.

The Gaussian distribution used to be known as the "law of frequency of error" in the old times. Its beauty is stated clearly in the following quote from the great statistician and biologist, Sir Francis Galton (*Natural Inheritance*, 1889):

> I know of scarcely anything so apt to impress the imagination as the wonderful form of cosmic order expressed by the "Law of Frequency of Error." The Law would have been personified by the Greeks and deified, if they had known of it. It reigns with serenity and in complete self-effacement amidst the wildest confusion. The huger the mob and the greater the apparent anarchy, the more perfect is its sway. It is the supreme law of unreason. ... Whenever a large sample of chaotic elements are taken in hand and marshaled in the order of their magnitude, an unsuspected and most beautiful form of regularity proves to have been latent all along. The tops of the marshaled row form a flowing curve of invariable proportions and each element, as it is sorted into place, finds, as it were, a preordained niche accurately adapted to it.

The mystery of the Gaussian distribution is stated nicely by the great mathematician and astronomer, Henri Poincaré:

> There must be something mysterious about the normal law since mathematicians think it is a law of nature whereas physicists are convinced that it is a mathematical theorem.

The error function is related to $\Phi(\cdot)$ by

$$\Phi(x) = \begin{cases} 0.5(1+\mathrm{erf}(x/\sqrt{2})) & x \geq 0 \\ 0.5(1-\mathrm{erf}(|x|/\sqrt{2})) & x < 0 \end{cases}$$

MATLAB has an function `erf` that returns the value of the error function.

Example 3.12: Calculation of Gaussian Probability and Percentile by P&R

Suppose we want to calculate the value of the CDF $F(x)$ of Gaussian RV: $X \sim \mathcal{N}(2.4, 0.65^2)$ at $x = 2.77$. This can be done easily using the companion software P&R as follows:

S1. Click "Tables" in the main window of P&R.
S2. Click "Normal." The "**Normal Distribution Table**" window will appear.
S3. Enter the mean and variance as shown in Fig. 3.10. Click "Ok." The value of the Gaussian CDF $F(x)$ at $x = 2.77$ with the specified mean $\bar{x}$ and variance σ^2 is then shown as 0.7154 in the box.

Following the same procedure, we obtained that $\Phi(1) = F(1)|_{\mathcal{N}(0,1)} = 0.84134$. Similarly, all the probabilities of Examples 3.13 and 3.14 can be obtained.

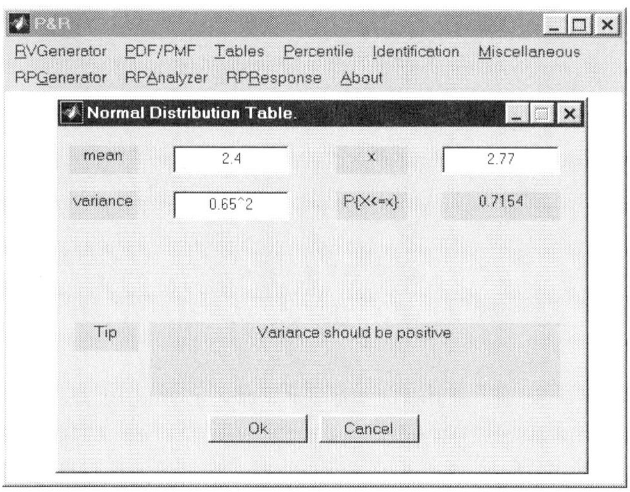

Figure 3.10: Calculation of Gaussian probability and percentile by P&R.

The point x that satisfies $P\{X < x\} = \alpha\%$ is called the $\alpha\%$ ***percentile*** point. For example, the above example shows that $x = 2.77$ is the 71.54% percentile of the $\mathcal{N}(2.4, 0.65^2)$ distribution. Following a similar procedure (except replacing "Tables" with "Percentile"), we found that the 22% percentile of the $\mathcal{N}(2.4, 0.65^2)$ distribution is 1.8981. See Appendix A for details.

Example 3.13: Common Gaussian Probabilities

Consider a RV $X \sim \mathcal{N}(\bar{x}, \sigma^2)$, where $\sigma > 0$ and $\bar{x}$ is a given real number.

(a) Find $P\{X - \bar{x} < \sigma\}$:

$$
\begin{aligned}
P\{X - \bar{x} < \sigma\} &= P\{X < \bar{x} + \sigma\} \stackrel{?}{=} P\{X \leq \bar{x} + \sigma\} = F(\bar{x} + \sigma) \\
&\stackrel{(3.28)}{=} \Phi\Big(\frac{(\bar{x} + \sigma) - \bar{x}}{\sigma}\Big) = \Phi(1) \stackrel{\mathcal{N}(0,1)\text{ table}}{=} 84.1\%
\end{aligned}
$$

(b) Find $P\{|X \quad \bar{x}| < \sigma\}$:

$$
\begin{aligned}
P\{|X - \bar{x}| < \sigma\} &= P\{-\sigma < X - \bar{x} < \sigma\} = \{\bar{x} - \sigma < X \leq \bar{x} + \sigma\} \\
&= F(\bar{x} + \sigma) - F(\bar{x} - \sigma) \\
&= \Phi\Big(\frac{(\bar{x} + \sigma) - \bar{x}}{\sigma}\Big) - \Phi\Big(\frac{(\bar{x} - \sigma) - \bar{x}}{\sigma}\Big) = \Phi(1) - \Phi(-1) \\
&\stackrel{(3.27)}{=} \Phi(1) - [1 - \Phi(1)] = 2\Phi(1) - 1 = 68.3\%
\end{aligned}
$$

Similarly,

$$
\begin{array}{ll}
P\{X - \bar{x} < 2\sigma\} = 97.7\% & P\{X - \bar{x} < 3\sigma\} = 99.9\% \\
P\{|X - \bar{x}| < 2\sigma\} = 95.4\% & P\{|X - \bar{x}| < 3\sigma\} = 99.7\%
\end{array}
$$

These probabilities are commonly used and are better learned by heart.

Example 3.14: Probability of Power Consumption

A voltage $V \sim \mathcal{N}[0, (10\text{V})^2]$ is applied across a 5Ω resistor. What is the probability that the power P consumed by the resistor is in [10W, 40W]?

$$
\begin{aligned}
P\{10 \leq P \leq 40\} &= P\{10 \leq V^2/5 \leq 40\} = P\{50 \leq V^2 \leq 200\} \\
&= P\{\sqrt{50} \leq V \leq \sqrt{200}\} + P\{-\sqrt{200} \leq V \leq -\sqrt{50}\} \\
&= P\{7.071 < V \leq 14.14\} + P\{-14.14 < V \leq -7.071\} \\
&= F_V(14.14) - F_V(7.071) + F_V(-7.071) - F_V(-14.14) \\
&= \Phi\Big(\frac{14.14 - 0}{10}\Big) - \Phi\Big(\frac{7.071}{10}\Big) \\
&\quad + \Big[1 - \Phi\Big(\frac{7.071}{10}\Big)\Big] - \Big[1 - \Phi\Big(\frac{14.14}{10}\Big)\Big] \\
&= 2[\Phi(1.414) - \Phi(0.7071)] = 0.322
\end{aligned}
$$

Importance of Gaussian RV — The Central Limit Theorem

The ***central limit theorem*** states that, for n independent RVs $X_1, \cdots, X_n$ with mean $\bar{x}_i$ and variance σ_i^2, respectively,

$$Y_n = \frac{\sum_{i=1}^n (X_i - \bar{x}_i)}{\sqrt{\sum_{i=1}^n \sigma_i^2}} \quad \xrightarrow{n\to\infty} \quad Y \sim \mathcal{N}(0,1) \tag{3.29}$$

The central limit theorem has the following **interpretation**:

> The properly normalized sum of many "uniformly" small and negligible independent RVs tends to be a standard Gaussian RV. If a random phenomenon is the *cumulative* effect of many "uniformly" small sources of uncertainty, it can be reasonably modeled as a Gaussian RV.

Fig. 3.11 illustrates how fast the sum of n independent uniform RVs tends to be a Gaussian RV: X_1 has a rectangular PDF; $X_1 + X_2$ has a triangular PDF; $X_1 + X_2 + X_3$ has a PDF with three parabolic pieces that is already quite close to a Gaussian PDF.

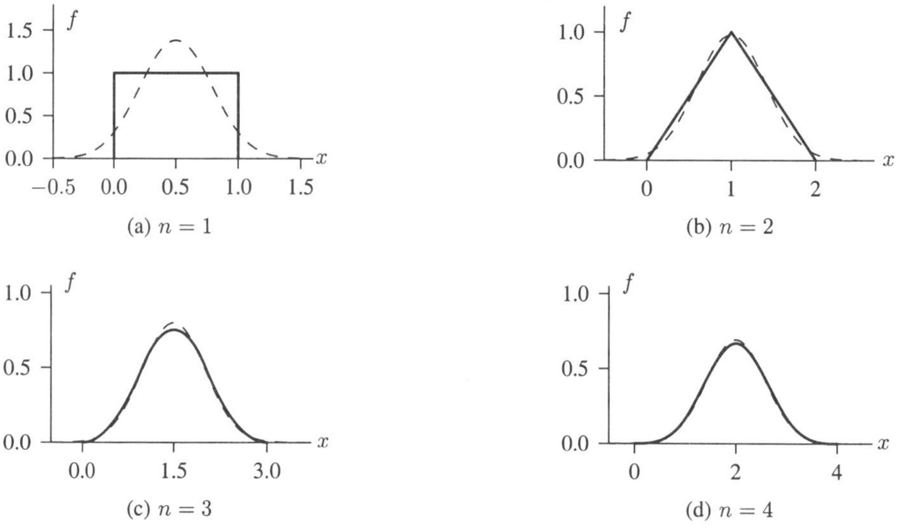

Figure 3.11: Convergence of sum of uniform RVs to Gaussian distribution.

Central limit theorem is probably the single most important result in the entire probability theory. It suggests that the Gaussian assumption is good when the random phenomenon under consideration is a cumulative effect of (a large number of independent) sources of uncertainty, whose individual effects are "uniformly" small and "negligible." Various versions of the central limit theorem give a variety of conditions under which a properly normalized sum of a large number of "uniformly" small and negligible independent RVs tends to be a standard Gaussian RV. This is the primary justification for the omnipresence of the Gaussian distribution.

A very nice property of Gaussian RVs is that the weighted sum of two independent Gaussian RVs is a Gaussian RV:

$$\left.\begin{array}{r} X \sim \mathcal{N}(\bar{x}, \sigma_x^2) \\ Y \sim \mathcal{N}(\bar{y}, \sigma_y^2) \\ X, Y \text{ independent} \end{array}\right\} \quad \Longrightarrow \quad Z = aX + bY \sim \mathcal{N}(a\bar{x} + b\bar{y}, a^2\sigma_x^2 + b^2\sigma_y^2)$$

Were this property not true, the central limit theorem would be in an awkward position: The normalized sum of RVs tends to be Gaussian but it is not true for the sum of Gaussian RVs.

Central limit theorem is for the convergence of the sum, irrespective of convergence rate. If the individual uncertainties are independent and identically distributed, Gaussian approximation is good in practice even when only a small number of random effects are involved. A good example is the case of Fig. 3.11. In general, this is typically the case for RVs with smooth and symmetric PDFs.

Independence of RVs will be studied in the next chapter.

Consider the occurrence of event A in N Bernoulli trials. Suppose that the probability of event A in a single trial is p. Define a RV Y that is equal to k if A occurs exactly k times. It is clear from (2.27) that Y has the PMF:

$$P\{Y = k\} = P\{A \text{ occurs } k \text{ times in } N \text{ trials}\} = \binom{N}{k} p^k (1-p)^{N-k}, \qquad k = 0, 1, 2, \ldots, N \tag{3.30}$$

Such a RV is known as to have a ***binomial distribution***. If p is not very close to either 0 or 1, then a properly normalized Y can be approximated by a standard Gaussian RV X, for large N,

$$\frac{Y - Np}{\sqrt{Np(1-p)}} \cong X, \qquad N \text{ large} \tag{3.31}$$

This result is known as the ***De Moivre-Laplace theorem***. This relation between the Gaussian distribution and the repeated coin-tossing explains why not surprisingly it was De Moivre who first introduced the Gaussian distribution.

In fact, the central limit theorem and the Gaussian law are not limited to random phenomena. For example, it can be shown that if the deterministic functions $\cos\omega_1 t, \cos\omega_2 t, \ldots, \cos\omega_n t$ are "independent," the proportion of time during which the normalized sum[1]

$$x_n(t) = \frac{\cos\omega_1 t + \cos\omega_2 t + \cdots + \cos\omega_n t}{\sqrt{n}}$$

[1]See M. Mac, *Enigmas of Chance*, University of California Press, 1987.

satisfies $a < x_n(t) < b$ approaches as n increases the area under the normal curve $f(x) = \frac{\sqrt{2}}{\sqrt{\pi}}e^{-2x^2}$ over the interval (a, b). The deterministic functions $\cos\omega_1 t, \cos\omega_2 t, \ldots, \cos\omega_n t$ are "independent" in the sense that the proportion of time during which the following is satisfied *jointly*

$$\cos\omega_1 t < \alpha_1, \cos\omega_2 t < \alpha_2, \ldots, \cos\omega_n t < \alpha_n$$

is equal to the product of the proportions of times during which the above is satisfied *individually* for each function. For example, $\cos\omega_1 t$ and $\cos\omega_2 t$ are independent if ω_1/ω_2 is a irrational number since in this case, the proportion of time during which both $\cos\omega_1 t < \alpha_1$ *and* $\cos\omega_2 t < \alpha_2$ is equal to the product of proportion of time during which $\cos\omega_1 t < \alpha_1$ and proportion of time during which $\cos\omega_2 t < \alpha_2$.

Example 3.15: System Reliability

A sophisticated system consists of a large number of n independent parts. Each part has an operational probability (i.e., reliability) of 0.9. If the system works normally only when at least 80% of its parts are in operation, find the minimum number of n such that the system's reliability (i.e., the probability that the system is operational) is at least 0.95.

Let RV Y be the number of parts (out of n) that are operational. Note that Y is binomial. Then, system is operational if and only if $Y \geq 0.8n$ and thus the sought-after reliability is simply

$$P\{Y \geq 0.8n\} \geq 0.95$$

Note that

$$\begin{aligned} P\{Y \geq 0.8n\} &= 1 - P\{Y < 0.8n\} \\ &= 1 - P\left\{\frac{Y - np}{\sqrt{np(1-p)}} \leq \frac{0.8n - np}{\sqrt{np(1-p)}}\right\} \\ &\overset{(3.31)}{\cong} 1 - P\left\{X \leq \frac{0.8n - np}{\sqrt{np(1-p)}}\right\} \\ &= 1 - \Phi\left(\frac{0.8n - np}{\sqrt{np(1-p)}}\right) \\ &= 1 - \Phi\left(\frac{-0.1n}{0.3\sqrt{n}}\right) \\ &= \Phi(\sqrt{n}/3) \end{aligned}$$

The requirement becomes

$$\Phi(\sqrt{n}/3) \geq 0.95$$

From the table of the standard Gaussian RV, the above requirement is satisfied if $\sqrt{n}/3 \geq 1.64$ and thus $n \geq 25$. Therefore, at least 25 parts are needed to have system reliability of at least 0.95.

3.6 Some Other Commonly Used Distributions

Some other commonly used distributions are described here. Their PDF or PMF curves with various parameter values can be obtained by the companion software P&R.

3.6.1 Chi-Square Distribution

A RV with the following PDF:

$$f(x) = \begin{cases} \frac{1}{2^{n/2}\sigma^n\Gamma(n/2)} x^{n/2-1}\, e^{-x/(2\sigma^2)} & x > 0 \\ 0 & x \leq 0 \end{cases} \tag{3.32}$$

with $\sigma > 0$ is said to have a ***chi-square distribution*** with n degrees of freedom, denoted by χ_n^2, where $\Gamma(\cdot)$ is the gamma function, defined by

$$\Gamma(x) = \int_0^\infty u^{x-1} e^{-u} du$$

Fig. 3.12 depicts the PDFs of chi-square distributed RVs with different degrees of freedom.

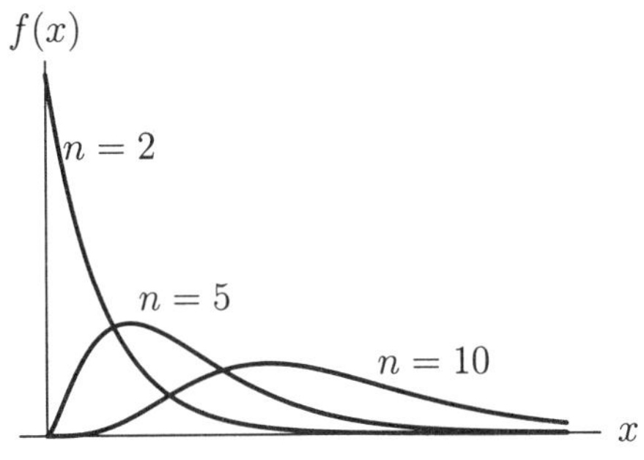

Figure 3.12: PDF of chi-square distribution.

A chi-square RV has a close relation with Gaussian RVs. It is actually the sum of the squares of independent zero-mean Gaussian RVs with a common variance; that is, given n independent and identically distributed Gaussian RVs X_i's with $X_i \sim \mathcal{N}(0, \sigma^2)$, then

$$X = X_1^2 + X_2^2 + \cdots + X_n^2$$

is a χ_n^2 RV with the PDF given by (3.32). As such, a chi-square χ_n^2 RV is often a good model for the total energy or power of a signal that have n independent components.

Chi-square RVs are often used in the so-called ***chi-square test***, which provides a check of the *goodness of fit* to judge if a RV contradicts its underlying assumptions made regarding its distribution. This test is among the most popular statistical tests.

The mean and variance, studied later, of a χ_n^2 RV are equal to $n\sigma^2$ and $2n\sigma^4$, respectively.

3.6.2 Rayleigh Distribution

A ***Rayleigh RV*** is one with the following PDF:

$$f(x) = \begin{cases} \frac{x}{\sigma^2} e^{-x^2/2\sigma^2} & x > 0 \\ 0 & x \leq 0 \end{cases} \tag{3.33}$$

where σ^2 is given. Fig. 3.13 illustrates the PDFs of two Rayleigh distributed RVs.

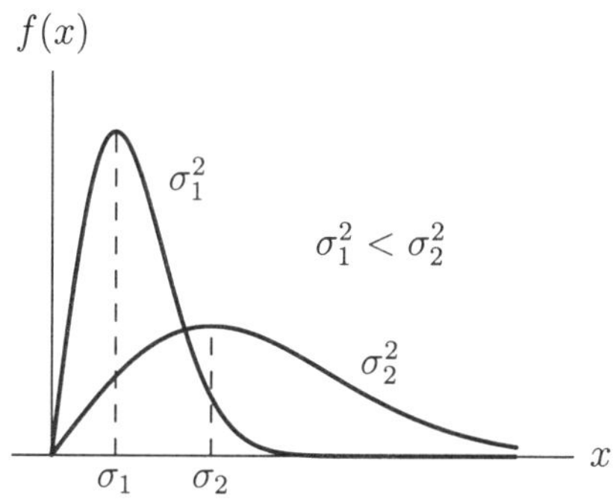

Figure 3.13: PDF of Rayleigh distribution.

A Rayleigh RV is the magnitude of the sum of two independent zero-mean Gaussian RV squares with a common variance; that is, if $X = \sqrt{X_1^2 + X_2^2}$ with independent $X_i \sim \mathcal{N}(0, \sigma^2)$, $i = 1, 2$, then X is Rayleigh distributed with PDF given by (3.33). As such, a Rayleigh RV is often a good model for measurement error (as the square-root of the sum of squares of two independent measurement errors) and the amplitude or envelope of a signal with two orthogonal components (in-phase and quadrature components).

Example 3.16: Plotting Rayleigh PDF Curve by P&R

The companion software P&R can be used to plot the PDF curve of any distribution covered in this book. We illustrate the procedure using the Rayleigh distribution with $\sigma^2 = 5$ as follows:

S1. Click "PDF/PMF" in the main window of P&R.
S2. Click "Rayleigh." The "**Rayleigh PDF**" window will appear.
S3. Enter the parameter σ as `sqrt(5)` as shown in Fig. 3.14 because $\sigma = \sqrt{5}$. Click "Ok." The Rayleigh PDF curve is then shown as in Fig. 3.14.

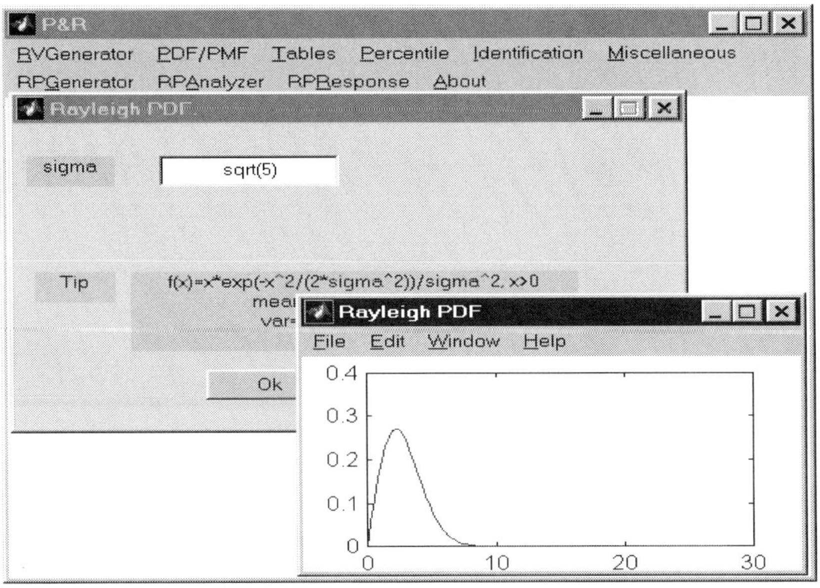

Figure 3.14: Plotting PDF curve by P&R.

3.6.3 Exponential Distribution

The RV in Example 3.21 is an example of an ***exponential RV***. More generally, an exponential RV is one with a PDF of the following exponential function:

$$f(x) = \begin{cases} \lambda e^{-\lambda(x-a)} & x > a \\ 0 & x \leq a \end{cases} \tag{3.34}$$

where $\lambda > 0$ is a scale parameter. Fig. 3.15 depicts the PDF of an exponential distributed RV.

An exponential RV is often a good model for the time between two consecutive occurrences of independent events, which can be modeled as uniform RVs, such as the time between failures (or customer visits) at a constant rate. It has many applications in such areas as reliability and random service systems (e.g., computer systems and communication systems) for the calculation of mean-time-to-failure, mean-time-between-failures, expected life time of a part, etc., as well as many other quantities in queuing theory and queuing networks.

The exponential distribution with $a = 0$ is a "memoryless" distribution: For $s, t \geq 0$,

$$P\{X > t+s | X > s\} = \frac{P\{X > t+s, X > s\}}{P\{X > s\}} = \frac{P\{X > t+s\}}{P\{X > s\}} = \frac{e^{-\lambda(t+s)}}{e^{-\lambda s}} = e^{-\lambda t} = P\{X > t\}$$

where use has been made of the fact that $P\{X > x\} = 1 - P\{X \leq x\} = 1 - (1 - e^{-\lambda x})$. Hence, if X is the service time of a device (or life expectancy of a person) and s is the time already

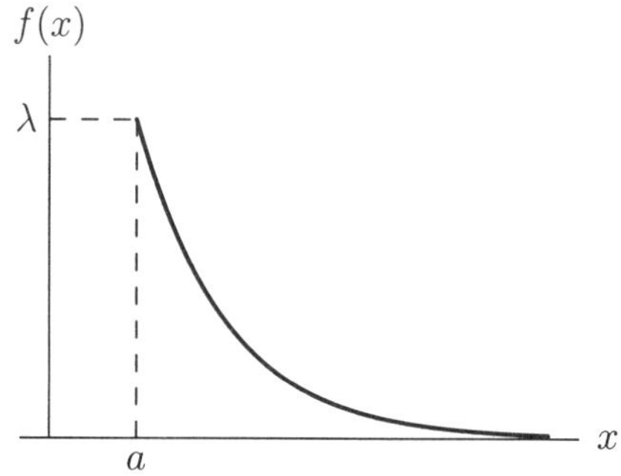

Figure 3.15: PDF of exponential distribution.

served (or the current age of the person), then the above property indicates that the device is always "like new" (or the person is "young forever") because the probability of living t more years is the same whether the person is s years old $P\{X > t + s|X > s\}$ or a new-born baby $P\{X > t\}$. It can be shown that the exponential distribution is the only continuous distribution that has this interesting "memoryless" property. Although this cannot be true for any real-world quantity, it is sometimes a good approximation for some practical problems.

3.6.4 Weibull Distribution

Closely related with the exponential distribution is the so-called ***Weibull distribution***, whose PDF is given by

$$f(x) = \left\{ \begin{array}{ll} abx^{a-1}e^{-bx^a} & x > 0 \\ 0 & x \leq 0 \end{array} \right\} = \left\{ \begin{array}{ll} \frac{\alpha}{\beta}\left(\frac{x}{\beta}\right)^{\alpha-1} e^{-(x/\beta)^\alpha} & x > 0 \\ 0 & x \leq 0 \end{array} \right. \tag{3.35}$$

where $\alpha, \beta > 0$ are shape and scale parameters, respectively, and $a = \alpha$, $b = (1/\beta)^\alpha$. Fig. 3.16 shows the PDFs of two Weibull distributed RVs with $b = 1$ and $a = 1, 2$, corresponding to exponential and Rayleigh distributions, respectively. Its CDF is given by

$$F(x) = \left\{ \begin{array}{ll} 1 - e^{-bx^a} & x > 0 \\ 0 & x \leq 0 \end{array} \right\} = \left\{ \begin{array}{ll} 1 - e^{-(x/\beta)^\alpha} & x > 0 \\ 0 & x \leq 0 \end{array} \right.$$

A Weibull RV is often a good model for a failure time with a time-invariant or time-varying failure rate or the time to complete a task. The popularity of the Weibull model is a result of the fact that it includes several popular failure models as special cases. Specifically, it can be shown that the failure rate is given by $h(t) = \frac{\alpha}{\beta}\left(\frac{t}{\beta}\right)^{\alpha-1}$. It thus reduces to the exponential model when $\alpha = 1$, which corresponds to a constant failure rate. For $\alpha > 1$, the failure rate is an increasing

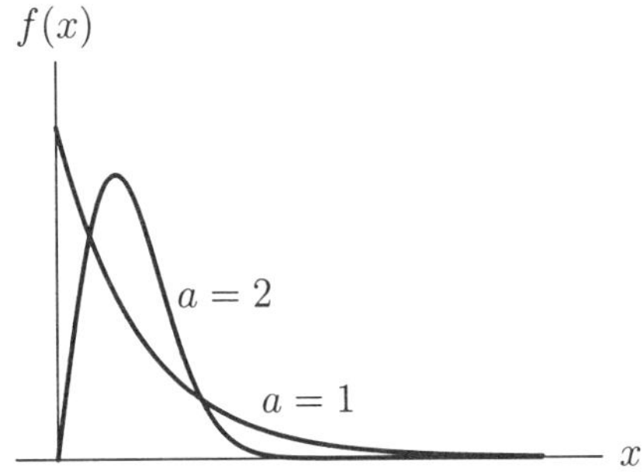

Figure 3.16: PDF of Weibull distribution.

function of time. For example, it is a linearly and quadratically increasing function for $\alpha = 2$ and 3, respectively. As a result, Weibull distribution has many applications in reliability, and in some statistical tests, in particular, the *Weibull plots*.

3.6.5 Log-Normal Distribution

If the logarithm of a RV is Gaussian (normal), the RV is known as to have the ***log-normal distribution***; that is, a log-normal RV is the exponential of a Gaussian RV:

$$X = e^Y \qquad \text{with} \qquad Y \sim \mathcal{N}(0, \sigma^2) \tag{3.36}$$

Its PDF is given by

$$f(x) = \begin{cases} \frac{1}{\sqrt{2\pi}\sigma} x^{-1} e^{-(\ln x - \mu)^2/2\sigma^2} & x > 0 \\ 0 & x \leq 0 \end{cases} \tag{3.37}$$

where $\sigma > 0$ and $-\infty < \mu < \infty$ are shape and scale parameters, respectively. Fig. 3.17 sketches the PDF of a log-normal distributed RV (with $\mu = 0, \sigma = 1$).

A log-normal RV is often a good model for many problems involving random effects that are Gaussian when measured in a logarithmic scale. For instance, if a quantity is the product of a large number of terms, such as multiplicative measurement errors, then the lognormal distribution is probably a good model.

3.6.6 Poisson Distribution

An extremely important and popular discrete distribution is the so-called ***Poisson distribution***. Its PMF is given by, for $\lambda > 0$,

$$P\{X = k\} = \left\{ \begin{array}{ll} e^{-\lambda} \frac{\lambda^k}{k!} & k \geq 0 \\ 0 & k < 0 \end{array} \right\}, \qquad k = 0, 1, 2, \ldots \tag{3.38}$$

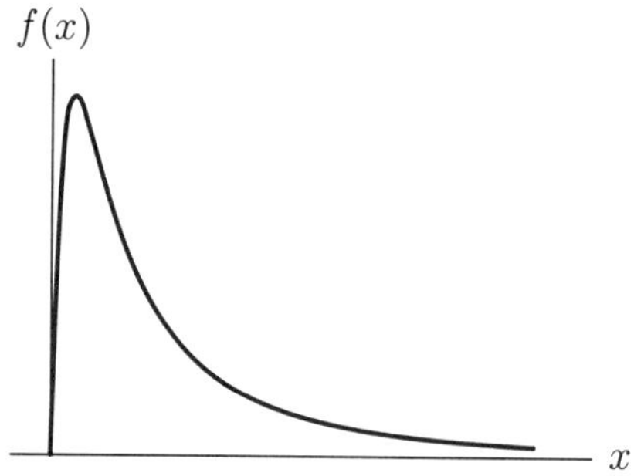

Figure 3.17: PDF of log-normal distribution.

It is a good model for rare events (i.e., those occur rarely). It is very useful in counting the number of the occurrences of certain events. This has wide-spread applications, ranging from various service systems to physics. In service systems, many discrete random quantities, such as the number of the customers of a store, are Poisson distributed. That is why the Poisson distribution is probably the most important probability model in operations research and management science, especially in queueing theory and queueing networks.

The Poisson distribution is one of the few most important distributions in probability theory. It is, in fact, one of the "elementary particles" that constructs various random phenomena, in particular those that are described by random processes.

The Poisson distribution is also the limit and thus a good approximation of the binomial distribution for large N when pN is small.

Example 3.17: Plotting Poisson PMF Curve by P&R

The companion software P&R can be used to plot the PMF curve of a distribution. We illustrate the procedure using the Poisson distribution with $\lambda = 3.8$ as follows:

S1. Click "PDF/PMF" in the main window of P&R.
S2. Click "Poisson." The "**Poisson PMF**" window will appear.
S3. Enter the parameter λ as `3.8` as shown in Fig. 3.18. Click "Ok." The Poisson PMF curve is then shown as in Fig. 3.18.

3.6.7 Some Other Discrete Distributions

There are many other distributions with well-known names. There are, of course, even more distributions that arise from practical problems that do not have generally accepted names.

A ***binary RV*** or ***Bernoulli RV*** is one that can take only on two possible values with probabilities p and $1 - p$, respectively. Only binary RVs may be defined on a coin tossing experiment.

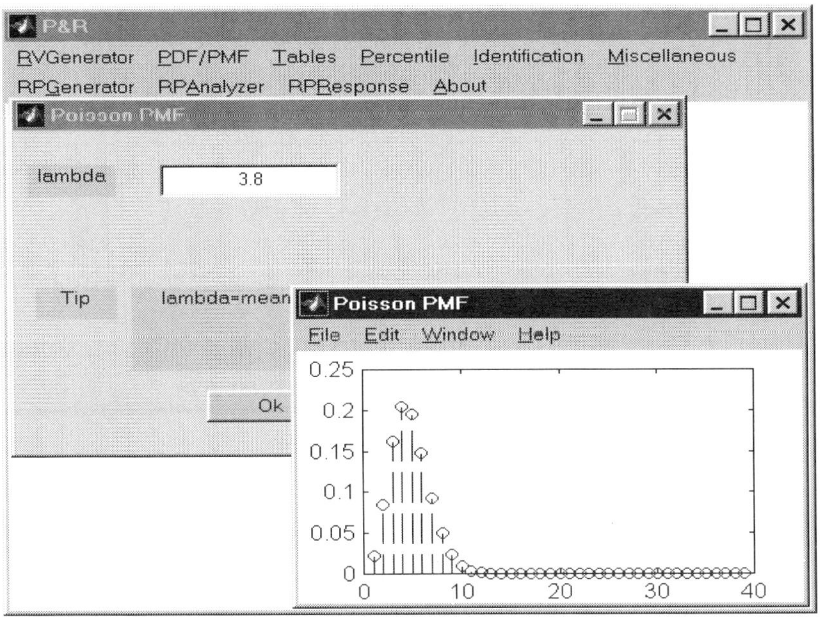

Figure 3.18: Plotting PMF curve by P&R.

The point masses of a ***binomial RV*** were given by (3.30) with parameters N and p. It can be applied to many problems of repeated independent trials with only two possible outcomes on each trial (Bernoulli trials). Examples of the applications include games of chance and signal detection in radar and sonar systems.

The sum of N independent and identical binary RVs with parameter p has a binomial distribution with parameters N and p.

In Bernoulli trials, the number of the occurrences of $\overline{A}$ (i.e., nonoccurrences of A) before the *first* occurrence of event A (with a probability p on each trial) is said to have a ***geometric distribution***. It has the PMF

$$P\{X = m\} = (1-p)^{m-1}p \qquad m = 1, 2, 3, \ldots \tag{3.39}$$

It can be applied to, for instance, the number of failures before the first success, the number of items inspected before the first defective one is encountered, the number of items in a batch of random size, and the number of items demanded from an inventory.

The geometric distribution is in some sense the discrete counterpart of the exponential distribution. As is the exponential distribution for continuous RV, it is the only discrete distribution that possesses the "memoryless" (or "like-new," "young for ever") property: $P\{X = n+m | X > n\} = P\{X = m\}$.

3.7 Expectation and Moments

3.7.1 Introduction and Motivation

Recall that given mass M distributed over $x_1, \ldots, x_n$ with point masses $m_1, \ldots, m_n$, respectively, the center of mass (i.e., centroid) is given by

$$x_c = \frac{\sum_{i=1}^n x_i m_i}{\sum_{i=1}^n m_i} = \frac{\sum_{i=1}^n x_i m_i}{M} \stackrel{\text{if } M=1}{=} \sum_{i=1}^n x_i m_i$$

If M is distributed continuously over an interval $[a, b]$ with a mass density function $m(x)$, then the center of mass (centroid) is given by

$$x_c = \frac{\int_a^b x m(x) dx}{\int_a^b m(x) dx} = \frac{\int_a^b x m(x) dx}{M} \stackrel{\text{if } M=1}{=} \int_a^b x m(x) dx$$

Since probability of a RV X can be viewed as mass with point masses $P\{X = x_i\} = p_i$ that sum up to unity or with a density function $f(x)$ that integrates up to unity, its "***centroid***" may be defined as

$$x_c = \begin{cases} \sum_{i=1}^n x_i p_i & X \text{ discrete} \\ \int_a^b x f(x) dx & X \text{ continuous} \end{cases} \tag{3.40}$$

Example 3.18: Consider Example 3.2 of die rolling. For the RV X, defined by $X(F_i) = i$, its ***average*** value $\bar{x}$ over all possible outcomes F_i's is

$$\bar{x} = \frac{1+2+3+4+5+6}{6} = \sum_{i=1}^{6} X(F_i) P\{X = X(F_i)\} = \sum_{i=1}^{6} \frac{i}{6} = 3.5$$

Define another RV W by the following assignments:

$$W(F_1) = 1, \quad W(F_3) = 2, \quad W(F_5) = 3, \quad W(\text{an even number shows}) = 4$$

Its average value $\bar{w}$ is given by

$$\bar{w} = 1 \cdot \frac{1}{6} + 2 \cdot \frac{1}{6} + 3 \cdot \frac{1}{6} + 4 \cdot \frac{1}{2} = 3 \neq \frac{1+2+3+4}{4} = \frac{10}{4}$$

Consequently, the ***average*** of a discrete RV X should be defined by

$$\bar{x} = \sum_i x_i P\{X = x_i\} = \sum_i x_i p_i \tag{3.41}$$

which is a probabilistically weighted sum over all possible outcomes. Note that it is identical to (3.40) and thus the average of a RV is the same as the centroid of its probability mass. This introduces the ***expected value***.

All children are above average.

Humorist Garrison Keillor, *Lake Wokegon*

3.7.2 Definitions

The ***expected value*** of a RV X with PDF $f_X(x)$ or PMF $P\{X = x_i\}$ is defined as

$$\bar{x} = E[X] = \begin{cases} \int_{-\infty}^{\infty} x f_X(x) dx & X \text{ continuous} \\ \sum_i x_i P\{X = x_i\} & X \text{ discrete} \end{cases} \tag{3.42}$$

where $\sum_i$ stands for sum over *all* i. Note the following aliases:

expected value = ***mean*** = ***average*** = ***expectation*** = ***first moment***

The expected value of a function $g(X)$ of a RV X is given by

$$\boxed{E[g(X)] = \begin{cases} \int_{-\infty}^{\infty} g(x) f_X(x) dx & X \text{ continuous} \\ \sum_i g(x_i) P\{X = x_i\} & X \text{ discrete} \end{cases}} \tag{3.43}$$

Note that $\sum_i g(x_i) P\{X = x_i\}$ is the weighted sum (or average) of $g(x)$.

Other quantities associated with expectation can then be defined, including:

- The mth ***moment*** of a RV X is defined as $E[X^m]$, given by (3.43) with $g(x) = x^m$.
- The mth ***central moment*** of X is defined as $E[(X - \bar{x})^m]$, given by (3.43) with $g(x) = (x - \bar{x})^m$.

In the above, m is a positive integer.

Second moments are of particular interest:

$$\begin{aligned} \textbf{\textit{mean square value}}\ E[X^2] &= \text{second moment} \\ \textbf{\textit{variance}}\ \sigma_x^2 = \text{var}(X) \triangleq E[(X - \bar{x})^2] &= \text{second central moment} \\ \textbf{\textit{standard deviation}}\ \sigma_x &\triangleq \sqrt{\sigma_x^2} \end{aligned}$$

Although the CDF and PDF are complete and better descriptions of a RV, they are difficult to obtain in many practical situations. Expectation and moments are much easier to obtain. This is a major reason why expectation and moments are important.

It should be clear from Example 3.18 that the expected value of a discrete RV X can be interpreted naturally as the common-sense average value of X. Based on this, the definition of the expected value of a continuous RV is also well supported by the ***law of large numbers***, which, loosely speaking, states that the expected value of a RV is equal to the average value of a large number of its realizations x_i's:

$$\text{common-sense average} = \frac{1}{n}\sum_{i=1}^{n} x_i = \sum_{i=1}^{n} x_i P\{X = x_i\} \stackrel{n\to\infty}{\longrightarrow} \text{expected value}$$

The law of large numbers is one of the best representatives of statistical laws. It is a primary foundation of probability and statistics. Without it, there would be no probability or statistics. Everyone has the experience that the relative frequency of the occurrence of an event under "identical" conditions is quite stable. In other words, the stability of relative frequency reflects the regularity about the occurrence of an event, usually imbedded in the contingency (uncertainties) involved in the event occurrence. The regularity will overcome the contingency only through a large number of trials since the uncertainties tend to cancel out each other. As a result, probability laws exhibit themselves only through a large number of trials. For example, the Bernoulli law of large numbers states that as the number of independent trials increases, the relative frequency of the occurrence of an event approaches the probability of the event.

The top part of (3.42) is valid also for discrete RVs if the delta functions are allowed.

Expectation or moments may turn out to be infinity. For example, a RV X with the following PDF

$$f(x) = \begin{cases} \frac{1}{x^2} & x \geq 1 \\ 0 & x < 1 \end{cases}$$

has an expected value

$$E[X] = \int_{-\infty}^{\infty} x f(x) dx = \int_{-\infty}^{1} x \frac{1}{x^2} dx = \infty$$

When this is the case, some other measure should be considered.

The integral of (3.42) is mathematically legitimate only if it is absolutely integrable; that is,

$$\int_{-\infty}^{\infty} |x f(x)| dx < \infty$$

The reason for this requirement is that if this is not true, then the integral (summation) may depend on the order of its terms, which does not make any sense in practice, and thus the expected value would be ambiguous.

There are two fundamentally different methods to calculate the expected value of a function of a RV $Y = g(X)$: The first one is to find the PDF or pmf of Y and then use (3.42). The second is to use (3.43) directly. Fortunately, the results are identical — if one integral converges so does the other and they are equal if the integral converges.

3.7.3 Interpretations

The expected value of a RV X can be interpreted naturally as

- the common-sense ***average value*** of X over all possible outcomes, where the average is weighted by the probability of the occurrence of the outcome
- the ***center of gravity*** (i.e., the centroid or the balance point).

The standard deviation σ is a measure of the dispersion of a RV from its mean: A smaller σ indicates that the distribution is more concentrated around its mean. A RV X with $\sigma = 0$ must have $P\{X = \bar{x}\} = 1$, meaning that X is actually not random. Thus, the standard deviation can be thought of as a measure of how random (uncertain) a RV is.

X has a small variance $\Longrightarrow$ X concentrates more around its mean

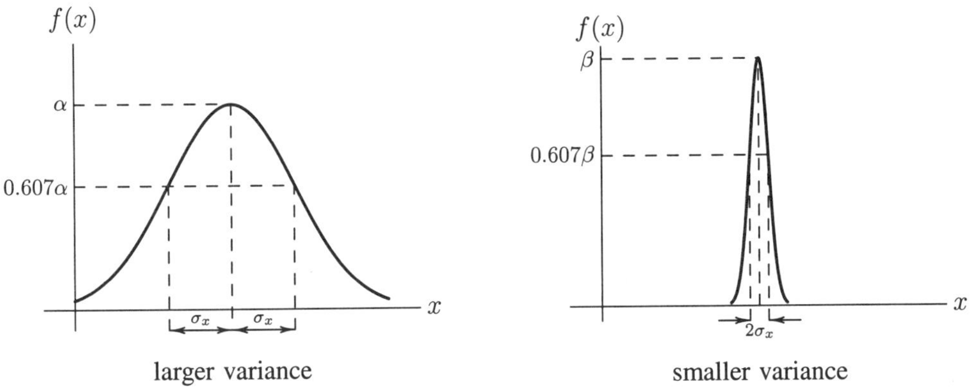

Figure 3.19: The PDF of a Gaussian RV with different variances.

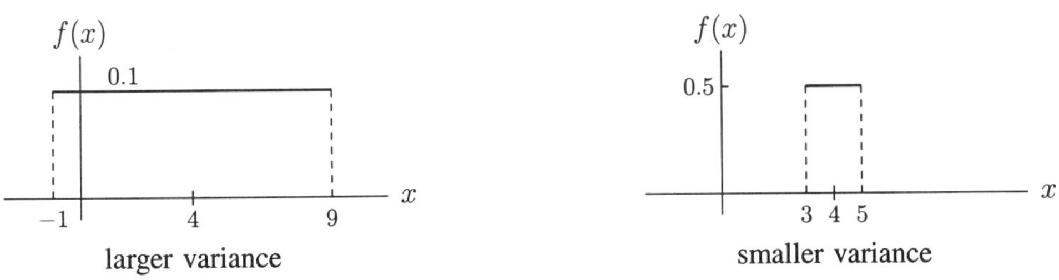

Figure 3.20: The PDF of a uniform RV with different variances.

It is clear from the introduction that thinking of probability as of (unity) mass on the real line, the expected value can be naturally interpreted as the centroid (or the balance point): The value of the PDF $f(x_i)$ (or point mass p_i) acts as the weight located at $x = x_i$ and the line will balance (i.e., not tip to the left or right) if a pivot is placed under the line at the expected value $\bar{x}$. Similarly, the mean-square value $E[X^2]$ can be interpreted as the moment of inertia with respect to the origin; the variance σ^2 as the central moment of inertia; and the standard deviation as the radius of gyration.

The expected value is a quantity that attempts to summarize a distribution function by a single value. The standard deviation (or variance) tries to measure the spread of the distribution about the expected value by means of a single scalar. It can be shown that $\sigma_x^2 = E[(x-\bar{x})^2] = \min_y E[(x-y)^2]$ for any y, meaning that a RV is most concentrated in the minimum mean-square sense around its expected value.

The following joke demonstrates the significance of the variance: A convention was held in a resort near a lake. Informed that the lake has an average depth of 1.2 meters, many people attending the convention jumped into the water for fun and were drowned because they did not known that the variance of the water depth is 5 meters!

There are other (scalar) measures of the central tendency of a RV. One of the most important is the ***mode***, defined as the value of the RV at which the PDF is a maximum. It is that value of a RV that is most probable (i.e., observed with the greatest frequency).

Another important measure is the ***median***, defined as the value of a RV at which the CDF is equal to 0.5. In other words, the median of a RV is a "center" point in the domain of the CDF that divides its range into two "equal" parts.

The popularity of expected value (mean) is a result of its nice mathematical properties, its clear physical interpretations and the law of large numbers.

For a RV with a PDF of multiple peaks, the standard deviation depends primarily on the separation between the peaks of the PDF.

For a RV with a PDF that is symmetrical about $x = a$, all its odd central moments vanish if they exist (i.e., are finite), and its mean, mode, and median are all equal to a. However, a continuous distribution need not be symmetrical even if all its odd central moments vanish. Typically, the third central moment provides a measure of the skewness of a PDF. Specifically, the quantity $\frac{E[(X-\bar{x})^3]}{\sigma^3}$ is known as ***skewness***, which provides a partial measure of how skew (i.e., asymmetrical) a PDF is.

Most probability inequalities are related with the law of large numbers. The most well-known is the following ***Chebyshev inequality***:

$$P\{|X - E(X)| \geq \epsilon\} \leq \frac{\text{var}(X)}{\epsilon^2}, \qquad \forall \epsilon > 0 \tag{3.44}$$

which provides a bound in probability for how close the RV is to its mean.

A useful result for a zero-mean Gaussian RV with variance σ^2 is

$$E(X^4) = 3\sigma^4 \tag{3.45}$$

3.7.4 Important Properties of Expectation and Variance

1. Expectation (average) of a constant is equal to the constant itself:
$$E[c] = \int_{-\infty}^{\infty} c f_X(x)dx = c\int_{-\infty}^{\infty} f_X(x)dx = c$$
This makes sense: a constant will show up the same value in every trial.
2. Expectation (average) of a constant times a RV is equal to the constant times the expectation (average) of the RV:
$$E[ch(X)] = \int_{-\infty}^{\infty} ch(x)f_X(x)dx = c\int_{-\infty}^{\infty} h(x)f_X(x)dx = cE[h(X)] \quad (3.46)$$
This makes sense: average of scaled numbers = scaled average.
3. Expectation of the sum of RVs is equal to the sum of expectations:
$$E\Big[\sum_i h_i(X)\Big] = \int_{-\infty}^{\infty} \Big(\sum_i h_i(x)\Big) f_X(x)dx = \sum_i \int_{-\infty}^{\infty} h_i(x)f_X(x)dx = \sum_i E[h_i(X)] \quad (3.47)$$
This makes sense: average of sets of numbers = sum of averages.
4.
$$\boxed{\text{variance} = \text{mean-square value} - (\text{mean})^2} \quad (3.48)$$
$$\sigma_x^2 \triangleq E[(X-\bar{x})^2] = E[X^2 - 2\bar{x}X + (\bar{x})^2] = E[X^2] - 2\bar{x}E[X] + (\bar{x})^2$$
$$= E[X^2] - 2(\bar{x})^2 + (\bar{x})^2 = E[X^2] - (\bar{x})^2$$
5. Variance of a nonrandom constant times a RV plus a constant is equal to the first constant squared times the variance of the RV:
$$\text{var}(aX+b) = E[(aX+b-(a\bar{x}+b))^2] = E[a^2(X-\bar{x})^2] = a^2E[(X-\bar{x})^2]$$
$$= a^2\text{var}(X) \quad (3.49)$$
This makes sense: Adding a constant in every trial does not alter randomness.
6. Variance of a constant is zero: $\text{var}(c) = 0$ because c is not random.

(3.46)–(3.47) implies that expectation E is a ***linear operator***:

$$\boxed{\begin{array}{c}\text{expectation of weighted sum} = \text{weighted sum of expectations}\\ E[ah_1(X) + bh_2(X)] = aE[h_1(X)] + bE[h_2(X)]\end{array}} \quad (3.50)$$

However, $E[h(X)] \neq h(E[X])$, if $h(\cdot)$ is nonlinear. If $y = h(x) \neq ax+b$ for some a and b, then $h(x)$ is nonlinear, such as $h(x) = \cos x$ and $h(x) = e^{-x}$.

3.7.5 Examples

Example 3.19: Mean and Variance of Uniform Distribution

Find the mean and variance of a uniform RV $X \sim \mathcal{U}(a, b)$:

$$\begin{aligned}
\bar{x} = E[X] &= \int_{-\infty}^{\infty} x f_X(x)dx = \int_a^b x\frac{1}{b-a}dx = \frac{1}{2}(a+b) \\
\sigma^2 = E[(X-\bar{x})^2] &= \int_{-\infty}^{\infty} (x-\bar{x})^2 f_X(x)dx \\
&= \frac{1}{b-a}\int_a^b (x-\bar{x})^2 dx = \frac{1}{b-a}\frac{1}{3}(x-\bar{x})^3\Big|_a^b \\
&= \frac{1}{b-a}\frac{1}{3}\Big\{[b-\frac{1}{2}(a+b)]^3 - [a-\frac{1}{2}(a+b)]^3\Big\} \\
&= \frac{1}{b-a}\frac{1}{3}\Big[\frac{1}{8}(b-a)^3 - \frac{1}{8}(a-b)^3\Big] = \frac{1}{12}(b-a)^2
\end{aligned}$$

Alternatively, the variance can be calculated using (3.48):

$$\begin{aligned}
\sigma^2 = E[X^2] - (\bar{x})^2 &= \int_a^b x^2\frac{1}{b-a}dx - \Big[\frac{1}{2}(a+b)\Big]^2 \\
&= \frac{1}{b-a}\frac{1}{3}x^3\Big|_a^b - \frac{1}{4}(a+b)^2 = \frac{1}{b-a}\frac{1}{3}(b^3-a^3) - \frac{1}{4}(a+b)^2 \\
&= \frac{1}{3}(b^2+ab+a^2) - \frac{1}{4}(a+b)^2 \\
&= \frac{1}{12}(4b^2+4ab+4a^2-3a^2-6ab-3b^2) \\
&= \frac{1}{12}(b-a)^2
\end{aligned}$$

In summary, for any $\mathcal{U}(a, b)$ RV,

$$\text{mean} = \frac{1}{2}(a+b) = \text{center point} \tag{3.51}$$

$$\text{variance} = \frac{1}{12}(b-a)^2 = \frac{(\text{length})^2}{12} \tag{3.52}$$

Since the PDF of a uniform RV is nonzero only over a finite interval, caution should be taken when using standard deviation (or variance). See Example 3.37 and problems 3.18 and 3.19.

Example 3.20: Mean and Variance of Gaussian Distribution

Find the mean and variance of a Gaussian RV $X \sim \mathcal{N}(\mu, \sigma^2)$:

$$E[X] = \int_{-\infty}^{\infty} x f_X(x)dx = \int_{-\infty}^{\infty}(x-\mu)f_X(x)dx + \mu \underbrace{\int_{-\infty}^{\infty} f_X(x)dx}_{\stackrel{?}{=}1}$$

$$= \int_{-\infty}^{\infty}(x-\mu)\frac{1}{\sqrt{2\pi}\sigma}e^{-\frac{(x-\mu)^2}{2\sigma^2}}dx + \mu$$

Since

$$\left.\begin{array}{r}\frac{1}{\sqrt{2\pi}\sigma}e^{-\frac{(x-\mu)^2}{2\sigma^2}} \text{ is even about } x=\mu \\ (x-\mu) \text{ is odd about } x=\mu\end{array}\right\} \Longrightarrow \frac{x-\mu}{\sqrt{2\pi}\sigma}e^{-\frac{(x-\mu)^2}{2\sigma^2}} \text{ is odd about } x=\mu$$

by odd symmetry,

$$\int_{-\infty}^{\infty}(x-\mu)\frac{1}{\sqrt{2\pi}\sigma}e^{-\frac{(x-\mu)^2}{2\sigma^2}}dx = 0$$

The mean is then given by

$$E[X] = \bar{x} = \mu = \text{peak point} \tag{3.53}$$

In general, *a RV X with a symmetric PDF about a point b has $E[X] = b$.*

Taking derivative on both sides with respect to σ of the identity

$$\int_{-\infty}^{\infty}\frac{1}{\sqrt{2\pi}}\,e^{-\frac{(x-\bar{x})^2}{2\sigma^2}}\,dx = \sigma\int_{-\infty}^{\infty}\frac{1}{\sqrt{2\pi}\sigma}\,e^{-\frac{(x-\bar{x})^2}{2\sigma^2}}\,dx = \sigma\int_{-\infty}^{\infty}\mathcal{N}(x;\bar{x},\sigma^2)dx = \sigma$$

yields

$$\frac{1}{\sigma^2}\int_{-\infty}^{\infty}(x-\bar{x})^2\frac{1}{\sqrt{2\pi}\sigma}\,e^{-\frac{(x-\bar{x})^2}{2\sigma^2}}\,dx = 1$$

where $\frac{\partial}{\partial\sigma}e^{-(x-\bar{x})^2/2\sigma^2} = e^{-(x-\bar{x})^2/2\sigma^2}[\frac{1}{2}(x-\bar{x})^2](-2)\sigma^{-3}$. Thus, the variance is given by

$$\text{var}(X) = E[(X-\bar{x})^2] = \int_{-\infty}^{\infty}(x-\bar{x})^2\frac{1}{\sqrt{2\pi}\sigma}\,e^{-\frac{(x-\bar{x})^2}{2\sigma^2}}\,dx = \sigma^2 \tag{3.54}$$

Consequently, *the two parameters μ and σ^2 of a Gaussian distribution are the mean and variance, respectively.*

Example 3.21: Mean and Variance of Exponential Distribution

A RV X has the ***exponential PDF***

$$f(x) = \lambda e^{-\lambda x} u(x) \qquad \lambda > 0 \qquad u(x) = \text{unit step function} \tag{3.55}$$

where the factor λ is necessary to guarantee $\int_{-\infty}^{\infty} f(x)dx = 1$.

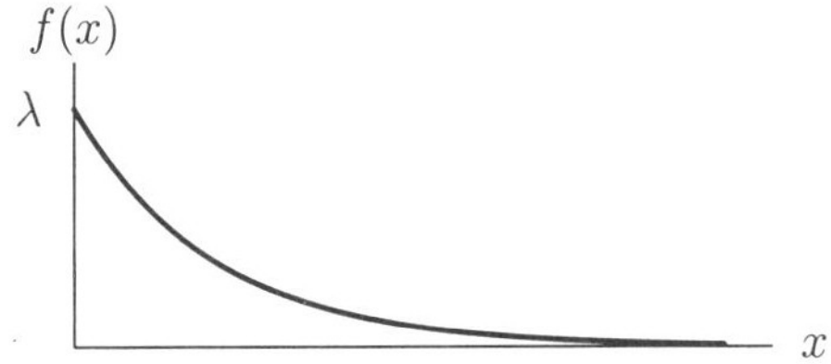

Figure 3.21: The PDF of an exponentially-distributed RV.

(a) Find $\bar{x}$ and σ_x^2: By integration by parts, we have

$$\bar{x} = \int_0^{\infty} x\lambda e^{-\lambda x} dx = -xe^{-\lambda x}\Big|_0^{\infty} + \int_0^{\infty} e^{-\lambda x} dx = 1/\lambda$$

$$E[X^2] = \int_0^{\infty} \lambda x^2 e^{-\lambda x} dx \overset{\text{similar to above}}{=} \frac{2}{\lambda^2} \quad \Longrightarrow \quad \sigma_x^2 \overset{(3.48)}{=} \frac{1}{\lambda^2}$$

Note that a small λ implies a flat PDF and a large mean and variance.

(b) Find the mean and variance of $Y = c_1 X + c_2$ (c_1, c_2 are constants):

$$\bar{y} = E[c_1 X + c_2] \overset{(3.50)}{=} c_1 \bar{x} + c_2 = c_1/\lambda + c_2$$

$$\sigma_y^2 \overset{(3.49)}{=} c_1^2 \sigma_x^2 = c_1^2/\lambda^2$$

(c) Find the mean and variance of $Z = e^{-aX}$ (a is a constant):

$$\bar{z} = E[e^{-aX}] = \int_{-\infty}^{\infty} e^{-ax} f_X(x)dx = \lambda \int_0^{\infty} e^{-ax} e^{-\lambda x} dx = \frac{\lambda}{a+\lambda}$$

$$E[Z^2] = E[(e^{-aX})^2] = E\left[e^{-2aX}\right] \overset{?}{=} \frac{\lambda}{2a+\lambda} \qquad \text{(from the above equation)}$$

$$\sigma_z^2 = E[Z^2] - (\bar{z})^2 = \frac{\lambda}{2a+\lambda} - \frac{\lambda^2}{(a+\lambda)^2} = \frac{a^2\lambda}{(2a+\lambda)(a+\lambda)^2}$$

Note the difference:

$$Y = c_1 X + c_2 \quad \text{(a linear function)} \Longrightarrow \bar{y} = c_1 \bar{x} + c_2$$

$$Z = e^{-aX} \quad \text{(a nonlinear function)} \not\Longrightarrow \bar{z} = e^{-a\bar{x}}$$

Example 3.22: Mean and Variance of Binary Distribution

A ***binary RV*** or ***Bernoulli RV*** X is one with the PMF $P\{X=1\}=p, P\{X=0\}=1-p$. Its mean and variance are

$$\begin{aligned} E[X] &\overset{(3.42)}{=} \sum_{k=-\infty}^{\infty} kP\{X=k\} = \sum_{k=0}^{1} kP\{X=k\} = p \\ E[X^2] &\overset{(3.43)}{=} \sum_{k=-\infty}^{\infty} k^2P\{X=k\} = \sum_{k=0}^{1} k^2P\{X=k\} = p \\ \sigma_x^2 &= E[X^2]-(E[X])^2 = p-p^2 = p(1-p) \end{aligned}$$

Example 3.23: Mean and Variance of Poisson Distribution

A ***Poisson RV*** X has the PMF given by (3.38).

(a) Find the mean of X:

$$\begin{aligned} E[X] &\overset{(3.42)}{=} \sum_{k=-\infty}^{\infty} kP\{X=k\} = \sum_{k=0}^{\infty} ke^{-\lambda}\frac{\lambda^k}{k!} = \lambda e^{-\lambda}\sum_{k=1}^{\infty}\frac{\lambda^{k-1}}{(k-1)!} \\ &\overset{n=k-1}{=} \lambda e^{-\lambda}\sum_{n=0}^{\infty}\frac{\lambda^n}{n!} = \lambda e^{-\lambda}e^{\lambda} = \lambda \qquad \left(\text{note: } e^{\lambda} = \sum_{n=0}^{\infty}\frac{\lambda^n}{n!}\right) \end{aligned}$$

(b) Find the variance of X:

$$\begin{aligned} E[X^2] &\overset{(3.43)}{=} \sum_{k=-\infty}^{\infty} k^2P\{X=k\} = \sum_{k=0}^{\infty} k^2e^{-\lambda}\frac{\lambda^k}{k!} \\ &= \lambda e^{-\lambda}\sum_{k=1}^{\infty}[(k-1)+1]\frac{\lambda^{k-1}}{(k-1)!} \\ &= \lambda e^{-\lambda}\left[\sum_{k=1}^{\infty}\frac{(k-1)\lambda^{k-1}}{(k-1)!} + \sum_{k=1}^{\infty}\frac{\lambda^{k-1}}{(k-1)!}\right] \\ &= \lambda e^{-\lambda}\left[\lambda\sum_{k=2}^{\infty}\frac{\lambda^{k-2}}{(k-2)!} + e^{\lambda}\right] = \lambda e^{-\lambda}(\lambda e^{\lambda}+e^{\lambda}) = \lambda^2+\lambda \\ \sigma_x^2 &= E[X^2]-(E[X])^2 = \lambda^2+\lambda-\lambda^2 = \lambda \end{aligned}$$

Thus, *both mean and variance of a Poisson RV are equal to the parameter* λ.

The expected values and variances of some commonly used distributions are given in the following table. For an engineering student, it is beneficial to study and try to remember important results. Memory will often provide a connection that may prove to be useful in e.g. developing or inventing a new product. You never know what you missed if you do not remember the results.

Table 3.1: Mean and variance of some commonly used distributions.

Distribution	*PDF or PMF*	*Mean*	*Variance*
Continuous:			
Chi-square	(3.32)	$n\sigma^2$	$2n\sigma^4$
Exponential	(3.34)	$a + \frac{1}{\lambda}$	$\frac{1}{\lambda^2}$
Gaussian	(3.23)	μ	σ^2
Log-normal	(3.37)	$e^{\mu+\frac{1}{2}\sigma^2}$	$(e^{\sigma^2} - 1)e^{2\mu+\sigma^2}$
Rayleigh	(3.33)	$\sigma\sqrt{\pi/2}$	$(2 - \pi/2)\sigma^2$
Uniform	(3.22)	$\frac{1}{2}(a + b)$	$(b - a)^2/12$
Weibull	(3.35)	$b^{-1/a}\Gamma(1 + \frac{1}{a})$	$b^{-2/a}\{\Gamma(1 + \frac{2}{a}) - [\Gamma(1 + \frac{1}{a})]^2\}$
Discrete:			
Binary	Example 3.22	p	$p(1 - p)$
Binomial	(3.30)	Np	$Np(1 - p)$
Geometric	(3.39)	$1/p$	$(1 - p)/p^2$
Poisson	(3.38)	λ	λ

3.8 Function of a Random Variable

Given a RV X with the PDF $f_X(x)$ and a deterministic function $g(\cdot)$, $Y = g(X)$ is another RV, what is the PDF $f_Y(y)$ of Y?

A general procedure for obtaining $f_Y(y)$ from $f_X(x)$ is to express first $F_Y(y)$ in terms of $F_X(\cdot)$ and then find $f_Y(y) = \frac{d}{dy}F_Y(y)$.

Example 3.24: PDF of a Linear Function of a RV

Find $f_Y(y)$ of $Y = aX + b$, where X is continuous and $a \neq 0$:

$$F_Y(y) = P\{Y \le y\} = P\{aX + b \le y\}$$
$$= \begin{cases} P\{X \le \frac{y-b}{a}\} = F_X(\frac{y-b}{a}) & a > 0 \\ P\{X \ge \frac{y-b}{a}\} \stackrel{?}{=} P\{X > \frac{y-b}{a}\} = 1 - F_X(\frac{y-b}{a}) & a < 0 \end{cases}$$

$$f_Y(y) = \frac{d}{dy}F_Y(y) = \left\{ \begin{matrix} \frac{1}{a}f_X(\frac{y-b}{a}) & a > 0 \\ -\frac{1}{a}f_X(\frac{y-b}{a}) & a < 0 \end{matrix} \right\} = \frac{1}{|a|}f_X\Big(\frac{y-b}{a}\Big) \tag{3.56}$$

An alternative and usually better method is to use the following theorem.

Theorem: The PDF $f_Y(y)$ of $Y = g(X)$ can be determined from the PDF $f_X(x)$ of X by:

$$f_Y(y) = \frac{f_X(x_1)}{|g'(x_1)|} + \cdots + \frac{f_X(x_n)}{|g'(x_n)|} + \cdots = \sum_i \frac{f_X(x_i)}{|g'(x_i)|} \tag{3.57}$$

where $g'(x) = \frac{d}{dx}g(x)$ and x_i's are the (distinct) real roots of the equation $y = g(x)$ in terms of y: $y = g(x_1) = \cdots = g(x_n) = \cdots$.

Example 3.25: PDF of the Output of a Square-Law Device

The output Y of a square-law device is related to its input X by $Y = X^2$. Find $f_Y(y)$ if $X \sim \mathcal{N}(0,1)$. Note first $g(x) = x^2 \Longrightarrow g'(x) = 2x$. The two roots of $y = x^2$ are $x_{1,2} = \pm\sqrt{y}$. Thus, $f_Y(y) = 0$ for $y < 0$. For $y \ge 0$, we have

$$f_Y(y) = \frac{f_X(x_1)}{|g'(x_1)|} + \frac{f_X(x_2)}{|g'(x_2)|} = \frac{f_X(\sqrt{y})}{|2\sqrt{y}|} + \frac{f_X(-\sqrt{y})}{|2(-\sqrt{y})|}$$
$$= \frac{1}{2\sqrt{y}}\left[\frac{1}{\sqrt{2\pi}}e^{-\frac{1}{2}(\sqrt{y})^2} + \frac{1}{\sqrt{2\pi}}e^{-\frac{1}{2}(-\sqrt{y})^2}\right] = \frac{1}{\sqrt{2\pi y}}e^{-y/2}$$

Clearly, a deterministic function of a RV X is a RV since it is a function of the outcome of the random experiment on which the RV X is defined.

There are many situations in practice in which we need to find the PDF or CDF of a function of a RV. For instance, if the resistors in a circuit is random due to say, uncertainties in the manufacturing process or aging, the current, voltage, and power are all functions of these random resistances and are therefore random. Another popular example is that a sinusoidal signal with a random phase, frequency, or amplitude (due to noise, etc.) is also random. The study of such signals is fundamental for communications. A third popular example is the kinetic energy of a molecule $Y = \frac{1}{2}mX^2$, where m is the mass of the molecule and X its random velocity.

For a derivation and discussion of Theorem (3.57), see A. Papoulis, *Probability, Random Variables, and Stochastic Processes*, McGraw-Hill, New York, 1991.

A deterministic function of a discrete RV is a discrete RV. Its PMF can usually be found quite easily, as the following examples illustrate.

Example 3.26: PMF of Output of a Square-Law Device with Discrete Input

Find the PMF of the output $Y = X^2$ of a square-law detector when the input X has the following PMF:

X	-2	-1	0	1	2	3
$P\{X = x_i\}$	0.2	0.1	0.1	0.3	0.2	0.1

Clearly, Y has the following possible values: 0, 1, 4, and 9. Its probabilities of taking on these values are, respectively:

$$P\{Y = 0\} = P\{X = 0\} = 0.1 \qquad P\{Y = 1\} = P\{X = 1\} + P\{X = -1\} = 0.4$$
$$P\{Y = 9\} = P\{X = 3\} = 0.1 \qquad P\{Y = 4\} = P\{X = 2\} + P\{X = -2\} = 0.4$$

Thus the PMF of Y is

Y	0	1	4	9
$P\{Y = y_i\}$	0.1	0.4	0.4	0.1

Square-law devices are quite popular in engineering. For example, it is often used in communications for demodulation or in signal detection.

Example 3.27: PMF of a Function of Binary Random Variable

Suppose that a binary RV X has PMF $P\{X = 1\} = 0.3$ and $P\{X = 0\} = 0.7$. Find the PMF of the RV $Y = e^X$.

Clearly, the PMF of the RV $Y = e^X$ is given by

$$P\{Y = e\} = P\{Y = e^X|_{X=1}\} = P\{X = 1\} = 0.3$$
$$P\{Y = 1\} = P\{Y = e^X|_{X=0}\} = P\{X = 0\} = 0.7$$

Example 3.28: PDF of a Trigonometric Function with a Random Phase

Given a random phase $X \sim \mathcal{U}(-\pi, \pi)$, find $f_Y(y)$ of $Y = a \tan X, a > 0$:

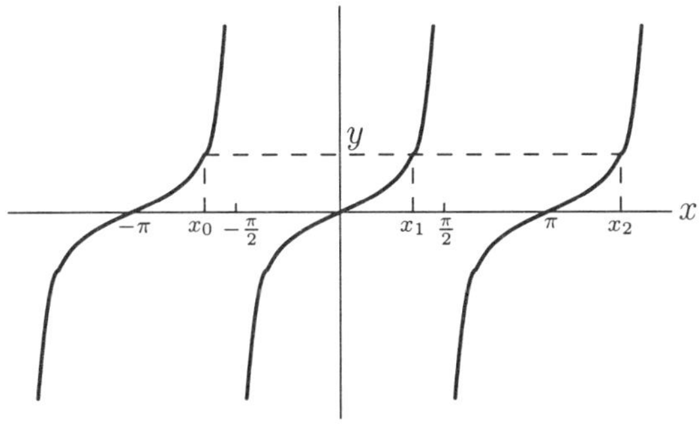

$$g(x) = a \tan x = y \Longrightarrow \begin{cases} g'(x) = \frac{a}{\cos^2 x} \\ x_i = \tan^{-1}(y/a), \quad i = \ldots, -1, 0, 1, \ldots \end{cases}$$

$$|g'(x_i)| = \frac{a}{\cos^2 x_i} = \frac{a(\cos^2 x_i + \sin^2 x_i)}{\cos^2 x_i} = \frac{a(1 + \tan^2 x_i)}{1} = \frac{a^2 + y^2}{a}$$

$$X \sim \mathcal{U}(-\pi, \pi) \quad \Longrightarrow \quad f_X(x) = \begin{cases} \frac{1}{2\pi} & -\pi < x < \pi \\ 0 & \text{elsewhere} \end{cases}$$

There are two roots x_0 and x_1 in $(-\pi, \pi)$ [$f_X(x) = 0$ outside $(-\pi, \pi)$]. Thus,

$$f_Y(y) \overset{(3.42)}{=} \sum_{i=-\infty}^{\infty} \frac{a f_X(x_i)}{a^2 + y^2} = \frac{a f_X(x_0)}{a^2 + y^2} + \frac{a f_X(x_1)}{a^2 + y^2} = \frac{a/2\pi}{a^2 + y^2} + \frac{a/2\pi}{a^2 + y^2} = \frac{a/\pi}{a^2 + y^2} \tag{3.58}$$

This is the PDF of the ***Cauchy distribution***.

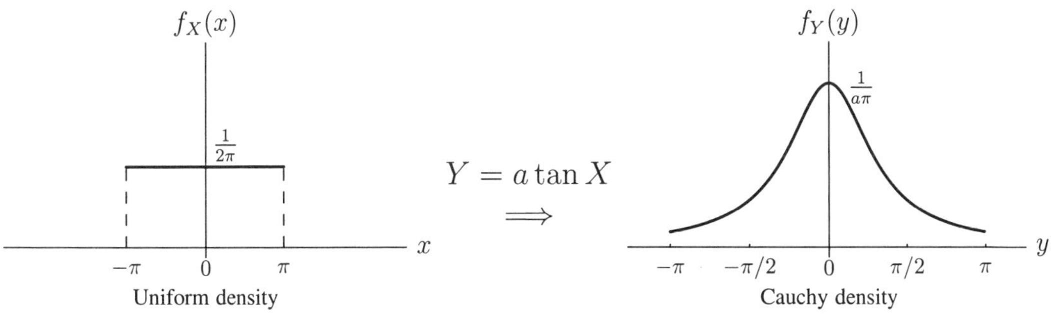

3.9 Conditional Distributions

Recall that the probability of event A conditioned on the occurrence of event B is defined by

$$P\{A|B\} = \frac{P\{A \cap B\}}{P\{B\}}$$

Since $\{X \leq x\}$ is an event, let $A = \{X \leq x\}$. Then the ***conditional CDF*** given event B is defined by

$$F(x|B) = F_X(x|B) = P\{X \leq x|B\} = \frac{P\{(X \leq x) \cap B\}}{P\{B\}} \tag{3.59}$$

The ***conditional PDF*** given event B is thus defined by

$$f(x|B) = \frac{dF(x|B)}{dx} = \frac{\frac{d}{dx}P\{X \leq x, B\}}{P\{B\}} \tag{3.60}$$

Delta functions may be introduced at discontinuous points of $F(x|B)$.

A conditional CDF possesses all the properties of an unconditional CDF, including

$$\begin{aligned}
F(-\infty|B) &= 0 \\
F(\infty|B) &= 1 \\
0 \leq F(x|B) &\leq 1 \\
P\{x_1 < X \leq x_2|B\} &= F(x_2|B) - F(x_1|B) \\
F(x_2|B) &\geq F(x_1|B) \quad \forall x_2 > x_1 \qquad \text{(nondecreasing)} \\
\lim_{\epsilon \to 0} F(x + \epsilon|B) &= F(x|B) \qquad \text{(continuous from right)}
\end{aligned}$$

All the properties of a PDF also apply to a conditional PDF, including

$$\begin{aligned}
f(x|B) &\geq 0 \\
\int_{-\infty}^{\infty} f(x|B)dx &= 1 \\
F_X(x|B) &= \int_{-\infty}^{x} f_X(y|B)dy \\
P\{x_1 < X \leq x_2|B\} &= \int_{x_1}^{x_2} f_X(x|B)dx
\end{aligned}$$

Example 3.29: Truncated Gaussian

Let $B = \{a < X < b\}$. The following so-called ***truncated PDF***

$$f(x|B) = \frac{d}{dx}\frac{P\{X \le x, a < X < b\}}{P\{a < X < b\}} = \frac{d}{dx}\begin{cases} 0 & x \le a \\ \frac{P\{a<X\le x\}}{P\{a<X<b\}} = \frac{F(x)-F(a)}{P\{a<X<b\}} & a < x < b \\ \frac{P\{a<X<b\}}{P\{a<X<b\}} = 1 & x \ge b \end{cases}$$

$$= \begin{cases} \frac{f(x)}{P\{a<X<b\}} & a < x < b \\ 0 & \text{elsewhere} \end{cases} \tag{3.61}$$

is the original PDF restricted to $a < X < b$ and *renormalized*, where $P\{a < X < b\}$ acts as a scaling factor to scale up the PDF so that the conditional PDF $f(x|B)$ integrates to unity. For example, if $X \sim \mathcal{N}(\bar{x}, \sigma^2)$, then

$$f(x|a < X < b) = \begin{cases} \dfrac{\mathcal{N}(x; \bar{x}, \sigma^2)}{P\{a < X < b\}} & a < x < b \\ 0 & \text{elsewhere} \end{cases} \tag{3.62}$$

which is a ***truncated (censored) Gaussian***.

The conditional PDF $f(x|a < X < b)$ of (3.62) and the unconditional (original) PDF $\mathcal{N}(x; \bar{x}, \sigma^2)$ are plotted in Fig. 3.22.

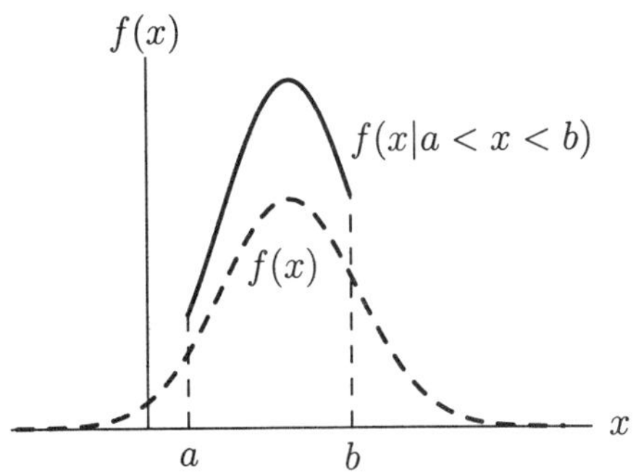

Figure 3.22: Gaussian PDF $f(x)$ and truncated Gaussian PDF $f(x|a < X < b)$.

Most Gaussian models of real-world problems are actually truncated Gaussian, such as test score, and height and weight of people.

3.10 Generation of Random Numbers

Uniform random number generators are available in many software packages.

Given a uniform random number generator, there are several approaches to generate other random numbers. The most popular one is the following. Random numbers of some popular distributions can be generated by the companion software P&R.

3.10.1 Generation of Continuous Random Numbers

Samples of many continuous RVs can be generated by the following ***inverse-transform method*** based on the following **theorem**:

> If $U \sim \mathcal{U}(0,1)$, then $X = F^{-1}(U)$ has the CDF $F(x)$ provided $F(x)$ is continuous and strictly increasing in x, where $F^{-1}(\cdot)$ is the inverse function of $F(\cdot)$.

Thus, if $U_1, \ldots, U_n$ are $\mathcal{U}(0,1)$ random numbers, then $(X_1, \ldots, X_n) = [F^{-1}(U_1), \ldots, F^{-1}(U_n)]$ are random numbers with the distribution F.

Example 3.30: Generation of Exponential Random Numbers

An ***exponential RV*** has the PDF given by (3.55). The corresponding CDF is

$$F(x) = \begin{cases} 1 - e^{-\lambda x} & x > 0 \\ 0 & x < 0 \end{cases}$$

Its inverse function $F^{-1}(u)$ can be obtained by expressing x in terms of u:

$$u = 1 - e^{-\lambda x} \quad \Longrightarrow \quad x = -\frac{1}{\lambda}\ln(1-u) \quad \Longrightarrow \quad x = F^{-1}(u) = -\frac{1}{\lambda}\ln(1-u)$$

Since $F(\cdot)$ is continuous and strictly increasing, by the above theorem, given $\mathcal{U}(0,1)$ random numbers $U_1, \ldots, U_n$, the random numbers

$$(X_1, \ldots, X_n) = \left[-\frac{1}{\lambda}\ln(1-U_1), \ldots, -\frac{1}{\lambda}\ln(1-U_n)\right]$$

are exponentially distributed with the CDF given above. In fact, $(1-U_i)$ can be replaced by U_i since both U_i and $(1-U_i)$ are $\mathcal{U}(0,1)$ random numbers.

One of the most important tools for probabilistic analysis is the so-called ***Monte Carlo simulation***. In this method, it is required that random numbers with specified distributions be generated. A common approach is to generate random numbers that are uniformly distributed over $(0, 1)$ first. Then, random numbers with various other distributions are generated as functions of these uniform random numbers.

A routine that generates random numbers uniformly distributed over $(0, 1)$ is called a ***uniform random number generator***. How such a generator can be developed will not be covered here. What is important is that such generators are available in most software packages that enable us to use the Monte Carlo method. However, most such packages provide a generator for *uniform* random numbers only. It is thus important to know how to generate other random numbers based on this generator.

The inverse-transform method, also known as the *percentile transform method*, for the generation of continuous random numbers described here is convenient for any distribution for which the inverse of its CDF can be found analytically (i.e., in a closed form). This is the case for many important continuous RVs. For the most important distribution, the Gaussian distribution, however, the inverse cannot be determined analytically. There are several methods to generate Gaussian random numbers. They will not be covered here because Gaussian random number generators are available in more and more software packages. For example, MATLAB has an internal function "`randn`" that generates vector-valued Gaussian random numbers that are independent.

Note that the inverse-transform method works only if the CDF $F(x)$ is strictly increasing in x. Since $F(x)$ is always nondecreasing, it is not strictly increasing only if it is a constant (say, c) over some interval. Thus, when $y = F(x) = c$, x is ambiguous, which can be any value in the interval and the inverse function does not exist.

In general, if a generator generates random numbers $U_1, \ldots, U_n$ with zero mean and unity variance, then random numbers $X_1, \ldots, X_n$ will have mean $\bar{x}$ and variance σ_x^2 if $X_i = \sigma_x U_i + \bar{x}$.

An advantage of the inverse-transform method is its ease of generating random numbers with truncated distributions. For example, it can be shown that random numbers $X_1, \ldots, X_n$ with the truncated PDF (3.61) or equivalently the following CDF

$$F(x|a < x < b) = \begin{cases} 0 & x \le a \\ \frac{F(x)-F(a)}{P\{a<X<b\}} & a < x < b \\ 1 & x \ge b \end{cases}$$

can be generated by the following procedure

S1. Generate $U_i \sim \mathcal{U}(0, 1)$
S2. Let $V_i = F(a) + P\{a < X < b\}U_i$
S3. Return $X_i = F^{-1}(V_i)$.

Another common approach to random number generation is the so-called ***acceptance-rejection method***. Its basic idea is to select (accept) among all the candidate random numbers generated only those that satisfy various conditions of the specified distribution. It is particularly good for random numbers with complex distributions for which the inverses do not exist or the inverse-transform method is inefficient.

Example 3.31: Generation of Truncated Gaussian Random Numbers

Random numbers $X_1, \dots, X_n$ that are truncated Gaussian (3.62) over (a, b) can be generated by the following procedure based on the acceptance-rejection method

S1. Generate $U_i \sim \mathcal{N}(\bar{x}, \sigma^2)$

S2. If $a < U_i < b$ then return $X_i = U_i$; otherwise go back to Step 1 and generate another U_i.

3.10.2 Generation of Discrete Random Numbers

Given $\mathcal{U}(0, 1)$ random numbers $U_1, \dots, U_n$ and a discrete RV X with PMF:

$$P\{X = x_i\} = p_i$$

where $p_i \geq 0, \sum_i p_i = 1$, if each X_i is obtained by the following logic, with $p_0 = 0$,

$$\text{if } \sum_{j=0}^{k-1} p_j \leq U_i < \sum_{j=0}^{k} p_j \quad \text{then} \quad X_i = x_k$$

then $X_1, \dots, X_n$ are random numbers with the distribution of X. This is illustrated below.

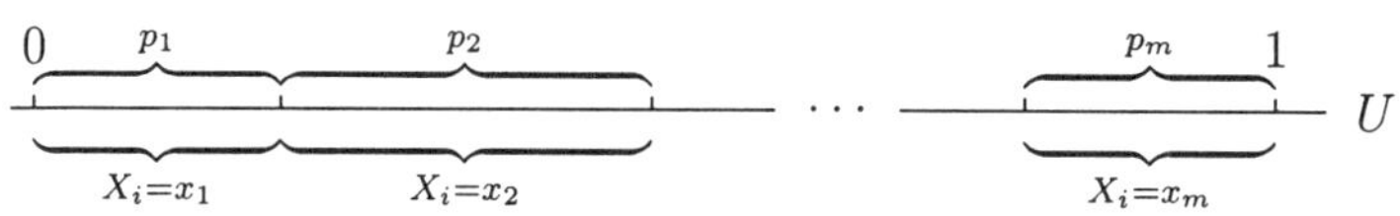

Figure 3.23: Generation of discrete random numbers X_i from $U_i \sim \mathcal{U}(0, 1)$.

Example 3.32: Generation of Binary Random Numbers

Given $\mathcal{U}(0, 1)$ random numbers $U_1, \dots, U_n$, a sequence $X_1, \dots, X_n$ of binary random numbers taking on value 1 with p and 0 with $1 - p$ can be obtained by

$$X_i = \begin{cases} 1 & 0 \leq U_i < p \\ 0 & p < U_i \leq 1 \end{cases}$$

Note that $Y = \sum_{i=1}^{N} X_i$ is binomial distributed with parameters N and p.

Example 3.33: Generation of Chi-Square Random Numbers by P&R

Random numbers can be generated easily using the companion software P&R. We use chi-square distribution as an example to illustrate the procedure as follows:

S1. Click "RVGenerator" in the main window of P&R.

S2. Click "ChiSquare." The "**Chi-Square Random Number Generator**" window will appear.

S3. Enter the parameters (i.e., 1.44 for "sigma" and 20 for "number"), as shown in Fig. 3.24. Click "Ok." 20 random numbers that are chi-square distributed with parameter $\sigma = 1.44$ are then generated and saved to a data file `e3_33.dat` as specified and plotted as shown in Fig. 3.24.

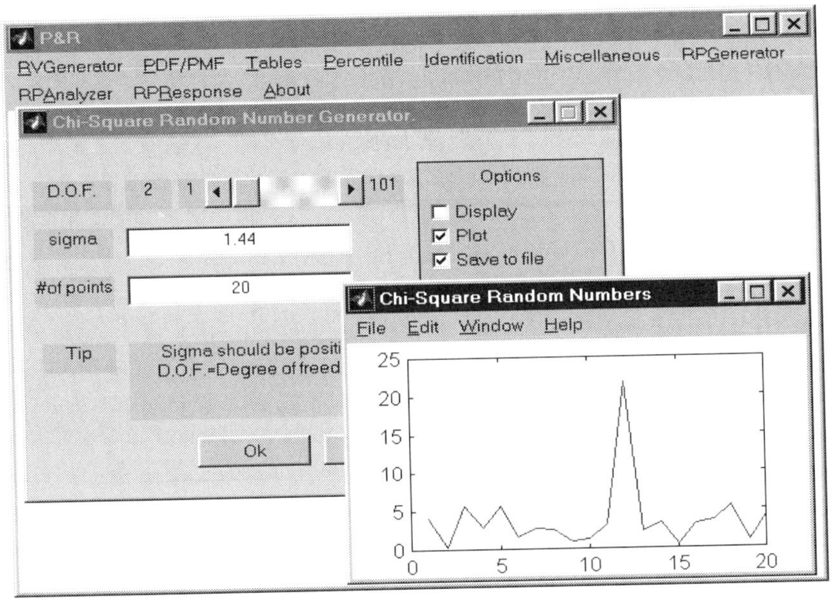

Figure 3.24: Generation of random numbers by P&R.

3.11 Determination of Distribution from Data

Given a set of numbers (data) that are the realizations of a RV X, how to determine the distribution of X?

A *frequency ratio* or relative frequency of X is the fraction (or percentage) of the number of values (realizations) of X falling within a specified interval. A ***histogram*** is a plot of the frequency ratios versus the intervals (also called ***bins***).

A histogram is a useful tool for the determination of the distribution from data. It can usually be plotted as follows.

S1. Find the maximum $x_{\max}$ and the minimum $x_{\min}$ of the data.

S2. Let a be the largest integer that is not larger than $x_{\min}$ and let b be the smallest integer that is not smaller than $x_{\max}$. If $x_{\max} - x_{\min}$ is much smaller than $b-a$, then let $(a, b) = (x_{\min}, x_{\max})$. Then the interval for plotting is (a, b).

S3. Divide the interval (a, b) by n to obtain n equally spaced frequency intervals, $t_i = (c_{i-1}, c_i]$, $i = 1, \ldots, n$, called ***bins***, where $c_0 = a$ and $c_n = b$.
S4. Examine all data points one by one. If the jth data point X_j is in the interval t_i; that is, if $c_{i-1} < X_j \leq c_i$, then increase the counter f_i of the interval t_i by 1. Repeat this process until all data points are examined.
S5. Divide f_i's by the total number of data points times the length of the interval $(c_{i-1}, c_i]$ and then plot f_i's against t_i's. This yields the histogram.

The MATLAB function "`hist`" plot the histogram automatically. Only the data set and the number of bins need to be provided. However, make sure the vertical value of the histogram is the frequency ratio.

Given a record of data, its histogram can be obtained by the companion software P&R.

Example 3.34: Generating Histogram of Data by P&R

Given the data contained in the data file `e3_34.dat` in the companion software P&R, its histogram can be obtained by P&R as follows[2].

S1. Click "Identification" in the main window of P&R.
S2. Choose "Histogram." You will be prompted for the name of a data file.
S3. Choose "data" subdirectory, enter "`e3_34.dat`" and click "Ok." The histogram of the data is then plotted, as shown in Fig. 3.25.

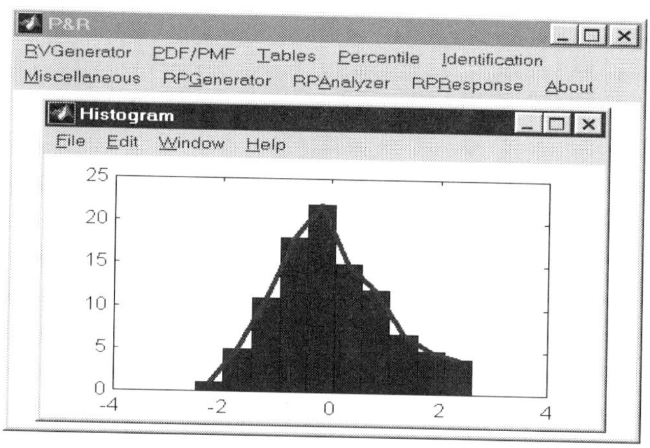

Figure 3.25: Generation of a histogram by P&R.

Fig. 3.26 gives the histogram obtained from P&R by using 10 bins.

[2]All data files used by a P&R example are included in P&R. You are strongly encouraged to repeat all these examples.

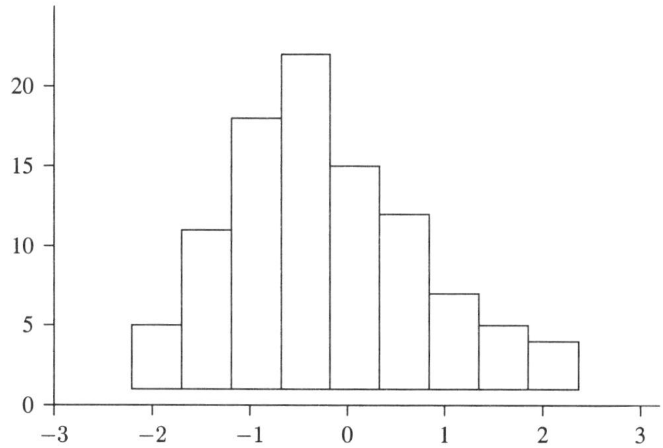

Figure 3.26: A histogram using 10 bins.

For example, the second highest bar indicates that 20% data have the value in between 0 and 0.5.

If the histogram of the data set matches a known PDF, then we conclude that the RV from the set of data was drawn has the known distribution. For example, the above set of data was actually 100 realizations of a $\mathcal{N}(0, 1)$ RV and thus (the envelope of) the histogram looks similar to the standard Gaussian PDF curve. This is more evident if more data are available, as shown in Fig. 3.27 with 20 bins and 400 data points (in `e3_34_2.dat`).

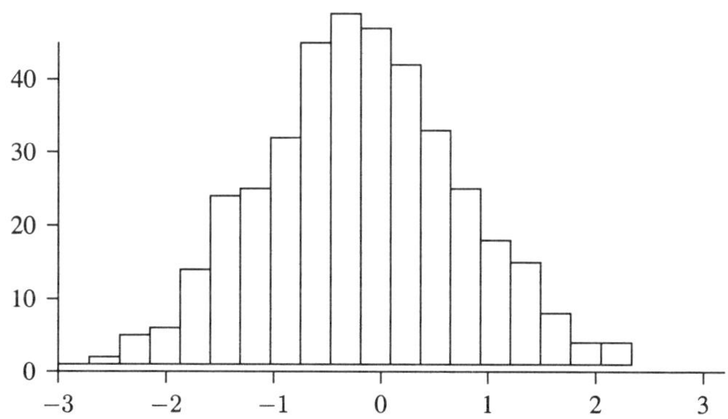

Figure 3.27: A histogram using 20 bins.

It is sometimes not easy to determine the parameters, e.g., the mean and variance, of a distribution from a histogram. This can be compensated by calculating the ***sample mean*** and

sample variance of the data set, given by

$$\text{sample mean:} \quad \hat{X} = \frac{1}{N}\sum_{i=1}^{N} X_i \tag{3.63}$$

$$\text{sample variance:} \quad \hat{V} = \frac{1}{N-1}\sum_{i=1}^{N}(X_i - \hat{X})^2 \tag{3.64}$$

where N is the number of data points in the data set and X_i is the value of the ith data point. Sample mean and sample variance can be used to replace the unknown mean and variance of X approximately.

3.12 Characteristic Functions

The ***characteristic function*** of a RV X is defined by

$$M_x(\omega) = E[e^{j\omega X}] = \int_{-\infty}^{\infty} e^{j\omega x} f(x)dx = \int_{-\infty}^{\infty} e^{-j(-\omega)x} f(x)dx \tag{3.65}$$

This function with $j\omega$ replaced by s is also known as the ***moment generating function*** because its ith derivative at $s = 0$ gives the ith moment:

$$M_x(s) = E[e^{sX}] \Longrightarrow \frac{d^n}{ds^n} M_x(s)\bigg|_{s=0} = E[X^n] \tag{3.66}$$

For example,

$$E[X] = \frac{d}{ds} M_x(s)\bigg|_{s=0}, \qquad E[X^2] = \frac{d^2}{ds^2} M_x(s)\bigg|_{s=0}$$

Clearly, a characteristic (or moment generating) function and a PDF are a Fourier (or two-sided Laplace) transform pair:

$$M_x(-\omega) = \mathcal{F}[p(x)], \qquad M_x(-s) = \mathcal{L}[p(x)]$$

where two-sided Laplace transform is defined exactly the same as the Laplace transform except that the integral is over the entire real line. Since the characteristic (or moment generating) function and the PDF are a Fourier (or two-sided Laplace) transform pair, they carry the same information about the RV: they are both complete descriptions of the RV.

Thus they have the same properties as a Fourier (or Laplace) transform does, e.g.,

$$M_{ax+b}(s) = e^{sb} M_x(as)$$

It can be easily shown that the magnitude of the characteristic (or moment generating) function is bounded by unity:

$$|M_x(s)| \leq M(0) = 1$$

Example 3.35: Moments of Gaussian RV

The moment generating function of a $\mathcal{N}(0, \sigma^2)$ RV X is

$$M_x(s) = \int_{-\infty}^{\infty} e^{sx} \frac{1}{\sqrt{2\pi}\sigma} e^{-\frac{x^2}{2\sigma^2}} dx = e^{\frac{1}{2}s^2\sigma^2} \int_{-\infty}^{\infty} \frac{1}{\sqrt{2\pi}\sigma} e^{-\frac{(x-\sigma^2 s)^2}{2\sigma^2}} dx = e^{\frac{1}{2}s^2\sigma^2}$$

Hence:

$$\begin{aligned}
E[X] &= \frac{d}{ds} M_x(s)|_{s=0} = s\sigma^2 e^{\frac{1}{2}s^2\sigma^2}\Big|_{s=0} = 0 \\
E[X^2] &= \frac{d^2}{ds^2} M_x(s)\Big|_{s=0} = [\sigma^2 + (s\sigma^2)^2]\, e^{\frac{1}{2}s^2\sigma^2}\Big|_{s=0} = \sigma^2 \\
E[X^3] &= \frac{d^3}{ds^3} M_x(s)\Big|_{s=0} = [2s\sigma^4 + (\sigma^2 + s^2\sigma^4)s\sigma^2]\, e^{\frac{1}{2}s^2\sigma^2}\Big|_{s=0} = [3s\sigma^4 + s^3\sigma^6]\, e^{\frac{1}{2}s^2\sigma^2}\Big|_{s=0} = 0 \\
E[X^4] &= \frac{d^4}{ds^4} M_x(s)\Big|_{s=0} = [3\sigma^4 + 3s^2\sigma^6 + (3s\sigma^4 + s^3\sigma^6)s\sigma^2]\, e^{\frac{1}{2}s^2\sigma^2}\Big|_{s=0} = 3\sigma^4
\end{aligned}$$

Since the moment generating function $M_x(s)$ for this RV is even in s, it is clear that all odd moments vanish, which can be seen also from the fact that the PDF of the $\mathcal{N}(0, \sigma^2)$ is even in x.

The moment generating function of a $\mathcal{N}(\bar{y}, \sigma^2)$ RV Y is

$$M_Y(s) = e^{s\bar{y} + \frac{1}{2}s^2\sigma^2}$$

Its moments follow easily from the moments of X in Example 3.35:

$$E[Y^n] = E[(X + \bar{y})^n]$$

For example, $\sigma_y^2 = E[(Y - \bar{y})^2] = E[X^2] = \sigma^2$.

3.13 Summary and Requirements

A *random variable* (RV) is defined as a numerical function of the outcome of a random experiment. It is *discrete* if it can take on only discrete values. It is *continuous* if its possible values form a continuous range. A RV is completely described by its *cumulative distribution function* (CDF) or *probability density function* (PDF), or for a discrete RV by its *probability mass function* (PMF). The CDF of a RV X is the probability $F(x) = P\{X \leq x\}$ as a function of the value x. Its derivative is the PDF $f(x)$, which is the density of probability as a function of x. The PMF of a discrete RV X taking on values $x_1, x_2, \ldots$ is the sequence of the probabilities $p(x_i) = P\{X = x_i\}$, which is defined only over points $x_1, x_2, \ldots$. The probability masses $p_i = P\{X = x_i\}$ located at these points are called the point masses. The PDF of a discrete RV requires the use of a delta function.

The CDF, PDF and PMF possess many important properties. The most important one concerning the direct use of a CDF or PDF is the following:

$$P\{x_1 < X \le x_2\} = F(x_2) - F(x_1) = \int_{x_1}^{x_2} f(x)dx$$

The probability of a continuous RV taking on any given value is always zero.

Although not a complete description of a RV, *expectation* and *moments* are usually easier to obtain in practice. The *expected value* of a RV $g(X)$ is

$$E[g(X)] = \begin{cases} \int_{-\infty}^{\infty} g(x) f_X(x) dx & X \text{ continuous} \\ \sum_i g(x_i) P\{X = x_i\} & X \text{ discrete} \end{cases}$$

The expected value or *mean* of a RV can be interpreted either as the common-sense *average* value or as the center of mass. The *variance* of a RV X is $\text{var}(X) = E[(X - E[X])^2]$. It is a measure of the dispersion of a RV from its mean (or how random the RV is).

Expectation and moments have many important properties. The most commonly used ones are:

$$E[ah_1(X) + bh_2(X)] = aE[h_1(X)] + bE[h_2(X)]$$
$$\text{var}(X) = E[X^2] - (E[X])^2$$
$$\text{var}(aX + b) = a^2\text{var}(X)$$

A *Gaussian* (or *normal*) RV X, denoted by $X \sim \mathcal{N}(\bar{x}, \sigma^2)$, is one having the following PDF:

$$f(x) = \mathcal{N}(x; \bar{x}, \sigma^2) \triangleq \frac{1}{\sqrt{2\pi}\sigma} e^{-(x-\bar{x})^2/2\sigma^2}$$

It is the most important distribution. Its mean is $\bar{x}$ and variance is σ^2. If a random quantity is the cumulative effect of many "uniformly" small sources of uncertainty, then it can be modeled as a Gaussian RV. This is the *central limit theorem.*

A RV X *uniformly distributed* over an interval (a, b), denoted by $X \sim \mathcal{U}(a, b)$, is one having a constant PDF over the interval (a, b). Its mean is the center $(a+b)/2$ and its variance is (length of the interval)$^2/12$.

A deterministic function $Y = g(X)$ of a RV X is a RV. Its PDF can be determined from the PDF of the original RV X by:

$$f_Y(y) = \frac{f_X(x_1)}{|g'(x_1)|} + \cdots + \frac{f_X(x_n)}{|g'(x_n)|} + \cdots = \sum_i \frac{f_X(x_i)}{|g'(x_i)|}$$

where $g'(x) = \frac{d}{dx} g(x)$ and x_i's are the (distinct) real roots of the equation $y = g(x)$ in terms of y.

A random number X with a continuous and strictly increasing CDF $F(x)$ can be generated by $X = F^{-1}(U)$, where $U \sim \mathcal{U}(0, 1)$ and $F^{-1}(\cdot)$ is the inverse function of $F(\cdot)$.

Basic Requirements

- The emphasis of the chapter is on expectation, PDF, CDF, and Gaussian RVs.
- Have a clear understanding of the CDF, PDF, and PMF. Be familiar with their relationships and properties.
- Comprehend the concepts of expected value and variance and be familiar with their properties.
- Be skillful in calculating the expected value and variance of a RV and their functions, making full use of the properties of the expectation and variance.
- Be skillful in calculating various probabilities using the CDF, PDF, or PMF.
- Be familiar with the PDF or PMF, expected values and variances of popular distributions, especially of Gaussian, uniform, exponential, Poisson, and binary distributions.
- Know how to find the CDF, PDF, and/or PMF of simple problems.
- Master the procedure to generate a random number from one that is uniformly distributed over $(0, 1)$.
- Know how to find the PDF of a simple function of a RV.

3.14 Additional Examples

3.36 *Probability calculation.* Consider Example 3.8. Calculate the following probabilities.

(a) $P\{2 < X \leq 7|3 < X < 5\}$
(b) $P\{X = 5|3 < X < 8\}$
(c) $P\{X = 5|3 < X < 5\}$
(d) $P\{X = 0|X \leq 0\}$
(e) $P\{X \leq 0|X = 0\}$
(f) $P\{2 < X \leq 7|X = 1\}$.

Solution:

(a) $P\{2 < X \leq 7|3 < X < 5\} = \frac{P\{(2<X\leq 7)\cap(3<X<5)\}}{P\{3<X<5\}} = \frac{P\{3<X<5\}}{P\{3<X<5\}} = 1$, which makes sense since $\{3 < X < 5\}$ implies $\{2 < X \leq 7\}$.
(b) $P\{X = 5|3 < X < 8\} = \frac{P\{(X=5)\cap(3<X<8)\}}{P\{3<X<8\}} = \frac{P\{X=5\}}{P\{3<X<8\}} = \frac{0}{1-0.5} = 0$.
(c) $P\{X = 5|3 < X < 5\} = \frac{P\{(X=5)\cap(3<X<5)\}}{P\{3<X<8\}} = \frac{P\{\emptyset\}}{P\{3<X<8\}} = 0$.
(d) This problem is not well-defined. It depends on how the "infinitesimal" (if not absolute zero) probability mass is distributed over the interval $(-\infty, 0]$ and thus could have more than one solution. For example, if negative values of x is not considered for the probability or if $\{X < 0\} = \emptyset$ (impossible event), then $P\{X = 0|X \leq 0\} = P\{X = 0|X = 0\} = \frac{P\{X=0\}}{P\{X=0\}} = 1$.
(e) $P\{X \leq 0|X = 0\} = \frac{P\{(X\leq 0)\cap(X=0)\}}{P\{X=0\}} = \frac{P\{X=0\}}{P\{X=0\}} = 1$, which make sense since we have already known that $X = 0$ and therefore it must have $X \leq 0$.
(f) $P\{2 < X \leq 7|X = 1\} = \frac{P\{(2<X\leq 7)\cap(X=1)\}}{P\{X=1\}} = \frac{P\{\emptyset\}}{P\{X=1\}} = 0$ since $X = 1$ is not impossible even though it has zero probability.

3.37 *Common probabilities for uniform distribution.* For a RV $X \sim \mathcal{U}(4,6)$, find its mean $\bar{x} = \int_{-\infty}^{\infty} x f(x)dx$ and standard deviation $\sigma = \sqrt{\int_{-\infty}^{\infty}(x-\bar{x})^2 f(x)dx}$ and the following probabilities: (a) $P\{X-\bar{x} < \sigma\}$; (b) $P\{X-\bar{x} < 2\sigma\}$; (c) $P\{|X-\bar{x}| < \sigma\}$; (d) $P\{|X-\bar{x}| < 2\sigma\}$.

Solution: From Example 3.19,

$$\bar{x} = \frac{1}{2}(x_1 + x_2) = \frac{1}{2}(4+6) = 5 \qquad \text{(the center point)}$$
$$\sigma^2 = \frac{(x_2 - x_1)^2}{12} = \frac{(6-4)^2}{12} = \frac{1}{3} \qquad \Longrightarrow \sigma = \frac{1}{\sqrt{3}}$$

(a)

$$\begin{aligned} P\{X - \bar{x} < \sigma\} &= P\{X - 5 < 1/\sqrt{3}\} = P\{4 \le X < 5 + 1/\sqrt{3}\} = \int_4^{5+1/\sqrt{3}} f(x)dx \\ &= \int_4^{5+1/\sqrt{3}} \frac{1}{6-4}dx \quad (\text{since } (4, 5+1/\sqrt{3}) \subset [4,6]) \\ &= \frac{1+1/\sqrt{3}}{6-4} = 0.789 \end{aligned}$$

(b)

$$\begin{aligned} P\{X - \bar{x} < 2\sigma\} &= P\{X - 5 < 2/\sqrt{3}\} = \int_4^{5+2/\sqrt{3}} f(x)dx \\ &= \int_4^{6} \frac{1}{6-4}dx \quad (\text{since } [4,6] \subset [4, 5+2/\sqrt{3})) \\ &= 1 \end{aligned}$$

(c)

$$\begin{aligned} P\{|X - \bar{x}| < \sigma\} &= P\{|X - 5| < 1/\sqrt{3}\} = P\{5 - 1/\sqrt{3} \le X < 5 + 1/\sqrt{3}\} \\ &= \int_{5-1/\sqrt{3}}^{5+1/\sqrt{3}} \frac{1}{6-4}dx \quad (\text{since } (5-1/\sqrt{3}, 5+1/\sqrt{3}) \subset [4,6]) \\ &= \frac{2/\sqrt{3}}{6-4} = \frac{1}{\sqrt{3}} = 0.577 \end{aligned}$$

(d)

$$\begin{aligned} P\{|X - \bar{x}| < 2\sigma\} &= P\{|X - 5| < 2/\sqrt{3}\} = P\{5 - 2/\sqrt{3} \le X < 5 + 2/\sqrt{3}\} \\ &= \int_{5-2/\sqrt{3}}^{5+2/\sqrt{3}} f(x)dx \\ &= \int_4^{6} \frac{1}{6-4}dx \quad (\text{since } [4,6] \subset (5-2/\sqrt{3}, 5+2/\sqrt{3}) \\ &= 1 \end{aligned}$$

3.38 *Gaussian percentiles.* For a RV $X \sim \mathcal{N}(\bar{x}, \sigma^2)$, find the values α_i's that satisfy the following equations, respectively:

(a) $P\{|X - \bar{x}| < \alpha_1\sigma\} = 90\%$
(b) $P\{|X - \bar{x}| < \alpha_2\sigma\} = 95\%$
(c) $P\{|X - \bar{x}| < \alpha_3\sigma\} = 99\%$
(d) $P\{X - \bar{x} < \alpha_4\sigma\} = 90\%$
(e) $P\{X - \bar{x} < \alpha_5\sigma\} = 95\%$
(f) $P\{X - \bar{x} < \alpha_6\sigma\} = 99\%$.

These probabilities are commonly used and are better learned by heart.

Solution: From the above problem, it is clear

$$\begin{aligned} P\{|X - \bar{x}| < \alpha\sigma\} &= F(\bar{x} + \alpha\sigma) - F(\bar{x} - \alpha\sigma) \\ &= \Phi\left(\frac{(\bar{x} + \alpha\sigma) - \bar{x}}{\sigma}\right) - \Phi\left(\frac{(\bar{x} - \alpha\sigma) - \bar{x}}{\sigma}\right) \\ &= \Phi(\alpha) - \Phi(-\alpha) = 2\Phi(\alpha) - 1 \\ P\{X - \bar{x} < \alpha\sigma\} &= P\{X < \bar{x} + \alpha\sigma\} = F(\bar{x} + \alpha\sigma) = \Phi(\alpha) \end{aligned}$$

Then, in general,

$$\begin{aligned} P\{|X - \bar{x}| < \alpha\sigma\} = \beta \quad &\Longrightarrow \quad \Phi(\alpha) = \frac{1+\beta}{2} \\ P\{X - \bar{x} < \alpha\sigma\} = \beta \quad &\Longrightarrow \quad \Phi(\alpha) = \beta \end{aligned}$$

Thus:

(a) $P\{|X - \bar{x}| < \alpha_1\sigma\} = 90\%$: $\Phi(\alpha_1) = \frac{1+0.9}{2} \Longrightarrow \alpha_1 = 1.645$
(b) $P\{|X - \bar{x}| < \alpha_2\sigma\} = 95\%$: $\Phi(\alpha_2) = \frac{1+0.95}{2} \Longrightarrow \alpha_2 = 1.96$
(c) $P\{|X - \bar{x}| < \alpha_3\sigma\} = 99\%$: $\Phi(\alpha_3) = \frac{1+0.99}{2} \Longrightarrow \alpha_3 = 2.575$
(d) $P\{X - \bar{x} < \alpha_4\sigma\} = 90\%$: $\Phi(\alpha_4) = 0.9 \Longrightarrow \alpha_4 = 1.28$
(e) $P\{X - \bar{x} < \alpha_5\sigma\} = 95\%$: $\Phi(\alpha_5) = 0.95 \Longrightarrow \alpha_5 = 1.645$
(f) $P\{X - \bar{x} < \alpha_6\sigma\} = 99\%$: $\Phi(\alpha_6) = 0.99 \Longrightarrow \alpha_6 = 2.33$

3.39 *Computer maintenance.* A computer will be damaged with probabilities 0.01, 0.001, and 0.05, respectively, if its power supply has a voltage X below 100V, in between 100V and 120V, and above 120V. Suppose that $X \sim \mathcal{N}(110\text{V}, (10\text{V})^2)$. Find

(a) the probability that the computer will be damaged
(b) the probability that the voltage is above 120V if the computer is damaged

Solution: Let $D = \{\text{computer damaged}\}$, $V_1 = \{X < 100\}$, $V_2 = \{100 \le X \le 120\}$, $V_3 = \{X > 120\}$. Clearly,

$$P\{V_1\} = P\{X < 100\} = P\left\{\frac{X - 110}{10} < \frac{100 - 110}{10}\right\} = P\{Y < 1\}|_{Y \sim \mathcal{N}(0,1)} = \Phi(-1)$$

$$
\begin{aligned}
&= 0.159 \\
P\{V_2\} &= P\{100 \le X \le 120\} = P\left\{\frac{100-110}{10} \le \frac{X-110}{10} \le \frac{120-110}{10}\right\} \\
&= \Phi(1) - \Phi(-1) = 0.682 \\
P\{V_3\} &= P\{X > 120\} = P\left\{\frac{X-110}{10} > \frac{120-110}{10}\right\} = 1 - \Phi(1) = 0.159
\end{aligned}
$$

Note that $P\{V_1\} + P\{V_2\} + P\{V_3\} = 1$.

(a) Since V_1, V_2, V_3 form a partition, by total probability theorem,

$$
\begin{aligned}
P\{D\} &= \underbrace{P\{D|V_1\}}_{=0.01} P\{V_1\} + \underbrace{P\{D|V_2\}}_{=0.001} P\{V_2\} + \underbrace{P\{D|V_3\}}_{=0.05} P\{V_3\} \\
&= (0.01)(0.159) + (0.001)(0.682) + (0.05)(0.159) = 0.01022
\end{aligned}
$$

(b) By Bayes' rule

$$
P\{V_3|D\} = \frac{P\{D|V_3\}P\{V_3\}}{P\{D\}} = \frac{(0.05)(0.159)}{0.01022} = 0.7777
$$

3.40 *Mean and variance of a linear function of a Gaussian RV.* For a RV $X \sim \mathcal{N}(\bar{x}, \sigma^2)$, find the mean and variance of $Y = aX + b$ for all $a \neq 0$.

Solution: Since the mean and variance of X are $\bar{x}$ and σ^2, respectively, we have, by the linearity of the expectation (3.50),

$$
\begin{aligned}
\bar{y} &= E[aX + b] = aE[X] + b = a\bar{x} + b \\
\sigma_y^2 &= E[(Y - \bar{y})^2] = E[(aX + b - (a\bar{x} + b))^2] = E[a^2(X - \bar{x})^2] = a^2\sigma^2
\end{aligned}
$$

Note that the above two equations are always valid, not just limited to Gaussian RVs.

3.41 *Inventory analysis.* An inventory has three machines that are operational with probabilities 0.8, 0.9, and 0.95, respectively. Whether a machine is operation is independent of the other machines. Let X be the number of machines that are operational. Find the mean and variance of X.

Solution: Let M_i = {machine i operational}. Note that M_1, M_2, M_3 are independent; X can take only on 4 possible values: 0, 1, 2, 3 and

$$
\begin{aligned}
P\{X = 0\} &= P\{\overline{M_1}\,\overline{M_2}\,\overline{M_3}\} = (1 - 0.8)(1 - 0.9)(1 - 0.95) = 0.001 \\
P\{X = 1\} &= P\{M_1\overline{M_2}\,\overline{M_3}\} + P\{\overline{M_1}M_2\overline{M_3}\} + P\{\overline{M_1}\,\overline{M_2}M_3\} \\
&= (0.8)(0.1)(0.05) + (0.2)(0.9)(0.05) + (0.2)(0.1)(0.95) = 0.032 \\
P\{X = 2\} &= P\{M_1M_2\overline{M_3}\} + P\{\overline{M_1}M_2M_3\} + P\{M_1\overline{M_2}M_3\} \\
&= (0.8)(0.9)(0.05) + (0.2)(0.9)(0.95) + (0.8)(0.1)(0.95) = 0.283 \\
P\{X = 3\} &= P\{M_1M_2M_3\} = (0.8)(0.9)(0.95) = 0.684
\end{aligned}
$$

Thus,

$$\begin{aligned} E[X] &= (1)P\{X=1\} + (2)P\{X=2\} + (3)P\{X=3\} = 2.65 \\ E[X^2] &= (1)^2P\{X=1\} + (2)^2P\{X=2\} + (3)^3P\{X=3\} = 7.32 \\ \text{var}(X) &= 7.32 - (2.65)^2 = 0.2975 \end{aligned}$$

3.42 *Expected GPA*. A senior student has to take *four* more 3-credit courses to complete his undergraduate study. Suppose that he has completed 120 credits with a Grade Point Average (GPA) of 2.7 and will obtain a grade in any of these four courses with the following probabilities:

$$P\{\text{“}A\text{”}\} = 0.15 \quad P\{\text{“}B\text{”}\} = 0.25 \quad P\{\text{“}C\text{”}\} = 0.4 \quad P\{\text{“}D\text{”}\} = 0.15 \quad P\{\text{“}F\text{”}\} = 0.05$$

(a) What is the expected GPA of this student for these four courses?
(b) What is the expected GPA of this student for his entire undergraduate study?

Use the following for GPA calculation: $A = 4.0, B = 3.0, C = 2.0, D = 1.0, F = 0$.

Solution: Denote G the grade point of a course and

$$G_1 = A = 4.0, G_2 = B = 3.0, G_3 = C = 2.0, G_4 = D = 1.0, G_5 = F = 0.0$$

(a) For any of the four courses, the expected grade point of the student is then

$$\begin{aligned} E[G] &= \sum_{i=1}^{5} G_i P\{G_i\} \\ &= (4.0)(0.15) + (3.0)(0.25) + (2.0)(0.4) + (1.0)(0.15) + (0)(0.05) = 2.3 \end{aligned}$$

Since the probability distributions of the grade point for the four courses are identical, the expected GPA for these courses are $E[G] = 2.3$.

(b) Since the student has a GPA of 2.7 over 120 credits and an expected GPA of 2.3 over the last four courses (12 credits), the total expected GPA for the entire undergraduate study is

$$\text{Total expected GPA} = \frac{(120)(2.7) + (12)(2.3)}{120 + 12} = 2.66$$

3.43 *Computer systems analysis*. The number of downs of a mainframe computer over a time period of x days is Poisson distributed with parameter λx. Find

(a) the PDF of the time between two consecutive downs
(b) the probability that it will be up for 10 more days after it has been up for 10 days

Solution: Let X be the up time and let $N(x)$ be the number of downs over x days.

(a) Since X is nonnegative, $F_X(x) = 0, \forall x < 0$. For $x > 0$, two events $\{X > x\}$ and $\{N(x) = 0\}$ are equivalent and

$$F_X(x) = P\{X \leq x\} = 1 - P\{X > x\} = 1 - P\{N(x) = 0\} = 1 - e^{-\lambda x}$$

Hence,

$$f(x) = \begin{cases} \lambda e^{-\lambda x} & x > 0 \\ 0 & x \leq 0 \end{cases}$$

which indicates that the up time X is exponentially distributed with parameter λ.

(b) The sought-after probability is

$$P\{X > 20|X > 10\} = \frac{P\{X > 20, X > 10\}}{P\{X > 10\}} = \frac{P\{X > 20\}}{P\{X > 10\}} = \frac{e^{-20\lambda}}{e^{-10\lambda}} = e^{-10\lambda}$$

Note that

$$P\{X > 20|X > 10\} = e^{-10\lambda} = P\{X > 10\}$$

which indicates that the computer is always "like new" after knowing it has been operational for some time. Of all continuous distributions, only the exponential distribution has this "memoryless" property.

3.15 Problems

3.1 *From CDF to probability.* Consider Example 3.7. Find the probability that the total power consumption is (a) not greater than 3 kW; (b) 1 kW, 2 kW or 3 kW; (c) greater than 2 kW.

3.2 *Sum of CDFs.* Two RVs X and Y have CDFs $F_X(x)$ and $F_Y(x)$. Can $F(x) = F_X(x) + F_Y(x)$ be the CDF of some RV?

3.3 *Point mass determination.* Find the point masses for RVs X, Y, and Z of Example 3.2, respectively.

3.4 *From CDF to PDF.* Find the PDF of the total power consumption for Example 3.7.

3.5 *Determination of CDF and PDF.* Find the CDF and PDF of the RV X of Example 3.1.

3.6 *Calculation based on CDF.* The CDF of a random variable X is

$$F(x) = \begin{cases} 1 - e^{-3x} & x \geq 0 \\ 0 & x < 0 \end{cases}$$

Find

(a) the PDF of X

(b) the probability that $X > 3$
(c) the probability $P\{X > 3| - 3 < X \leq 4\}$

3.7 *Determination of PDF.* Find the PDF of the RV X for Example 3.8.

3.8 *Conditions to be a CDF.* Can any of the following functions be a CDF? Justify your answer.

$$F_1(x) = \begin{cases} 2/3 + (1 - e^{-x})/3 & x > 0 \\ 1/3 & x \leq 0 \end{cases}$$
$$F_2(x) = \begin{cases} 2/3 + (1 - e^{-x})/3 & x > 0 \\ 0 & x \leq 0 \end{cases}$$
$$F_3(x) = \begin{cases} 1/3 + (1 - e^{-x})/3 & x > 0 \\ 0 & x \leq 0 \end{cases}$$
$$F_4(x) = \begin{cases} 1 - e^{-x} & x > 0 \\ 0 & x \leq 0 \end{cases}$$

3.9 *Conditions to be a PDF.* Can any of the following functions be a PDF? Justify your answer.

$$f_1(x) = \begin{cases} \sin x & 0 < x < \pi \\ 0 & \text{elsewhere} \end{cases}$$
$$f_2(x) = \begin{cases} \sin x & 0 < x < \pi/2 \\ 0 & \text{elsewhere} \end{cases}$$
$$f_3(x) = \begin{cases} \sin x & 0 < x < 2\pi \\ 0 & \text{elsewhere} \end{cases}$$
$$f_4(x) = \begin{cases} \cos x & 0 < x < \pi/2 \\ 0 & \text{elsewhere} \end{cases}$$
$$f_5(x) = \begin{cases} \cos x & -\pi/2 < x < \pi/2 \\ 0 & \text{elsewhere} \end{cases}$$

3.10 *Conditions to be a PDF.* Can the following function be a PDF? Justify your answer.

$$f(x) = \begin{cases} x & 0 < x \leq 1 \\ 9/4 - x & 1 < x \leq 3 \\ 0 & \text{elsewhere} \end{cases}$$

3.11 *Conditions to be a PDF.* Given a continuous RV X, can the following two functions be its PDF? Justify your answer.

$$h_1(x) = \begin{cases} 0.4 \sin x & 0 < x < \pi \\ 0 & \text{elsewhere} \end{cases}$$
$$h_2(x) = \begin{cases} \sin x & -\pi/2 < x < \pi/2 \\ 0 & \text{elsewhere} \end{cases}$$

3.12 *From PDF to CDF.* A RV has the following PDF

$$f(x) = \begin{cases} x/4 & 0 < x \leq 2 \\ 3 - x & 2 < x \leq 3 \\ 0 & \text{elsewhere} \end{cases}$$

(a) Sketch $f(x)$. Is X continuous?
(b) Is the following calculation of the CDF $F(x)$ correct?

$$\begin{aligned} x \leq 0: \; F(x) &= \int_{-\infty}^{x} 0 dx = 0 \\ 0 < x \leq 2: \; F(x) &= \int_{0}^{x} x/4 dx = x^2/2 \\ 2 < x \leq 3: \; F(x) &= \int_{2}^{x} (3 - x) dx = 3(x-2) - x^2/2 + 2 = 3x - 4 + x^2/2 \\ x > 3: \; F(x) &= \int_{3}^{x} 0 dx = 0 \end{aligned}$$

If not, give the correct calculation of $F(x)$.

3.13 *Determination of PDF and CDF.* The following waveform is sampled at a random time τ over a period $T = 1.5$. Let $X = \{\text{sampled value of } v(\tau)\}$. Find (a) the PDF $f_X(x)$ of X; (b) the CDF $F_X(x)$ of X; (c) $P\{0 \leq X < 3\}$; and (d) $P\{0.1 \leq \tau < 0.3 \mid X \geq 2\}$.

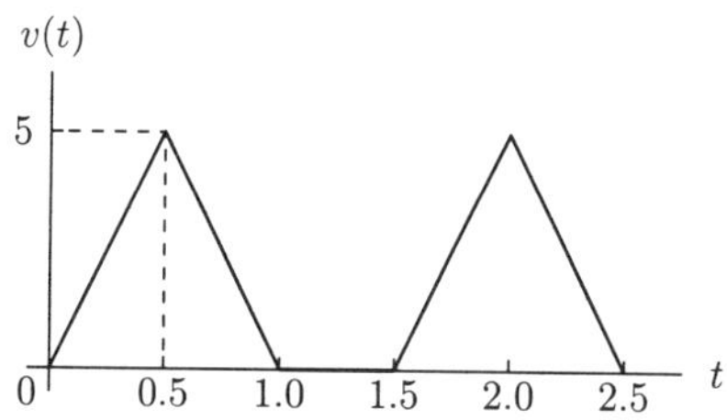

3.14 *Determination of parameters of a PDF.* Determine the value of constants c and d such that the following function is a PDF of some RV

$$h(x) = (c + de^{-x^2/2})u(x)$$

where u is the unit step function.

3.15 *Determination of a parameter of a PDF.* Determine the value of constant c such that $h(x) = \frac{c}{e^x + e^{-x}}$ is a PDF of some continuous RV.

3.16 *Laplace distribution.* A ***Laplace distributed*** RV X has the following PDF

$$f(x) = ce^{-|x|}$$

Find (a) constant c; (b) the CDF of X; and (c) $P\{0 < X \leq 1\}$.

3.17 *Random capacitance.* Consider Example 3.4. Assume the capacitance C is a RV uniformly distributed over $(9, 11]$. Find

(a) thc CDF and PDF of X
(b) the probability that the discrete RV X is in level 1
(c) the probability that X is in the upper two levels (3 or 4)
(d) the probability that the capacitance is in between 9.3 and 9.5 given that X is in level 1
(e) the average capacitance
(f) the average level for the discrete RV X

3.18* *Percentiles of uniform distribution.* Consider a RV $X \sim \mathcal{U}(a, b)$, where $a < b$.

(a) Find $P\{X - \bar{x} < \alpha\sigma\}$ in terms of α, where σ is the standard deviation.
(b) Find the minimum value of α such that $P\{X - \bar{x} < \alpha\sigma\} = 1$.

3.19* *Percentiles of uniform distribution.* Given a RV $X \sim \mathcal{U}(a, b)$, where $a < b$, show that

$$P\{|x - \bar{x}| < \alpha\sigma\} = \begin{cases} 1 & \alpha \geq \sqrt{3} \\ \frac{\alpha}{\sqrt{3}} & 0 \leq \alpha < \sqrt{3} \\ 0 & \alpha < 0 \end{cases}$$

3.20 *Probability of a solution.* Find the probability that the following equation has a real-valued solution

$$x^2 + 3xY + 1 = 0$$

where Y is a $\mathcal{U}(-4, 4)$ RV.

3.21 *Gaussian percentiles.* For a RV $X \sim \mathcal{N}(\bar{x}, \sigma^2)$, find the following probabilities using only the results of Example 3.13 (i.e., you are not allowed to look up any probability table because you may not have it at hand): (a) $P\{|X - \bar{x} - \sigma| < \sigma\}$; (b) $P\{|X - \bar{x} - \sigma| < 2\sigma\}$.

3.22 *Probability of a Gaussian RV.* Given a RV $X \sim \mathcal{N}(3, \sigma^2)$ and $P\{0 < X < 3\} = 0.4$ but σ is unknown, find $P\{X < 0\}$.

3.23 *Probability of a Gaussian RV.* Find $P\{2 < X \leq 10\}$ for the random variable $X \sim \mathcal{N}(4, 2^2)$.

3.24 *Price for quality.* A production line manufactures 1 $k\Omega$ resistors that must satisfy 10% tolerance. Suppose that the resistance X is adequately described by $\mathcal{N}[1k\Omega, (40\Omega)^2]$. What percent of resistors is expected to be rejected?

3.25 *Confidence intervals.* In radar detection, assume that the true measurement is a RV $X \sim \mathcal{N}(105, 3^2)$.

(a) Find the one-sided threshold c such that $P\{X > c\} = 0.01$.

(b) Find the two-sided threshold d such that $P\{|X - 105| > d\} = 0.01$.

3.26 *Life time probability.* The life time of a light bulb is a RV $X \sim \mathcal{N}(1000, \sigma^2)$. Find σ such that $P\{800 < X < 1200\} = 0.9$.

3.27 *Mean and variance of standardized RV.* A standardized RV is defined by $\tilde{X} \triangleq \frac{X-\bar{x}}{\sigma}$, where $\bar{x}$ and σ^2 are the mean and variance of X. Show that a standardized RV has zero mean and unity variance.

3.28 *Mean and variance.* A RV X has the following PDF

$$f(x) = \begin{cases} x & 0 < x \leq 1 \\ 3/4 - x/4 & 1 < x \leq 3 \\ 0 & \text{elsewhere} \end{cases}$$

Find the mean and variance of X.

3.29 *Mean and variance of exponential distribution.* An exponentially distributed RV X has the PDF given by (3.34). Find the mean and variance of X.

3.30 *Mean, variance, and probability of Poisson distribution.* Given a Poisson RV X with parameter λ, as defined by (3.38), and $P\{X = 0\} = 1/3$, find (a) the mean and variance of X and (b) the probability $P\{X > 2\}$.

3.31 *Mean, variance, and probability of Poisson distribution.* Given a Poisson RV X with parameter λ, as defined by (3.38), and $P\{X = 1\} = P\{X = 2\}$, find (a) the mean and variance of X and (b) the probability $P\{X = 3\}$.

3.32 *Mean and variance of discrete uniform distribution.* Find the mean and variance of the discrete uniform distribution over possible values $x = n, n+1, \ldots, m$.

3.33 *Median of exponential distribution.* Find the median of the exponential distribution, as defined by (3.55).

3.34 *Mean of a function of a RV.* Given an exponential distributed RV X with parameter $\lambda = 1$, as defined by (3.55), find the mean of $Y = 2X + e^{-3X}$.

3.35* *Mean and variance of a mixture.* A RV X has the mixture PDF

$$f(x) = \begin{cases} 0.1 + 0.5\mathcal{N}(x; 0, 1) & 0 \leq x < 5 \\ 0.5\mathcal{N}(x; 0, 1) & \text{elsewhere} \end{cases}$$

where $\mathcal{N}(x; 0, 1)$ was defined by (3.23). Find the mean and variance of X.

3.36 *Lot size estimation.* The lot size of a real estate property is measured with a measurement error that is Gaussian distributed with mean 1 and variance 4. Suppose that the true lot size is equal to 7000.

(a) What is the average measured lot size?
(b) Find the probability that the measured lot size is in between 6997 and 7005?

3.37 *Linear function of a uniform RV.* Given a RV $X \sim \mathcal{U}(0,1)$, determine the constants c and d such that $Y = cX + d$ is uniformly distributed over (a, b).

3.38* *PDF of functions of a uniform RV.* Consider a RV $X \sim \mathcal{U}(1,6)$.

(a) Find the PDF of $Y = X^2$? Is it uniform?
(b) Find the PDF of $Y = X^3$? Is it uniform?
(c) What are the PDFs of $Y = X^2$ and $Z = X^3$ if $X \sim \mathcal{U}(-1,6)$? Are they uniform?
(d) What are the PDFs of $Y = X^{2n}$ and $Z = X^{2n+1}$ if $X \sim \mathcal{U}(-1,6)$ (n is a positive integer)? Are they uniform?

3.39 *PDF of the magnitude of a RV.* Find the PDF $f_Y(y)$ of the RV $Y = |X|$ in terms of the PDF of the RV X. Write down the explicit expression for $f_Y(y)$ if $X \sim \mathcal{N}(0, \sigma^2)$. In this case, Y is a *Rayleigh* distributed RV.

3.40 *PDF of the absolute value function of a RV.* Find the PDF $f_Y(y)$ of the RV $Y = |X - 3|$ in terms of the PDF of a continuous RV X. Write down the explicit expression for $f_Y(y)$ if (a) $X \sim \mathcal{N}(3, \sigma^2)$; or (b) $X \sim \mathcal{U}(1,9)$. Find the corresponding variances of Y for the two cases.

3.41 *Expected values of a circle.* The radius of a circle is measured four times to be $r = 99, 100, 101, 102$, with the probabilities of correct measurement 0.2, 0.3, 0.4, and 0.1, respectively. Find the expected value of (a) the radius; (b) the circumference of the circle; and (c) the area of the circle.

3.42 *PDF of a linear function.* Use the theorem in Section 3.8 to find the PDF $f_Y(y)$ of $Y = aX + b$ in terms of the PDF $f_X(x)$ of X.

3.43 *PDF of a linear function of a Gaussian RV.* Find the PDF $f_Y(y)$ of $Y = aX + b$, where $X \sim \mathcal{N}(\bar{x}, \sigma^2)$.

3.44* *Exponential function of an exponential RV.* For $Z = e^{aX}$, where $a > 0$ and X is an exponentially distributed RV with the PDF given by (3.55). Find (a) the PDF of Z and (b) the mean and variance of Z.

3.45* *Unique position of the uniform distribution.* Consider a continuous RV X with a strictly increasing CDF $F(x)$. Let $Y = F(X)$. Find the PDF of Y.

3.46 *Continuous function of a RV.* Given a continuous function $h(x)$, is $Y = h(X)$ always discrete if X is discrete? Can $Y = h(X)$ be discrete or mixed if X is continuous?

3.47 *Like-new property of exponential distribution.* It is known that the life X of a TV set is exponentially distributed with an expected life of 7000 hours.

(a) Write down the PDF and CDF of X.
(b) What is the variance of X?
(c) Find the probability that a TV set has a life longer than 5000 hours.
(d) Find the probability that a TV set is operational for at least another 5000 hours knowing that it has been operational for 5000 hours.
(e) Do you get the same probabilities for part (c) and (d)? Is it true that $P\{X > x_2\} = P\{X > x_2 + x_1 | X > x_1\}$ for every possible $x_1, x_2 > 0$ for this RV X? Justify your answer.

3.48 *Conditional exponential PDF.* The failure time X of a solid-state device is an exponential RV with PDF given by (3.55). It is observed that it is operational at time t_0. Find the conditional PDF of its failure time $f(x|X \geq t_0)$.

3.49 *Machine operation.* There are five identical machines in a laboratory. Each of them has a probability of 0.8 in operation at any time and consumes 1000 watts if in operation. The operation of each machine is independent of other machines. Find

(a) the probability that at least four machines are in operation at a particular time
(b) the probability that these five machines will consume at least 4500 watts at a particular time
(c) the average power consumption of these five machines

3.50 *Generation of arbitrary uniform random numbers.* Given a random number generator for $U \sim \mathcal{U}(0,1)$, how to generate $X \sim \mathcal{U}(a,b)$ for $b > a$? Give a specific formula. Use your formula to generate and list 100 random numbers that are uniformly distributed over $(-1, 6)$.

3.51 *Generation of chi-square random numbers.* Given a random number generator for independent $U \sim \mathcal{N}(0,1)$, how to generate χ_n^2 random numbers? Give a specific formula.

3.52 *Generation of Rayleigh random numbers.* Given a uniform random number generator for $U \sim \mathcal{U}(0,1)$, how to generate Rayleigh random numbers with PDF given by (3.33)? Give a specific formula.

3.53 *Generation of Rayleigh random numbers.* Given a Gaussian random number generator for $X \sim \mathcal{N}(0,1)$, how to generate Rayleigh random numbers with PDF given by (3.33)? Give a specific formula.

3.54 *Parameter determination for uniform distribution.* Given mean $\bar{x}$ and variance σ^2 of a uniformly distributed RV X, determine the interval over which X is uniformly distributed; that is, determine a and b in terms of $\bar{x}$ and σ such that $X \sim \mathcal{U}(a,b)$.

3.55 *Puzzling questions.* Raise two technical questions of your own that you are puzzled by and that are closely related to the material covered in this chapter. These questions should be important, not too simple, and not too involved.

3.16 Computer Exercises

3.1 *PDF and PMF plots.* Use the companion software P&R to plot (a) the PDF curve of the chi-square distribution with $\sigma = 2.7$ and 5 degrees of freedom; and (b) the PMF curve of the binomial distribution with $N = 30$ and $p = 0.32$.

3.2 *Log-normal probability.* Use the companion software P&R to calculate the probabilities $P\{X \leq 4.4\}$ and $P\{2.5 < X \leq 5.7\}$, where X is a RV having the log-normal distribution with $\sigma = 1.2$ and $a = 2.9$.

3.3 *Calculation of Poisson probability.*

(a) Write a computer subroutine `poisson_pmf.m` to calculate the Poisson probability

$$P\{X = m\} = e^{-\lambda}\frac{\lambda^m}{m!}$$

for an arbitrary m and λ.

(b) Write a computer subroutine `poisson_cdf.m` to calculate the probability $P\{X \leq m\}$ of a Poisson RV X given above for an arbitrary m and λ. Utilize the subroutine in (a).

(c) Run both subroutines in (a) and (b) for $\lambda = 2.4$ and $m = 0, 2, 50$, respectively. Compare the results with your theoretical calculation for $m = 0, 2$.

(d) Repeat (c) by using the companion software P&R and compare the results with those in (c).

3.4 *Chi-square percentile.* Use the companion software P&R to calculate the 37% and 89% percentile points of the chi-square distribution with $\sigma = 3.4$ and 6 degrees of freedom.

3.5 *Identify distribution.* The data files `m3_5x.dat`, `m3_5y.dat` and `m3_5z.dat` in the companion software P&R contain three records of data, which are the realizations of three RVs X, Y, and Z, respectively.

(a) Write a computer program to plot three histograms of the three data sets, respectively, and compute the sample means and sample variances of the three data sets, respectively.

(b) Give your guess what the distributions of X, Y and Z are, respectively. Justify.

(c) Repeat (a) and (b) using P&R.

3.6 *Generation of a random phase.*

(a) Show that if $X \sim \mathcal{U}(0,1)$ then $Y = aX + b$ is uniformly distributed over $(b, b + a)$.

(b) What should a and b be such that $Y \sim \mathcal{U}(\alpha, \beta)$?

(c) Give a formula and procedure to generate a random phase $\phi \sim \mathcal{U}(-\pi, \pi)$.

(d) Generate 1000 random numbers that are the realizations of the random phase $\phi \sim \mathcal{U}(-\pi, \pi)$. Plot the values of these random numbers. Compute the sample mean and sample variance of these random numbers generated.

3.7 *Generation and testing of exponential random numbers.*

(a) Use MATLAB $\mathcal{U}(0, 1)$ random number generator "`rand`" and the results of Example 3.30 to generate 10,000 random numbers that are exponentially distributed with parameter $\lambda = 10$.
(b) Compute the average value of the 10,000 random numbers generated.
(c) Compare the average value obtained in (b) with the true mean of an exponential RV with parameter $\lambda = 10$.
(d) Compute the average value of the first 1,000 of the 10,100 random numbers generated and compare them with the true mean obtained in (c).
(e) Use MATLAB function "`hist`" (with 50 bins) to plot the histogram based on the 10,000 random numbers generated. The vertical value of the histogram should be frequency ratio.
(f) Generate 10,000 exponentially distributed random numbers with $\lambda = 10$ using the companion software P&R and repeat parts (b) and (e) for these numbers.
(g) Plot the PDF of an exponentially distributed RV with parameter $\lambda = 10$ using the same x-y scales as the histogram. Overlay the two histograms and the PDF plot in a single figure. The height of the histogram should be divided by the total number of data points times the length of each bin. Comment on the similarity and difference.

3.8 *Binary random number generation.* Given a sequence of $\mathcal{U}(0, 1)$ random numbers, obtain and describe an algorithm that generate binary random numbers with the following point masses:

$$P\{X = 0\} = 0.3, \quad P\{X = 1\} = 0.7$$

(a) Find the (theoretical) mean and variance of X.
(b) Use MATLAB $\mathcal{U}(0, 1)$ random number generator "`rand`" to generate 10,000 random numbers with the above distribution.
(c) Compute the average value of the 10,000 random numbers generated. Compare it with the theoretical mean.
(d) Give the percentages of the numbers generated that are equal to -1, 0, and 1, respectively. Compare them with the point masses and make a comment.
(e) Generate 10,000 random numbers with the above distribution using the companion software P&R by taking advantage of the relationship between the binomial and binary distributions and repeat (c)–(d) for these numbers.

3.9 *Ternary random number generation.* Given a sequence of $\mathcal{U}(0, 1)$ random numbers, obtain and describe an algorithm that generate random numbers with the following point masses:

$$P\{X = -1\} = 0.3, \quad P\{X = 0\} = 0.2, \quad P\{X = 1\} = 0.5$$

(a) Find the (theoretical) mean and variance of X.
(b) Use MATLAB $\mathcal{U}(0,1)$ random number generator "`rand`" to generate 10,000 random numbers with the above distribution.
(c) Compute the average value of the 10,000 random numbers generated. Compare it with the theoretical mean.
(d) Give the percentages of the numbers generated that are equal to -1, 0, and 1, respectively. Compare them with the point masses and make a comment.

3.10 *Approximations to binomial distributions.* Use the companion software P&R to do the following.

(a) Generate 200 random numbers of binomial distribution with $N = 100$ and $p = 0.005$ and save to data file `m3_10a.dat`. Overlay the histogram of the random numbers generated and the PMF of Poisson distribution with the sample mean and sample variance of the random numbers as its mean and variance, respectively.
(b) Generate 400 random numbers of binomial distribution with $N = 100$ and $p = 0.55$ and save to data file `m3_10b.dat`. Overlay the histogram of the random numbers generated and the PDF of Gaussian distribution with the sample mean and sample variance of the random numbers as its mean and variance, respectively.
(c) Comment on the conditions under which the binomial distribution can be approximated by either Poisson or Gaussian distributions.

3.17 Self-Test Problems

3.1 Answer the following questions briefly.

(a) What are the domain and range of a RV?
(b) Is it possible to define a continuous RV over the outcomes of a die-rolling experiment?
(c) Can the CDF of a continuous RV have a discontinuous point?
(d) Is it true that the probability of a RV taking on any given value is zero?
(e) Given the mean and variance of a Gaussian RV, is the PDF uniquely determined?
(f) Given the peak value of the PDF of a Gaussian RV and the location of the peak, is the PDF uniquely determined?
(g) Does a larger variance imply a larger or smaller mean?
(h) If a RV X has mean $\bar{x}$ and zero variance, what can you say about this RV?
(i) If $Y = aX + b$, where $a \neq 0$ and b are constants and X is a discrete RV, can Y be a continuous RV?
(j) What is the relationship between the chi-square and Gaussian RVs?

3.2 *Understanding variance.* For two RVs $X \sim \mathcal{N}(a, \sigma_x^2)$ and $Y \sim \mathcal{N}(b, \sigma_y^2)$, where $\sigma_x < \sigma_y$, which of the following is true? Justify your answer.

(a) $P\{|X - a| < 1\} < P\{|Y - b| < 1\}$ for any a and b.

(b) $P\{|X-a|<1\}>P\{|Y-b|<1\}$ for any a and b.
(c) $P\{|X-a|<1\}>P\{|Y-b|<1\}$ for some a and b values; for other values, the inequality should be reversed.

3.3 *Mean and variance.* Given a RV X with PDF

$$f(x)=\begin{cases}1-|x| & -1\leq x\leq 1\\ 0 & \text{elsewhere}\end{cases}$$

find the mean and variance of X.

3.4 *Roundoff errors.* Roundoff errors X are uniformly distributed. suppose that the sixth decimal place of a calculator will be rounded (e.g., 0.00004573 rounded to 0.00005 and 0.50004486 rounded to 0.50004). Find

(a) the PDF and CDF of X
(b) the probability that the numerical error is in between 0.000001 and 0.000003
(c) the mean and variance of X

3.5 *GRE test model.* Assume the score of a student in a GRE test is a Gaussian RV $X\sim\mathcal{N}(\mu,\sigma^2)$. Suppose it is known that the average score is 1440 and 15.9% students' scores are above 1650. Find μ, σ, and $P\{1600<X<2000\}$.

3.6 *Test score.* A student's test score of an exam is assumed to be a random variable $X\sim\mathcal{N}(18,2^2)$ (assume that a negative score is also possible — tough?!).

(a) What are the mean and variance of X? Find the probability that the student's score is higher than 18 points.
(b) If the score is to be converted to 100-point scale by $Y=4X+c$, where c is a constant, find the mean and variance of Y in terms of c. Can you express σ_y^2 in terms of σ_x^2? If you can, what is the expression? How much should c be if the mean of Y is required to be 75?
(c) Find the PDF of Y. Is Y a Gaussian random variable?
(d) Suppose that you will answer parts (a), (b), (c), and (d) correctly with probabilities 0.9, 0.5, 0.7, and 0.7, respectively, what is your expected score on this problem if a total of 10 points are distributed, respectively, over the parts with 20%, 40%, 20%, and 20%?

3.7 *Weibull random number generation.* For $\mathcal{U}(0,1)$ random numbers $U_1,\ldots,U_n$, derive a formula $X_i=g(U_i)$ such that X_i's are random numbers with ***Weibull distribution*** whose PDF is given by (3.35) and thus the CDF is given by

$$F(x)=\begin{cases}1-e^{-ax^b} & x>0\\ 0 & x<0\end{cases}\tag{3.67}$$

where $a>0, b>0$.

3.18 Solutions to Self-Test Problems

3.1 (a) The domain is the sample space (i.e., the set of all outcomes) and range is the set of all real numbers (real line).
(b) No, because a die-rolling experiment has only 6 possible outcomes and thus a RV defined over these outcomes can have at most 6 different values and therefore cannot be continuous.
(c) No, a continuous RV must have a continuous CDF.
(d) It is true for a continuous RV but not true in general for a discrete or mixed RV, which must have nonzero probability of taking on some specific values.
(e) Yes, it is uniquely determined as $\mathcal{N}(x;\bar{x},\sigma^2)$, where $\bar{x}$ and σ^2 are the mean and variance, respectively.
(f) Yes, it is uniquely determined because the peak locates at $x=\bar{x}$ and the peak value is equal to $\frac{1}{\sqrt{2\pi}\sigma}$ and thus the mean and variance can be obtained, which then determine the PDF uniquely.
(g) No. A larger variance implies that the RV is more uncertain in the sense that it is not so concentrated around its mean but the mean could be any value.
(h) Practically speaking, we have $X=\bar{x}$; that is, the RV is actually not random but equal to a constant that is its mean.
(i) No. Since $Y=aX+b$ is simply the scaling and translation of X, Y will have the same number of possible values to take on as X does. Thus, if X is discrete, so is Y; if X is continuous, so is Y.
(j) A chi-square RV of n degrees of freedom is the sum of n independent zero-mean Gaussian RVs of a common variance.

3.2 Since X has a smaller variance than Y, X will be more concentrated around its mean, which is a, than Y around b. This means that X will have a higher probability of taking on a value close to a than Y taking on a value close to b. As a result, $P\{|X-a|<1\}>P\{|Y-b|<1\}$ for any a and b (since the above argument is true for any a and b).

3.3 Clearly, $f(x)$ is symmetrical about $x=0$. Thus, the mean of X is zero. Note that

$$1-|x|=\begin{cases}1+x & -1\le x\le 0\\ 1-x & 0<x\le 1\end{cases}$$

Hence,

$$E[X^2]=\int_{-1}^{0}x^2(1+x)dx+\int_{0}^{1}x^2(1-x)dx=x^3/3+x^4/4\Big|_{-1}^{0}+x^3/3-x^4/4\Big|_{0}^{1}=1/6$$
$$\text{var}(X)=E[X^2]-(\bar{x})^2=1/6-0=1/6$$

3.4 Clearly X is uniformly distributed over $[-0.000005, 0.000005]$. Then

(a)

$$f(x) = \begin{cases} \frac{1}{0.00001} & -0.000005 < x < 0.000005 \\ 0 & \text{elsewhere} \end{cases}$$

$$F(x) = \int_{-\infty}^{x} f(x)dx = \begin{cases} 0 & x \leq -0.000005 \\ \frac{x+0.000005}{0.00001} & -0.000005 < x < 0.000005 \\ 1 & x \geq 0.000005 \end{cases}$$

(b)

$$P\{0.000001 < X < 0.000003\} = \frac{0.000003 - 0.000001}{0.00001} = 0.2$$

(c) $\bar{x}$ = center point $= (-0.000005 + 0.000005)/2 = 0.$
$\sigma_x^2 = (\text{length})^2/12 = (0.00001)^2/12 = 8.33 \times 10^{-12}.$

3.5 Clearly, $\mu = 1440$, and $P\{X > 1650\} = 0.159$. Hence

$$P\left\{\frac{X - 1440}{\sigma} > \frac{1650 - 1440}{\sigma}\right\} = 0.159$$

From $\mathcal{N}(0, 1)$ table, $\Phi(1) = 0.159$. This means $\frac{1650-1440}{\sigma} = 1$ and thus $\sigma = 210$.

$$\begin{aligned} P\{1600 < X < 2000\} &= P\left\{\frac{1600 - 1440}{210} < \frac{X - 1440}{210} < \frac{2000 - 1440}{210}\right\} \\ &= \Phi(2.667) - \Phi(0.762) = 0.230 \end{aligned}$$

3.6 (a) $\bar{x} = 18$ and $\sigma^2 = 2^2 = 4$.

$$P\{X > 18\} \overset{\text{by symmetry}}{=} P\{X < 18\} = \frac{1}{2}P\{-\infty < X < \infty\} = \frac{1}{2}$$

(b)

$$\begin{aligned} \bar{y} &= E[Y] = E[4X + c] = 4\bar{x} + c \\ \sigma_y^2 &= E[((4X + c) - (4\bar{x} + c))^2] = E[(4(X - \bar{x}))^2] = 16\sigma_x^2 \\ 75 &= \bar{y} = 4\bar{x} + c = 4 \times 18 + c \quad \Longrightarrow \quad c = 3 \end{aligned}$$

(c)

$$g(x) = 4x + c \quad \Longrightarrow \quad g'(x) = 4$$

$y = 4x + c$ has one solution only: $x_1 = \frac{y-c}{4}$.
Thus

$$\begin{aligned} f_Y(y) &= \frac{f_X(x_1)}{|g'(x_1)|} = \frac{1}{4}\mathcal{N}(x_1; 18, 2^2) = \frac{1}{4}\frac{1}{\sqrt{2\pi}\cdot 2}e^{-(\frac{y-c}{4}-18)^2/(2\times 2^2)} \\ &= \frac{1}{\sqrt{2\pi}\cdot 8}e^{-(y-72-c)^2/(2\times 8^2)} = \mathcal{N}(y; 72 + c, 8^2) \end{aligned}$$

Thus it is a Gaussian RV.

(d) Note that the total point for this problem is 10. The expected score is

$$E(S) = 10[(0.9)(0.2) + (0.5)(0.4) + (0.7)(0.2) + (0.7)(0.2)] = 6.6$$

3.7 Solving the equation $u = 1 - e^{-ax^b}$ for x yields

$$x = \left[-\frac{1}{a} \ln(1-u) \right]^{1/b}$$

Thus, given $\mathcal{U}(0,1)$ random numbers $U_1, \ldots, U_n$, $(X_1, \ldots, X_n)$ are Weibull distributed random numbers, where $X_i = [-\frac{1}{a} \ln(1-U_i)]^{1/b}$. Since $(1-U_i)$ and U_i have the same distribution, $(1-U_i)$ can be replaced by U_i in random number generation and finally, $X_i = [-\frac{1}{a} \ln U_i]^{1/b}$.

4

MULTIPLE RANDOM VARIABLES

This branch of mathematics [probability] is the only one, I believe, in which good writers frequently get results entirely erroneous.

Charles Sanders Peirce

In many practical situations, the outcome of a random experiment is better labeled by two or more RVs, rather than just one RV.

This chapter studies the probabilistic tools for problems that involve more than one RV.

Main Topics

- Joint Distribution Functions
- Joint Probability Density Functions
- Independence of Random Variables
- Expectation and Moments
- Relation between Two Random Variables
- Jointly Gaussian Random Variables
- Functions of Random Variables

These topics are all extensions of the single RV case.

4.1 The Joint Distribution Function

The ***joint cumulative distribution function*** (***joint CDF***) of two RVs X and Y is defined by

$$\boxed{F(x,y) = F_{X,Y}(x,y) = P\{(X \le x) \cap (Y \le y)\} = P\{X \le x, Y \le y\}} \quad (4.1)$$

That is, $F_{X,Y}(x,y)$ is the joint probability of $(X \le x)$ and $(Y \le y)$, which is often of major interest to us. As shown later, probabilities in the form of $P\{a < X \le b, c < Y \le d\}$ can also be obtained from the above probabilities.

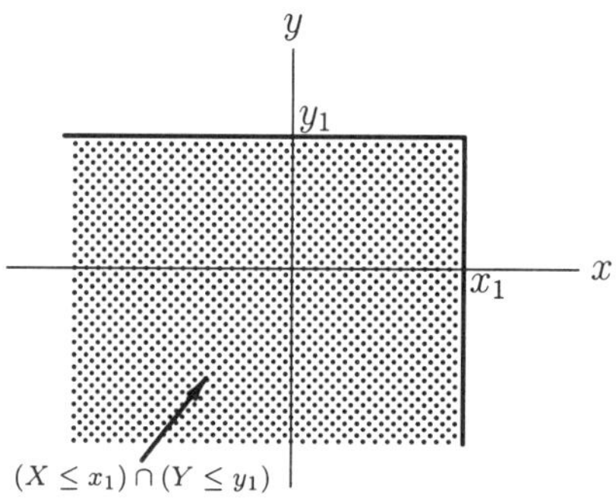

Figure 4.1: Area of $\{X \le x, Y \le y\}$.

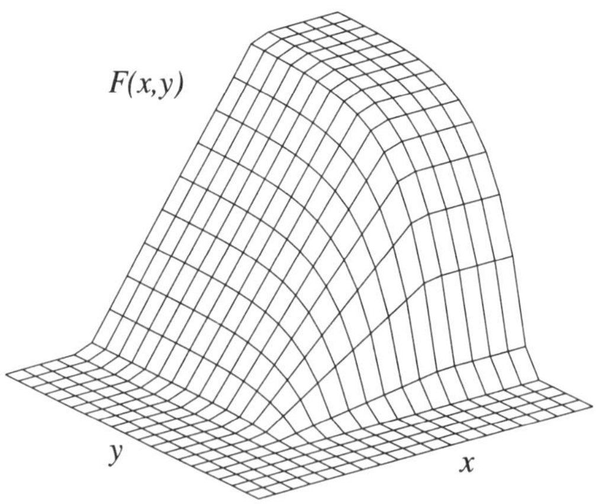

Figure 4.2: A joint cumulative distribution function.

In many practical situations, use of only one RV is not enough; that is, assigning a single number to an outcome is not sufficient. Examples include

- Three numbers are needed to specify the instantaneous position of a moving object.
- We would like to investigate the relation between the response of a system to an input that is random due to the existence of noise. We need at least two RVs to model the input and output at the same or different time instants. In fact, most often we will need to have two *random time functions*, to be studied later.
- We would like to investigate the relation between the measured values at different time instants of a random time function.

Extension from a single RV to two RVs is not trivial in many cases. Extension from two RVs to three or more RVs is usually not hard and thus will not be studied in detail.

A simple and concrete example in which two RVs are needed is the combined experiment of Subsection 2.6.1 for die-rolling and coin-tossing. We may define X by the assignment $X(\text{head}) = 1$, $X(\text{tail}) = 0$ and define Y by the assignment $Y(F_i) = i$.

A pair of RVs can be treated as a ***random vector*** whose components are the RVs. Likewise, we may define an n-dimensional random vector $\mathbf{X}$ as a vector whose components are all RVs: $\mathbf{X} = [X_1, \ldots, X_n]'$. Such a random vector may be thought of as a random point in an n-dimensional hyperspace. The (joint) CDF of a random vector (i.e., a set of RVs) $\mathbf{X} = [X_1, \ldots, X_n]'$ is defined by, using notation $\mathbf{x} = [x_1, \ldots, x_n]'$,

$$F_{\mathbf{X}}(\mathbf{x}) = P\{\mathbf{X} \leq \mathbf{x}\}$$

meaning that

$$F_{X_1,\ldots,X_n}(x_1, \ldots, x_n) = P\{X_1 \leq x_1, \ldots, X_n \leq x_n\}$$

where $\{\mathbf{X} \leq \mathbf{x}\}$ means $\{X_1 \leq x_1, \ldots, X_n \leq x_n\}$ or

$$\{\mathbf{X} \leq \mathbf{x}\} \triangleq \begin{bmatrix} X_1 \leq x_1 \\ \vdots \\ X_n \leq x_n \end{bmatrix}$$

Similarly as for the single RV case, a discrete random vector is one that can take on only discrete (finitely or countably many) values. These values are called their ***point masses***. Its CDF consists only of products of one-dimensional unit step functions. A continuous random vector is one having a continuous range of values and a mixed random vector is one with both discrete and continuous possible values. More rigorously, (X, Y) is continuous if there exists a nonnegative function $f(x, y)$ such that $F(x, y) = \int_{-\infty}^{x} \int_{-\infty}^{y} f(u, v)dudv$, where $F(x, y)$ is the joint CDF of (X, Y) and such a function $f(x, y)$ is known as the ***joint probability density function*** of (X, Y), to be studied later.

Note that $\{X \leq x, Y \leq y\} = \{X \leq x\} \cap \{Y \leq y\}$ represents the joint event of $\{X \leq x\}$ and $\{Y \leq y\}$. However, the following is not true in general

$$P\{X \leq x, Y \leq y\} = P\{X \leq x\}P\{Y \leq y\}$$

unless RVs X and Y are *independent*, to be discussed later.

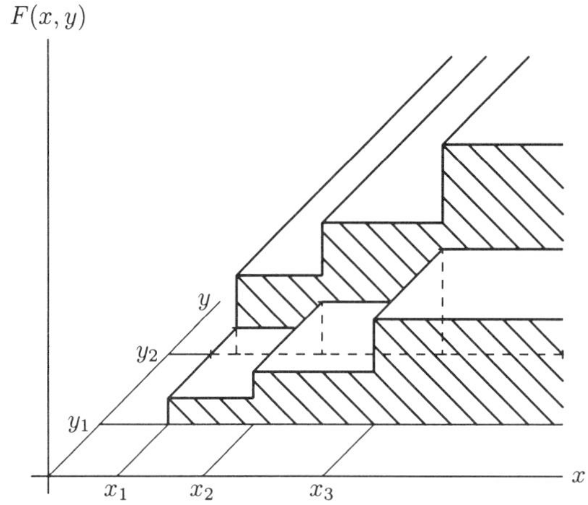

Figure 4.3: The joint cumulative distribution function of two discrete RVs.

A RV X may be thought of as a random point on the real line. Likewise, a pair of RVs (X, Y) can be thought of as a random point in the x-y plane.

A pair of RVs (X, Y) is

- ***discrete*** if it can take on only discrete (finitely many or countably many) values;
- ***continuous*** if it has a continuous range of values;
- ***mixed*** if it is neither discrete nor continuous.

If X and Y are discrete RVs, taking on discrete values $x_1, x_2, \ldots, x_n$ and $y_1, y_2, \ldots, y_m$, respectively, then, similarly to (3.7),

$$F(x,y) = \sum_{i=1}^{n}\sum_{j=1}^{m} P\{X = x_i, Y = y_j\}u(x - x_i)u(y - y_j)$$
$$= \sum_{x_i \le x}\sum_{y_j \le y} P\{X = x_i, Y = y_j\} = \sum_{x_i \le x}\sum_{y_j \le y} p_{i,j} \tag{4.2}$$

where u is the unit step function and $p_{i,j} \triangleq P\{X = x_i, Y = y_j\}$ is called the ***point mass*** at (x_i, y_j).

Note that if there are N x_i's such that $x_i \le x$ and M y_j's such that $y_j \le y$, then the sum in (4.2) has $N \times M$ terms.

Important Properties of Joint CDF

1. $F(x, y)$ is nondecreasing as x or y or both increase: Note that

$$\begin{aligned}\{(X \le x_2) \cap (Y \le y_2)\} &= \{X \le x_1, Y \le y_1\} \uplus \{X \le x_1, y_1 < Y \le y_2\} \\ &\uplus \{x_1 < X \le x_2, Y \le y_1\} \uplus \{x_1 < X \le x_2, y_1 < Y \le y_2\}\end{aligned}$$

   Since these areas are disjoint, we have

$$\begin{aligned}F(x_2, y_2) = P\{x_1 < X \le x_2, Y \le y_1\} + P\{x_1 < X \le x_2, y_1 < Y < y_2\} \\ + F(x_1, y_1) + P\{X \le x_1, y_1 < Y \le y_2\}\end{aligned} \tag{4.3}$$

   Thus,

$$F(x_2, y_2) \ge F(x_1, y_1), \qquad \forall x_2 \ge x_1, y_2 \ge y_1 \tag{4.4}$$

2. Note that

$$(X \le x_1) \cap (Y \le y_2) = \{(X \le x_1) \cap (Y \le y_1)\} \uplus \{(X \le x_1) \cap (y_1 < Y \le y_2)\}$$

   We have

$$\begin{aligned}F(x_1, y_2) &= F(x_1, y_1) + P\{X \le x_1, y_1 < Y \le y_2\} \\ P\{X \le x_1, y_1 < Y \le y_2\} &= F(x_1, y_2) - F(x_1, y_1) \\ P\{x_1 < X \le x_2, Y \le y_1\} &= F(x_2, y_1) - F(x_1, y_1) \qquad \text{(by symmetry)}\end{aligned}$$

   Substituting into (4.3) yields

$$\boxed{\begin{aligned}P\{x_1 < X \le x_2, y_1 < Y \le y_2\} \;=\; & F(x_2, y_2) + F(x_1, y_1) \\ & - F(x_1, y_2) - F(x_2, y_1)\end{aligned}} \tag{4.5}$$

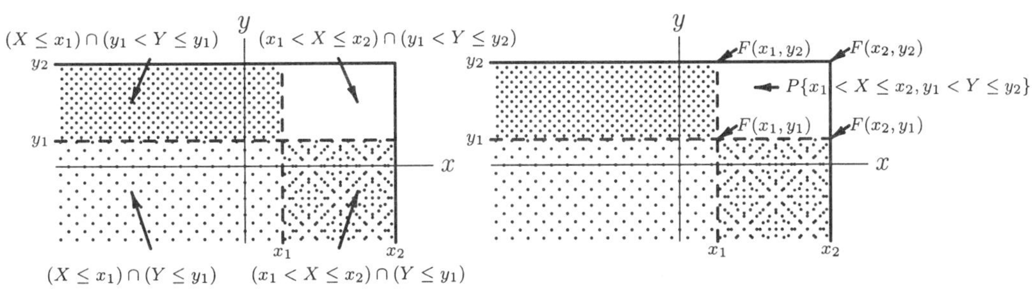

Figure 4.4: Breakdown of $\{X \le x_2, Y \le y_2\}$ and $F(x_2, y_2)$.

3. $0 \leq F(x, y) \leq 1$, since $F(x, y)$ is a probability.
4. Since $P\{X \leq -\infty\} = P\{Y \leq -\infty\} = 0$,

$$F(-\infty, y) = F(x, -\infty) = F(-\infty, -\infty) = 0 \tag{4.6}$$

5. $F(\infty, \infty) = 1$, since $P\{X \leq \infty, Y \leq \infty\} = P\{S\} = 1$.
6. The ***marginal CDFs*** of X and Y can be obtained from their joint CDF:

$$\boxed{\begin{aligned} F_X(x) &= P\{X \leq x\} = P\{X \leq x, Y \leq \infty\} = F_{X,Y}(x, \infty) \\ F_Y(y) &= P\{Y \leq y\} = P\{X \leq \infty, Y \leq y\} = F_{X,Y}(\infty, y) \end{aligned}} \tag{4.7}$$

If X and Y are continuous then the probability of taking on any given value is zero:

$$\begin{aligned} P\{X = a, Y = b\} &= P\{X = a, y_1 < Y \leq y_2\} = P\{x_1 < X \leq x_2, Y = b\} \\ &= P\{g(X, Y) = c\} = 0 \end{aligned} \tag{4.8}$$

since a single point, a straight line and a curve all have zero area, where $g(X, Y) = c$ is a curve in x-y plane.

(4.7) has the following geometric interpretation: $F_X(x)$ is the curve determined by the intersection of the surface $F(x, y)$ and the plane $y = \infty$ and similarly, $F_Y(y)$ is the curve determined by the intersection of the surface $F(x, y)$ and the plane $x = \infty$.

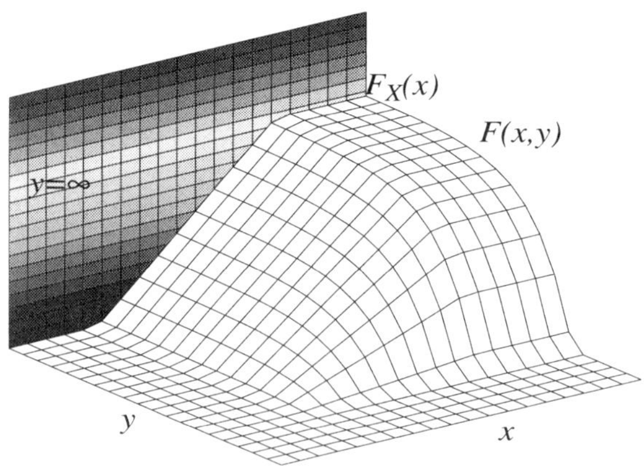

Figure 4.5: Geometric interpretation of (4.7): intersection of $F(x, y)$ and $y = \infty$.

An additional property of the joint CDF is that it is continuous from right in x and y:

$$\lim_{\epsilon \to 0} F(x+\epsilon, y) = F(x,y), \qquad \lim_{\epsilon \to 0} F(x, y+\epsilon) = F(x,y)$$

A bivariate function can be the joint CDF of a pair of RVs if and only if it satisfies the properties 1, 3, 4, 5, the one above, and the following property:

$$F(x_2, y_2) + F(x_1, y_1) - F(x_1, y_2) - F(x_2, y_1) \geq 0$$

for every quadruple of real numbers $x_1 \leq x_2, y_1 \leq y_2$. Note that this last property is not implied by the other properties (see problem 4.3).

(4.7) makes sense: $F_X(x) = P\{X \leq x\}$ is the probability that X takes on a value not greater than x irrespective of the value of Y (i.e., $Y \leq \infty$).

Note that $F_X(x_0)$ [or $F_Y(y_0)$] is not necessarily continuous at $x = x_0$ [or $y = y_0$] even if $F(x_0, y_0)$ is continuous at $(x,y) = (x_0, y_0)$. This should be clear from the fact that $F_X(x_0) = F(x_0, \infty)$ [or $F_Y(y_0) = F(\infty, y_0)$].

The ***probability mass function*** (***PMF***) of discrete RVs (X, Y) is

$$p(x_i, y_j) \triangleq P\{X = x_i, Y = y_j\} = \sum_m \sum_n p_{m,n} \delta_{m-i} \delta_{n-j} \tag{4.9}$$

where δ_{m-i} is the Kronecker delta function, defined by (3.12).

The joint distribution of a pair of discrete RVs (X, Y) is sometimes tabulated as follows

$X \backslash Y$	y_1	$\cdots$	y_j	$\cdots$
x_1	p_{11}	$\cdots$	p_{1j}	$\cdots$
$\vdots$	$\vdots$	$\cdots$	$\vdots$	$\cdots$
x_i	p_{i1}	$\cdots$	$p_{i,j}$	$\cdots$
$\vdots$	$\vdots$	$\cdots$	$\vdots$	$\cdots$

For discrete (X, Y), their point masses $p_{i,j} = P\{X = x_i, Y = y_j\}$ satisfy, denoting by $\sum_i$ sum over all i,

$$\begin{aligned} \text{nonnegativity property:} &\quad p_{i,j} \geq 0, \quad \forall i, j \\ \text{normalization property:} &\quad \sum_i \sum_j p_{i,j} = 1 \end{aligned}$$

Conversely, any set of numbers that satisfies these two conditions can be the joint PMF of some pair of discrete RVs.

The marginal PMFs and CDFs of RVs X and Y are

$$\begin{aligned} p_i &\triangleq P\{X = x_i\} = \sum_j p_{i,j}, & p_j &\triangleq P\{Y = y_j\} = \sum_i p_{i,j} \\ F_X(x) &= \sum_{x_i \leq x} \sum_j p_{i,j} = \sum_{x_i \leq x} p_i, & F_Y(y) &= \sum_{y_j \leq y} \sum_i p_{i,j} = \sum_{y_j \leq y} p_j \end{aligned}$$

Example 4.1: Consider two RVs X and Y having the joint CDF

$$F(x,y) = \begin{cases} 1 - e^{-y} & x > 4,\ y \geq 0 \\ \frac{x}{4}(1 - e^{-y}) & 0 \leq x \leq 4,\ y \geq 0 \\ 0 & \text{elsewhere} \end{cases}$$

(a) Find $P\{X < 2, Y < 2\}$:

$$P\{X < 2, Y < 2\} = P\{X \leq 2, Y \leq 2\} = F(2,2) = \frac{2}{4}(1 - e^{-2}) = 0.4323$$

(b) Find $P\{1 < X < 2, Y < 2\}$:

$$P\{1 < X < 2, Y < 2\} = F(2,2) - F(1,2) = \frac{2}{4}(1 - e^{-2}) - \frac{1}{4}(1 - e^{-2}) = 0.216$$

(c) Find $P\{0 < X < 4, Y = 1\}$: Since Y is continuous, it has zero probability of taking on any given value. Hence, $P\{0 < X < 4, Y = 1\} \stackrel{(4.8)}{=} 0$.

(d) Find $P\{X < 4, Y > 2\}$:

$$\begin{aligned} P\{X < 4, Y > 2\} &= P\{-\infty < X < 4, 2 < Y < \infty\} \\ &= F(4,\infty) + F(-\infty,2) - F(-\infty,\infty) - F(4,2) \\ &= \frac{4}{4}(1 - e^{-\infty}) + 0 - 0 - \frac{4}{4}(1 - e^{-2}) = e^{-2} = 0.1353 \end{aligned}$$

(e) Find $P\{-1 < X < 0, Y < 2\}$: It is clear from the given joint CDF that the probability is zero whenever X is negative. Thus, $P\{-1 < X < 0, Y < 2\} = 0$. This can also be obtained formally as below:

$$P\{-1 < X < 0, Y < 2\} = F(0,2) - F(-1,2) = 0 - 0 = 0$$

(f) Find the marginal CDFs of X and Y: $F(x)$ is the curve determined by the intersection of the surface $F(x,y)$ and the plane $y = \infty$ and likewise for $F(y)$:

$$F_X(x) = F(x,\infty) = \left\{ \begin{matrix} 1 & x > 4 \\ \frac{x}{4} & 0 \leq x \leq 4 \\ 0 & x < 0 \end{matrix} \right\} \neq \left\{ \begin{matrix} 1 & x > 4 \\ \frac{x}{4} & x \leq 4 \end{matrix} \right\}$$

$$F_Y(y) = F(\infty,y) = \left\{ \begin{matrix} 1 - e^{-y} & y \geq 0 \\ 0 & y < 0 \end{matrix} \right\} \neq 1 - e^{-y}$$

Note that $F(y) \neq 1 - e^{-y}$ because

$$F(x,y) = 0,\ \text{"elsewhere"} \quad \Longrightarrow \quad F(x) = 0, \forall x < 0, \quad F(y) = 0, \forall y < 0$$

4.2 The Joint Density Function

The ***joint probability density function*** (***joint PDF*** for two RVs X and Y is defined as

$$\boxed{f(x,y) = f_{X,Y}(x,y) = \frac{\partial^2}{\partial x \partial y} F(x,y) = \frac{\partial^2}{\partial y \partial x} F(x,y)} \tag{4.10}$$

Important Properties of the Joint PDF:

1. $f(x,y) \geq 0$, $\forall x, y$, since $F(x,y)$ is nondecreasing.
2. $F(x,y) = \int_{u=-\infty}^{x} \int_{v=-\infty}^{y} f(u,v) dv du$ (true from definition).
3. By $F(\infty,\infty) = 1$ and property 2 above, $\int_{-\infty}^{\infty} \int_{-\infty}^{\infty} f(u,v) du dv = 1$.
4. By property 2 above and (4.7),
$$\int_{u=-\infty}^{x} \int_{v=-\infty}^{\infty} f(u,v) dv du = F(x,\infty) = F_X(x)$$
$$\int_{v=-\infty}^{y} \int_{u=-\infty}^{\infty} f(u,v) du dv = F(\infty,y) = F_Y(y)$$
5. Note that
$$f_X(x) = \frac{d}{dx} F_X(x) = \frac{d}{dx} \left\{ \int_{-\infty}^{x} \left[\int_{-\infty}^{\infty} f(u,v) dv \right] du \right\} \stackrel{?}{=} \int_{-\infty}^{\infty} f(x,v) dv \tag{4.11}$$
and similarly for $f_Y(y)$; that is, the ***marginal PDFs*** of X and Y can be obtained from their joint PDF:
$$\boxed{f_X(x) = \int_{-\infty}^{\infty} f(x,v) dv, \qquad f_Y(y) = \int_{-\infty}^{\infty} f(u,y) du} \tag{4.12}$$
6. Since
$$\begin{aligned} &P\{x_1 < X \leq x_2, y_1 < Y \leq y_2\} \\ &= F(x_2,y_2) + F(x_1,y_1) - F(x_1,y_2) - F(x_2,y_1) \\ &= \left(\int_{-\infty}^{x_2} \int_{-\infty}^{y_2} + \int_{-\infty}^{x_1} \int_{-\infty}^{y_1} - \int_{-\infty}^{x_1} \int_{-\infty}^{y_2} - \int_{-\infty}^{x_2} \int_{-\infty}^{y_1} \right) f(u,v) dv du \\ &= \left(\int_{x_1}^{x_2} \int_{-\infty}^{y_2} - \int_{x_1}^{x_2} \int_{-\infty}^{y_1} \right) f(u,v) dv du = \int_{x_1}^{x_2} \int_{y_1}^{y_2} f(u,v) dv du \end{aligned}$$
we have
$$\boxed{P\{x_1 < X \leq x_2, y_1 < Y \leq y_2\} = \int_{x=x_1}^{x_2} \int_{y=y_1}^{y_2} f(x,y) dx dy} \tag{4.13}$$

Similarly as in the single RV case, joint PDF is a density function that indicates where the RV values are more (or less) consolidated.

The marginal CDF (or PDF, PMF) of a RV is simply the CDF (or PDF, PMF) of the RV itself. The only difference is that we say marginal CDF (or PDF, PMF) of a single RV when we have more than one RV in mind and we are interested in their relations; while we say CDF (or PDF, PMF) of the RV if we are not concerned with its relation to other RVs.

If the joint CDF $F(x, y)$ [or joint PDF $f(x, y)$, joint PMF $p(x, y)$] of (X, Y) is symmetrical about x and y, that is, $F(x, y) = F(y, x)$ [or $f(x, y) = f(y, x)$, $p(x_i, y_j) = p(y_j, x_i)$], X and Y have an identical marginal CDF (or PDF, PMF). This should be clear from (4.7) and (4.12) since the formulas for the two marginal CDFs (or PDFs, PMFs) are also symmetrical about x and y.

The last equation $\overset{?}{=}$ in (4.11) follows from the following result in calculus:

$$\frac{d}{dx}\int_{-\infty}^{x} g(u)du = g(x)$$

where $g(x) = \int_{-\infty}^{\infty} f(x, v)dv$.

A geometric interpretation of (4.12) will be given later in terms of conditional distribution.

The joint PMF and joint PDF of a random vector (i.e., a set of RVs) $\mathbf{X} = [X_1, \ldots, X_n]'$ are defined by

$$\begin{aligned} p_{\mathbf{X}}(\mathbf{x}) &= P\{\mathbf{X} = \mathbf{x}\} \triangleq P\{X_1 = x_1, \ldots, X_n = x_n\} \\ f_{\mathbf{X}}(\mathbf{x}) &= \frac{\partial}{\partial \mathbf{x}} F_{\mathbf{X}}(\mathbf{x}) \end{aligned}$$

The last equation is a compact notation of the following:

$$f_{X_1,\ldots,X_n}(x_1, \ldots, x_n) = \frac{\partial^n}{\partial x_1 \cdots \partial x_n} F_{X_1,\ldots,X_n}(x_1, \ldots, x_n)$$

Similarly as for the single RV case, a discrete random vector is one that has a PDF consisting only of products of (one-dimensional) delta functions, as illustrated by (4.15). The PDF of a continuous random vector contains no delta functions and a mixed random vector has a PDF with both delta functions and other regular functions.

All properties of the joint CDF (or PDF, PMF) of two RVs can be extended to the joint CDF (or PDF, PMF) of an arbitrary number of RVs. For example,

$$\begin{aligned} &F_{X_1,\ldots,X_n}(x_1, \ldots, x_{i-1}, -\infty, x_{i+1}, \ldots, x_n) = 0 \\ &F_{X_1,\ldots,X_m}(x_1, \ldots, x_m) = F_{X_1,\ldots,X_n}(x_1, \ldots, x_m, \underbrace{\infty, \ldots, \infty}_{n-m \text{ fold}}) \\ &f_{X_1,\ldots,X_m}(x_1, \ldots, x_m) = \underbrace{\int_{-\infty}^{\infty} \cdots \int_{-\infty}^{\infty}}_{n-m \text{ fold}} f_{X_1,\ldots,X_n}(x_1, \ldots, x_n) dx_{m+1} \cdots dx_n \end{aligned} \tag{4.14}$$

If X and Y are discrete RVs, taking on discrete values $x_1, x_2, \cdots, x_n$ and $y_1, y_2, \cdots, y_m$, respectively, then

$$f(x,y) = \sum_{i=1}^{n}\sum_{j=1}^{m} P\{X = x_i, Y = y_j\}\delta(x - x_i)\delta(y - y_j)$$
$$= \sum_{i=1}^{n}\sum_{j=1}^{m} p_{i,j}\delta(x - x_i)\delta(y - y_j) \tag{4.15}$$

where δ is the delta (impulse) function. This follows from

$$\begin{aligned} f(x,y) &= \frac{\partial^2}{\partial x \partial y} F(x,y) \\ &\overset{(4.2)}{=} \frac{\partial^2}{\partial x \partial y}\Big[\sum_{i=1}^{n}\sum_{j=1}^{m} P\{X = x_i, Y = y_j\}u(x - x_i)u(y - y_j)\Big] \\ &= \sum_{i=1}^{n}\sum_{j=1}^{m}\Big[P\{X = x_i, Y = y_j\}\frac{\partial^2}{\partial x \partial y}[u(x - x_i)u(y - y_j)]\Big] \\ &= \sum_{i=1}^{n}\sum_{j=1}^{m} P\{X = x_i, Y = y_j\}\delta(x - x_i)\delta(y - y_j) \end{aligned}$$

The corresponding PMF is given by

$$p(x_i, y_j) = \sum_{k=1}^{n}\sum_{l=1}^{m} p_{k,l}\delta_{k-i}\delta_{l-j} \tag{4.16}$$

where δ_{k-i} and δ_{l-j} are the Kronecker delta functions.

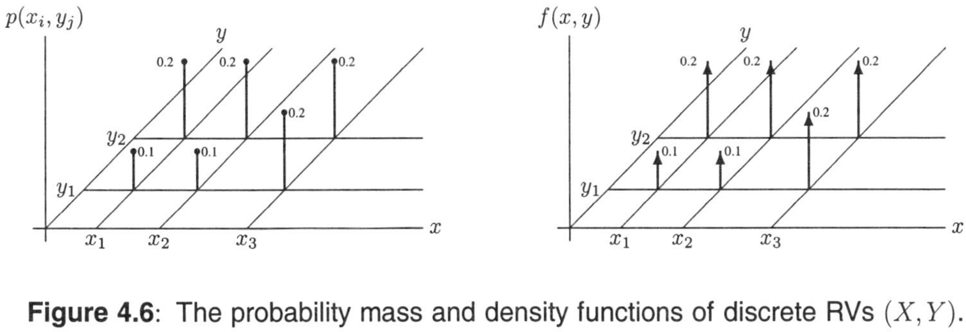

Figure 4.6: The probability mass and density functions of discrete RVs (X, Y).

Example 4.2: Find PDF from CDF

Consider again Example 4.1. The joint CDF of RVs X and Y were given before by

$$F(x,y) = \begin{cases} 1-e^{-y} & x>4,\ y\geq 0 \\ \frac{x}{4}(1-e^{-y}) & 0\leq x\leq 4,\ y\geq 0 \\ 0 & \text{elsewhere} \end{cases}$$

(a) Find the joint PDF of X and Y:

$$f(x,y) = \frac{\partial^2 F(x,y)}{\partial x \partial y} = \begin{cases} \frac{1}{4}e^{-y} & 0<x<4,\ y>0 \\ 0 & \text{elsewhere} \end{cases}$$

(b) Find the marginal PDFs of X and Y: Since the marginal CDFs are, from Example 4.1,

$$F_X(x) = \left\{ \begin{matrix} 1 & x>4 \\ \frac{x}{4} & 0\leq x\leq 4 \\ 0 & x<0 \end{matrix} \right\}, \qquad F_Y(y) = \left\{ \begin{matrix} 1-e^{-y} & y\geq 0 \\ 0 & y<0 \end{matrix} \right\}$$

the marginal PDFs are

$$f_X(x) = \frac{dF_X(x)}{dx} = \begin{cases} \frac{1}{4} & 0<x<4 \\ 0 & \text{elsewhere} \end{cases}$$

$$f_Y(y) = \frac{dF_Y(y)}{dy} = \begin{cases} e^{-y} & y>0 \\ 0 & y<0 \end{cases}$$

Alternatively,

$$f_X(x) = \int_{-\infty}^{\infty} f(x,y)dy = \left\{ \begin{matrix} \int_0^{\infty} \frac{1}{4}e^{-y}dy & 0<x<4 \\ 0 & \text{elsewhere} \end{matrix} \right\} = \begin{cases} \frac{1}{4} & 0<x<4 \\ 0 & \text{elsewhere} \end{cases}$$

$$f_Y(y) = \int_{-\infty}^{\infty} f(x,y)dx = \left\{ \begin{matrix} \int_0^{4} \frac{1}{4}e^{-y}dx & y>0 \\ 0 & y<0 \end{matrix} \right\} = \begin{cases} e^{-y} & y>0 \\ 0 & y<0 \end{cases}$$

(c) Identify the distribution and the associated parameters of RVs X and Y: Clearly, X is uniform over $[0,4]$ [i.e., $X \sim \mathcal{U}(0,4)$] and Y is an exponential RV (3.34) with parameter $\lambda = 1, a = 0$.

4.3 Independence of Random Variables

Recall that two events A and B are independent if and only if $P\{A, B\} = P\{A\}P\{B\}$. Let $A = \{X \leq x\}$ and $B = \{Y \leq y\}$. Then A and B are independent if

$$P\{X \leq x, Y \leq y\} = P\{X \leq x\}P\{Y \leq y\}$$

that is

$$F_{X,Y}(x, y) = F_X(x)F_Y(y) \tag{4.17}$$

Note that $F(x,y_1) = F(x)F(y_1)$ for all x implies that the event $\{X \leq x\}$ for every x (i.e., RV X) and event $\{Y \leq y_1\}$ for a given y_1 are independent. $F(x_1, y_1) = F(x_1)F(y_1)$ implies the independence of $\{X \leq x_1\}$ and $\{Y \leq y_1\}$ but not the independence of $\{X \leq x_2\}$ and $\{Y \leq y_2\}$.

Thus, if (4.17) holds for every x and y [i.e., *function* $F(x, y)$ equals the product of *functions* $F(x)$ and $F(y)$], then the RVs X and Y are said to be ***independent***. Also, $\frac{\partial^2}{\partial x \partial y}$[(4.17)] yields

$$f(x, y) = \frac{\partial^2 F(x, y)}{\partial x \partial y} = \frac{\partial^2}{\partial x \partial y} F_X(x)F_Y(y) = \frac{\partial}{\partial x} F_X(x) \frac{\partial}{\partial y} F_Y(y) = f_X(x)f_Y(y)$$

Consequently,

$$\boxed{\begin{aligned} X, Y \textbf{\textit{ independent}} \quad &\Longleftrightarrow \quad F_{X,Y}(x, y) = F_X(x)F_Y(y), \quad \forall x, y \\ &\Longleftrightarrow \quad f_{X,Y}(x, y) = f_X(x)f_Y(y), \quad \forall x, y \end{aligned}} \tag{4.18}$$

The independence of RVs X and Y means that the probability of X taking on any value (within any interval) is *independent* of the value of Y and vice versa.

Example 4.3: The RVs X and Y of Example 4.1 are independent because

$$\begin{aligned} F_X(x)F_Y(y) &= \begin{Bmatrix} 1 & x > 4 \\ \frac{x}{4} & 0 \leq x \leq 4 \\ 0 & \text{elsewhere} \end{Bmatrix} \begin{Bmatrix} 1 - e^{-y} & y \geq 0 \\ 0 & \text{elsewhere} \end{Bmatrix} \\ &= \begin{Bmatrix} 1 - e^{-y} & x > 4,\ y \geq 0 \\ \frac{x}{4}(1 - e^{-y}) & 0 \leq x \leq 4,\ y \geq 0 \\ 0 & \text{elsewhere} \end{Bmatrix} = F(x, y) \end{aligned}$$

Example 4.4: Two RVs X and Y have the joint PDF

$$f(x,y) = \begin{cases} 2xy & 0 < x \leq 1,\ 0 < y \leq \sqrt{2} \\ 0 & \text{elsewhere} \end{cases}$$

(a) Determine if X and Y are independent:

$$f_X(x) = \int_{-\infty}^{\infty} f(x,v)dv = \int_0^{\sqrt{2}} 2xvdv = 2x\int_0^{\sqrt{2}} vdv = 2x,\ 0 < x \leq 1$$
$$f_Y(y) = \int_{-\infty}^{\infty} f(u,y)du = \int_0^1 2uydu = y,\ 0 < y \leq \sqrt{2}$$

That is

$$f_X(x) = \begin{cases} 2x & 0 < x \leq 1 \\ 0 & \text{elsewhere} \end{cases}$$
$$f_Y(y) = \begin{cases} y & 0 < y \leq \sqrt{2} \\ 0 & \text{elsewhere} \end{cases}$$

Thus

$$f(x,y) = f_X(x)f_Y(y) \quad \Longrightarrow \quad X \text{ and } Y \text{ are independent}$$

(b) Find $F(x,y)$:

$$F_X(x) = \int_{-\infty}^{x} f_X(x)dx = \begin{cases} 0 & x \leq 0 \\ x^2 & 0 < x \leq 1 \\ 1 & x > 1 \end{cases}$$
$$F_Y(y) = \int_{-\infty}^{y} f_Y(y)dy = \begin{cases} 0 & y \leq 0 \\ \frac{1}{2}y^2 & 0 < y \leq \sqrt{2} \\ 1 & y > \sqrt{2} \end{cases}$$

Since X and Y are independent,

$$F(x,y) = F_X(x)F_Y(y) = \begin{cases} \frac{1}{2}x^2y^2 & 0 < x \leq 1, 0 < y \leq \sqrt{2} \\ x^2 & 0 < x \leq 1, y > \sqrt{2} \\ \frac{1}{2}y^2 & x > 1, 0 < y \leq \sqrt{2} \\ 1 & x > 1, y > \sqrt{2} \\ 0 & \text{elsewhere} \end{cases}$$

This approach is simpler than using property 2 of the joint PDF.

Two discrete RVs X and Y are ***independent*** if their joint PMF is the product of their marginal PMFs:

$$p_{X,Y}(x_i, y_j) = p_X(x_i)p_Y(y_j)$$

In general, a set of RVs $X_1, \ldots, X_n$ is said to be ***independent*** if any of the following equivalent conditions holds for every group of values $x_1, \ldots, x_n$

$$F_{X_1,\ldots,X_n}(x_1, \ldots, x_n) = F_{X_1}(x_1) \cdots F_{X_n}(x_n)$$
$$f_{X_1,\ldots,X_n}(x_1, \ldots, x_n) = f_{X_1}(x_1) \cdots f_{X_n}(x_n)$$

To check if X and Y are independent given their joint PDF, it is often easier to use the following result than based on (4.18):

Theorem: X and Y are independent if and only if their joint PDF $f(x, y)$ can be factorized as (i.e., variables can be separated):

$$f(x, y) = f_1(x)f_2(y) \quad \text{or} \quad F(x, y) = F_1(x)F_2(y) \tag{4.19}$$

where $f_i(x)$ are *nonnegative* functions and $F_i(x)$ are nondecreasing *nonnegative* functions.

Note that such factorization implies that the region over which $f(x, y)$ is nonzero is also separated in the form of $x_1 < x < x_2, y_1 < y < y_2$, where x_1, x_2, y_1, y_2 are constants.

A similar result holds true for discrete RVs (for PMF).

Using this criterion, RVs X and Y in Examples 4.1 and 4.4 are clearly independent.

Example 4.5: Independence Check by Factorization

If (X, Y) has the joint PDF

$$F(x, y) = \begin{cases} 1 - e^{-x} - e^{-2y} + e^{-(x+2y)} & x \geq 0, y \geq 0 \\ 0 & \text{elsewhere} \end{cases}$$

then X and Y are independent because $1 - e^{-x} - e^{-2y} + e^{-(x+2y)} = (1 - e^{-x})(1 - e^{-2y})$ and $x \geq 0, y \geq 0 = (x \geq 0) \cap (y \geq 0)$. Similarly, X and Y are independent if their joint PDF is

$$f(x, y) = e^{-x^2/2} \frac{1}{\sqrt{2\pi^3}(1 + y^2)}$$

because $f_1(x) = \frac{1}{\sqrt{2\pi^3}} e^{-x^2/2}, \quad f_2(x) = \frac{1}{1+y^2}$.

X and Y are not independent if their joint PDF is

$$f(x, y) = \begin{cases} \frac{1}{24}(1 + xy) & 0 \leq x \leq 2, 0 \leq y \leq 4 \\ 0 & \text{elsewhere} \end{cases}$$

or

$$f(x, y) = \begin{cases} 2 & 0 < x < 1, y < 1 - x \\ 0 & \text{elsewhere} \end{cases}$$

because either the function (former case) or the region (latter case) cannot be factorized.

4.4 Expectation and Moments

As for the single RV case, although quite often not a complete description of the RVs, expectation and moments are numbers that are essential characterizations and can usually be obtained easily.

The ***expected value***, ***expectation***, ***mean***, or ***average*** of a function $g(x, y)$ of two RVs X and Y is defined by, denoting by $\sum_i$ sum over all i,

$$E[g(X,Y)] = \begin{cases} \int_{-\infty}^{\infty}\int_{-\infty}^{\infty} g(x,y)f(x,y)dxdy & (X,Y) \text{ continuous} \\ \sum_i \sum_j g(x_i,y_j)P\{X=x_i, Y=y_j\} & (X,Y) \text{ discrete} \end{cases} \tag{4.20}$$

In particular, for $p, q \geq 0$,

$$(p,q)\text{th } \boldsymbol{moment} \triangleq E[X^pY^q]$$

$$(p,q)\text{th } \boldsymbol{central\ moment} \triangleq E[(X-\bar{x})^p(Y-\bar{y})^q]$$

For example,

$$E[X] = \bar{x} = \begin{cases} \int_{-\infty}^{\infty} xf_X(x)dx & \text{(method 1)} \\ \int_{-\infty}^{\infty}\int_{-\infty}^{\infty} xf(x,y)dxdy & \text{(method 2)} \end{cases}$$

$$E[X^2] = \begin{cases} \int_{-\infty}^{\infty} x^2f_X(x)dx & \text{(method 1)} \\ \int_{-\infty}^{\infty}\int_{-\infty}^{\infty} x^2f(x,y)dxdy & \text{(method 2)} \end{cases}$$

$$\sigma_x^2 = \begin{cases} \int_{-\infty}^{\infty} (x-\bar{x})^2f_X(x)dx & \text{(method 1)} \\ \int_{-\infty}^{\infty}\int_{-\infty}^{\infty} (x-\bar{x})^2f(x,y)dxdy & \text{(method 2)} \end{cases}$$

Similar formulas are valid for discrete X (or Y). For example,

$$E[X] = \bar{x} = \begin{cases} \sum_i x_iP\{X=x_i\} & \text{(method 1)} \\ \sum_i \sum_j x_iP\{X=x_i, Y=y_j\} & \text{(method 2)} \end{cases}$$

$$\sigma_x^2 = \begin{cases} \sum_i (x_i-\bar{x})^2P\{X=x_i\} & \text{(method 1)} \\ \sum_i \sum_j (x_i-\bar{x})^2P\{X=x_i, Y=y_j\} & \text{(method 2)} \end{cases}$$

Symmetric formulas are valid for $E[Y]$, $E[Y^2]$, and σ_y^2.

Example 4.6: Given the PDF of a uniform RV for measurement errors:

$$f(x,y) = \begin{cases} 1/12 & 0 < x \le 6, 0 < y \le 2 \\ 0 & \text{elsewhere} \end{cases}$$

(a) Find $\bar{x}, \bar{y}$, and $E(XY)$:

$$\begin{aligned} \bar{x} &= \int_{-\infty}^{\infty} x f_X(x)dx = \int_{-\infty}^{\infty} x \int_{-\infty}^{\infty} f(x,y)dydx = \frac{1}{12}\int_0^6 x \int_0^2 dydx = 3 \\ \bar{y} &= \int_{-\infty}^{\infty} y f_Y(y)dx = \int_{-\infty}^{\infty} y \int_{-\infty}^{\infty} f(x,y)dxdy = \frac{1}{12}\int_0^2 y \int_0^6 dxdy = 1 \\ E[XY] &= \int_{-\infty}^{\infty}\int_{-\infty}^{\infty} xyf(x,y)dxdy = \int_0^6\int_0^2 \frac{xy}{12}dxdy = \frac{1}{12}\int_0^6 xdx \int_0^2 ydy = 3 \end{aligned}$$

Note that $E[XY] = \bar{x}\bar{y}$ for this example. When this is the case, we say that X and Y are ***uncorrelated***. Clearly, X and Y are independent by the theorem on page 159.

(b) Find $P\{X \ge Y\}$: Let $g(x,y) = \begin{Bmatrix} 1 & x \ge y \\ 0 & x < y \end{Bmatrix}$. Then

$$\begin{aligned} E[g(X,Y)] &= 1 \cdot P\{g(X,Y) = 1\} + 0 \cdot P\{g(X,Y) = 0\} \\ &= 1 \cdot P\{X \ge Y\} = P\{X \ge Y\} \end{aligned}$$

$$\begin{aligned} P\{X \ge Y\} = E[g(X,Y)] &= \int_{y=0}^2 \int_{x=0}^6 g(x,y)\frac{1}{12}dxdy = \int_{y=0}^2\int_{x=y}^6 1 \cdot \frac{1}{12}dxdy \\ &= \frac{1}{12}\int_0^2 [x|_{x=y}^6]dy = \frac{1}{12}(6-y)dy = \frac{5}{6} \end{aligned}$$

For this function g, $E[g(X,Y)]$ is equal to the area of the shaded region since (X,Y) is uniformly distributed over the rectangle.

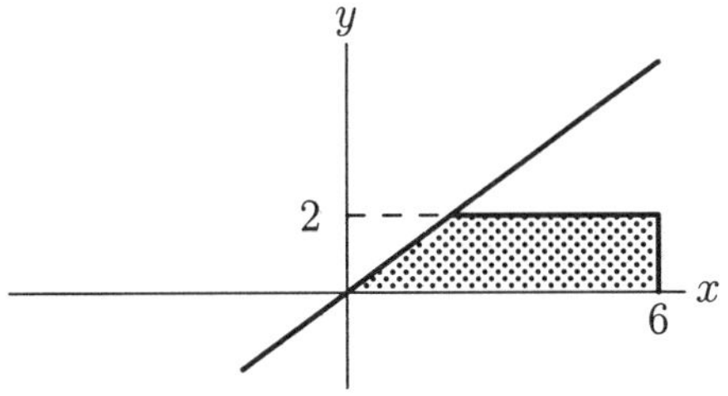

4.5 Relation between Two Random Variables

The ***correlation*** of two RVs X and Y is the $(1,1)$th joint moment:

$$R_{xy} \triangleq E[XY] = E[g(X,Y)]|_{g(x,y)=xy} \tag{4.21}$$

The ***covariance*** of two RVs X and Y is the $(1,1)$th joint central moment:

$$\begin{aligned} C_{xy} &\triangleq \text{cov}(X,Y) \triangleq \sigma_{xy}^2 \triangleq E[(X-\bar{x})(Y-\bar{y})] \\ &= E[g(X,Y)]|_{g(x,y)=(x-\bar{x})(y-\bar{y})} \\ &= E[XY] - \bar{x}\bar{y} = R_{xy} - \bar{x}\bar{y} \end{aligned} \tag{4.22}$$

Note that σ_{xy}^2 *could be negative*. This notation is somewhat misleading.

The ***correlation coefficient*** of two RVs X and Y is defined as the *correlation* of the ***standardized*** RVs $\tilde{X} \triangleq \frac{X-\bar{x}}{\sigma_x}$ and $\tilde{Y} \triangleq \frac{Y-\bar{y}}{\sigma_y}$:

$$\rho_{xy} = E[\tilde{X}\tilde{Y}] = E\Big[\frac{(X-\bar{x})}{\sigma_x}\frac{(Y-\bar{y})}{\sigma_y}\Big] = \frac{C_{xy}}{\sigma_x\sigma_y} \qquad -1 \le \rho \le 1 \tag{4.23}$$

which is a normalized *covariance* rather than a normalized correlation.

The correlation coefficient ρ_{xy} is a measure of ***linear correlation***:

- $\rho_{xy} = 1$ indicates that X and Y have a positive linear relation (almost) *surely*: $\frac{Y-\bar{y}}{\sigma_y} = \frac{X-\bar{x}}{\sigma_x}$.
- $\rho_{xy} = -1$ indicates that X and Y have a negative linear relation (almost) *surely*: $\frac{Y-\bar{y}}{\sigma_y} = -\frac{X-\bar{x}}{\sigma_x}$.
- $\rho_{xy} = 0$ indicates that X and Y are (almost) *surely not* related *linearly* (i.e., by a *linear* equation), which is called ***uncorrelated***.
- A large (or small) $|\rho_{xy}|$ indicates a strong (or weak) likelihood that X and Y have a linear relation.

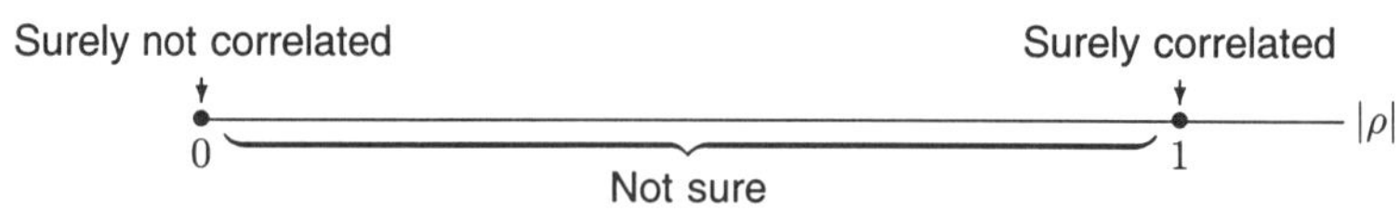

Figure 4.7: Interpretation of correlation coefficient.

To study the interdependence of two RVs X and Y, we may use the following ***scatter diagram***: Plot the values of (X, Y) as points in the x-y plane according to the outcomes of many performances of the random experiment on which X and Y are defined, such as Fig. 4.8. The interdependence of X and Y can be well explained using such a scatter diagram:

(a) If the points are scattered throughout the plane without a pattern, then X and Y are independent.
(b) If all points align over a straight line, then X and Y are linearly correlated and thus $|\rho_{xy}| = 1$ ($\rho_{xy} = 1$ corresponds to a line with a positive slope and $\rho_{xy} = -1$ with a negative slope).
(c) If the points are scattered closely (or loosely) around a straight line, then X and Y have a strong (or weak) linear correlation and thus $|\rho_{xy}|$ is large (or small).
(d) If the points align over a certain curve, then X and Y are related by the equation that describes the curve. Note, however, that this does not necessarily imply that X and Y are uncorrelated even if the curve is not a straight line. See also Example 4.9.
(e) If the points are scattered closely around a certain curve, then this indicates a strong interdependence between X and Y defined by the equation that describes the curve.

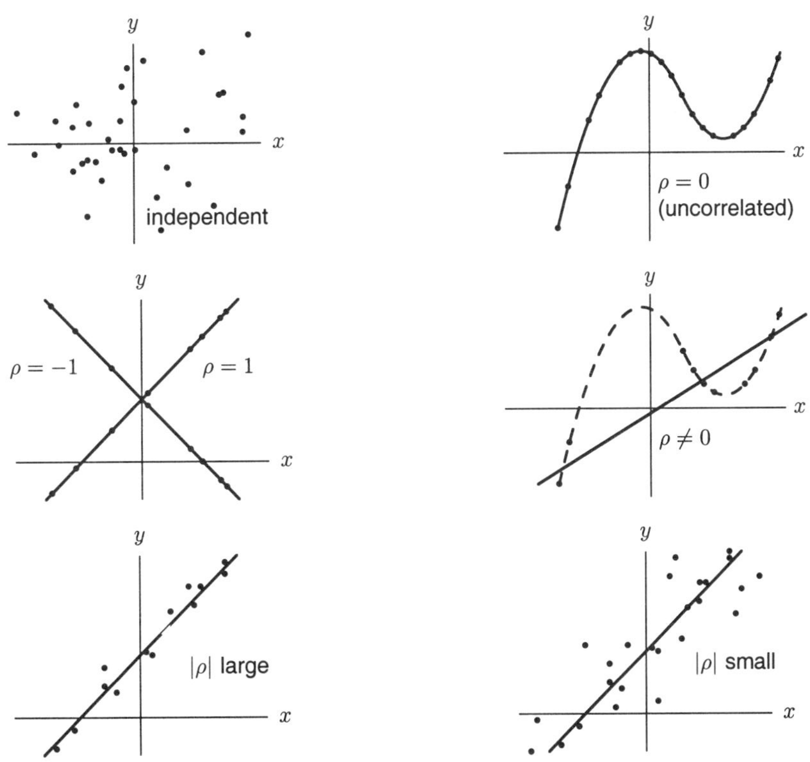

Figure 4.8: Comparison of independence and correlation by scatter diagrams.

Suppose that X and Y have a strong linear correlation, say $Y = aX + b$ in some statistical sense. The slope a and intercept b are unknown and may be determined such that the predicted value $\check{Y}$ of Y based on a sample value of X is optimal in the sense of having the smallest mean square error $E[(Y - \check{Y})^2] = E[(Y - (aX + b))^2]$. It can be shown that this leads to

$$\check{Y} = \frac{C_{xy}}{\sigma_x^2}(X - \bar{x}) + \bar{y} = \sigma_y \frac{C_{xy}}{\sigma_y \sigma_x} \frac{(X - \bar{x})}{\sigma_x} + \bar{y} \quad \Longrightarrow \quad \frac{(\check{Y} - \bar{y})}{\sigma_y} = \rho \frac{(X - \bar{x})}{\sigma_x}$$

It is thus clear that the straight line passes through the point $(\bar{x}, \bar{y})$ and the ***correlation coefficient*** ρ is the slope of the line that gives best standardized predicted value $\tilde{\check{Y}} = \frac{(\check{Y} - \bar{y})}{\sigma_y}$ using a sample point of the standardized RV $\tilde{X} = \frac{(X - \bar{x})}{\sigma_x}$.

The magnitude of the covariance C_{xy} or correlation R_{xy} itself is not a direct measure of the linear correlation of X and Y since it does not take into account how random the RVs X and Y themselves are.

An unknown correlation coefficient may be approximated by the ***sample correlation coefficient***, given by for a data set $\{(X_1, Y_1), (X_2, Y_2), \ldots, (X_n, Y_n)\}$

$$\hat{\rho}_{xy} = \frac{1}{(n-1)\sigma_x \sigma_y} \sum_{i=1}^{n} (X_i - \hat{X})(Y_i - \hat{Y}) \approx \frac{1}{(n-1)\sqrt{\hat{V}_x \hat{V}_y}} \sum_{i=1}^{n} (X_i - \hat{X})(Y_i - \hat{Y}) \qquad (4.24)$$

where $\hat{X}$, $\hat{Y}$, $\hat{V}_x$ and $\hat{V}_y$ are the sample means and sample variances of X and Y, respectively, given by (3.63)–(3.64).

(4.23) can be shown easily as follows, noting that $E[\tilde{X}^2] = E[\tilde{Y}^2] = 1$,

$$0 \le E[(\tilde{X} \pm \tilde{Y})^2] = E[\tilde{X}^2] \pm 2E[\tilde{X}\tilde{Y}] + E[\tilde{Y}^2] \overset{(4.23)}{=} 2(1 \pm \rho) \quad \Longrightarrow \quad |\rho| \le 1 \qquad (4.25)$$

We emphasize that two RVs are uncorrelated only if it is (almost) *sure* that they are not related *linearly* — they are correlated if not (almost) sure.

A RV and its function are not necessarily correlated, as demonstrated below.

Example 4.7: Uncorrelatedness of a RV and Its Function

Consider $X \sim \mathcal{N}(0, 1)$ and $Y = X^2$. Note that

$$C_{xy} = E[XY] - \bar{x}\bar{y} = E[X^3] - 0 = \int_{-\infty}^{\infty} \frac{x^3}{\sqrt{2\pi}} e^{-x^2/2} dx \overset{\text{odd symmetry}}{=} 0$$

where the integral vanishes because the integrand $\frac{x^3}{\sqrt{2\pi}} e^{-x^2/2}$ is odd symmetrical about $x = 0$ (and goes to zero as $x \to \infty$). Then the correlation coefficient $\rho = 0$ (because $C_{xy} = 0$) and thus X and Y are uncorrelated. However, note that

$$\left.\begin{aligned} f(x)f(y) &\overset{\text{Example 3.25}}{=} \mathcal{N}(x; 0, 1) \tfrac{1}{\sqrt{2\pi y}} e^{-y/2} \\ f(x, y) &\overset{(4.45)}{=} f(y|x) f(x) = \delta(y - x^2) \mathcal{N}(x; 0, 1) \end{aligned}\right\} \quad \Longrightarrow \quad f(x, y) \neq f(x)f(y)$$

where $f(y|x) = \delta(y - x^2)$ since $Y = a^2$ if $X = a$ is known. Thus they are dependent.

Example 4.8: Determination of Correlation of Two RVs by P&R

Data file `e4_8.dat` in P&R stores realizations of a pair of RVs (X, Y) in the following format:

$$\begin{array}{cc} x_1 & y_1 \\ x_2 & y_2 \\ \vdots & \vdots \\ x_n & y_n \end{array}$$

The correlation coefficient as well as some other parameters of the RVs X and Y can be obtained by P&R following the following steps (see Fig. 4.9):

S1. Click "Identification" in the main window of P&R.
S2. Point to "Estimation and Validation" and then click "2-D Distribution."
S3. You will be prompted to enter the name of a data file. Choose "data" subdirectory and enter "`e4_8.dat`" and click "Ok." The computed sample correlation coefficient as well as covariance, correlation, means and variances of X and Y are then shown as in Fig. 4.9.
S4. Click "Scatter Diagram." A scatter diagram is then plotted as shown in Fig. 4.9.

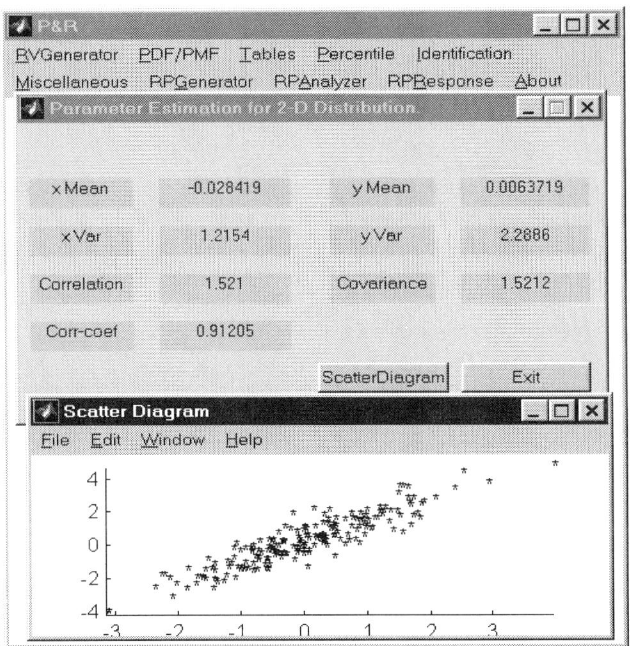

Figure 4.9: Determination of correlation by P&R.

Example 4.9: Correlation of Nonlinearly Related RVs

Nonlinearly related RVs are not necessarily uncorrelated. Let $Y = (X - a)^2$. If $f(x)$ is symmetrical about its mean $\bar{x} = a$, e.g., $X \sim \mathcal{N}(a, 1)$ or $X \sim \mathcal{U}(a - 1, a + 1)$, then X and Y are uncorrelated because

$$\begin{aligned} C_{xy} &= E[(X - \bar{x})(Y - \bar{y})] = E\{(X - a)[(X - a)^2 - E[(X - a)^2]]\} \\ &= E[(X - a)^3] - E[X - a]E[(X - a)^2] \overset{\text{odd symmetry}}{=} 0 - 0 \times \sigma_x^2 = 0 \\ &\Longrightarrow \rho = 0 \end{aligned}$$

The above is not true if $f(x)$ is not symmetrical or it is symmetrical but $\bar{x} \neq a$. For example, assume $X \sim \mathcal{U}(a, a + 1)$. Then

$$\begin{aligned} \sigma_x^2 &= (\text{length})^2/12 = (a + 1 - a)^2/12 = 1/12 \\ \sigma_y^2 &= E[Y^2] - (\bar{y})^2 = E[(X - a)^4] - (E[(X - a)^2])^2 \\ &= \int_a^{a+1} (x - a)^4 dx - \left(\int_a^{a+1} (x - a)^2 dx\right)^2 = 1/5 - (1/3)^2 = 4/45 \\ C_{xy} &= E[(X - \bar{x})(Y - \bar{y})] = E[(X - \bar{x})(X - a)^2] - E[X - \bar{x}]\bar{y} \\ &= E[(X - a + a - \bar{x})(X - a)^2] = E[(X - a)^3] + (a - \bar{x})E[(X - a)^2] \\ &= 1/4 - (1/2)(1/3) = 1/12 \quad \Longrightarrow \quad \rho = \frac{C_{xy}}{\sigma_x \sigma_y} = \frac{1/12}{\sqrt{1/12}\sqrt{4/45}} = 0.968 \end{aligned}$$

This indicates a close-to-perfect linear correlation although X and Y are actually related nonlinearly! This is because $Y = (X - a)^2$ is *statistically* close to the line $\frac{Y - \bar{y}}{\sigma_y} = \rho \frac{X - \bar{x}}{\sigma_x}$ over $a < x < a + 1$ but surely not close to $Y = \alpha X + \beta$ for any $\alpha, \beta \neq 0$ over $a - 1 < x < a + 1$, as shown in Fig. 4.10. Note that the curve in (b) can be reasonably approximated by a straight line of a nonzero slope.

(a) $\rho = 0$: uncorrelated

(b) $\rho = 0.968$: highly correlated

Figure 4.10: Correlation between X and $Y = (X - a)^2$.

Example 4.10: Numerical Evaluation of Correlation between RVs by P&R

The correlation of a RV and its linear or nonlinear function can be evaluated numerically by the companion software P&R. We demonstrate how this is done by verifying the correlation in the last example using 100 pairs of random numbers with $a = 2$.

S1. Generate pairs of random numbers (X, Y). For the above example, this can be done by first generating 100 random numbers $X_i \sim \mathcal{U}(a-1, a+1)$ with $a = 2$, obtaining the corresponding random numbers for Y using $Y_i = (X_i - a)^2$ and then saving them in pairs to data file `e4_10.dat`. These are accomplished by the following MATLAB commands at the MATLAB prompt (of P&R) (see Appendix B for an explanation):

```
cd data
a=2;
x=2*rand(100,1)+a-1;
y=(x-a).^2;
z=[x, y];
save e4_10.dat z -ascii
cd ..
```

Alternatively, we may use P&R to generate 100 random numbers $X_i \sim \mathcal{U}(a-1, a+1)$ with $a = 2$; save them as `e4_10x.dat` under `data` subdirectory; and execute the following MATLAB commands at the MATLAB prompt (of P&R):

```
a=2;
cd data
load e4_10x.dat
y=(e4_10x-a).^2;
z=[e4_10x, y];
save e4_10.dat z -ascii
cd ..
```

This latter approach is more convenient if X has a more complex distribution, such as log-normal and Weibull distributions because the generation of random numbers of these distributions is not easy.

S2. Calculate the correlation between the pairs of random numbers. Following the procedure described in Example 4.8, use P&R to obtain the correlation of the data contained in the data file `e4_10.dat`. The sample correlation results, along with the scatter diagram, are shown in Fig. 4.11. Note that the correlation coefficient is indeed approximately zero.

The above procedure can be repeated for $X \sim \mathcal{U}(a, a+1)$ with only one modification: Replace the MATLAB command `x=2*rand(100,1)+a-1` with `x=rand(100,1)+a` or use P&R to generate $X_i \sim \mathcal{U}(a, a+1)$ rather than $X_i \sim \mathcal{U}(a-1, a+1)$. The results of the correlation are shown in Fig. 4.12. Note that the correlation coefficient for this case is 0.966, which is very close to the theoretical value of 0.968.

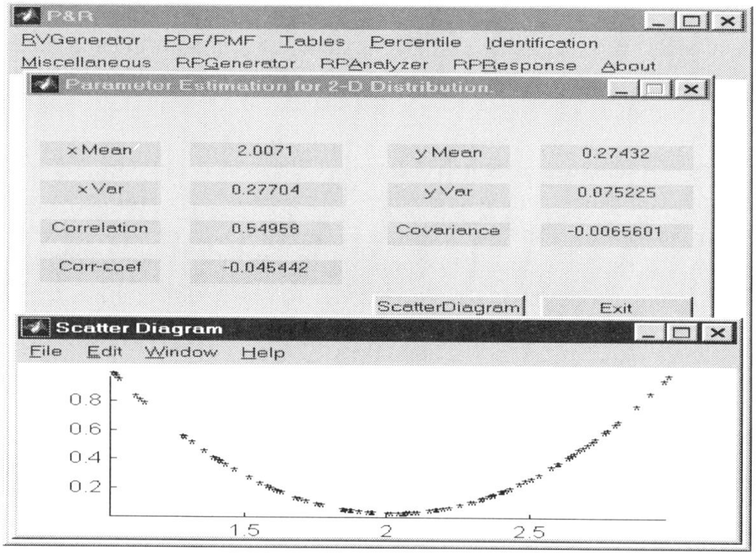

Figure 4.11: Numerical evaluation of correlation of $X \sim \mathcal{U}(a-1, a+1)$ and $(X-a)^2$ by P&R.

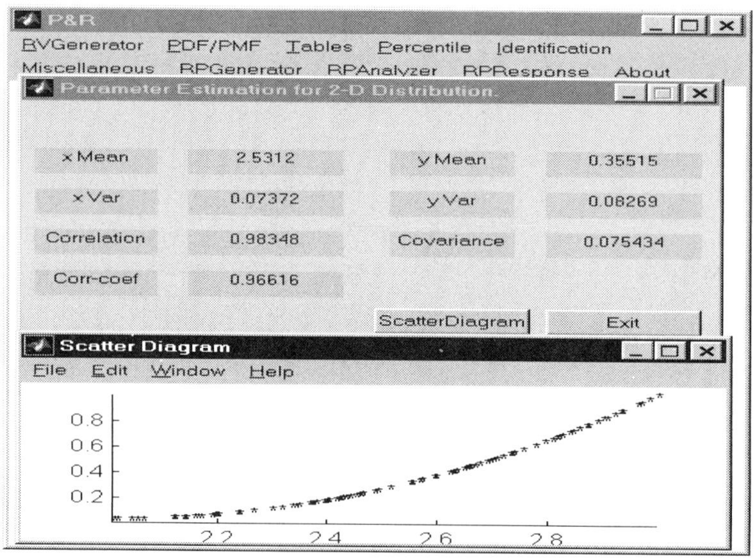

Figure 4.12: Numerical evaluation of correlation of $X \sim \mathcal{U}(a, a+1)$ and $(X-a)^2$ by P&R.

Two RVs X and Y are said to be

- ***uncorrelated*** if

$$E[XY] = E[X]E[Y] \tag{4.26}$$

or equivalently

$$C_{xy} = 0 \qquad \text{or} \qquad \rho_{xy} = 0 \tag{4.27}$$

- ***orthogonal*** if

$$E[XY] = 0$$

- ***independent*** if, for every pair (x, y),

$$F_{X,Y}(x, y) = F_X(x)F_Y(y)$$

or equivalently

$$f_{X,Y}(x, y) = f_X(x)f_Y(y)$$

Note that

$$\text{uncorrelated} \triangleq \text{surely not } \textit{linearly} \text{ related}$$

Clearly, if either X or Y has zero mean, then

$$X \text{ and } Y \text{ are orthogonal} \iff X \text{ and } Y \text{ are uncorrelated}$$

In general,

$$X \text{ and } Y \text{ are independent} \;\underset{\not\Longleftarrow}{\Longrightarrow}\; X \text{ and } Y \text{ are uncorrelated}$$

This can be shown easily:

$$E[XY] = \int_{-\infty}^{\infty}\int_{-\infty}^{\infty} xyf(x,y)dxdy$$
$$E[X]E[Y] = \int_{-\infty}^{\infty} xf(x)dx \int_{-\infty}^{\infty} yf(y)dy = \int_{-\infty}^{\infty}\int_{-\infty}^{\infty} xyf(x)f(y)dxdy$$

Thus

$$f(x, y) = f(x)f(y) \quad \Longrightarrow \quad E[XY] = E[X]E[Y]$$

It clearly makes sense:

$$\text{independence} = \text{no dependence} \Longrightarrow \text{no linear correlation} = \text{uncorrelatedness}$$

In fact, for two functions g and h,

$$\begin{aligned} X, Y \text{ independent} \quad &\Longrightarrow \quad g(X) \text{ and } h(Y) \text{ independent} \\ &\Longrightarrow \quad E[g(X)h(Y)] = E[g(X)]E[h(Y)] \end{aligned} \tag{4.28}$$

The definitions of expectation, correlation, covariance, independent, uncorrelatedness, and orthogonality can be extended to an n-dimensional random vector $\mathbf{X} = [X_1, \ldots, X_n]'$ in a straightforward way as follows. Let $\mathbf{x} = [x_1, \ldots, x_n]'$ denote the value of the random vector $\mathbf{X}$.

The ***mean vector*** of $\mathbf{X}$ is defined as the n-fold integral

$$\bar{\mathbf{x}} \triangleq E[\mathbf{X}] = \int_{-\infty}^{\infty} \cdots \int_{-\infty}^{\infty} x_1 \cdots x_n \, f_{X_1,\ldots,X_n}(x_1, \ldots, x_n) dx_1 \cdots dx_n \overset{\text{shorthand}}{=} \int \mathbf{x} f_{\mathbf{X}}(\mathbf{x}) d\mathbf{x} \tag{4.29}$$

The ***covariance matrix*** of $\mathbf{X}$ is the n-fold integral

$$C_{\mathbf{x}} = \text{cov}(\mathbf{X}) \triangleq E[(\mathbf{X} - \bar{\mathbf{x}})(\mathbf{X} - \bar{\mathbf{x}})'] = \int (\mathbf{x} - \bar{\mathbf{x}})(\mathbf{x} - \bar{\mathbf{x}})' f_{\mathbf{X}}(\mathbf{x}) d\mathbf{x} \tag{4.30}$$

The ***correlation matrix*** of a random n-vector x is the n-fold integral

$$R_{\mathbf{x}} \triangleq E[\mathbf{X}\mathbf{X}'] = \int \mathbf{x}\mathbf{x}' f_{\mathbf{X}}(\mathbf{x}) d\mathbf{x}$$

Covariance and correlation matrices are at least positive semidefinite; that is,

$$\begin{aligned} \mathbf{x}' C_{\mathbf{x}} \mathbf{x} &\geq 0, \quad \forall \mathbf{x} \\ \mathbf{x}' R_{\mathbf{x}} \mathbf{x} &\geq 0, \quad \forall \mathbf{x} \end{aligned}$$

The ***crosscovariance matrix*** of two random vectors $\mathbf{X}$ and $\mathbf{Y}$ is

$$\text{cov}(\mathbf{X}, \mathbf{Y}) \triangleq E[(\mathbf{X} - \bar{\mathbf{x}})(\mathbf{Y} - \bar{\mathbf{y}})'] = \int (\mathbf{x} - \bar{\mathbf{x}})(\mathbf{y} - \bar{\mathbf{y}})' f_{\mathbf{X},\mathbf{Y}}(\mathbf{x}, \mathbf{y}) d\mathbf{x} = E[\mathbf{X}\mathbf{Y}'] - \bar{\mathbf{x}}\bar{\mathbf{y}}'$$

A set of RVs $X_1, \ldots, X_n$, is said to be

- (mutually) ***independent*** if, for every possible point $(x_1, \ldots, x_n)$,

$$F_{X_1,\ldots,X_n}(x_1, \ldots, x_n) = F_{X_1}(x_1) F_{X_2}(x_2) \cdots F_{X_n}(x_n)$$

  or equivalently

$$f_{X_1,\ldots,X_n}(x_1, \ldots, x_n) = f_{X_1}(x_1) f_{X_2}(x_2) \cdots f_{X_n}(x_n)$$

- (mutually) ***uncorrelated*** if

$$E[X_i X_j] = E[X_i] E[X_j] \qquad \forall i \neq j \qquad \text{or} \qquad \text{cov}(\mathbf{X}) \text{ is diagonal}$$

- ***orthogonal*** if

$$E[X_i X_j] = 0 \qquad \forall i \neq j \qquad \text{or} \qquad R_{\mathbf{x}} \text{ is diagonal}$$

These definitions are similar to those for the scalar case, except that the scalars are replaced by vectors and the scalar multiplications are replaced by vector *outer* products.

Example 4.11: Interpretation of Correlation Coefficient

Find the correlation coefficient of $Y = \cos X$ and $Z = \cos(X + \theta)$, where the random phase $X \sim \mathcal{U}(-\pi, \pi)$ and θ is a constant. Are Y and Z uncorrelated, independent? Clearly,

$$f(x) = \begin{cases} \dfrac{1}{2\pi} & -\pi < x \leq \pi \\ 0 & \text{elsewhere} \end{cases}$$

$$\bar{y} = \int_{-\pi}^{\pi} (\cos x)\frac{1}{2\pi}dx = \frac{1}{2\pi}\sin x\Big|_{-\pi}^{\pi} = 0 \qquad \text{(average over a period = 0)}$$

$$\bar{z} = \int_{-\pi}^{\pi} \cos(x+\theta)\frac{1}{2\pi}dx = \frac{1}{2\pi}\sin(x+\theta)\Big|_{-\pi}^{\pi} = 0$$

$$\sigma_y^2 = E[Y^2] - (\bar{y})^2 = \int_{-\pi}^{\pi}(\cos^2 x)\frac{1}{2\pi}dx = \frac{1}{2\pi}\int_{-\pi}^{\pi}\frac{1}{2}(1+\cos 2x)dx$$

$$= \frac{1}{4\pi}\left(x + \frac{1}{2}\sin 2x\right)\Big|_{-\pi}^{\pi} = \frac{1}{2}$$

$$\sigma_z^2 = E[Z^2] - (\bar{z})^2 = \int_{-\pi}^{\pi}\cos^2(x+\theta)\frac{1}{2\pi}dx = \frac{1}{2\pi}\int_{-\pi}^{\pi}\frac{1}{2}[1 - \cos 2(x+\theta)]dx$$

$$= \frac{1}{4\pi}\left[x + \frac{1}{2}\sin 2(x+\theta)\right]_{-\pi}^{\pi} = \frac{1}{2}$$

$$C_{yz} = E[YZ] - \bar{y}\bar{z} = E[YZ] = \int_{-\pi}^{\pi}\cos x\cos(x+\theta)\frac{1}{2\pi}dx$$

$$= \int_{-\pi}^{\pi}\frac{\cos\theta + \cos(2x+\theta)}{4\pi}dx = \frac{1}{4\pi}\left[2\pi\cos\theta + \frac{\sin(2x+\theta)}{2}\right]_{-\pi}^{\pi} = \frac{1}{2}\cos\theta$$

$$\rho_{yz} = \frac{C_{yz}}{\sigma_y\sigma_z} = \frac{(1/2)\cos\theta}{\sqrt{(1/2)(1/2)}} = \cos\theta$$

Thus,

- $\theta = 0 \implies \rho = 1 \implies Y = Z$ (perfect linear relation).
- $\theta = \pi \implies \rho = -1 \implies Y = -Z$ (perfect linear relation).
- $\theta = \pm\pi/2 \implies \rho = 0 \implies Y, Z$ are uncorrelated, meaning that Y and Z are (almost) surely not related *linearly*. In fact, they are related *nonlinearly*: $Y^2 + Z^2 = \cos^2 X + \sin^2 X = 1$, and thus are dependent. This makes sense since Y and Z are related through the same RV X.
- For other θ, $0 < |\rho| < 1 \implies Y, Z$ are correlated but not perfectly.

See also Example 4.29 and problems 4.20 and 4.30.

4.6 Mean and Variance of Weighted Sum of RVs

The mean of a weighted sum of two RVs is equal to the weighted sum of the individual means:

$$\begin{aligned}
E[aX+bY] &= \int_{-\infty}^{\infty}\int_{-\infty}^{\infty}(ax+by)f(x,y)dxdy \\
&= \int_{-\infty}^{\infty}\int_{-\infty}^{\infty}axf(x,y)dxdy + \int_{-\infty}^{\infty}\int_{-\infty}^{\infty}byf(x,y)dxdy \\
&= \int_{x=-\infty}^{\infty}ax\int_{y=-\infty}^{\infty}f(x,y)dydx + \int_{-\infty}^{\infty}by\int_{x=-\infty}^{\infty}f(x,y)dxdy \\
&= a\int_{-\infty}^{\infty}xf_X(x)dx + b\int_{-\infty}^{\infty}yf_Y(y)dy \\
&= aE[X]+bE[Y]
\end{aligned}$$

This is in fact the same as (3.47). This can be easily extended to a weighted sum of n RVs:

mean of weighted sum of RVs = the weighted sum of means

$$\boxed{E\Big[\sum_{i=1}^{n} a_i X_i\Big] = a_1E[X_1] + a_2E[X_2] + \cdots + a_nE[X_n]} \tag{4.31}$$

The variance of a weighted sum of two RVs can be obtained as

$$\begin{aligned}
\text{var}[aX+bY] &= E\left[[aX+bY-(a\bar{x}+b\bar{y})]^2\right] \\
&= E\left[[a(X-\bar{x})+b(Y-\bar{y})]^2\right] \\
&= E\left[a^2(X-\bar{x})^2 + b^2(Y-\bar{y})^2 + 2ab(X-\bar{x})(Y-\bar{y})\right] \\
&= a^2\sigma_x^2 + b^2\sigma_y^2 + 2abC_{xy}
\end{aligned}$$

Thus,

$$\text{var}(aX+bY) \overset{\text{if } X,Y \text{ uncorrelated}}{=} a^2\text{var}(X) + b^2\text{var}(Y)$$

This can also be extended to a weighted sum of *uncorrelated* RVs:

variance of weighted sum of RVs = sum of variances with weight squared

$$\boxed{\text{var}\Big(\sum_{i=1}^{n} a_i X_i\Big) \overset{\text{if uncorrelated}}{=} a_1^2\text{var}(X_1) + a_2^2\text{var}(X_2) + \cdots + a_n^2\text{var}(X_n)} \tag{4.32}$$

(4.31) can be shown easily as follows:

$$\begin{aligned}
E\left[\sum_{i=1}^{n} a_i X_i\right] &= E\left[a_1 X_1 + \sum_{i=2}^{n} a_i X_i\right] \\
&= a_1 E[X_1] + E\left[\sum_{i=2}^{n} a_i X_i\right] \\
&= a_1 E[X_1] + E\left[a_2 X_2 + \sum_{i=3}^{n} a_i X_i\right] \\
&= a_1 E[X_1] + a_2 E[X_2] + \cdots + a_n E[X_n]
\end{aligned}$$

Likewise, (4.32) can be shown. In general, if an expression is true for two arguments, then it will be true for n arguments provided the conditions for the n arguments are really the generalization of that for the two arguments.

In general, the following holds no matter X_i and X_j are uncorrelated or not:

$$\begin{aligned}
\text{var}\left(\sum_{i=1}^{n} a_i X_i\right) &= \sum_{i=1}^{n}\sum_{j=1}^{n} a_i a_j \text{cov}(X_i, X_j) \\
&= \sum_{i=1}^{n} a_i^2 \text{var}(X_i) + 2\sum_{i=1}^{n}\sum_{j>i} a_i a_j \text{cov}(X_i, X_j)
\end{aligned}$$

which reduces to (4.32) if X_i and X_j are uncorrelated for every $i \neq j$.

Example 4.12: Mean and Variance of Chi-Square Distribution

The PDF of the chi-square distribution is defined by (3.32). Its mean and variance can be found as follows. Note that an n-degrees of freedom chi-square RV X is the sum of the squares of independent zero-mean Gaussian RVs X_i's with a common variance σ^2: $X = X_1^2 + X_2^2 + \cdots + X_n^2$, where $X_i \sim \mathcal{N}(0, \sigma^2)$. Thus the mean is

$$\begin{aligned}
E[X] &= E[X_1^2 + X_2^2 + \cdots + X_n^2] \\
&= E[X_1^2] + E[X_2^2] + \cdots + E[X_n^2] \\
&= [\text{var}(X_1) + (\bar{x}_1)^2] + \cdots + [\text{var}(X_n) + (\bar{x}_n)^2] \\
&= \sigma^2 + \cdots + \sigma^2 = n\sigma^2
\end{aligned}$$

Since the independence of $X_1, X_2, \ldots, X_n$ implies that they are uncorrelated, the variance is

$$\begin{aligned}
\text{var}(X) &= \text{var}(X_1^2 + X_2^2 + \cdots + X_n^2) \\
&= \text{var}(X_1^2) + \text{var}(X_2^2) + \cdots + \text{var}(X_n^2) \\
&= \{E[(X_1^2)^2] - (E[X_1^2])^2\} + \cdots + \{E[(X_n^2)^2] - (E[X_n^2])^2\} \\
&\overset{(3.45)}{=} [3\sigma^4 - \sigma^4] + \cdots + [3\sigma^4 - \sigma^4] \\
&= 2n\sigma^4
\end{aligned}$$

Example 4.13: Moments of Sum and Difference

Consider Example 4.6 again. Let $U = X + Y$ and $V = X - 2Y$.

(a) Find $E[U]$ and $E[V]$:

$$E[U] = E[X + Y] = E[X] + E[Y] = 3 + 1 = 4$$
$$E[V] = E[X - 2Y] = E[X] - 2E[Y] = 3 - 2 \times 1 = 1$$

(b) Find σ_u^2 and σ_v^2:

$$E[X^2] = \int_{-\infty}^{\infty}\int_{-\infty}^{\infty} x^2 f(x,y)dydx = \frac{1}{12}\int_0^6 x^2 \int_0^2 dydx = 12$$
$$E[Y^2] = \int_{-\infty}^{\infty}\int_{-\infty}^{\infty} y^2 f(x,y)dydx = \frac{1}{12}\int_0^6\int_0^2 y^2 dydx = \frac{4}{3}$$
$$E[U^2] = E[(X+Y)^2] = E[X^2] + 2R_{xy} + E[Y^2]$$
$$= 12 + 2 \times 3 + \frac{4}{3} = 19\frac{1}{3}$$
$$E[V^2] = E[(X-2Y)^2] = E[X^2] - 4R_{xy} + 4E[Y^2]$$
$$= 12 - 4 \times 3 + 4 \times \frac{4}{3} = \frac{16}{3}$$
$$\sigma_u^2 = E[U^2] - (E[U])^2 = 19\frac{1}{3} - (4)^2 = \frac{10}{3}$$
$$\sigma_v^2 = E[V^2] - (E[V])^2 = 16/3 - 1^2 = 13/3$$

Since X and Y are uncorrelated, a better alternative is to use (4.32), which leads to, noting that $\sigma_x^2 = E[X^2] - (\bar{x})^2$ and $\sigma_y^2 = E[Y^2] - (\bar{y})^2$,

$$\sigma_u^2 = (1)^2 \cdot (12 - 3^2) + 1^2 \cdot \left(\frac{4}{3} - 1^2\right) = \frac{10}{3}$$
$$\sigma_v^2 = (1)^2 \cdot (12 - 3^2) + (-2)^2 \cdot \left(\frac{4}{3} - 1^2\right) = 13/3$$

(c) Find the $(1,1)$th moment R_{uv} and the $(1,1)$th central moment C_{uv}:

$$R_{uv} = E[UV] = E[(X+Y)(X-2Y)]$$
$$= E[X^2] - 2R_{xy} + R_{xy} - 2E[Y^2] = 12 - 3 - 2 \times \frac{4}{3} = 19/3$$
$$C_{uv} = E[(U - \bar{u})(V - \bar{v})] = E[UV + \bar{u}\bar{v} - U\bar{v} - V\bar{u}]$$
$$= R_{uv} - E[U]E[V] = 19/3 - (4)(1) = 7/3$$

4.7 Jointly Gaussian Random Variables

Two RVs X and Y are said to be ***jointly Gaussian*** or ***jointly normal*** if their joint PDF is given by

$$f(x,y) = \frac{1}{2\pi\sigma_x\sigma_y\sqrt{1-\rho^2}} e^{-\frac{1}{2(1-\rho^2)}\left[\frac{(x-\bar{x})^2}{\sigma_x^2} + \frac{(y-\bar{y})^2}{\sigma_y^2} - \frac{2\rho(x-\bar{x})(y-\bar{y})}{\sigma_x\sigma_y}\right]} \tag{4.33}$$

where $\bar{x}, \bar{y}, \sigma_x^2, \sigma_y^2$ and ρ are the mean, variances and correlation coefficient of X and Y. This is sometimes denoted by the shorthand

$$(X, Y) \sim \mathcal{N}(\bar{x}, \bar{y}; \sigma_x^2, \sigma_y^2, \rho)$$

Gaussian RVs have many nice and important properties. For example,

- If (X, Y) are jointly Gaussian, then X and Y are both Gaussian RVs.
- For *jointly* Gaussian RVs X, Y, their weighted sum $aX + bY$ is another Gaussian RV. Specifically,

$$\left.\begin{array}{r} X \sim \mathcal{N}(\bar{x}, \sigma_x^2) \\ Y \sim \mathcal{N}(\bar{y}, \sigma_y^2) \\ X,\ Y \text{ independent} \end{array}\right\} \implies \begin{array}{l} aX + bY + c \sim \mathcal{N}(a\bar{x} + b\bar{y} + c, a^2\sigma_x^2 + b^2\sigma_y^2) \\ \text{for all } a, b \text{ and } c \end{array} \tag{4.34}$$

  Also, $Z = aX + bY + c$ and $W = \alpha X + \beta Y + \gamma$ are jointly Gaussian.
- Two *jointly* Gaussian RVs are independent if and only if they are uncorrelated. This is clear since $(\bar{x}, \bar{y}; \sigma_x^2, \sigma_y^2, \rho)|_{\rho=0} = \mathcal{N}(\bar{x}, \sigma_x^2)\mathcal{N}(\bar{y}, \sigma_y^2)$. It is, however, not true if Gaussian RVs X, Y are not *jointly* Gaussian.

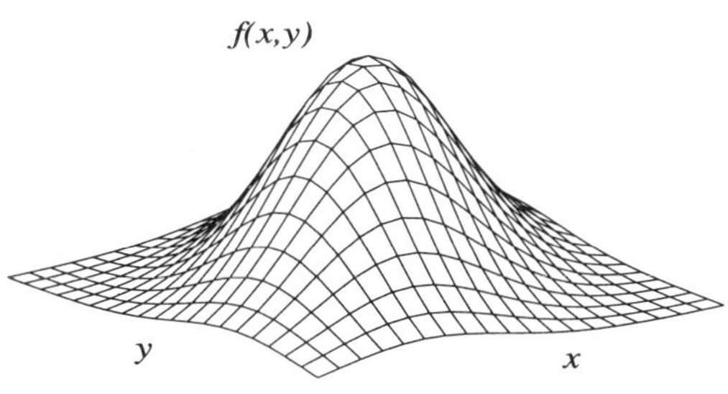

Figure 4.13: The PDF of jointly Gaussian RVs.

Fig. 4.13 corresponds to $\rho = 0.5$. Fig. 4.14 shows three Gaussian PDF surfaces corresponding to $\rho = 0, \pm 0.9$, respectively.

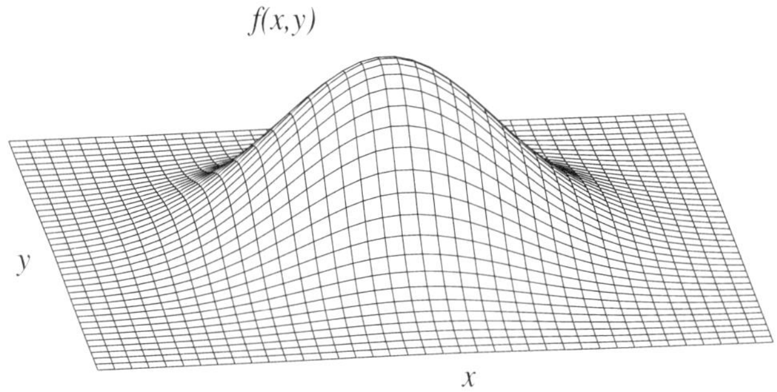

(a) $\rho = 0$

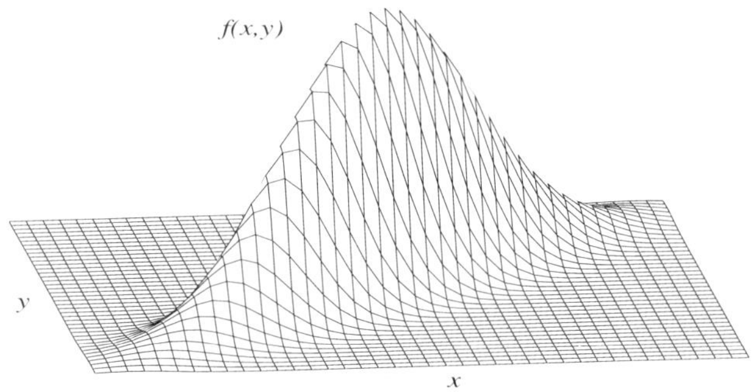

(b) $\rho = 0.9$

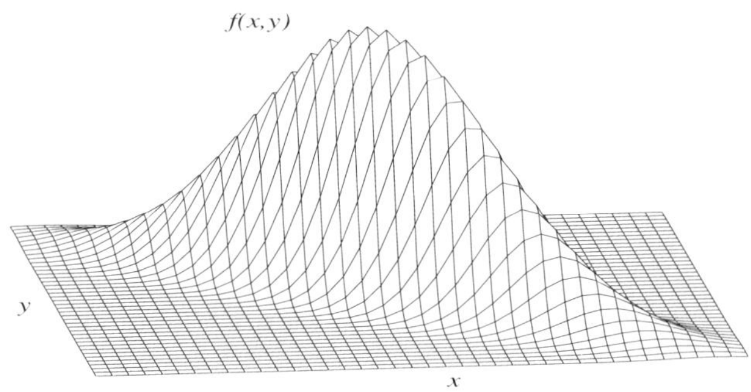

(c) $\rho = -0.9$

Figure 4.14: Illustration of the joint Gaussian PDF with various correlation coefficients.

The ***joint Gaussian (normal) distribution*** is the most popular and important joint distribution.

Note that

- Not every pair of Gaussian RVs (X, Y) is *jointly* Gaussian.
- Every pair of *independent* Gaussian RVs (X, Y) is *jointly* Gaussian.
- Marginal PDFs of X, Y do not depend on the correlation coefficient ρ.

Example 4.14: Dependence of Weighted Sums of Gaussian RVs

Consider two independent RVs $X \sim \mathcal{N}(\bar{x}, \sigma_x^2)$ and $Y \sim \mathcal{N}(\bar{y}, \sigma_y^2)$. Let $U = aX + bY$ and $V = cX + dY$, where $\bar{x}, \sigma_x^2, \bar{y}, \sigma_y^2, a, b, c, d$ are known constants.

(a) Find the means and variances of U and V: Similar to Example 4.31,

$$\bar{u} = a\bar{x} + b\bar{y}, \qquad \bar{v} = c\bar{x} + d\bar{y}$$

$$\text{var}(U) = \text{var}(aX + bY) \overset{(4.32)}{=} a^2\text{var}(X) + b^2\text{var}(Y) = a^2\sigma_x^2 + b^2\sigma_y^2$$

$$\text{var}(V) = c^2\sigma_x^2 + d^2\sigma_y^2$$

(b) Find the covariance, correlation, and correlation coefficient of U and V:

$$\begin{aligned}
C_{uv} &\overset{(4.49)}{=} ac\sigma_x^2 + bd\sigma_y^2 + (ad + bc)C_{xy} \\
&= ac\sigma_x^2 + bd\sigma_y^2 + 0 \qquad \text{(because } X, Y \text{ are independent)} \\
R_{uv} &\overset{(4.22)}{=} C_{uv} + \bar{u}\bar{v} = ac\sigma_x^2 + bd\sigma_y^2 + (a\bar{x} + b\bar{y})(c\bar{x} + d\bar{y}) \\
&= ac(\sigma_x^2 + \bar{x}^2) + bd(\sigma_y^2 + \bar{y}^2) + (ad + bc)\bar{x}\bar{y} \\
\rho_{uv} &= \frac{C_{uv}}{\sigma_u\sigma_v} = \frac{ac\sigma_x^2 + bd\sigma_y^2}{\sqrt{(a^2\sigma_x^2 + b^2\sigma_y^2)(c^2\sigma_x^2 + d^2\sigma_y^2)}}
\end{aligned}$$

(c) Find the conditions under which U and V are uncorrelated and the conditions under which U and V are independent:

Since U and V are weighted sum of Gaussian RVs, they are jointly Gaussian. Thus they are independent if and only if they are uncorrelated (i.e., $C_{uv} = 0$); that is, if and only if $\sigma_x^2, \sigma_y^2, a, b, c, d$ satisfy

$$ac\sigma_x^2 + bd\sigma_y^2 = 0$$

If X and Y are uncorrelated, from Example 4.31, the condition for uncorrelatedness of U and V is (4.50).

The loci of constant values of the Gaussian PDF are ellipses since the following describes an ellipse

$$\frac{(x-\bar{x})^2}{\sigma_x^2} + \frac{(y-\bar{y})^2}{\sigma_y^2} - \frac{2\rho(x-\bar{x})(y-\bar{y})}{\sigma_x \sigma_y} = r^2$$

A ***vector-valued Gaussian RV*** or a ***Gaussian vector*** **X** is one with the following joint PDF

$$f(\mathbf{x}) = \frac{1}{|2\pi P|^{1/2}} e^{-\frac{1}{2}(\mathbf{x}-\bar{\mathbf{x}})' P^{-1}(\mathbf{x}-\bar{\mathbf{x}})} \tag{4.35}$$

where $\bar{\mathbf{x}}$ and P are the mean vector and covariance matrix of $\mathbf{X}$, defined by (4.29) and (4.30). (4.35) is sometimes denoted compactly as

$$f(\mathbf{x}) = \mathcal{N}(\mathbf{x}; \bar{\mathbf{x}}, P) \qquad \text{or} \qquad \mathbf{X} \sim \mathcal{N}(\bar{\mathbf{x}}, P)$$

Components of a Gaussian vector are said to be ***jointly Gaussian***.

As an extension of two-dimensional case, an n-dimensional Gaussian vector also has many nice properties. For example,

- The sum of many "uniformly" small and negligible random effects tends to be Gaussian — central limit theorem.
- The weighted sums of jointly Gaussian RVs are jointly Gaussian: $Y_1, \ldots, Y_m$ as defined by

  $$Y_j = a_{1j}X_1 + a_{2j}X_2 + \cdots + a_{1j}X_n + b_j \qquad j = 1, 2, \ldots, m$$

  are jointly Gaussian. In matrix-vector language, that is: For every matrix $A = [a_{ij}]$ and vector $\mathbf{b} = [b_1, \ldots, b_m]'$ of suitable dimensions,

  $$\mathbf{X} \sim \mathcal{N}(\bar{\mathbf{x}}, P) \Longrightarrow A\mathbf{X} + \mathbf{b} \sim \mathcal{N}(A\bar{\mathbf{x}} + \mathbf{b}, APA') \tag{4.36}$$

  (4.34) is actually a special case of (4.36).
- Independence $\Longleftrightarrow$ uncorrelatedness $\Longleftrightarrow$ $\text{cov}(\mathbf{X}) = \text{diagonal}$
- It is fully characterized by its mean vector and covariance matrix.
- Joint conditional density of jointly Gaussian RVs (i.e., some components of a Gaussian vector) given the values of some other components is also Gaussian.

Example 4.15: Jointly Gaussian vs. Marginal Gaussian

A well-known example demonstrating that *not every pair of Gaussian RVs X and Y is jointly Gaussian* is the one with the following PDF:

$$f(x, y) = \frac{1}{2\pi} e^{-\frac{x^2+y^2}{2}} (1 + \sin x \sin y)$$

Clearly, (X, Y) is not jointly Gaussian. But both X and Y are Gaussian RVs since clearly

$$f_X(x) = \frac{1}{\sqrt{2\pi}} e^{-\frac{x^2}{2}}, \qquad f_Y(y) = \frac{1}{\sqrt{2\pi}} e^{-\frac{y^2}{2}}$$

Example 4.16: Generation of Jointly Gaussian Random Numbers by P&R

Jointly Gaussian random numbers can be generated easily using the companion software P&R. The procedure is illustrated as follows:

S1. Click "RVGenerator" in the main window of P&R.
S2. Click "Normal." The "**Normal Random Number Generator**" window will appear.
S3. Click "2" for "dimension" and enter the parameters as shown in Fig. 4.15. Click "Ok." 20 random numbers that are jointly Gaussian distributed with parameters $(\bar{x}, \bar{y}, \sigma_x^2, \sigma_y^2, \rho) = (1.2, 0.7, 0.66, 1.48, 0.799)$ are then generated (and saved to a data file if specified in a two-column format). A scatter diagram of the random numbers is then plotted as shown in Fig. 4.15.

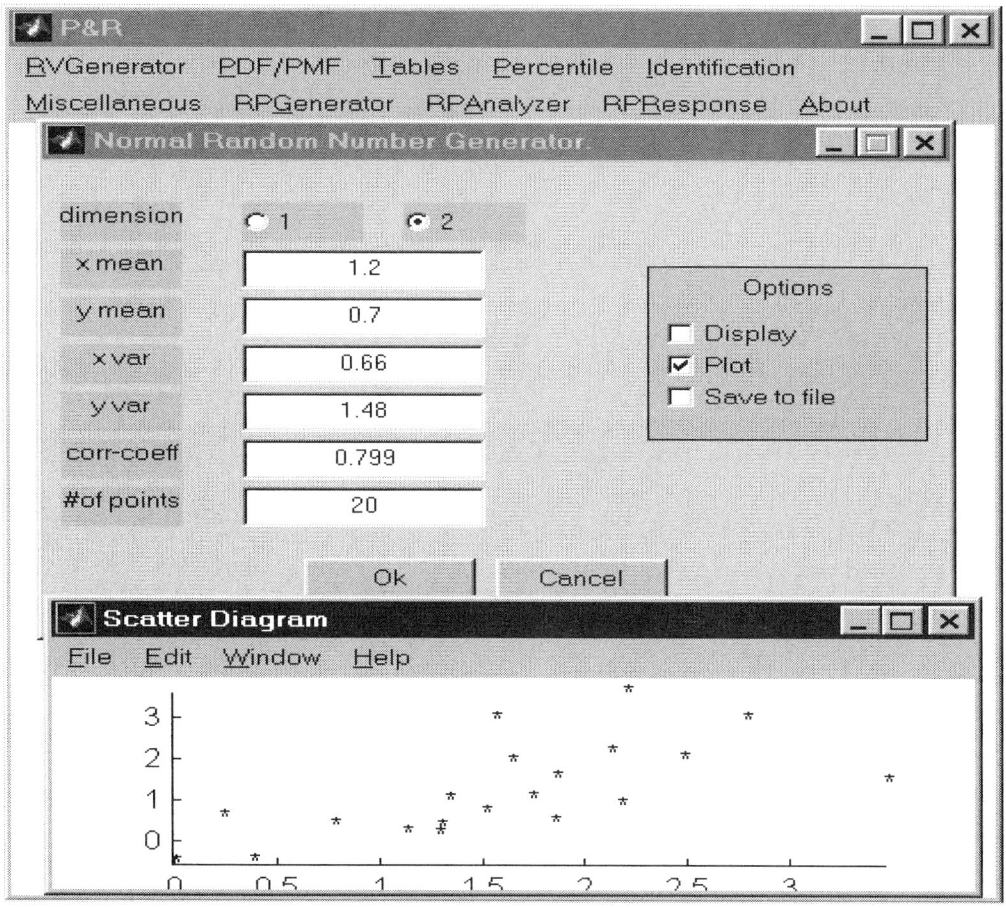

Figure 4.15: Generation of jointly Gaussian random numbers by P&R.

4.8 Functions of Random Variables

4.8.1 PDF of Sum of Random Variables

The PDF of the ***sum of two RVs*** $Z = X + Y$ is in general given by

$$f_Z(z) = \int_{-\infty}^{\infty} f_{X,Y}(z-y, y)dy = \int_{-\infty}^{\infty} f_{X,Y}(x, z-x)dx$$

If X, Y are independent, then it reduces to the ***convolution*** $f_X * f_Y$:

$$f_Z(z) = f_X(z) * f_Y(z) = \int_{-\infty}^{\infty} f_X(z-y)f_Y(y)dy = \int_{-\infty}^{\infty} f_X(x)f_Y(z-x)dx \tag{4.37}$$

Example 4.17: Sum of Independent Gaussian RVs

Given independent RVs $X \sim \mathcal{N}(0,1)$ and $Y \sim \mathcal{N}(0,1)$, the PDF of the sum $Z = X + Y$ is

$$f_Z(z) \overset{(4.37)}{=} \int_{-\infty}^{\infty} f_X(z-y)f_Y(y)dy = \int_{-\infty}^{\infty} \frac{1}{\sqrt{2\pi}}e^{-(z-y)^2/2}\frac{1}{\sqrt{2\pi}}e^{-y^2/2}dy$$

Note that

$$\frac{(z-y)^2}{2} + \frac{y^2}{2} = \frac{2y^2 - 2zy + z^2}{2} = \frac{2(y-z/2)^2 + z^2/2}{2} = \frac{(y-z/2)^2}{2 \times 0.5} + \frac{z^2}{4}$$

Then

$$\begin{aligned} f_Z(z) &= \frac{1}{\sqrt{2\pi}}e^{-z^2/4}\int_{-\infty}^{\infty}\frac{1}{\sqrt{2\pi}}e^{-(y-z/2)^2/(2\times 0.5)}dy \\ &= \frac{1}{\sqrt{2\pi}\sqrt{2}}e^{-z^2/(2\times 2)}\int_{-\infty}^{\infty}\frac{1}{\sqrt{2\pi}\sqrt{0.5}}e^{-(y-z/2)^2/(2\times 0.5)}dy \\ &\overset{(3.23)}{=} \mathcal{N}(z;0,2)\int_{-\infty}^{\infty}\mathcal{N}(y;z/2,0.5)dy \overset{(3.19)}{=} \mathcal{N}(z;0,2) \end{aligned}$$

Thus, symbolically, $\mathcal{N}(0,1) + \mathcal{N}(0,1) = Z \sim \mathcal{N}(0,2)$.

In general, it can be shown that the weighted sum $Z = \sum_{i=1}^n a_i X_i$ of *independent* Gaussian RVs $X_1 \sim \mathcal{N}(\bar{x}_1, \sigma_{x_1}^2), \ldots, X_n \sim \mathcal{N}(\bar{x}_n, \sigma_{x_n}^2)$ is a Gaussian RV with mean $\sum_{i=1}^n a_i\bar{x}_i$ and variance $\sum_{i=1}^n a_i^2\sigma_{x_i}^2$; that is, symbolically:

$$\sum_{i=1}^n a_i\mathcal{N}(\bar{x}_i, \sigma_{x_i}^2) = Z \sim \mathcal{N}\Big(\sum_{i=1}^n a_i\bar{x}_i, \sum_{i=1}^n a_i^2\sigma_{x_i}^2\Big) \tag{4.38}$$

For discrete (X,Y), the PMF of a function of two RVs $Z = g(X,Y)$ is equal to the sum of the point masses on the curve $g(x,y) = z$:

$$P\{Z = z_k\} = \sum_{g(x_i,y_j)=z_k} P\{X = x_i, Y = y_j\} \overset{\text{if } X,Y \text{ independent}}{=} \sum_{g(x_i,y_j)=z_k} P\{X = x_i\}P\{Y - y_j\}$$

This can clearly be extended to the general case $Z = g(X_1,\ldots,X_n)$. In particular, if $Z = X+Y$, then

$$\begin{aligned}
P\{Z = z_k\} &= \sum_{x_i+y_j=z_k} P\{X = x_i, Y = y_j\} \\
&\overset{\text{if } X,Y \text{ independent}}{=} \sum_{x_i+y_j=z_k} P\{X - x_i\}P\{Y = y_j\} \\
&= \sum_{x_i} P\{X = x_i\}P\{Y = z_k - x_i\} \\
&= \sum_{y_j} P\{X = z_k - y_j\}P\{Y = y_j\}
\end{aligned}$$

where $\sum_{x_i}$ denotes "sum over all points x_i's where X has a nonzero point mass." The last three summations are known as ***convolution sum***.

The study of the sum of RVs is very useful with many applications. For example, we often are facing the problem of filtering out the noise in the signal-plus-noise case.

Similarly to (4.37), the PDF of the ***sum of* n *independent RVs*** $X_1,\ldots,X_n$ is simply the ***convolution*** of the PDFs of $X_1,\ldots,X_n$:

$$f_{X_1+\cdots+X_n}(x) = f_{X_1}(x) * f_{Y_2}(x) = f_{X_1}(x) * f_{X_2}(x) * f_{Y_3}(x) = f_{X_1}(x) * \cdots * f_{X_n}(x)$$

where

$$\begin{aligned}
Y_1 &= X_1 + X_2 + \cdots + X_n \\
&= X_1 + \underbrace{(X_2 + \cdots + X_n)}_{Y_2} \\
&= X_1 + X_2 + \underbrace{(X_3 + \cdots + X_n)}_{Y_3} = \cdots
\end{aligned}$$

One of the most popular probability laws deals with the sum of a large number of RVs — the ***central limit theorem***. It states that under some fairly nonrestrictive conditions, the sum of a large number of "negligible" random effects that are "uniformly small," meaning that there are no dominant ones, is a Gaussian RV.

Although some versions of the central limit theorem are applicable to sum of dependent RVs, a sum of independent RVs is easier to handle and converges to Gaussian more quickly.

A powerful version of the central limit theorem states that the properly normalized sum $\hat{X} = \sum_{i=1}^{n} \frac{X_i - \bar{x}_i}{\sigma(n)}$ of independent RVs $X_1,\ldots,X_n,\ldots$ with mean $\bar{x}_i$ and variance σ_i^2, where $\sigma^2(n) = \sum_{i=1}^{n} \sigma_i^2$, tends to be a standard Gaussian RV: $(\lim_{n\to\infty} \hat{X}) \sim \mathcal{N}(0,1)$, if the random terms are "uniformly small" (i.e., no one stands out) and the sum is over a large number of terms that are sufficiently small to be "negligible."

4.8.2 Extrema of Random Variables

Given two independent RVs X and Y with CDFs $F_X(x)$ and $F_Y(y)$, what are the distributions of $U = \max(X, Y)$ and $V = \min(X, Y)$?

Note that

$$\max(X, Y) \le u \Longleftrightarrow (X \le u) \cap (Y \le u)$$

Thus

$$\begin{aligned} F_{\max}(u) &\triangleq P\{\max(X, Y) \le u\} = P\{X \le u, Y \le u\} \\ &= F_{X,Y}(u, u) \overset{X,Y \text{ independent}}{=} F_X(u)F_Y(u) \end{aligned}$$

Note that the same value is used in $F_X(\cdot)$ and $F_Y(\cdot)$ and that $F_{\max}(u)$ is not larger than the smaller of $F_X(u)$ and $F_Y(u)$, which makes sense.

For the minimum $V = \min(X, Y)$, note that

$$\min(X, Y) \le v \not\Longleftrightarrow (X \le v) \cap (Y \le v)$$

but

$$\min(X, Y) > v \Longleftrightarrow (X > v) \cap (Y > v)$$

Thus, the CDF of $V = \min(X, Y)$ is

$$\begin{aligned} F_{\min}(v) &\triangleq P\{\min(X, Y) \le v\} = 1 - P\{\min(X, Y) > v\} \\ &= 1 - P\{X > v, Y > v\} \overset{X,Y \text{ independent}}{=} 1 - P\{X > v\}P\{Y > v\} \\ &= 1 - [1 - F_X(v)][1 - F_Y(v)] \end{aligned}$$

The above results can be easily extended to the case of n independent RVs. Clearly, the CDFs of $U = \max(X_1, \ldots, X_n)$ and $V = \min(X_1, \ldots, X_n)$ are

$$\begin{aligned} F_{\max}(u) &\triangleq P\{\max(X_1, \ldots, X_n) \le u\} \\ &= F_{X_1}(u)F_{X_2}(u) \cdots F_{X_n}(u) \qquad (4.39) \\ F_{\min}(v) &\triangleq P\{\min(X_1, \ldots, X_n) \le v\} \\ &= 1 - [1 - F_{X_1}(v)][1 - F_{X_2}(v)] \cdots [1 - F_{X_n}(v)] \qquad (4.40) \end{aligned}$$

These results are very useful in order statistics, where RVs are ranked.

Example 4.18: System Reliability

A system S consists of two independent subsystems S_1 and S_2 with one of the following connections: (a) in series; (b) in parallel; and (c) a subsystem in operation while the other in standby. The service times (i.e., times to failure) of the subsystems S_1 and S_2 are exponential RVs X and Y with CDFs

$$F_X(x) = \begin{Bmatrix} 1 - e^{-ax} & x \geq 0 \\ 0 & x < 0 \end{Bmatrix}, \qquad F_Y(y) = \begin{Bmatrix} 1 - e^{-by} & y \geq 0 \\ 0 & y < 0 \end{Bmatrix}$$

with $a > 0, b > 0$. Find the time to failure Z of the system S.

(a) Connected in series. In this case, $Z = \min(X, Y)$. Hence,

$$F_Z(z) \overset{(4.40)}{=} 1 - [1 - F_X(z)][1 - F_Y(z)] = \begin{cases} 1 - e^{-(a+b)z} & z \geq 0 \\ 0 & z < 0 \end{cases}$$

$$f_Z(z) = \frac{dF_Z(z)}{dz} = \begin{cases} (a+b)e^{-(a+b)z} & z > 0 \\ 0 & z < 0 \end{cases}$$

Thus, Z is exponentially distributed with parameter $a + b$ in this case.

(b) Connected in parallel. In this case, $Z = \max(X, Y)$. Hence,

$$F_Z(z) \overset{(4.39)}{=} F_X(z)F_Y(z) = \begin{cases} (1 - e^{-az})(1 - e^{-bz}) & z \geq 0 \\ 0 & z < 0 \end{cases}$$

$$f_Z(z) = \frac{dF_Z(z)}{dz} = \begin{cases} ae^{-az} + be^{-bz} - (a+b)e^{-(a+b)z} & z > 0 \\ 0 & z < 0 \end{cases}$$

There is no special name for such a RV Z.

(c) In a standby connection, the second subsystem is put into operation at the moment the first subsystem fails, which means $Z = X + Y$. Note that the PDFs of the exponential RVs X and Y are

$$f_X(x) = \frac{dF_X(x)}{dx} = \begin{Bmatrix} ae^{-ax} & x > 0 \\ 0 & x < 0 \end{Bmatrix}, \qquad f_Y(y) = \begin{Bmatrix} be^{-by} & y > 0 \\ 0 & y < 0 \end{Bmatrix}$$

Hence, $f_Z(z) = 0$ for $z \leq 0$, and for $z > 0$,

$$\begin{aligned} f_Z(z) &\overset{(4.37)}{=} \int_{-\infty}^{\infty} f_X(z-y)f_Y(y)dy = \int_0^z ae^{-a(z-y)}be^{-by}dy \\ &= abe^{-az}\int_0^z e^{(a-b)y}dy = \begin{cases} \frac{ab}{a-b}[e^{-bz} - e^{-az}] & a \neq b \\ a^2ze^{-az} & a = b \end{cases} \end{aligned}$$

4.8.3 Joint PDF of Functions of a Random Vector

Consider m functions of n RVs $X_1, \ldots, X_n$:

$$\begin{aligned} Y_1 &= g_1(X_1, \ldots, X_n) \\ Y_2 &= g_2(X_1, \ldots, X_n) \\ &\vdots \\ Y_m &= g_m(X_1, \ldots, X_n) \end{aligned}$$

If $m > n$, then $Y_{n+1}, \ldots, Y_m$ can be expressed in terms of $Y_1, \ldots, Y_n$ and thus their joint PDF can be expressed in terms of the joint PDF of $Y_1, \ldots, Y_n$. If $m < n$, then we could find the joint PDF of $Y_1, \ldots, Y_m, X_{m+1}, \ldots, X_n$ first and then use (4.14) to determine the joint PDF of $Y_1, \ldots, Y_m$. Therefore, we may reasonably assume $n = m$.

Let $\mathbf{Y} = [Y_1, \ldots, Y_n]'$, $\mathbf{X} = [X_1, \ldots, X_n]'$. Then as for the single RV case (p. 111), depending on the solution of the following system of equations

$$\begin{cases} y_1 &= g_1(x_1, \ldots, x_n) \\ y_2 &= g_2(x_1, \ldots, x_n) \\ &\vdots \\ y_n &= g_n(x_1, \ldots, x_n) \end{cases} \tag{4.41}$$

the joint PDF of $\mathbf{Y}$ is given by

$$f_{\mathbf{Y}}(y_1, \ldots, y_n) = \begin{cases} 0 & \text{if (4.41) has no solution} \\ \dfrac{f_{\mathbf{X}}(x_1, \ldots, x_n)}{|J(x_1, \ldots, x_n)|} & \text{if (4.41) has unique solution} \\ \displaystyle\sum_i \dfrac{f_{\mathbf{X}}(x_1^i, \ldots, x_n^i)}{|J(x_1^i, \ldots, x_n^i)|} & \text{if (4.41) has more than one solution} \end{cases}$$

where $(x_1^i, \ldots, x_n^i)$ is the ith solution of (4.41) and the sum is over all solutions; and the Jacobian of the transformation (4.41) is given by the determinant

$$J(x_1, \ldots, x_n) = \begin{vmatrix} \frac{\partial g_1}{\partial x_1} & \cdots & \frac{\partial g_1}{\partial x_n} \\ \vdots & \ddots & \vdots \\ \frac{\partial g_n}{\partial x_1} & \cdots & \frac{\partial g_n}{\partial x_n} \end{vmatrix}$$

Example 4.19: Independence of Functions of Random Variables

Example 4.31 shows that two *linear* combinations of two correlated RVs X and Y with distinct variances can always be made uncorrelated. Moreover, *two functions $U = g(X, Y)$ and $V = h(X, Y)$ of the same pair of RVs (X, Y) are not even necessarily dependent.* For example, consider independent RVs $X \sim \mathcal{N}(0, \sigma^2)$ and $Y \sim \mathcal{N}(0, \sigma^2)$. Let $U = \sqrt{X^2 + Y^2}$ and $V = X/Y$. Then, solving the system of equations:

$$\begin{cases} u &= \sqrt{x^2 + y^2} \\ v &= x/y \end{cases}$$

yields two solutions:

$$(x_1, y_1) = \Big(\frac{uv}{\sqrt{1+v^2}}, \frac{u}{\sqrt{1+v^2}}\Big)$$
$$(x_2, y_2) = \Big(-\frac{uv}{\sqrt{1+v^2}}, -\frac{u}{\sqrt{1+v^2}}\Big)$$

Then

$$|J(x_2, y_2)| \stackrel{\text{it turns out}}{=} |J(x_1, y_1)| = \left|\det \begin{bmatrix} \frac{\partial}{\partial x}\sqrt{x^2+y^2} & \frac{\partial}{\partial y}\sqrt{x^2+y^2} \\ \frac{\partial}{\partial x}(x/y) & \frac{\partial}{\partial y}(x/y) \end{bmatrix}\right|_{x=x_1, y=y_1} = \frac{1+v^2}{|u|}$$

Thus,

$$\begin{aligned} f_{U,V}(u,v) &= \begin{cases} 0 & u < 0 \\ \dfrac{f_{X,Y}(x_1,y_1)}{|J(x_1,y_1)|} + \dfrac{f_{X,Y}(x_2,y_2)}{|J(x_2,y_2)|} & u \geq 0 \end{cases} \\ &= \begin{cases} 0 & u < 0 \\ \dfrac{u}{1+v^2}\Big[f_{X,Y}\Big(\dfrac{uv}{\sqrt{1+v^2}}, \dfrac{u}{\sqrt{1+v^2}}\Big) + f_{X,Y}\Big(-\dfrac{uv}{\sqrt{1+v^2}}, -\dfrac{u}{\sqrt{1+v^2}}\Big)\Big] & u \geq 0 \end{cases} \end{aligned}$$

Since

$$f_{X,Y}(x,y) = \frac{1}{\sqrt{2\pi}\sigma}e^{-x^2/2\sigma^2}\frac{1}{\sqrt{2\pi}\sigma}e^{-y^2/2\sigma^2} = \frac{1}{2\pi\sigma^2}e^{-(x^2+y^2)/2\sigma^2}$$

and note that $f_{X,Y}(x,y) = f_{X,Y}(-x,-y)$,

$$\begin{aligned} f_{U,V}(u,v) &= \begin{cases} 0 & u < 0 \\ \dfrac{u}{1+v^2}\dfrac{1}{\pi\sigma^2}\exp\Big[-\dfrac{1}{2\sigma^2}\Big(\dfrac{u^2v^2}{1+v^2} + \dfrac{u^2}{1+v^2}\Big)\Big] & u \geq 0 \end{cases} \\ &= \begin{cases} 0 & u < 0 \\ \dfrac{1}{\pi(1+v^2)}\dfrac{u}{\sigma^2}e^{-u^2/2\sigma^2} & u \geq 0 \end{cases} \\ &= \underbrace{\frac{1}{\pi(1+v^2)}}_{f_V(v)}\underbrace{\begin{Bmatrix} 0 & u < 0 \\ \dfrac{u}{\sigma^2}e^{-u^2/2\sigma^2} & u \geq 0 \end{Bmatrix}}_{f_U(u)} \end{aligned}$$

Thus, U and V are independent. In fact, in view of (3.33) and (3.58) U is Rayleigh and V is Cauchy.

See also problems 4.40 and 4.41.

4.9 Uniform Distribution

A pair of RVs (X, Y) is said to be ***uniformly distributed*** over a region D if its PDF is constant over D and zero elsewhere:

$$f(x,y) = \begin{cases} \frac{1}{a(D)} & (x,y) \in D \\ 0 & \text{elsewhere} \end{cases} \tag{4.42}$$

where $a(D)$ = area of D; that is, the probability of (X, Y) falling inside any region C inside D is proportional to the area of C. Note that the PDF integrates up to unity. This is an example that the extension from one-dimension to two-dimension is not trivial. Uniform distribution is one of the simplest joint distributions

Example 4.20: Measurement Errors — Uniform over a Rectangle

Find the PDF, means, variances, and correlation coefficient of measurement error (X, Y) that is uniformly distributed over a rectangle $x_1 < x < x_2, y_1 < y < y_2$: Since $a(D) = (x_2 - x_1)(y_2 - y_1)$, we have

$$f(x,y) = \begin{cases} \frac{1}{(x_2-x_1)(y_2-y_1)} & x_1 < x < x_2, y_1 < y < y_2 \\ 0 & \text{elsewhere} \end{cases}$$

$$\text{mean} = \text{center} = [\bar{x}, \bar{y}] = \left[\frac{1}{2}(x_1 + x_2), \frac{1}{2}(y_1 + y_2)\right]$$

$$\text{variance} = [\sigma_x^2, \sigma_y^2] = \left[\frac{1}{12}(x_2 - x_1)^2, \frac{1}{2}(y_2 - y_1)^2\right]$$

and $\rho_{xy} = 0$ since it is easy to show that X and Y are independent.

Example 4.21: Radar Clutter — Uniform over an Ellipse

Find the PDF of the location (X, Y) of radar clutter that is uniformly distributed over the ellipse $\frac{(x-\alpha)^2}{a^2} + \frac{(y-\beta)^2}{b^2} \leq 1$: Since the lengths of the semi-axes of this ellipse are a and b, respectively, its area is πab. Thus, the PDF is

$$f(x,y) = \begin{cases} \frac{1}{\pi ab} & \frac{(x-\alpha)^2}{a^2} + \frac{(y-\beta)^2}{b^2} \leq 1 \\ 0 & \text{elsewhere} \end{cases}$$

The mean $(\bar{x}, \bar{y}) = (\alpha, \beta)$, but σ_x^2, σ_y^2 and ρ_{xy} are not easy to find.

4.10 Conditional Distributions

As an extension of the conditional probability, the ***conditional distribution*** of a RV X given (the value of) another RV Y provides a powerful tool for the study of dependent RVs.

The ***conditional PDF*** of a continuous RV X given another continuous RV $Y = y$ is defined by, if $f_Y(y) > 0$,

$$f_{X|Y}(x|y) \triangleq \frac{f_{X,Y}(x,y)}{f_Y(y)} \tag{4.43}$$

which has a clear geometric interpretation: For a specific $y = y_1$, the function $f(x, y_1)$ is a profile of $f(x, y)$ (i.e., the intersection of the surface $f(x, y)$ and the plane $y = y_1$). Thus, the conditional PDF $f(x|y)$ is just this profile normalized by the factor $1/f_Y(y)$ such that $f(x|y)$ integrates up to unity.

The marginal PDF can be obtained from the conditional PDF:

$$f_X(x) = \int_{-\infty}^{\infty} f(x,y)dy = \int_{-\infty}^{\infty} f_{X|Y}(x|y)f_Y(y)dy = E[f_{X|Y}(x|y)]$$

It indicates that a marginal PDF is an average conditional PDF.

It is clear from (4.43) that if X and Y are independent, then the conditional PDFs are equal to the marginal PDFs (this makes perfect sense):

$$f_{X|Y}(x|y) = f_X(x), \qquad \forall y \tag{4.44}$$

$$f_{Y|X}(y|x) = f_Y(y), \qquad \forall x \tag{4.45}$$

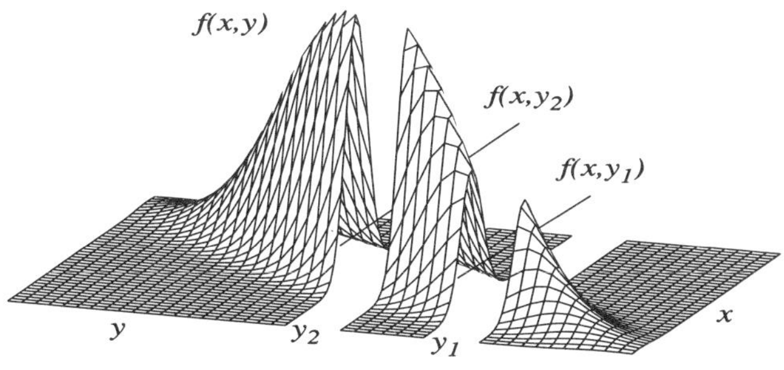

Figure 4.16: $f(x, y_1)$ and $f(x, y_2)$ as the profiles of the surface $f(x, y)$.

For discrete (X, Y) with joint PMF $P\{X = x_i, Y = y_j\} = p_{i,j}$ and marginal PMFs $P\{Y = y_j\} = p_j$, the ***conditional PMF*** of X given $Y = y_j$ is defined as the joint PMF divided by the marginal PMF of Y:

$$p_{i|j} \triangleq P\{X = x_i | Y = y_j\} = \frac{P\{X = x_i, Y = y_j\}}{P\{Y = y_j\}} = \frac{p_{i,j}}{p_j}, \quad \forall i, j \tag{4.46}$$

This definition is not valid for continuous (X, Y) since $P\{Y = y\} = 0, P\{X = x\} = 0$ for every given point (x, y).

The *conditional CDF* of a RV X given $Y = y$ is defined by

$$\begin{aligned} F_{X|Y}(x|y) &\triangleq \lim_{\epsilon \to 0} P\{X \leq x | y - \epsilon < Y \leq y + \epsilon\} & \epsilon > 0 \\ &= \lim_{\epsilon \to 0} \frac{P\{X \leq x, y - \epsilon < Y \leq y + \epsilon\}}{P\{y - \epsilon < Y \leq y + \epsilon\}} & \epsilon > 0 \end{aligned}$$

Conditional distributions given an event with a nonzero probability has been defined by (3.59) and (3.60).

Note that

$$\sum_i p_{i|j} = 1, \qquad \sum_j p_{i|j} \neq 1 \tag{4.47}$$

which should not come as a surprise.

It is clear from (4.46) that if X and Y are independent, then the conditional PMFs are equal to the marginal PMFs:

$$\begin{aligned} P\{X = x_i | Y = y_j\} = P\{X = x_i\} \\ P\{Y = y_j | X = x_i\} = P\{Y = y_j\} \end{aligned}$$

This and (4.44) indicate that the definitions of conditional distributions and independence of RVs are consistent.

Conditional Expectation

The ***conditional mean*** of a RV X given an event B is defined by

$$E[X|B] = \int_{-\infty}^{\infty} x f(x|B) dx$$

where the conditional PDF $f(x|B)$ was defined by (3.60). For example, if $B = \{x_1 < X \leq x_2\}$, then

$$E[X | x_1 < X \leq x_2] = \int_{-\infty}^{\infty} x f(x | x_1 < X \leq x_2) dx$$

The ***conditional mean*** of a RV X given the value of another RV Y is defined by

$$E[X|y] \triangleq E[X|Y = y] = \int_{-\infty}^{\infty} x f_{X|Y}(x|y) dx$$

The ***conditional variance*** of a RV X given the value of another RV Y is defined by

$$\text{var}(X|y) = \int_{-\infty}^{\infty} (x - E[X|y])^2 f(x|y)dx$$

The ***conditional expectations*** of a RV $g(X,Y)$ given the value of Y or event B are defined by

$$E[g(X,Y)|y] \triangleq E[g(X,Y)|Y=y] = \int_{-\infty}^{\infty} g(x,y) f_{X|Y}(x|y)dx$$
$$E[g(X,Y)|B] \triangleq \int_{-\infty}^{\infty}\int_{-\infty}^{\infty} g(x,y) f(x,y|B)dxdy$$

The most powerful result relating unconditional expectation and conditional expectation is the following ***total expectation theorem***

$$E[X] = E[E(X|Y)]$$

where the inner expectation is for X alone even though Y is also random, and the outer is for Y. Total expectation theorem is closely related with total probability theorem. It may simplify greatly the calculation of the unconditional expectation. It has rich applications in statistics, in particular, estimation and filtering.

Example 4.22: Conditional Distribution of Jointly Gaussian RVs

Given jointly Gaussian RVs

$$(X,Y) \sim \mathcal{N}(\bar{x},\bar{y};\sigma_x^2,\sigma_y^2,\rho)$$

the marginal PDF of Y is $\mathcal{N}(y;\bar{y};\sigma_y^2)$ and thus the conditional PDF of X given $Y=y$ is

$$\begin{aligned}
f(x|y) &= \frac{f(x,y)}{f(y)} = \frac{\mathcal{N}[(x,y);\bar{x},\bar{y};\sigma_x^2,\sigma_y^2,\rho]}{\mathcal{N}(y;\bar{y};\sigma_y^2)} \\
&= \frac{1}{\sqrt{2\pi}\sigma_x\sqrt{1-\rho^2}} e^{-\frac{1}{2(1-\rho^2)}\left[\frac{(x-\bar{x})^2}{\sigma_x^2} + \frac{(y-\bar{y})^2}{\sigma_y^2} - \frac{2\rho(x-\bar{x})(y-\bar{y})}{\sigma_x\sigma_y}\right] + \frac{(y-\bar{y})^2}{2\sigma_y^2}} \\
&= \frac{1}{\sqrt{2\pi}\sigma_x\sqrt{1-\rho^2}} e^{-\frac{1}{2(1-\rho^2)}\left[\frac{(x-\bar{x})^2}{\sigma_x^2} + \frac{(y-\bar{y})^2\rho^2}{\sigma_y^2} - \frac{2\rho(x-\bar{x})(y-\bar{y})}{\sigma_x\sigma_y}\right]} \\
&= \frac{1}{\sqrt{2\pi}\sigma_x\sqrt{1-\rho^2}} e^{-\frac{1}{2(1-\rho^2)}\left[\frac{(x-\bar{x})}{\sigma_x} - \rho\frac{(y-\bar{y})}{\sigma_y}\right]^2} \\
&= \frac{1}{\sqrt{2\pi}\sigma_x\sqrt{1-\rho^2}} e^{-\frac{1}{2(1-\rho^2)\sigma_x^2}\left[x-\bar{x}-\rho\frac{\sigma_x}{\sigma_y}(y-\bar{y})\right]^2} \\
&= \mathcal{N}\left(x;\bar{x}+\rho\frac{\sigma_x}{\sigma_y}(y-\bar{y});\sigma_x^2(1-\rho^2)\right)
\end{aligned}$$

That is, X *conditioned on* $Y = y$ *is also Gaussian distributed* with PDF

$$f(x|y) = \mathcal{N}\Big(x; \bar{x} + \rho\frac{\sigma_x}{\sigma_y}(y - \bar{y}), \sigma_x^2(1 - \rho^2)\Big)$$

The conditional mean $E[X|y]$ is given by

$$E[X|y] = \int_{-\infty}^{\infty} x f(x|y)dx = \int_{-\infty}^{\infty} x\mathcal{N}\Big(x; \bar{x} + \rho\frac{\sigma_x}{\sigma_y}(y - \bar{y}); \sigma_x^2(1 - \rho^2)\Big)dx = \bar{x} + \rho\frac{\sigma_x}{\sigma_y}(y - \bar{y})$$

Note that the conditional mean is a *linear function* of the value y of the RV Y. This is important in estimation and filtering theory.

The conditional variance var$(X|y)$ is given by

$$\begin{aligned}
\text{var}(X|y) &= \int_{-\infty}^{\infty} (x - E[X|y])^2 f(x|y)dx \\
&= \int_{-\infty}^{\infty} (x - E[X|y])^2 \mathcal{N}\Big(x; \bar{x} + \rho\frac{\sigma_x}{\sigma_y}(y - \bar{y}); \sigma_x^2(1 - \rho^2)\Big)dx \\
&= \sigma_x^2(1 - \rho^2)
\end{aligned}$$

4.11 Summary and Requirements

Many random phenomena can only be described by two or more RVs jointly. The *joint CDF* of two RVs X and Y is the joint probability $P\{(X \le x) \cap (Y \le y)\}$. Its second partial derivative is the *joint PDF*, which describes the density distribution of the joint probability over the x-y plane. Various joint probabilities can be determined in terms of the joint CDF or PDF. The following is the most important formula:

$$\begin{aligned}
P\{x_1 < X \le x_2, y_1 < Y \le y_2\} &= F(x_2, y_2) + F(x_1, y_1) - F(x_1, y_2) - F(x_2, y_1) \\
&= \int_{x=x_1}^{x_2} \int_{y=y_1}^{y_2} f(x, y)dxdy
\end{aligned}$$

The joint CDF (or PDF) and the *marginal CDF* (or PDF) are related by

$$\begin{aligned}
F_X(x) &= F_{X,Y}(x, \infty), & F_Y(y) &= F_{X,Y}(\infty, y) \\
f_X(x) &= \int_{-\infty}^{\infty} f(x, v)dv, & f_Y(y) &= \int_{-\infty}^{\infty} f(u, y)du
\end{aligned}$$

The expected value, mean or average of a function $g(x, y)$ of two RVs X, Y is defined by

$$E[g(X,Y)] = \begin{cases} \displaystyle\int_{-\infty}^{\infty}\int_{-\infty}^{\infty} g(x, y)f(x, y)dxdy & (X, Y) \text{ continuous} \\ \displaystyle\sum_i\sum_j g(x_i, y_j)P\{X = x_i, Y = y_j\} & (X, Y) \text{ discrete} \end{cases}$$

The mean of a weighted sum of RVs is equal to the weighted sum of the individual means:

$$E\left[\sum_{i=1}^{n} a_i X_i\right] = \left[\sum_{i=1}^{n} a_i E(X_i)\right]$$

If $X_1, \ldots, X_n$ are uncorrelated, then the variance of a weighted sum of RVs is equal to the squarely weighted sum of the individual variances:

$$\text{var}\left(\sum_{i=1}^{n} a_i X_i\right) = \sum_{i=1}^{n} a_i^2 \text{var}(X_i)$$

For two RVs X and Y, their *correlation* R_{xy} is the expectation $E[XY]$; their *covariance* C_{xy} is the central expectation $E[(X-\bar{x})(Y-\bar{y})]$; their *correlation coefficient* ρ is the normalized covariance $C_{xy}/(\sigma_x \sigma_y)$. They measure how likely if the two RVs are related *linearly.* If $|\rho| = 1$ then they are surely related linearly.

Two RVs X and Y are *independent* if $F_{X,Y}(x, y) = F_X(x)F_Y(y)$ or equivalently $f_{X,Y}(x, y) = f_X(x)f_Y(y)$. They are *uncorrelated* (i.e., surely not related linearly) if $E[XY] = E[X]E[Y]$, or their covariance $C_{xy} = 0$ or their correlation coefficient $\rho_{xy} = 0$. If they are independent then they are uncorrelated, but uncorrelatedness does not imply independence in general unless they are jointly Gaussian.

Two *jointly Gaussian* or *jointly normal* RVs X and Y have the following joint PDF

$$f(x, y) = \frac{1}{2\pi\sigma_x\sigma_y\sqrt{1-\rho^2}} e^{-\frac{1}{2(1-\rho^2)}\left[\frac{(x-\bar{x})^2}{\sigma_x^2} + \frac{(y-\bar{y})^2}{\sigma_y^2} - \frac{2\rho(x-\bar{x})(y-\bar{y})}{\sigma_x\sigma_y}\right]}$$

where $|\rho| \le 1$ is their correlation coefficient. A linear function (e.g., weighted sum) of jointly Gaussian RVs is still Gaussian. Two independent Gaussian RVs are jointly Gaussian.

The PDF of the sum of two independent RVs is the convolution of their two PDFs.

A pair of RVs is uniformly distributed if its joint PDF is constant over a single region and zero elsewhere.

Basic Requirements

- Have a clear understanding of the joint CDF, PDF, and PMF. Be clear about the similarity and difference between joint and marginal CDFs and PDFs. Be familiar with their properties, especially their relationships with marginal CDFs, PDFs, and PMFs.
- Know how to find marginal CDFs and PDFs from joint CDF and PDF.
- Comprehend the concepts of independence and uncorrelatedness of RVs and know how to check if two RVs are independent or uncorrelated.
- Understand the implication of correlation, covariance, and correlation coefficient of two RVs. Know how to calculate them.
- Know how to find the mean and variance of weighted sum of RVs.
- Be skillful in calculating various probabilities from joint CDF, PDF, or PMF.

- Be familiar with the various properties of joint Gaussian distribution.

The emphasis of the chapter is on the relations between RVs, between the joint and marginal distributions and the mean and variance of weighted sum of RVs.

4.12 Additional Examples

4.23 *From joint CDF to probability.* Two jointly exponential RVs X and Y have the joint CDF

$$F(x,y) = \begin{cases} 1 - e^{-4x} - e^{-3y} + e^{-(4x+3y)} & x \geq 0, y \geq 0 \\ 0 & \text{elsewhere} \end{cases}$$

(a) Find $P\{X \leq 0.5, Y \leq 1\}$.
(b) Find $P\{0.5 < X \leq 1, 0 \leq Y \leq 0.75\}$.
(c) Find $P\{1 < X \leq 2\}$.
(d) Find $f_X(x)$ and $f_Y(y)$.
(e) Find $f(x,y)$.
(f) Find $P\{Y \leq \frac{3}{2}X\}$.

Solution:

(a) $P\{X \leq 0.5, Y \leq 1\} \overset{(4.1)}{=} F(0.5,1) = 1 - e^{-4\times 0.5} - e^{-3} + e^{-(2+3)} = 0.6717$
(b)

$$\begin{aligned} & P\{0.5 < X \leq 1, 0 \leq Y \leq 0.75\} \\ \overset{(4.5)}{=} \; & F(1,0.75) + \underbrace{F(0.5,0)}_{0} - F(0.5,0.75) - \underbrace{F(1,0)}_{0} \\ = \; & (1 - e^{-4} - e^{-2.25} - e^{-6.25}) - (1 - e^{-2} - e^{-2.25} - e^{-4.25}) \\ = \; & 0.1047 \end{aligned}$$

(c)

$$P\{X \leq x\} = F_X(x) \overset{(4.7)}{=} F(x,\infty) = \begin{cases} 1 - e^{-4x} & x \geq 0 \\ 0 & x < 0 \end{cases}$$

Thus

$$\begin{aligned} P\{1 < X \leq 2\} &= F_X(2) - F_X(1) \\ &= F(2,\infty) - F(1,\infty) \\ &= (1 - e^{-8}) - (1 - e^{-4}) \\ &= 0.01798 \end{aligned}$$

Alternatively,

$$\begin{aligned} P\{1 < X \le 2\} &= P\{1 < X \le 2, -\infty < Y < \infty\} \\ &= F(2,\infty) + F(1,-\infty) - F(2,-\infty) - F(1,\infty) \\ &= F(2,\infty) - F(1,\infty) = 0.01798 \end{aligned}$$

(d) If $x \ge 0, y \ge 0$, then

$$f(x,y) = \frac{\partial^2}{\partial x \partial y} F(x,y) = \frac{\partial}{\partial x}\left[3e^{-3y} - 3e^{-(4x+3y)}\right] = 12e^{-(4x+3y)}$$

Hence

$$f(x,y) = \begin{cases} 12e^{-(4x+3y)} & x \ge 0, y \ge 0 \\ 0 & \text{elsewhere} \end{cases}$$

Check:

$$\int_{-\infty}^{\infty}\int_{-\infty}^{\infty} f(u,v)dudv = 12\int_0^{\infty}\int_0^{\infty} e^{-4x}e^{-3y}dydx = 1$$

(e)

$$f_X(x) = \int_{-\infty}^{\infty} f(x,v)dv = 12\int_0^{\infty} e^{-(4x+3v)}dv = 4e^{-4x},\ x \ge 0$$

Thus

$$f_X(x) = \begin{cases} 4e^{-4x} & x \ge 0 \\ 0 & x < 0 \end{cases}$$

Alternatively, $f_X(x) = \frac{d}{dx}F(x)$, where $F(x)$ was given in (c). Similarly, we have

$$f_Y(y) = \begin{cases} 3e^{-3y} & y \ge 0 \\ 0 & y < 0 \end{cases}$$

Check:

$$\begin{aligned} \int_{-\infty}^{\infty} f_X(x)dx &= \int_0^{\infty} 4e^{-4x}dx = 1 \\ \int_{-\infty}^{\infty} f_Y(y)dy &= \int_0^{\infty} 3e^{-3y}dy = 1 \end{aligned}$$

Note that since $f(x,y) = f_X(x)f_Y(y)$, we know X and Y are independent.

(f)

$$\begin{aligned} P\{Y \le \frac{3}{2}X\} &= P\{0 \le X < \infty, 0 \le Y \le \frac{3}{2}X\} \\ &= \int_{u=0}^{\infty}\int_{v=0}^{\frac{3}{2}u} f(u,v)dudv \end{aligned}$$

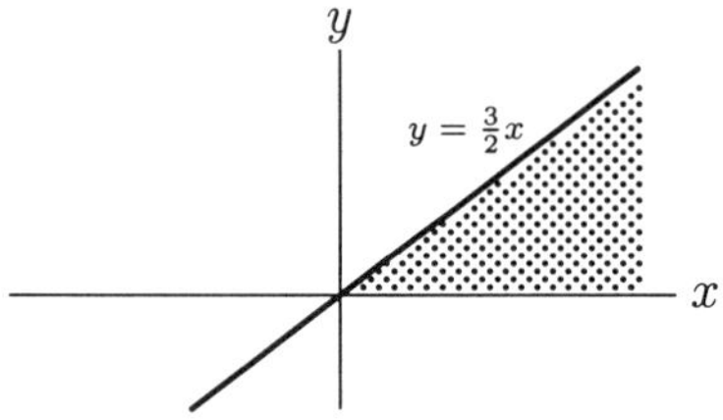

$$
\begin{aligned}
&= 12 \int_{u=0}^{\infty} e^{-4u} \left[\int_{v=0}^{\frac{3}{2}u} e^{-3v} dv \right] du \\
&= \frac{12}{3} \int_{0}^{\infty} e^{-4u} \left[e^{-3v} \Big|_{0}^{\frac{3}{2}u} \right] du \\
&= -4 \int_{0}^{\infty} \left(e^{-\frac{17}{2}u} - e^{-4u} \right) du \\
&= \frac{9}{17} = 0.5294
\end{aligned}
$$

4.24 *From joint PMF to others.* RVs (X, Y) have the following PMF

X \ Y	1	2	3
1	1/8	1/8	1/8
2	1/8	1/8	1/8
3	0	1/8	1/8

(a) Find the marginal PMFs of X and Y, respectively. Are X and Y independent?
(b) Find the expected values and correlation of X and Y. Are X and Y uncorrelated?
(c) Find the conditional PMF of X given Y.

Solution:

(a) The marginal PMFs can be obtained easily from (4.10) by summing up the corresponding joint PMF:

X \ Y	1	2	3	$P\{X = x_i\}$
1	1/8	1/8	1/8	3/8
2	1/8	1/8	1/8	3/8
3	0	1/8	1/8	2/8
$P\{Y = y_j\}$	2/8	3/8	3/8	1

Since $P\{X = 3, Y = 2\} = \frac{1}{8} \neq \frac{2}{8}\frac{3}{8} = P\{X = 3\}P\{Y = 2\}$, X and Y are not independent.

(b) The expected values and correlation of X and Y are

$$
\begin{aligned}
E[X] &= (1)(3/8) + (2)(3/8) + (3)(2/8) = 9/8 \\
E[Y] &= (1)(2/8) + (2)(3/8) + (3)(3/8) = 17/8
\end{aligned}
$$

$$\begin{aligned}E[XY] &= (1)(1)(1/8) + (1)(2)(1/8) + (1)(3)(1/8) \\ &\quad + (2)(1)(1/8) + (2)(2)(1/8) + (2)(3)(1/8) \\ &\quad + (3)(1)(0/8) + (3)(2)(1/8) + (3)(3)(1/8) \\ &= 33/8\end{aligned}$$

Since $E[XY] \neq E[X]E[Y]$, X and Y are not uncorrelated.

(c) The conditional PMF of X given Y is given by

$$p_{i|j} = \frac{p_{i,j}}{p_j}$$

which leads to

$p_{i\|j}$	$j=1$	$j=2$	$j=3$
$i=1$	1/2	1/3	1/3
$i=2$	1/2	1/3	1/3
$i=3$	0	1/3	1/3

Note that this verifies (4.47).

4.25 *Inventory analysis.* An inventory has three machines that are operational with probabilities 0.8, 0.9, and 0.95, respectively. Whether a machine is operational is independent of the other machines. Let X be the number of machines that are operational. Find the mean and variance of X.

Solution: Let $M_i = \{\text{machine } i \text{ operational}\}$ and

$$X_i = \begin{cases} 1 & \text{machine } i \text{ operational } (M_i) \\ 0 & \text{machine } i \text{ not operational } (\overline{M_i}) \end{cases}$$

Clearly

$$\begin{aligned}X &= X_1 + X_2 + X_3 \\ E[X_i] &= (0)P\{\overline{M_i}\} + (1)P\{M_i\} = P\{M_i\} \\ E[X_i^2] &= (0)^2P\{\overline{M_i}\} + (1)^2P\{M_i\} = P\{M_i\} \\ \text{var}(X_i) &= E[X_i^2] - (E[X_i])^2 = P\{M_i\}[1 - P\{M_i\}]\end{aligned}$$

Then,

$$\begin{aligned}E[X] &= E[X_1 + X_2 + X_3] \\ &= E[X_1] + E[X_2] + E[X_3] = 0.8 + 0.9 + 0.95 \\ &= 2.65 \\ \text{var}(X) &= P\{M_1\}[1 - P\{M_1\}] + P\{M_2\}[1 - P\{M_2\}] + P\{M_3\}[1 - P\{M_3\}] \\ &= (0.8)(0.2) + (0.9)(0.1) + (0.95)(0.05) \\ &= 0.2975\end{aligned}$$

Note an alternate solution provided in Example 3.41.

4.26 *Power consumption.* A random voltage $V \sim \mathcal{U}(-3\text{V}, 7\text{V})$ is applied to a resistor whose resistance R is a binary RV taking on a value of either 5Ω or 12Ω with equal probability and is independent of V.

(a) Find the expected power dissipated by R.
(b) Find R_{vp}, R_{rp}, and σ^2_{vp}.

Solution:

(a) Since

$$f_V(v) = \begin{cases} \frac{1}{10} & -3 \leq v \leq 7 \\ 0 & \text{elsewhere} \end{cases}$$

$$f_R(r) = \frac{1}{2}\delta(r-5) + \frac{1}{2}\delta(r-12)$$

$$E[V^2] = \int_{-\infty}^{\infty} v^2 f_V(v)dv = \int_{-3}^{7} v^2 \frac{1}{10} dv = \frac{37}{3}$$

$$E[1/R] = \sum_{i=1}^{2} \frac{1}{r_i} P\{R = r_i\} = \frac{1}{5}\cdot\frac{1}{2} + \frac{1}{12}\cdot\frac{1}{2} = \frac{17}{120}$$

$$V \text{ and } R \text{ are independent} \Longleftrightarrow V^2 \text{ and } \frac{1}{R} \text{ are independent}$$
$$\Longrightarrow V^2 \text{ and } \frac{1}{R} \text{ are uncorrelated}$$

the expected power is

$$E[P] = E[V^2/R] \stackrel{?}{=} E[V^2]E[1/R] = \frac{37}{3}\cdot\frac{17}{120} = 1.747 \text{ (watts)}$$

(b)

$$R_{vp} = E[VP] = E[V\cdot V^2/R] = \left[\int_{-3}^{7} \frac{v^3}{10} dv\right]\cdot\frac{17}{120} = 58 \times \frac{17}{120} \simeq 8.21$$

$$R_{rp} = E[R\cdot V^2/R] = E[V^2] = \frac{37}{3}$$

$$E[V] = \int_{-3}^{7} \frac{1}{10} dv = 2 \quad [\text{the center point of } (-3,7)]$$

$$\sigma^2_{vp} = R_{vp} - E[V]E[P] = 8.21 - 2\times 1.747 = 4.722$$

4.27* *Moments of discrete RVs.* RVs (X, Y) have the PDF

$$f(x,y) = 0.25\delta(x-c)\delta(y-c) + 0.4\delta(x+c)\delta(y-3) + 0.35\delta(x-1)\delta(y-1)$$

(a) Find c such that $\text{cov}(X, Y)$ is minimum. Find the minimum $\text{cov}(X, Y)$.

(b) When are X and Y uncorrelated?

Solution:

(a) (X, Y) has the following PMF

$X\backslash Y$	$y_1 = c$	$y_2 = 3$	$y_3 = 1$
$x_1 = c$	$p_{11} = 0.25$	$p_{12} = 0$	$p_{13} = 0$
$x_2 = -c$	$p_{21} = 0$	$p_{22} = 0.4$	$p_{23} = 0$
$x_3 = 1$	$p_{31} = 0$	$p_{32} = 0$	$p_{33} = 0.35$

Thus

$$\begin{aligned}(\bar{x}, \bar{y}) &= \sum_i \sum_j (x_i, y_j) p_{i,j} = \sum_i (x_i, y_i) p_{i,i} \\ &= p_{11}(x_1, y_1) + p_{22}(x_2, y_2) + p_{33}(x_3, y_3) \\ &= 0.25(c, c) + 0.4(-c, 3) + 0.35(1, 1) \\ &= (0.35 - 0.15c, 1.55 + 0.25c)\end{aligned}$$

That is,

$$\bar{x} = 0.35 - 0.15c, \qquad \bar{y} = 1.55 + 0.25c$$

Hence,

$$\begin{aligned}\text{cov}(X, Y) &= E[(X - \bar{x})(Y - \bar{y})] \\ &= \sum_i \sum_j (x_i - \bar{x})(y_j - \bar{y}) p_{i,j} \\ &= \sum_i (x_i - \bar{x})(y_i - \bar{y}) p_{ii} \\ &= 0.25(x_1 - \bar{x})(y_1 - \bar{y}) + 0.4(x_2 - \bar{x})(y_2 - \bar{y}) + 0.35(x_3 - \bar{x})(y_3 - \bar{y}) \\ &= 0.25[c - (0.35 - 0.15c)][c - (1.55 + 0.25c)] \\ &\quad +0.4[-c - (0.35 - 0.15c)][3 - (1.55 + 0.25c)] \\ &\quad +0.35[1 - (0.35 - 0.15c)][1 - (1.55 + 0.25c)] \\ &= 0.2875c^2 - 1.07657c - 0.1925\end{aligned}$$

Taking $\frac{d}{dc}[\text{cov}(X, Y)]$ and setting it to zero yields $c = 0.5341$. Since $\frac{d^2}{dc^2}[\text{cov}(X, Y)] = 0.575 > 0$, it is a minimum point. Thus the minimum covariance is

$$\min \text{cov}(X, Y) = 0.2875c^2 - 1.07657c - 0.1925|_{c=0.5341} = -0.6122$$

(b) X and Y are uncorrelated if and only if

$$\text{cov}(X, Y) = 0.2875c^2 - 1.07657c - 0.1925 = 0$$

The solutions of this equation are $c_1 = -5.311$ and $c_2 = -0.2812$. That is, X and Y are uncorrelated if and only if $c = -5.311$ or $c = -0.2812$.

4.28 *Circuit reliability*. An RLC circuit will be operational only if none of the R, L, C fails. Assume that the failures of the R, L, C are independent and each of their time-to-failure is an exponential RV with the following PDF (with identical parameter λ):

$$f(x) = \begin{cases} \lambda e^{-\lambda x} & x > 0 \\ 0 & x \le 0 \end{cases}$$

Find the CDF of the time-to-failure of the circuit.

Solution: Let X_R, X_L, X_C be the time-to-failure of R, L, C, respectively. Then, they are independent and have identical CDF

$$F(x) = \int_{-\infty}^{x} f(x)dx = \begin{cases} 1 - e^{-\lambda x} & x > 0 \\ 0 & x \le 0 \end{cases}$$

Hence, the CDF $F_Y(y)$ of the time-to-failure of the circuit is, for $y > 0$,

$$\begin{aligned} F_Y(y) &= P\{Y \le y\} = 1 - P\{Y > y\} \\ &= 1 - P\{X_R > y, X_L > y, X_C > y\} \\ &= 1 - P\{X_R > y\}P\{X_L > y\}P\{X_C > y\} \\ &= 1 - [1 - F_X(y)]^3 \\ &= 1 - e^{-3\lambda y} \end{aligned}$$

Note that Y is nonnegative. Thus,

$$F_Y(y) = \begin{cases} 1 - e^{-3\lambda y} & y > 0 \\ 0 & y \le 0 \end{cases}$$

which is the exponential distribution with parameter 3λ.

4.29* *Uniform distribution over a circle*. Suppose that a target with coordinates (X, Y) appears uniformly on a radar display screen of radius r; that is, (X, Y) is uniformly distributed over the circle $x^2 + y^2 \le r^2$.

(a) Find the correlation coefficient of X and Y. Are X and Y uncorrelated?
(b) Find the conditional PDF $f_{Y|X}(y|x)$.
(c) Are X and Y independent?

Solution: Clearly, the joint PDF of (X, Y) is

$$f(x, y) = \begin{cases} \dfrac{1}{\pi r^2} & x^2 + y^2 \le r^2 \\ 0 & \text{elsewhere} \end{cases} \tag{4.48}$$

(a) The marginal PDF of X is, $f_X(x) = 0$ for $|x| > r$ and for $|x| \leq r$, the upper and lower limits of y are $\sqrt{r^2 - x^2}$ and $-\sqrt{r^2 - x^2}$, respectively, and thus

$$f_X(x) = \int_{-\sqrt{r^2-x^2}}^{\sqrt{r^2-x^2}} \frac{1}{\pi r^2} dy = \frac{2}{\pi r^2}\sqrt{r^2 - x^2}$$

By symmetry,

$$f_Y(y) = \begin{cases} \frac{2}{\pi r^2}\sqrt{r^2 - y^2} & |y| \leq r \\ 0 & |y| > r \end{cases}$$

Note that $f_X(x)$ is evenly symmetrical in x about $x = 0$ and thus $E(X) = 0$. Likewise, $E(Y) = 0$. This makes sense since the center of mass (i.e., the balance point) of (X, Y) is $(0, 0)$. Hence,

$$C_{xy} = E(XY) = \iint_{x^2+y^2 \leq r^2} \frac{xy}{\pi r^2} dxdy = 0 \quad \Longrightarrow \quad \rho_{xy} = 0$$

Thus, X and Y are uncorrelated.

(b) For $|x| < r$,

$$\begin{aligned} f(y|x) &= \frac{f(x,y)}{f_X(x)} \\ &= \begin{cases} \frac{1}{\pi r^2} \frac{\pi r^2}{2\sqrt{r^2 - x^2}} = \frac{1}{2\sqrt{r^2 - x^2}} & -\sqrt{r^2 - x^2} < y < \sqrt{r^2 - x^2} \\ 0 & \text{elsewhere} \end{cases} \end{aligned}$$

Note that given x, $f_{Y|X}(y|x)$ is a constant and thus the conditional distribution of Y is also uniform, but the interval over which Y is uniform depends on the value of X (see Fig. 4.17).

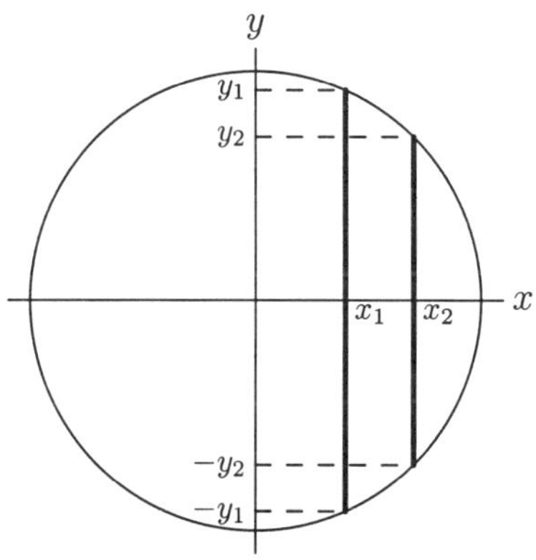

Figure 4.17: Uniform distribution over a circle and straight lines.

(c) X and Y are dependent since

$$f(x,y) \neq f_X(x)f_Y(y) \qquad \text{or} \qquad f(y|x) \neq f_Y(y)$$

By the theorem on page 159, it is also clear that X and Y are dependent because the region over which $f(x,y)$ is nonzero (i.e., the circle) cannot be separated into the form of $x_1 < x < x_2, y_1 < y < y_2$, where x_1, x_2, y_1, y_2 are constants. In other words, the dependence comes from the relation that defines the boundary: $x^2+y^2 \leq r^2$. This is made explicit in part (b): the interval over which Y (or X) is uniform depends on the value of X (or Y), as illustrated in Fig. 4.17. Specifically, if $X = x_1$, then $Y \sim \mathcal{U}(-y_1, y_1)$. If $X = x_2$, however, then $Y \sim \mathcal{U}(-y_2, y_2)$.

4.30 *Generation of uniform numbers over a circle.* Random numbers $(X_1, Y_i), \ldots, (X_n, Y_n)$ that are uniformly distributed over a circle $x^2+y^2 \leq r^2$ can be generated by the following procedure based on the ***acceptance-rejection method***

S1. Generate $U_i \sim \mathcal{U}(0,1)$ and $V_i \sim \mathcal{U}(0,1)$.
S2. If $U_i^2 + V_i^2 \leq 1$ then return $(X_i, Y_i) = (rU_i, rV_i)$; otherwise go back to Step 1 and generate another pair of (U_i, V_i).

4.31 *Decorrelation of Random Variables.* Given two RVs X and Y with distinct variances σ_x^2, σ_y^2, and correlation coefficient ρ_{xy}, it is always possible to introduce two RVs $U = aX+bY, V = cX+dY$ as their linear combinations such that U and V are uncorrelated. Clearly,

$$\begin{aligned}
\bar{u} &= a\bar{x} + b\bar{y} \qquad \bar{v} = c\bar{x} + d\bar{y} \\
C_{uv} &= E\{[(aX+bY) - (a\bar{x}+b\bar{y})][(cX+dY) - (c\bar{x}+d\bar{y})]\} \\
&= E\{[a(X-\bar{x}) + b(Y-\bar{y})][c(X-\bar{x}) + d(Y-\bar{y})]\} \\
&= E[ac(X-\bar{x})^2 + bd(Y-\bar{y})^2 + (ad+bc)(X-\bar{x})(Y-\bar{y})] \\
&= ac\sigma_x^2 + bd\sigma_y^2 + (ad+bc)C_{xy}
\end{aligned} \tag{4.49}$$

Thus, U and V are uncorrelated if a, b, c, d are chosen such that

$$ac\sigma_x^2 + bd\sigma_y^2 + (ad+bc)C_{xy} = 0 \tag{4.50}$$

In particular, if $a = d = \cos\theta, b = -c = \sin\theta$, then (4.50) becomes

$$0 = (\sigma_y^2 - \sigma_x^2)\cos\theta\sin\theta + (\cos^2\theta - \sin^2\theta)C_{xy} = (\sigma_y^2 - \sigma_x^2)\frac{\sin 2\theta}{2} + (\cos 2\theta)C_{xy}$$

Thus, U and V, given by

$$\begin{aligned}
U &= X\cos\theta + Y\sin\theta \\
V &= -X\sin\theta + Y\cos\theta
\end{aligned}$$

are uncorrelated if, since $C_{xy} = \rho_{xy}\sigma_x\sigma_y$,

$$\tan 2\theta = \frac{2\rho_{xy}\sigma_x\sigma_y}{\sigma_x^2 - \sigma_y^2} \quad\Longrightarrow\quad \theta = \frac{1}{2}\tan^{-1}\frac{2\rho_{xy}\sigma_x\sigma_y}{\sigma_x^2 - \sigma_y^2}$$

Note that X and Y can be expressed in terms of U and V as

$$X = U\cos\theta - V\sin\theta$$
$$Y = U\sin\theta + V\cos\theta$$

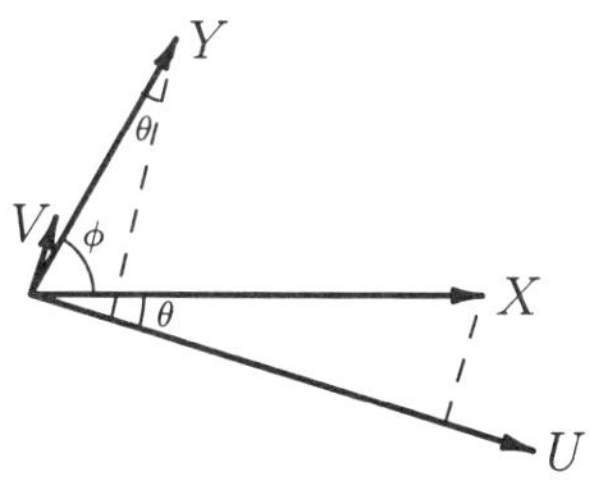

Figure 4.18: Conversion of correlated zero-mean RVs (X, Y) to uncorrelated RVs (U, V).

4.13 Problems

4.1 *From joint CDF to probability.* Consider the jointly exponential RVs of Example 4.23.

(a) Find $P\{X > 3, Y \leq 2\}$.
(b) Find $P\{X > 3 | Y \leq 2\}$.
(c) Find $P\{X > 3Y/2\}$.

4.2 *From joint PDF to probability.* Consider the RVs of Example 4.4.

(a) Find $P\{X > 1/2, Y \leq 1\}$.
(b) Find $P\{Y > 1 | X \leq 1/2\}$.
(c) Find $P\{X \leq Y\}$.

4.3* *Properties of CDF.* Consider a function

$$g(x, y) = \begin{cases} 1 & x + y > 0 \\ 0 & x + y \leq 0 \end{cases}$$

(a) Show that $g(x, y)$ is nondecreasing as x or y (or both) increases.
(b) Show that $g(x, y)$ is continuous from right in x and y.
(c) Show that $g(x, y)$ satisfies properties 1, 3, 4, 5 of the joint CDF.
(d) Find $g(1, 1) + g(0, 0) - g(1, 0) - g(0, 1)$.

(e) Can $g(x,y)$ be the joint CDF of some RV?

4.4 *From joint to marginal.* Find $F(x)$ and $F(y)$ if

$$F(x,y) = \begin{cases} 1 - e^{-x} & x > 0, y > 0 \\ 0 & \text{elsewhere} \end{cases}$$

Identify the RVs X and Y.

4.5 *Symmetric CDF.* If the joint CDF $F(x,y)$ of (X,Y) is symmetrical about x and y (i.e., $F_{X,Y}(x,y) = F_{X,Y}(y,x)$), do X and Y have an identical marginal PDF and an identical CDF? Justify your answer.

4.6 *Symmetric PDF.* If the joint PDF $f(x,y)$ of (X,Y) is symmetrical about x and y (i.e., $f_{X,Y}(x,y) = f_{X,Y}(y,x)$), do X and Y have an identical marginal PDF and an identical CDF? Justify your answer.

4.7 *Identically distributed RVs.* Two RVs X and Y have identical marginal CDFs. Are they always independent? Is it true that $X = Y$? Justify your answer.

4.8 *Problem given PDF.* Two RVs X and Y have the joint PDF

$$f(x,y) = \begin{cases} 3xy & 0 < x \le 1,\ 0 < y \le \frac{2}{\sqrt{3}} \\ 0 & \text{elsewhere} \end{cases}$$

(a) Determine if X and Y are independent.
(b) Find $F(x,y)$.

4.9* *Independence check with coupled PDF.* Given two RVs X and Y with the joint PDF

$$f(x,y) = \begin{cases} g(x,y) & 0 < x \le 1,\ 0 < y \le 1 - x \\ 0 & \text{elsewhere} \end{cases}$$

where $g(x,y)$ is nonzero everywhere over $0 < x \le 1,\ 0 < y \le 1 - x$, can X and Y be independent? Justify your answer.

4.10 *Independence check.* Are RVs X and Y independent if their joint PDF is

(a) $f(x,y) = \begin{cases} x + y & 0 < x < 1, 0 < y < 1 \\ 0 & \text{elsewhere} \end{cases}$

(b) $f(x,y) = \begin{cases} 1/\pi & 0 < x^2 + y^2 < 1 \\ 0 & \text{elsewhere} \end{cases}$

(c) $f(x,y) = \begin{cases} 2\lambda e^{-\lambda x} y & x > 0, 0 < y < 1 \\ 0 & \text{elsewhere} \end{cases}$

You should be able to answer these questions without obtaining the marginal PDFs.

4.11 *PDF of uniform distribution.* Find the joint PDF of (X, Y) which is uniformly distributed over the triangle $0 < x < 2, 0 < y < 1, x + 2y \leq 2$.

4.12 *From joint PDF to others.* Given the joint PDF of (X, Y)

$$f(x,y) = \begin{cases} 6(1-x-y) & 0 < x < 1,\ 0 < y < 1-x \\ 0 & \text{elsewhere} \end{cases}$$

(a) Find the marginal PDF of X. Check if $\int_{-\infty}^{\infty} f(x)dx = 1$. What is the marginal PDF of Y?
(b) Are X and Y independent?
(c) Find the mean $E[X]$ using the marginal PDF and the joint PDF, respectively.

4.13 *Probability of being larger.* Two RVs X and Y have the joint PDF

$$f(x,y) = \begin{cases} 2xy & 0 < x \leq 1,\ 0 < y \leq \sqrt{2} \\ 0 & \text{elsewhere} \end{cases}$$

Find $P\{X \geq Y\}$.

4.14 *Expectation and correlation.* Consider the RVs of Example 4.4.

(a) Find $E[X]$.
(b) Find $E[Y]$.
(c) Find R_{xy} and C_{xy}.
(d) Are X and Y uncorrelated?
(e) Find $E[X - Y]$.

4.15 *Expectation and correlation.* Consider the jointly exponential RVs of Example 4.23.

(a) Find $E[X]$.
(b) Find $E[Y]$.
(c) Find R_{xy} and C_{xy}.
(d) Are X and Y uncorrelated?

4.16 *Expectation and correlation.* Given the joint PDF of two random variables X and Y

$$f(x,y) = \begin{cases} xy/96 & 0 < x < 4, 1 < y < 5 \\ 0 & \text{elsewhere} \end{cases}$$

(a) Find $E[X]$, $E[Y]$, $E[XY]$, and $E[2X + 3Y]$.
(b) Are X and Y uncorrelated?
(c) Are X and Y independent?
(d) Find $\text{var}(2X + 3Y)$.

4.17 *Sum and difference of two RVs.* For two random variables X and Y with marginal PDFs $f_X(x) = 2e^{-2x}u(x)$ and $f_Y(y) = \delta(y+5)$, form two new random variables $U = X + Y$ and $V = X - Y$. Find the expected values of U and V.

4.18 *Correlation of discrete RVs.* RVs (X, Y) have the joint PDF

$$\begin{aligned} f(x,y) = {} & 0.15\delta(x)\delta(y) + 0.3\delta(x)\delta(y-2) + 0.35\delta(x-1)\delta(y+2) \\ & + 0.1\delta(x-1)\delta(y-1) + 0.2\delta(x-1)\delta(y-3) \end{aligned}$$

(a) Find $E(XY)$, and correlation coefficient ρ_{xy}.
(b) Are X and Y uncorrelated?
(c) Are X and Y orthogonal?

4.19 *Effect of scale and translation on correlation.* Find the correlation coefficient ρ_{uv} between $U = aX + b$ and $V = cY + d$, where a, b, c, d are given real numbers, X and Y are RVs with known correlation coefficient ρ_{xy}. When do ρ_{xy} and ρ_{uv} have the same sign?

4.20* *Correlation of a RV and its absolute value.* RV X has the so-called ***Laplace distribution*** with the following PDF:

$$f(x) = \frac{1}{2}e^{-|x|} \tag{4.51}$$

(a) Find the correlation coefficient between X and $|X|$.
(b) Are X and $|X|$ uncorrelated?
(c) Are X and $|X|$ independent?

4.21 *Perfectly correlated RVs.* Two RVs X and Y are defined over the same sample space. They have correlation coefficient $\rho = -1$ and $\bar{x} = 1, \bar{y} = 3, \sigma_x = 2, \sigma_y = 1$. If $\{Y = 2.3\}$ occurred, what is X equal to?

4.22* *Standard Gaussian RV and its nth power.* Let $X \sim \mathcal{N}(0,1)$ and $Y = X^n$, where n is a positive integer.

(a) Find the expected value of Y.
(b) Find the correlation coefficient of X and Y.

4.23 *From joint PDF to others.* Given the joint PDF of X and Y

$$f(x,y) = \begin{cases} 4.8y(2-x) & 0 < x < 1,\ 0 < y < x \\ 0 & \text{elsewhere} \end{cases}$$

(a) Find the marginal PDFs of X and Y.
(b) Are X and Y independent?
(c) Find the correlation R_{xy} and correlation coefficient ρ_{xy}.

4.24 *From joint PDF to others.* Given the joint PDF of X and Y

$$f(x,y) = \begin{cases} A\sin(x+y) & 0 < x \le \pi/2, 0 < y \le \pi/2 \\ 0 & \text{elsewhere} \end{cases}$$

(a) Determine constant A.
(b) Find the expected values of X and Y.
(c) Find the variances of X and Y.
(d) Find the correlation coefficient ρ_{xy}.

4.25* *Mean and variance of uniform RVs over a circle.* Find the expected values and variances of X and Y that are uniformly distributed over the circle $(x-a)^2 + (y-b)^2 \le r^2$.

4.26 *Correlation of sum and difference.* Consider two mutually independent zero-mean RVs X and Y with variances σ_x^2 and σ_y^2, respectively. Let $U = X + Y$ and $V = X - Y$.

(a) Find the correlation and correlation coefficient of U and V if $\sigma_x^2 = 1$ and $\sigma_y^2 = 0.25$. Are U and V perfectly correlated, strongly correlated, weekly correlated, or uncorrelated?
(b) Find the correlation and correlation coefficient of U and V if $\sigma_x^2 = 1$ and $\sigma_y^2 = 10$. Are U and V perfectly correlated, strongly correlated, weekly correlated, or uncorrelated?

4.27 *Sum of squares of Gaussian RVs.* For independent RVs $X_1, X_2, \ldots, X_n$, each being a $\mathcal{N}(0, \sigma^2)$ RV, form a new RV $X = X_1^2 + X_2^2 + \cdots + X_n^2$.

(a) Find the mean of X.
(b) Find the variance of X [use (3.45)].
(c) What are the mean and variance of a chi-square RV of n degrees of freedom?

4.28 *Mean and variance of sample mean.* From n independent RVs $X_1, \ldots, X_n$ with a common expected value $\bar{x}$ and a common variance σ^2, form a new RV

$$\hat{X} = \frac{1}{n}\sum_{i=1}^{n} X_i$$

which is known as ***sample average*** or ***sample mean***. Find the mean and variance of $\hat{X}$ in terms of the true mean $\bar{x}$, the sample size n and the variance σ^2. Is $\hat{X}$ a Gaussian RV if X_i's are independent or dependent Gaussian distributed?

4.29 *Mean of sample variance.* Given n independent RVs $X_1, \ldots, X_n$ with a common expected value $\bar{x}$ and a common variance σ^2, form a new RV

$$\hat{V} = \frac{1}{n-1}\sum_{i=1}^{n}(X_i - \hat{X})^2$$

where $\hat{X}$ is the sample average, defined by (3.63). This is known as ***sample variance***. Find the mean of $\hat{V}$ in terms of the true mean $\bar{x}$, the sample size n and the variance σ^2.

4.30 *Correlation of sum and difference.* Zero-mean RVs X and Y have equal variances. Show that $U = X + Y$ and $V = X - Y$ are uncorrelated.

4.31 *Orthogonality of sum and difference.* Find the real number c, in terms of the moments of X and Y, such that $U = X - cY$ and $V = X + cY$ are orthogonal.

4.32 *Marginal vs. joint Gaussian.* Two RVs X and Y have the following joint PDF

$$f(x,y) = \begin{cases} \frac{1}{2\pi}e^{-\frac{x^2+y^2}{2}} + \frac{1}{2\pi}e^{-\pi^2}\cos x \cos y & |x| < \pi, |y| < \pi \\ \frac{1}{2\pi}e^{-\frac{x^2+y^2}{2}} & \text{elsewhere} \end{cases}$$

(a) Find the marginal PDFs of X and Y.
(b) Are X and Y Gaussian RVs? Are they jointly Gaussian?

4.33 *Probability of sum of independent Gaussian RVs.* Given two independent RVs $X_1 \sim \mathcal{N}(2,1)$ and $X_2 \sim \mathcal{N}(1,4)$, find $P\{0 < X_1 + X_2 < 2\}$.

4.34 *Distribution of sum of independent uniform RVs.* A RV Y is the sum of 500 independent and identically distributed RVs $X_1, X_2, \ldots, X_{500}$. Each X_i is uniformly distributed over $(1,2)$. Find the mean and variance of Y. Give an approximate distribution or PDF of Y and justify your answer. What is the probability that Y is larger than 750?

4.35 *PDF of sum of two RVs.* Two resistors with independent random resistances R_1 and R_2 are connected in series.

(a) Find the PDF of the total resistance if $R_1 \sim \mathcal{N}(100, 3^2)$ and $R_2 \sim \mathcal{N}(500, 5^2)$.
(b) Find the PDF of the total resistance if R_1 and R_2 have the following identical PDF:

$$f_{R_i}(r) = \begin{cases} \frac{100-r}{400} & 0 < r < 100 \\ 0 & \text{elsewhere} \end{cases}$$

4.36 *Failure time analysis.* Three systems are connected as shown. Their times to failure X_1, X_2, X_3 have the following CDFs:

$$F_{X_i}(x) = (1 - e^{-a_i x})u(x), \qquad i = 1, 2, 3$$

where $u(x)$ is the unit step function. Find the CDF $F(x)$ and PDF $f(x)$ of the time to failure X of the total system S. If $a_1 = 1, a_2 = 2, a_3 = 3$, what are $F(x)$ and $f(x)$?

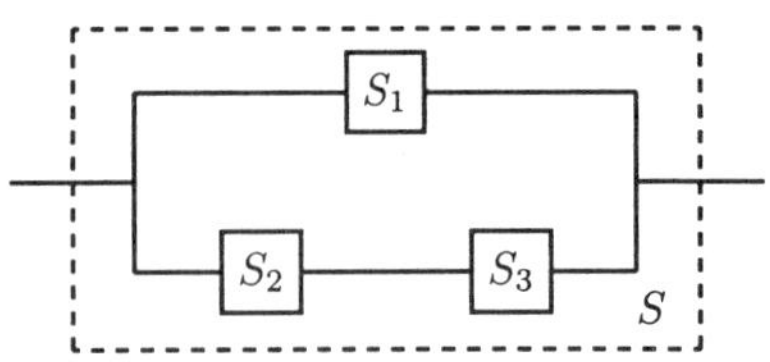

4.37 *PMF of functions of discrete RVs.* Consider two independent tossing of a fair die. Let $X(F_i^1) = i, Y(F_i^2) = i$, where F_i^1 and F_i^2 stand for the events that face i shows up in the first and second tossing, respectively. Find the PMF of (a) $Z = X + Y$; (b) $U = \max(X, Y)$; (a) $V = \min(X, Y)$.

4.38 *From joint PMF to others.* RVs (X, Y) have the following PMF

$Y\backslash X$	1	2	3	4	5	6
1	0.01	0.04	0.04	0.05	0.06	0.01
2	0.01	0.06	0.05	0.07	0.05	0.02
3	0.04	0.09	0.03	0.05	0.05	0.03
4	0.01	0.05	0.04	0.06	0.06	0.02

(a) Find the PMF of X.
(b) Find the PMF of Y.
(c) Find the PMF of $U = \max(X, Y)$.
(d) Find the PMF of $V = \min(X, Y)$.
(e) Find the PMF of $W = X + Y$.

4.39* *Correlation of product and difference of independent Gaussian RVs.* Consider two independent RVs $X \sim \mathcal{N}(0, 1)$ and $Y \sim \mathcal{N}(0, 1)$. Let RVs $U = XY$ and $V = X - Y$.

(a) Find the expected values of U and V.
(b) Find the correlation coefficient of U and V.

4.40* *Functions of independent RVs.* Consider two independent RVs X and Y with identical exponential PDF

$$f(x) = \begin{cases} e^{-x} & x > 0 \\ 0 & x < 0 \end{cases}$$

Let $U = X + Y$ and $V = X/Y$.

(a) Find the joint PDF of U and V.
(b) Find the marginal PDFs of U and V, respectively.
(c) Are U and V independent? uncorrelated?

4.41* *Functions of independent RVs.* Consider two independent RVs $X \sim \mathcal{N}(0, 1)$ and $Y \sim \mathcal{N}(0, 1)$. Let $U = X + Y$ and $V = \tan^{-1}(X/Y)$.

(a) Find the joint PDF of U and V.
(b) Find the marginal PDFs of U and V, respectively.
(c) Are U and V independent? uncorrelated?
(d) Identify the distributions of U and V, respectively.

4.42 *System reliability.* Three independent subsystems S_1, S_2, S_3 are connected in series to form a system S. The times to failure of the subsystems S_1, S_2 and S_3 are exponential RVs X_1, X_2, and X_3 with CDFs

$$F_{X_i}(x_i) = \left\{ \begin{array}{ll} 1 - e^{-a_i x_i} & x > 0 \\ 0 & x_i \leq 0 \end{array} \right\}, \qquad i = 1, 2, 3$$

with $a_i > 0$. Find the CDF and PDF of the time to failure of the system S. Can you extend the results to a system with n independent subsystems?

4.43 *Correction to a mistake.* For Example 4.1, a student calculates $P\{(X < 4) \cap (Y > 2)\}$ as follows: Let $Z = -Y$. Then

$$F_{X,Z}(x, z) = \begin{cases} 1 - e^z & x > 4,\ z \leq 0 \\ \frac{x}{4}(1 - e^z) & 0 \leq x \leq 4,\ z \leq 0 \\ 0 & \text{else} \end{cases}$$

Thus,

$$\begin{aligned} P\{(X < 4) \cap (Y > 2)\} &= P\{(X < 4) \cap (Z \leq -2)\} \\ &= F_{X,Z}(4, -2) = 1 - e^{-2} \\ &= 0.8647 \end{aligned}$$

Indicate and explain where the student made a mistake.

4.44 *Puzzling questions.* Raise two technical questions of your own that you are puzzled by and that are closely related to the material covered in this chapter. These questions should be important, not too simple, and not too involved.

4.14 Computer Exercises

4.1 *Scatter diagram.* The MATLAB command `randn(2,1)` generates a pair of two independent, standard (i.e., with zero mean and unity variance) Gaussian distributed random numbers X and Y in the form of $\begin{bmatrix} X \\ Y \end{bmatrix}$, that is, $(X, Y) \sim \mathcal{N}(0, 0; 1, 1, 0)$. Let

$$\begin{bmatrix} U \\ V \end{bmatrix} = \begin{bmatrix} 1 & 3 \\ 3 & 4 \end{bmatrix} \begin{bmatrix} X - 2 \\ Y - 3 \end{bmatrix}$$

(a) Generate 40 pairs of random numbers (U_i, V_i) for $i = 1, 2, \ldots, 40$ and plot them in a scatter diagram; that is, plot v_i vs. u_i for $i = 1, 2, \ldots, 40$.
(b) Are U and V dependent? Are they correlated? If yes, give your best guess of their correlation coefficient from the scatter diagram.

(c) Give your guess of the means $\bar{u}$ and $\bar{v}$, variances σ_u^2 and σ_v^2, and correlation coefficient ρ_{uv}.

You may use the companion software P&R as much as possible to complete the above tasks.

4.2 *Correlation determination.* Data files `m4_2a.dat`, `m4_2b.dat` and `m4_2c.dat` in the companion software P&R contain n values of (X, Y), (U, V) and (W, Z), respectively, in the following format:

$$\begin{array}{cc} x_1 & y_1 \\ x_2 & y_2 \\ \vdots & \vdots \\ x_n & y_n \end{array}$$

(a) Plot the data points $(x_1, y_1), \ldots, (x_n, y_n)$. Plot similarly the data points (u_i, v_i) and (w_i, z_i) in separate figures.
(b) Do you think that the two RVs in each of the pairs (X, Y), (U, V), (W, Z) are independent, uncorrelated or correlated? Justify your answer. For each pair that you think is correlated, give your best guess of how large the correlation coefficients, the expected values and the variances of all RVs are.
(c) Repeat parts (a) and (b) using the companion software P&R.

4.3 *Generation and testing of uncorrelated Gaussian random numbers.* A pair of independent standard Gaussian RVs X_1 and X_2 can be generated by the following formulas:

$$R = \sqrt{-2\ln U_1}$$
$$X_1 = R\cos(2\pi U_2)$$
$$X_2 = R\sin(2\pi U_2)$$

where $U_1 \sim \mathcal{U}(0,1)$ and $U_2 \sim \mathcal{U}(0,1)$ are independent RVs.

(a) Write a computer program implementing this method to generate independent Gaussian random numbers.
(b) Use the companion software P&R to plot the histogram of 500 random numbers generated in (a) to verify that they are indeed Gaussian distributed.
(c) Identify the mean and variance of the RV by calculating sample mean and sample variance (3.63)–(3.64) of the random numbers generated. Compare the results with those obtained by P&R.

4.4 *Identify parameters of jointly Gaussian distributions.* The files `m4_4a.dat`, `m4_4b.dat` and `m4_4c.dat` in the companion software P&R contain three records of data, which are the realizations of three pairs of jointly Gaussian RVs (X, Y), (U, V), (W, Z), respectively.

(a) Write a computer program to do the following:

- Plot six histograms of the six RVs using the three data sets.
- Compute the sample means, sample variances and sample correlation coefficients of the three pairs of RVs, respectively.

(b) Repeat the above using the companion software P&R.

(c) Give your guess of the distributions of RVs (X, Y), (U, V), (W, Z), respectively.

4.5 *Generation and testing of correlated Gaussian random numbers.* The vector form of the PDF (4.33) of jointly Gaussian RVs is

$$f(x, y) = \frac{1}{2\pi\sqrt{|C|}} e^{-\frac{1}{2}[x \;\; y]C^{-1}\left[\begin{smallmatrix} x \\ y \end{smallmatrix}\right]} \tag{4.52}$$

where $|C|$ is the determinant of the covariance matrix

$$C = \begin{bmatrix} \sigma_x^2 & \rho\sigma_x\sigma_y \\ \rho\sigma_x\sigma_y & \sigma_y^2 \end{bmatrix}$$

The MATLAB command `randn(2,1)` generates a pair of two independent, standard (i.e., with zero mean and unity variance) Gaussian distributed random numbers U, V in the form of $\begin{bmatrix} U \\ V \end{bmatrix}$, that is, $(U, V) \sim \mathcal{N}(0, 0; 1, 1, 0)$. If $C^{1/2}(C^{1/2})' = C$, where $C^{1/2}$ can be chosen as a lower triangular matrix $C^{1/2} = \begin{bmatrix} a & 0 \\ b & c \end{bmatrix}$, then

$$\begin{bmatrix} X \\ Y \end{bmatrix} = C^{1/2} \begin{bmatrix} U \\ V \end{bmatrix} + \begin{bmatrix} \bar{x} \\ \bar{y} \end{bmatrix}$$

will be jointly Gaussian with the PDF (4.33); i.e., $(X, Y) \sim \mathcal{N}(\bar{x}, \bar{y}, \sigma_x^2, \sigma_y^2, \rho_{xy})$. Give a procedure for the generation of the following pairs of random numbers:

$$(X, Y) \sim \mathcal{N}(\bar{x}, \bar{y}, \sigma_x^2, \sigma_y^2, \rho_{xy})$$

For each of the following three cases:

(i) $(\bar{x}, \bar{y}, \sigma_x^2, \sigma_y^2, \rho_{xy}) = (1, 2, 1.5, 2.3, 0)$,

(ii) $(\bar{x}, \bar{y}, \sigma_x^2, \sigma_y^2, \rho_{xy}) = (10, 1, 1.5, 2.3, 0.97)$,

(iii) $(\bar{x}, \bar{y}, \sigma_x^2, \sigma_y^2, \rho_{xy}) = (1, 2, 1.5, 2.3, 0.7)$,

do the following.

(a) Generate 500 pairs of RVs (X, Y) using the above procedure. Plot the histograms of X and Y, respectively, using these pairs of random numbers to verify that they are indeed Gaussian distributed. Identify $\bar{x}, \bar{y}, \sigma_x^2, \sigma_y^2, \rho_{xy}$ by calculating sample means (3.63), sample variances (3.64) and sample correlation coefficient (4.24) of X and Y.

(b) Use the last 100 pairs of the random numbers generated to plot scatter diagrams to check roughly the correlation of X and Y generated in (a).
(c) Repeat (a) and (b) using the companion software P&R.

4.6 *Generation and testing of correlated Gaussian random numbers.* An alternative method of generating correlated Gaussian RVs is based on Example 4.22. It was shown there that if $(X, Y) \sim \mathcal{N}(\bar{x}, \bar{y}; \sigma_x^2, \sigma_y^2, \rho)$, then the conditional PDF of X given Y is given by

$$f(x|y) = \mathcal{N}\Big(x; \bar{x} + \rho\frac{\sigma_x}{\sigma_y}(y - \bar{y}), \sigma_x^2(1 - \rho^2)\Big)$$

Thus a pair of $(X, Y) \sim \mathcal{N}(\bar{x}, \bar{y}; \sigma_x^2, \sigma_y^2, \rho)$ RVs can be generated in two steps:

S1. Generate a random number $Y \sim \mathcal{N}(\bar{y}; \sigma_y^2)$ using MATLAB function `randn`.
S2. Generate a random number $X \sim \mathcal{N}\left(\bar{x} + \rho\frac{\sigma_x}{\sigma_y}(y - \bar{y}), \sigma_x^2(1 - \rho^2)\right)$ using MATLAB function `randn`.

(a) For $(\bar{x}, \bar{y}, \sigma_x^2, \sigma_y^2, \rho) = (1, 2, 1.5, 2.3, 0.8)$, fix $y = 2.3$, using step S2 to generate 1000 random numbers X_i's.
(b) Plot the histogram of X_i's using the random numbers generated to verify that they are indeed Gaussian distributed.
(c) Identify the mean and variance of X given $Y = 2.3$ by calculating sample mean (3.63) and sample variance (3.64) of the random numbers X_i generated.
(d) Write a program to implement the above method of generating correlated Gaussian RVs.
(e) Generate random numbers for the following three cases, respectively:
(i) $(\bar{x}, \bar{y}, \sigma_x^2, \sigma_y^2, \rho_{xy}) = (1, 2, 1.5, 2.3, 0)$.
(ii) $(\bar{x}, \bar{y}, \sigma_x^2, \sigma_y^2, \rho_{xy}) = (10, 1, 1.5, 2.3, 0.97)$.
(iii) $(\bar{x}, \bar{y}, \sigma_x^2, \sigma_y^2, \rho_{xy}) = (1, 2, 1.5, 2.3, 0.7)$.
(f) Identify $\bar{x}, \bar{y}, \sigma_x^2, \sigma_y^2$, and ρ_{xy} by calculating sample means (3.63), sample variances (3.64) and sample correlation coefficient (4.24) of X and Y based on data generated in (e).
(g) Use three 100-point scatter diagrams to check roughly the correlation of X and Y based on data generated in (e).
(h) Repeat parts (e)–(g) using the companion software P&R.

4.7 *Numerical evaluation of correlation.* Following Example 4.8, use the companion software P&R to evaluate the correlation of (a) $Y = \sin(X + \pi/3)$ and X, and (b) $Z = X^4$ and X, where X has a log-normal distribution with $\sigma = 2$ and $a = 3.4$.

4.15 Self-Test Problems

4.1 Answer the following questions briefly:

(a) Is the marginal CDF of RV X equal to the joint CDF of (X, Y) with the value of Y fixed at an arbitrary value?
(b) Is the joint PDF of two RVs always equal to the product of the marginal PDFs of each RV? If not, under what condition is it true?
(c) If $g(x, y)$ describes a curve in the x-y plane, is it always true that $P\{g(X, Y) = 0\} = 0$? If not, under what condition is it true?
(d) Can a pair of jointly uniform RVs have two different nonzero PDF values?
(e) Is it true that X and Y have a stronger linear correlation than U and V if $|R_{xy}| > |R_{uv}|$? Is it true if $|C_{xy}| > |C_{uv}|$? Is it true if $|\rho_{xy}| > |\rho_{uv}|$?
(f) Can a RV and its nonlinear function be uncorrelated?
(g) Are two different functions of two RVs always correlated or uncorrelated?
(h) For two independent RVs, what is the convolution of their marginal PDFs equal to?
(i) For two arbitrary Gaussian RVs, Is it true that they are independent if and only if they are uncorrelated?
(j) What is the distribution of the sum of two jointly Gaussian RVs? What is the distribution of the sum of two independent Gaussian RVs?

4.2 Given the joint PDF of two RVs X and Y

$$f(x, y) = \begin{cases} \frac{1}{24}(1 + xy) & 0 \leq x \leq 2,\ 0 \leq y \leq 4 \\ 0 & \text{elsewhere} \end{cases}$$

(a) Find the joint CDF of X and Y.
(b) Show that the marginal PDFs of X and Y are, respectively:

$$f(x) = \begin{cases} \frac{1+2x}{6} & 0 \leq x < 2 \\ 0 & \text{elsewhere} \end{cases}$$

$$f(y) = \begin{cases} \frac{1+y}{12} & 0 \leq y \leq 4 \\ 0 & \text{elsewhere} \end{cases}$$

(c) Find $P\{X \leq 0.8, Y \leq 1\}$.
(d) Find $P\{X + Y > 1\}$.
(e) Find $P\{X - Y > 3\}$.
(f) Find the expected values and variances of X and Y.
(g) Find the correlation coefficient between X and Y.
(h) Determine if X and Y are independent. Justify your answer.
(i) Determine if X and Y are uncorrelated. Justify your answer.
(j) Find $E[2X + 5Y]$.

4.3 Given three independent RVs $X_1 \sim \mathcal{N}(1, 1)$, $X_2 \sim \mathcal{N}(2, 1)$, and $X_3 \sim \mathcal{N}(2, 4)$, find $P\{0 < X_1 + X_2 + X_3 < 6\}$.

4.4 Given two zero-mean unity-variance random variables X and Y with a correlation coefficient $\rho_{xy} = -0.5$, form a new random variable

$$V = (cX - 2Y)^2 - 3Y$$

where c is a real-valued constant.

(a) Find the value of c such that $E[V] = 4$.
(b) Find the value of c that minimizes the mean $E[V]$.
(c) Can $E[V]$ be smaller than 2.5? Justify your answer.

4.5 Given the joint CDF of X and Y

$$F_{X,Y}(x, y) = \begin{cases} 1 - e^{-x} & x > 0, y > 0 \\ 0 & \text{elsewhere} \end{cases}$$

(a) Find the marginal CDF and PDF of X. Identify RV X.
(b) Find the marginal CDF and PDF of Y. How many possible values can Y take on?
(c) Are X and Y independent?
(d) Find the correlation R_{xy}.

4.16 Solutions to Self-Test Problems

4.1 (a) No, the marginal PDF $F(x) = F(x, y)|_{y=\infty}$. It is equal to the joint PDF of (X, Y) with y fixed at infinity.
(b) No, the joint PDF of two RVs is equal to the product of the marginal PDFs of each RV if and only if the two RVs are independent.
(c) $P\{g(X, Y) = 0\} = 0$ if and only there is no discrete point mass on the curve $g(x, y) = 0$. Thus, it is true if (X, Y) are continuous because a curve has zero area.
(d) No, a pair of RVs is defined to be jointly uniform if their PDF is constant over only one area.
(e) Correlation R_{xy} depends on the expected values and variances as well as the correlation of X and Y; covariance R_{xy} depends on the variances as well as the correlation of X and Y. Thus neither $|R_{xy}| > |R_{uv}|$ nor $|C_{xy}| > |C_{uv}|$ necessarily implies that (X, Y) have a stronger linear correlation than (U, V). However, since the correlation coefficient ρ_{xy} is the correlation of the standardized RVs $\tilde{X}$ and $\tilde{Y}$ or the properly normalized covariance, it is the right thing (measure) for linear correlation between RVs and thus if $|\rho_{xy}| > |\rho_{uv}|$ then (X, Y) have a stronger linear correlation than (U, V).
(f) Yes, see Examples 4.7 and 4.9.
(g) No, they may or may not be correlated. See Example 4.31.
(h) The convolution is the PDF of their sum. See (4.37).
(i) It is true only if the two RVs are *jointly* Gaussian.

(j) Both are Gaussian distributed.

4.2 (a)

$$F(x,y) = \int_{-\infty}^{x}\int_{-\infty}^{y} f(x,y)dydx$$
$$= \begin{cases} 1 & x \geq 2,\ y \geq 4 \\ \int_0^2 \int_0^y \frac{1}{24}(1+xy)dydx & x \geq 2,\ 0 \leq y < 4 \\ \int_0^x \int_0^4 \frac{1}{24}(1+xy)dydx & 0 \leq x < 2,\ y \geq 4 \\ \int_0^x \int_0^y \frac{1}{24}(1+xy)dydx & 0 \leq x \leq 2,\ 0 \leq y \leq 4 \\ 0 & \text{elsewhere} \end{cases}$$
$$= \begin{cases} 1 & x \geq 2,\ y \geq 4 \\ (y^2+2y)/24 & x \geq 2,\ 0 \leq y < 4 \\ (x^2+x)/6 & 0 \leq x < 2,\ y \geq 4 \\ (xy + x^2y^2/4)/24 & 0 \leq x \leq 2,\ 0 \leq y \leq 4 \\ 0 & \text{elsewhere} \end{cases}$$

(b)

$$f(x) = \int_{-\infty}^{\infty} f(x,y)dy = \begin{cases} \int_0^4 \frac{1}{24}(1+xy)dy = \frac{1+2x}{6} & 0 \leq x < 2 \\ 0 & \text{elsewhere} \end{cases}$$
$$f(y) = \int_{-\infty}^{\infty} f(x,y)dx = \begin{cases} \int_0^2 \frac{1}{24}(1+xy)dx = \frac{1+y}{12} & 0 \leq y \leq 4 \\ 0 & \text{elsewhere} \end{cases}$$

(c)

$$P\{X \leq 0.8, Y \leq 1\} = F(0.8,1) = (xy + x^2y^2/4)/24\Big|_{x=0.8,y=1} = 0.04$$

(d)

$$\begin{aligned} P\{X+Y>1\} &= P\{Y > 1-X\} \\ &= \int_{x=-\infty}^{\infty}\int_{y=1-x}^{\infty} f(x,y)dxdy \\ &= \int_0^2\int_{1-x}^{4} \frac{1}{24}(1+xy)dxdy \\ &= \frac{1}{24}\int_0^2 \left[4-(1-x)+x\frac{1}{2}[16-(1-x)^2]\right]dx \\ &= \frac{1}{24}\left[6+\frac{9}{2}x^2\Big|_0^2 - \frac{1}{2}\int_0^2 (x-2x^2+x^3)dx\right] \\ &= \frac{1}{24}\left[24 - \frac{1}{2}\left(\frac{1}{2}x^2 - \frac{2}{3}x^3 + \frac{1}{4}x^4\right)\Big|_0^2\right] = \frac{71}{72} \\ &= 0.9861 \end{aligned}$$

(e) $P\{X - Y > 3\}$ can be obtained similarly as for $P\{X + Y > 1\}$. However, a better way is the following. Since the maximum value that X can take on with nonzero probability is 2 and the minimum value that Y can take on with nonzero probability is 0, the difference $X - Y > a$ will have a zero probability for any $a \geq 2$. Thus, $P\{X - Y > 3\} = 0$.

(f)

$$
\begin{aligned}
E[X] &= \int_{-\infty}^{\infty} x f(x) dx = \int_0^2 x \frac{1+2x}{6} dx = \frac{1}{6}(2 + \frac{2}{3} \times 8) = 1.2222 \\
E[Y] &= \int_{-\infty}^{\infty} y f(y) dx = \int_0^4 y \frac{1+y}{12} dy = \frac{1}{12}(8 + \frac{1}{3} \times 64) = 2.4444 \\
E[X^2] &= \int_{-\infty}^{\infty} x^2 f(x) dx = \int_0^2 x^2 \frac{1+2x}{6} dx = \frac{1}{6}(\frac{8}{3} + \frac{2}{4} \times 16) = 1.7778 \\
E[Y^2] &= \int_{-\infty}^{\infty} y^2 f(y) dx = \int_0^4 y^2 \frac{1+y}{12} dy = \frac{1}{12}(\frac{64}{3} + \frac{1}{4} \times 256) = 7.1111 \\
\sigma_x^2 &= E[X^2] - (E[X])^2 = 0.284 \\
\sigma_y^2 &= E[Y^2] - (E[Y])^2 = 1.136
\end{aligned}
$$

(g)

$$
\begin{aligned}
E[XY] &= \int_{-\infty}^{\infty} \int_{-\infty}^{\infty} xy f(x,y) dx dy \\
&= \int_0^2 \int_0^4 xy \frac{1}{24}(1 + xy) dx dy \\
&= \frac{1}{24}[\frac{4}{2}\frac{16}{2} + \frac{8}{3}\frac{64}{3}] \\
&= 3.037 \\
\rho_{xy} &= \frac{E[(X - \bar{x})(Y - \bar{y})]}{\sigma_x \sigma_y} = \frac{E[XY] - \bar{x}\bar{y}}{\sigma_x \sigma_y} \\
&= \frac{3.037 - (1.2222)(2.4444)}{\sqrt{(0.284)(1.136)}} \\
&= 0.08706
\end{aligned}
$$

(h) Since $f(x)f(y) \neq f(x,y)$, X and Y are not independent.

(i) Since $E[XY] \neq E[x]E[Y]$ (or equivalently, $\rho_{xy} \neq 0$), X and Y are not uncorrelated.

(j)

$$
E[2X + 5Y] = 2E[X] + 5E[Y] = (2)(1.2222) + (5)(2.4444) = 14.6664
$$

4.3 Let $Y = X_1 + X_2 + X_3$. Since Y is a weighted sum of independent Gaussian RVs, it is also Gaussian with the mean

$$
\begin{aligned}
E[Y] &\overset{(4.31)}{=} E[X_1 + X_2 + X_3] = \bar{x}_1 + \bar{x}_2 + \bar{x}_3 = 1 + 2 + 2 = 5 \\
\sigma_y^2 &\overset{(4.32)}{=} 1^2 + 1^2 + 4^2 = 18
\end{aligned}
$$

That is, $Y \sim \mathcal{N}(5, 18)$. Hence,

$$\begin{aligned} P\{0 < X_1 + X_2 + X_3 < 6\} &= P\{0 < Y < 6\} \\ &= P\left\{\frac{0-5}{\sqrt{18}} < \frac{Y-5}{\sqrt{18}} < \frac{6-5}{\sqrt{18}}\right\} \\ &= \Phi(0.236) - \Phi(-1.18) = 0.5933 - 0.119 \\ &= 0.4743 \end{aligned}$$

4.4

$$V = (cX - 2Y)^2 - 3Y = c^2X^2 + 4Y^2 - 4cXY - 3Y$$

(a)

$$\begin{aligned} E[V] &= c^2 E[X^2] + 4E[Y^2] - 4c \underbrace{E[XY]}_{=\sigma_x \sigma_y \rho_{xy}} - 3\bar{y} \\ &= c^2 + 4 - 4c\sigma_x\sigma_y\rho_{xy} - 0 \\ &= 4 + 2c + c^2 \end{aligned}$$

then

$$E[V] = 4 \implies c = 0$$

(b)

$$\left.\begin{aligned} \tfrac{d}{dc}E[V] = 2c + 2 = 0 \Rightarrow c = -1 \\ \tfrac{d^2}{dc^2}E[V] = 2 > 0 \end{aligned}\right\} \implies c = -1 \text{ minimizes } E[V]$$

(c) Since $\min E[V] = E[V]\Big|_{c=-1} = 4 + 2(-1) + (-1)^2 = 3 > 2.5$, the answer is no.

4.5 It is better to rewrite $F_{X,Y}(x, y)$ as

$$F_{X,Y}(x,y) = \begin{cases} 1 - e^{-x} & x > 0, y > 0 \\ 0 & x \le 0, y > 0 \\ 0 & x \le 0, y \le 0 \\ 0 & x > 0, y \le 0 \end{cases}$$

Note that the regions correspond to the four quadrants, respectively.

(a) Note that $F_X(x) = F_{X,Y}(x, \infty)$. For $x > 0$, the first piece of $F_{X,Y}(x, y)$ leads to $F_X(x) = 1 - e^{-x}$. For $x \le 0$, the second piece indicates that $F_X(x) = 0$. Thus, the marginal CDF of X is

$$F_X(x) = \begin{cases} 1 - e^{-x} & x > 0 \\ 0 & x \le 0 \end{cases}$$

The PDF of X is, by taking derivative,

$$f_X(x) = \begin{cases} e^{-x} & x > 0 \\ 0 & x < 0 \end{cases}$$

This indicates that X is exponentially distributed with parameter $\lambda = 1$.

(b) For $y > 0$, the first piece of $F_{X,Y}(x, y)$ leads to

$$F_Y(y) = F_{X,Y}(\infty, y) = \lim_{x \to \infty} (1 - e^{-x}) = 1$$

For $y \leq 0$, the last piece indicates that $F_Y(y) = F_{X,Y}(\infty, y) = 0$. Thus, the marginal CDF of Y is the unit step function; that is, $F_Y(y) = u(y)$. Taking derivative yields that the marginal PDF is the delta function $f_Y(y) = \delta(y)$, which indicates that all the probability mass of Y is located at $y = 0$ and thus Y is actually not random; it is equal to 0 (i.e., there is only one possible value for Y).

(c) It is easy to show that $F_{X,Y}(x, y) = F_X(x)F_Y(y)$ and thus X and Y are independent. The same conclusion could be arrived by showing $f_{X,Y}(x, y) = f_X(x)f_Y(y)$, which is, however, complicated due to the involvement of delta functions.

(d) Since X and Y are independent, the correlation $R_{xy} = \bar{x}\bar{y} = (1)(0) = 0$.

5

INTRODUCTION TO STATISTICS

Statistical thinking will one day be as necessary for efficient citizenship as the ability to read and write.

H. G. Wells

Statistics has been likened to a telescope. The latter enables one to see further and to make clear objects which were diminished or obscured by distance. The former enables one to discern structure and relationships which were distorted by other factors or obscured by random variation.

D. J. Hand, *Psychological Medicine* (1985)

This chapter covers some elementary concepts of statistics, which are necessary for probabilistic analysis when an appropriate probabilistic model is not given.

Main Topics

- Sampling, Sample Mean, and Sample Variance
- Empirical Distributions
- Parameter Estimation
- Hypothesis Testing
- Linear Regression and Curve Fitting

5.1 Introduction

Previous chapters are devoted to the *analysis* of a random phenomenon, where it is assumed that a probabilistic model of the phenomenon (e.g., the distribution of the random variable) is given. In reality, however, it is quite often the case that the probabilistic model is not known. In this case, statistics is essential in establishing the correct probabilistic model. In this chapter we study some elementary concepts of statistics.

Statistics is a science dealing with data subject to uncertainties. It has an extremely wide spectrum of application. Examples of the application areas are

- Quality control
- Instrumentation
- Insurance
- Poll taking
- Weather forecasting.

Statistics is a magic weapon: Given a collection of data/facts, one can arrive at almost any conclusion he/she wants by *abusing* statistics.

Terminology

- The entire collection of data being studied is called a ***population***. It is a RV whose possible values are the values of the data.
- A ***sample*** $(X_1, \ldots, X_n)$ is a subset of the population selected at random.
- The *population size* is the number of data pieces that make up the population.
- The ***sample size*** is the number of pieces of data that make up the sample.

For example, if we conduct a poll by asking 1,000 people to predict the outcome of a U.S. presidential election, then the population consists of all people in the U.S. who are eligible to vote. The sample for this poll is those people surveyed and the sample size is 1,000. The population size is the number of eligible people in the U.S.

It is almost always assumed that the data (i.e., RVs $X_1, X_2, \ldots, X_n$) making up a sample are ***independent and identically distributed*** **(*i.i.d.*)**.

Statistics has two fundamental objectives. The primary objective of *descriptive statistics* is to better understand the population; that is, it is concerned with the question of how to collect, assemble, and present data in a way that can be easily understood. On the other hand, *inductive statistics* or *statistical inference* aims at making inferences, such as decisions and estimates of the population based on a limited set of observations called a sample.

Probability and statistics may be viewed as an analysis-synthesis pair: Probability provides tools for the analysis of a random phenomenon while statistics deals with the various topics of establishing a probabilistic model of the random phenomenon from the observation (data) of the phenomenon. Almost all studies of synthesis are dependent on the corresponding analysis tools, so it is natural that the study of statistics relies heavily on the probabilistic tools.

The field of statistics is extremely vast. It has many areas, including the following.

- *Sampling theory* studies problems involving the selection of samples from a collection of data called the population. Two examples of such problems are how to select samples from a large collection of data and how to determine the properties of a sample.
- *Estimation theory* studies various methods of estimating unknowns based on available data and how these methods should be evaluated.
- *Hypothesis testing* is concerned with the problem of deciding which of a number of hypotheses is true, based on available data, and the properties and decision errors associated with the tests.
- *Curve fitting* fits curves (i.e., mathematical expressions) to the given data.
- *Analysis of variance* assesses the variation in the data (population).

A population usually has a large size. It is often unrealistic to study each and every element of the population. For example, if we want to know the average life of a batch of bulbs, it is clearly unwise to test each and every bulb for its life because not only is it time consuming but also all bulbs are ruined after the test. A natural idea is to study a (small) part of the population that represents the population. Such a representative part is called a ***sample*** in statistics. A good example of using such an idea is spot-checking wherein a conclusion is made based only on the examination of a sample of the products.

Mathematically, a population is a random variable X and a sample is a set of RVs: $\{X_1, \ldots,$ X_n, where n is the sample size. To "represent" the population truly and faithfully, a sample has three basic requirements: (a) it is random in the sense that each element of the population has an equal probability of being sampled (rather than in the sense of just being uncertain); (b) each of its elements has the same distribution as that of the population; and (c) its elements are independent of each other. More specifically, let X be the random variable that represents the population and let $X_1, \ldots, X_n$ be the elements of a (random) sample of X. Then the above requirements can be stated as: $X_1, \ldots, X_n$ are independent and they all have the same distribution as that of X. In short, X and $X_1, \ldots, X_n$ are i.i.d.

In practice, a random sample of independent elements is usually gathered by taking elements of the population at random in such a way that the elements taken have no clear relation. For example, if a telephone poll is to be conducted, then at least the phone numbers should be selected at random (e.g., from a random number generator).

The sample $(X_1, \ldots, X_n)$ is a set of RVs. Its particular value is denoted in lower case by $(x_1, \ldots, x_n)$.

Note that since the elements of a sample are independent, formulas that require the independence of RVs are applicable.

There are two types of sampling. In the *sampling with replacement*, after an item is drawn from the population, it is put back into the population and thus there is a chance it would be drawn again. As such, X_i and X_j have the same distribution even if the population size is finite. If the drawing of an item is done statistically independent of any other drawing, then the RVs $X_1, \ldots, X_n$ that make up a sample are truly i.i.d. On the other hand, in the *sampling without replacement*, an item is not put back into the population after it is drawn. If the population size is finite, the distribution of the (remaining) population is no longer the same as before. For example, if there is one defective resistor among N resistors, the probability that any one of them is defective is $1/N$. If, however, one is drawn and found not defective, then the probability that any one of the remaining is defective is now $1/(N-1)$. If N is not large, the difference would be significant. In this case, the distribution of X_2 would clearly depend on the value of X_1 and is not identical to that of X_1 and, therefore, $X_1, \ldots, X_n$ that make up a sample would not be i.i.d. However, they are approximately i.i.d. if the population size is large (and the defective resistors do not have a constant percentage in the population).

Although the RVs that make up a sample are almost always assumed to be i.i.d., it should be noted that this assumption is only approximately valid for sampling without replacement of a population of a *finite* size. In practice, it is almost always the case that sampling without replacement is used and thus it is important to apply the statistical tools only to problems with a population of a large size unless the i.i.d. assumption is given up.

Note that when the population size is infinite, sampling with or without replacement makes no difference.

Example 5.1: Biased Sample — The 1936 Literary Digest Poll on the Presidential Election

Before the presidential election between Franklin Roosevelt and Alf Landon in 1936, the *Literary Digest* polled 2.4 million people, the largest political poll ever conducted, and predicted a landslide victory for Landon, 57 percent to 43 percent, with a margin of error "within a fraction of 1 percent." However, it was Roosevelt who won in a landslide, 62 percent to 38 percent. The mistake was due to the fact that the sample used by the *Literary Digest* was biased. The *Literary Digest* mailed questionnaires to 10 million people. The primary source of their names was telephone books in the U.S. In 1936, telephone service was relatively new and expensive, and there were only 11 million households with phones. These homes were relatively better off financially and were in favor of the Republican candidate Landon, while those households without a phone were overwhelmingly for the Democrat Roosevelt.

In summary, quality is far more important than quantity in sampling — the elements in the sample should be independent and have an identical distribution as that of the population.

5.2 Sample Mean and Sample Variance

5.2.1 Sample Mean

The ***sample mean*** (or ***sample average***) of a sample $(X_1, \ldots, X_n)$ is defined as the average value of the sample:

$$\boxed{\hat{X} = \frac{1}{n} \sum_{i=1}^{n} X_i} \tag{5.1}$$

The sample mean is used in practice to estimate (approximate) the unknown true mean of the population:

$$\bar{x} = E[X] \approx \hat{X} \tag{5.2}$$

The sample mean is a RV since it is an (evenly weighted) sum of RVs $X_1, \ldots, X_n$. Given a realization $(x_1, \ldots, x_n)$ of the sample (i.e., given the values $x_1, \ldots, x_n$ of the RVs $X_1, \ldots, X_n$), the value (or realization)

$$\hat{x} = \frac{1}{n} \sum_{i=1}^{n} x_i$$

of the sample mean is a nonrandom number.

The expected value (mean) of the sample mean is equal to the true mean:

$$E[\hat{X}] = \frac{1}{n} \sum_{i=1}^{n} E[X_i] = \frac{1}{n} \sum_{i=1}^{n} \bar{x} = \bar{x} \tag{5.3}$$

which has the following important interpretation: *The sample mean equals the true mean on the average.*

The variance of the sample mean as a RV is, since $X_1, \ldots, X_n$ are i.i.d.,

$$\sigma_{\hat{x}}^2 = \text{var}(\hat{X}) = \text{var}\left[\frac{1}{n} \sum_{i=1}^{n} X_i\right] \overset{(4.32)}{=} \frac{1}{n^2} \sum_{i=1}^{n} \text{var}(X_i) = \frac{1}{n} \sigma_x^2 \tag{5.4}$$

where σ_x^2 is the variance of the population (i.e., of the RV X). $\sigma_{\hat{x}}^2$ provides a measure of the variation of the sample mean. Thus, the variation of the sample mean is reduced by increasing the sample size n.

Since the sample mean is the sum of independent RVs, if the size is large, the sample mean is approximately Gaussian distributed in view of the central limit theorem.

Example 5.2: Probabilistic Analysis of the Average Score of a Class

Suppose that the score of a student in a school is approximately a $\mathcal{N}(70, 10^2)$ RV. Consider the average score of a class of 20 students.

Here we are given a sample of size 20 and $\bar{x} = 70, \sigma_x = 10$.

(a) Find the expected value of the average score of this class: The average score is the sample mean and thus its expected value is

$$E[\hat{X}] = \bar{x} \text{ (true mean)} = 70$$

(b) Find the variance of the average score of the class: The variance of the average score is the variance of the sample mean

$$\sigma_{\hat{x}}^2 \overset{(5.4)}{=} \sigma_x^2/n = 10^2/20 = 5$$

(c) Is the average score of this class a Gaussian RV? The average score is the sample mean of the score of a student in this school. Since it is a weighted sum of independent Gaussian RVs for this example, from Example 3.19 it is a Gaussian RV.

(d) Find the probability that a student's score will be in $(60, 85)$: The score X of an arbitrary student is a $\mathcal{N}(70, 10^2)$ RV. Thus, $\tilde{X} = \frac{X-\bar{x}}{\sigma_x} = \frac{X-70}{10}$ is a standard Gaussian RV. Hence

$$P\{60 < X < 85\} = P\left\{\frac{60-70}{10} < \frac{X-70}{10} < \frac{85-70}{10}\right\} = P\{-1 < \tilde{X} < 1.5\}$$
$$= \Phi(1.5) - \Phi(-1) \overset{\Phi \text{ table}}{=} 0.9332 - (1 - 0.8413) = 77.45\%$$

(e) Find the probability that the average score of this class will be in $(65, 75)$: Since $E[\hat{X}] = 70, \sigma_{\hat{x}}^2 = 5$ and the average score $\hat{X}$ is a Gaussian RV, it is a $\mathcal{N}(70, 5)$ RV. Thus, $\tilde{\hat{X}} = \dfrac{\hat{X} - \bar{x}}{\sigma_{\hat{x}}} = \dfrac{\hat{X} - 70}{\sqrt{5}}$ is a standard Gaussian RV. Hence

$$P\{65 < \hat{X} < 75\} = P\left\{\frac{65-70}{\sqrt{5}} < \frac{\hat{X}-70}{\sqrt{5}} < \frac{75-70}{\sqrt{5}}\right\}$$
$$= P\{-2.236 < \tilde{\hat{X}} < 2.236\} = \Phi(2.236) - \Phi(-2.236)$$
$$= 0.9873 - (1 - 0.9873) = 97.46\%$$

From (d) and (e), the distribution of the average score is much more concentrated around its mean than the score of a student. Does this make sense?

5.2.2 Sample Variance

The ***sample variance*** of a sample $(X_1, X_2, \ldots, X_n)$ is defined as

$$\boxed{\hat{V} = \frac{1}{n-1}\sum_{i=1}^{n}(X_i - \hat{X})^2} \tag{5.5}$$

Sample variance is used in practice to estimate the variance of the population:

$$\sigma_x^2 \approx \hat{V} \tag{5.6}$$

Sample variance is also a RV. Its value, given the value of a sample, will be denoted by $\hat{v}$.

Note that

$$\begin{aligned}
E\left[(X_i - \hat{X})^2\right] &= E\left[[(X_i - \bar{x}) - (\hat{X} - \bar{x})]^2\right] \\
&= E\left[(X_i - \bar{x})^2 + (\hat{X} - \bar{x})^2 - 2(X_i - \bar{x})(\hat{X} - \bar{x})\right] \\
&= \sigma_x^2 + \sigma_{\hat{x}}^2 - 2E\left[(X_i - \bar{x})\Big(\frac{1}{n}\sum_{i=1}^{n}(X_j - \bar{x})\Big)\right] \\
&= \sigma_x^2 + \frac{1}{n}\sigma_x^2 - 2E\left[(X_i - \bar{x})\cdot\frac{1}{n}[(X_1 - \bar{x}) + \cdots + (X_n - \bar{x})]\right] \\
&\stackrel{?}{=} \sigma_x^2 + \frac{\sigma_x^2}{n} - \frac{2\sigma_x^2}{n} = \sigma_x^2 - \frac{\sigma_x^2}{n} = \frac{n-1}{n}\cdot\sigma_x^2
\end{aligned}$$

where $\stackrel{?}{=}$ follows from the fact that $X_1, X_2, \ldots, X_n$ are independent and thus uncorrelated. Thus, the expected value of $\hat{V}$ is

$$\begin{aligned}
E[\hat{V}] &= E\Big[\frac{1}{n-1}\sum_{i=1}^{n}(X_i - \hat{X})^2\Big] = \frac{1}{n-1}\sum_{i=1}^{n}E\left[(X_i - \hat{X})^2\right] \\
&= \frac{1}{n-1}\Big[n\cdot\frac{n-1}{n}\cdot\sigma_x^2\Big] = \sigma_x^2
\end{aligned}$$

That is, *the sample variance is equal to the true variance of the population (i.e., the RV) on the average.*

Note that were the sample variance defined as $S^2 = \frac{1}{n}\sum_{i=1}^{n}(X_i - \hat{X})^2$, it would not be equal to the true variance on the average, although it is sometimes also called the *sample variance*. If the true mean $\bar{x}$ of the population is known, however, the sample variance should be defined as $\hat{V} = \frac{1}{n}\sum_{i=1}^{n}(X_i - \bar{x})^2$.

The square root of the sample variance is called the ***standard error***.

Example 5.3: Determination of the Gaussian Curve of Test Scores

Suppose that the score of a student in a school is approximately a $\mathcal{N}(\bar{x}, \sigma^2)$ RV and that 30 students were surveyed whose scores are: 79, 43, 66, 99, 91, 88, 89, 74, 78, 83, 77, 68, 89, 85, 69, 58, 90, 84, 75, 77, 83, 94, 57, 63, 65, 79, 66, 68, 74, 75.

(a) The average score of the class is the sample mean, given by

$$\hat{x} = \frac{1}{30}[79 + 43 + 66 + \cdots + 68 + 74 + 75] = 76.2$$

(b) The sample variance is a measure of the deviation of an individual score from the average score of the class. It has the value

$$\hat{v} = \frac{1}{30-1}\left[(79 - 76.2)^2 + (43 - 76.2)^2 + \cdots + (75 - 76.2)^2\right] = 154.23$$

Hence, the *standard error* is $\sqrt{\hat{v}} = \sqrt{154.23} = 12.42$.

(c) The standard deviation of the sample mean is approximately given by

$$\sigma_{\hat{x}} = \sqrt{\sigma^2/n} \approx \sqrt{\hat{v}/30} = \sqrt{154.23/30} = 2.27$$

It is a measure of the variation (randomness) of the average score.

We may thus conclude that the test score of a student in this school is approximately a $\mathcal{N}(76.2, 154.23)$ RV and the average score is approximately a $\mathcal{N}(76.2, 2.27^2)$ RV. The Gaussian curve of the test scores provided by your professor may have been generated in this way.

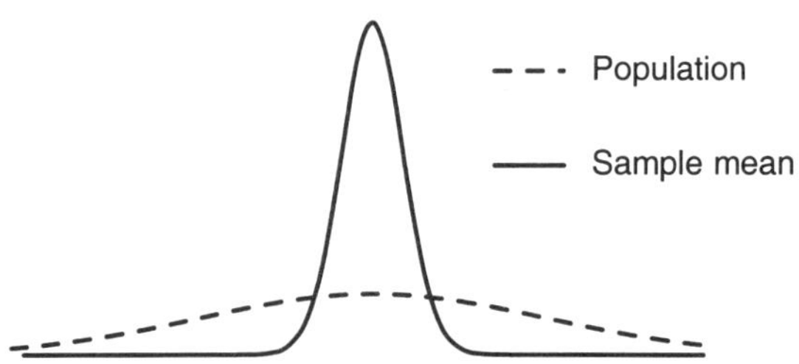

Figure 5.1: PDFs of the population and sample mean.

The ***sample variance*** is sometimes defined as

$$S^2 = \frac{1}{n}\sum_{i=1}^{n}(X_i - \hat{X})^2$$

although it is biased, as shown previously, in the sense that its expected (average) value is not equal to the true variance. Similarly, the ***sample moment*** and ***sample central moment*** are defined as

$$\text{sample } k\text{th moment} = \frac{1}{n}\sum_{i=1}^{n} X_i^k$$

$$\text{sample } k\text{th central moment} = \frac{1}{n}\sum_{i=1}^{n}(X_i - \hat{X})^k$$

It can be shown that the sample moments and sample central moments approach the true moment and true central moment, respectively, as sample size increases.

Although (5.3) is valid provided that $X_1, \ldots, X_n$ have the same mean $\bar{x}$, which is always the case for sampling of a population, (5.4) is valid only if $X_1, \ldots, X_n$ are i.i.d. (or more rigorously, uncorrelated and have the same variance), which is clearly not true for *sampling without replacement* of a population of a finite size. In such a case, (5.4) should be replaced by the following

$$\text{var}(\hat{X}) = \frac{\sigma_x^2}{n}\frac{N-n}{N-1}$$

where N is the population size. Clearly, this variance approaches (5.4) as the sample size n increases. Note that this variance vanishes when the sample size n is equal to the population size N. This makes sense since the sample mean in this case is exactly equal to the true mean of the population because all elements of the population have been drawn. Note also that this variance is identical to (5.4) for $n = 1$, which makes perfect sense since sampling with or without replacement makes no difference when only one item is drawn.

Since the sample variance itself is a RV, it has a variance. The ***variance of the sample variance*** (5.5) is given by, after a lengthy but straightforward derivation,

$$\sigma_{\hat{v}}^2 = \text{var}(\hat{V}) = E[(\hat{V} - E[\hat{V}])^2] = \frac{1}{n}\left(E[(X - \bar{x})^4] - \frac{n-3}{n-1}\sigma_x^4\right)$$

It provides a measure of the variation of the sample variance around its mean, which is the true variance of the population.

5.3 Empirical Distributions

In the previous chapter, it was assumed that the distribution of a RV X is given. In reality, it is more often the case that the distribution is not known but the values $x_1, \ldots, x_n$ of a sample of X are given. Note that the elements $X_1, \ldots, X_n$ of a sample are assumed to be independent. How do we obtain an approximate distribution of the RV X based on the data $x_1, \ldots, x_n$?

The ***empirical CDF*** of a RV X given the values $x_1, \ldots, x_n$ of a sample of X is defined as

$$\bar{F}(x|x_1, \ldots, x_n) = \frac{\text{number of sample values } x_1, \ldots, x_n \text{ not greater than } x}{n}$$

The empirical CDF can be used to approximate the true CDF. This is justified by the so-called Glivenko theorem which states that the empirical CDF uniformly converges to the true CDF with probability one as $n \to \infty$.

$\bar{F}(x|x_1, \ldots, x_n)$ is actually a histogram of a stairway type. Its value at x is the percentage of the points $x_1, \ldots, x_n$ that are not larger than x. It is often more convenient to use a histogram of the PDF type. This can be done based on the relation $f_X(x) \approx P\{x < X \leq x + \Delta x\}/\Delta x$ for small Δx. Thus the ***empirical PDF*** can be defined as

$$\bar{f}(x|x_1, \ldots, x_n) = \frac{\text{number of sample values } x_1, \ldots, x_n \text{ in } [x, x + \Delta x)}{n\Delta x}$$

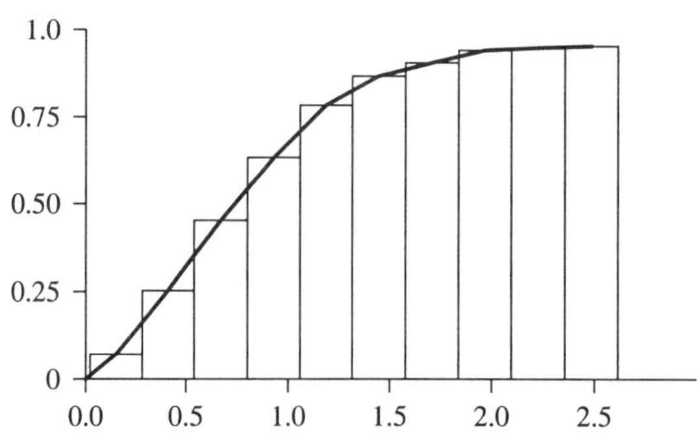

Figure 5.2: The empirical CDF of Example 5.4.

Example 5.4: Identification of Distribution From Data

The file `e5_4.dat` in the companion software P&R contains a record of 400 pieces of data. The distribution of the population from which the data was drawn can be identified as follows.

S1. Use P&R to plot the empirical PDF (i.e., histogram of the PDF type), as shown in Fig. 5.3. It can be observed that the empirical PDF resembles the PDF of a Weibull or Rayleigh RV. Since Weibull includes Rayleigh as a special case, let us assume it is a Weibull RV.

S2. Use P&R to compute the sample mean and sample variance of the data as $\hat{x} = 0.8891$ and $\hat{v} = 0.2299$. From Table 3.1, it is found that the parameters $a \approx 1.946$ and $b \approx 0.9930$.

S3. Use P&R to overlay the empirical PDF and the Weibull PDF with $a = 1.946$ and $b = 0.993$, as shown in Fig. 5.4. They match quite well and thus it can be concluded that the population is probably Weibull distributed with parameters $a = 1.946$ and $b = 0.993$.

In fact, the data was drawn from a population that is Weibull distributed with parameters $a = 2$ and $b = 1$ (i.e., Rayleigh distributed).

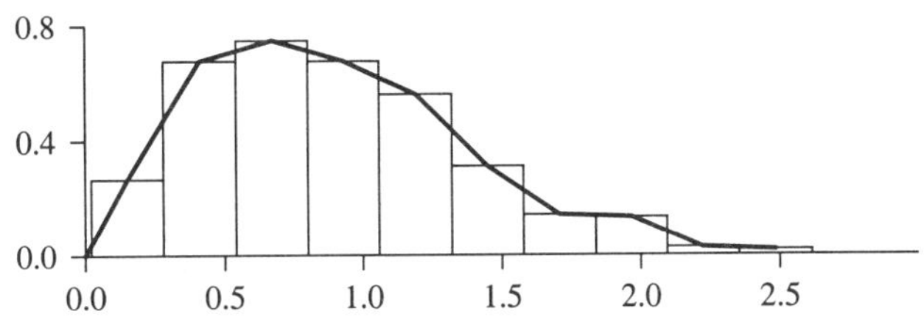

Figure 5.3: Empirical PDF of data `e5_4.dat`.

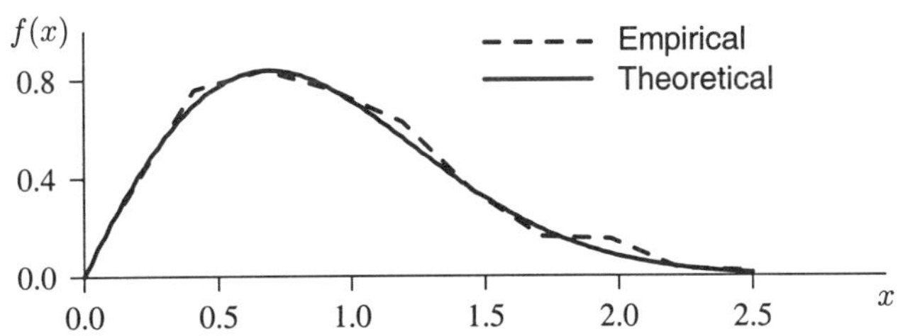

Figure 5.4: Comparison of the empirical and theoretical PDFs.

5.4 Statistical Inference

Statistical inference consists of two parts: Estimation and decision making concerning some unknown parameters of the population.

- ***Estimation*** provides an approximate value of the parameter that is close to the true value.
- ***Decision*** or ***hypothesis testing*** decides whether a given or hypothesized value of the parameter should be rejected as the true value or not.

Example 5.5: Estimation of Mean and Variance of a Gaussian Population

The mean and variance of a population $X \sim \mathcal{N}(\bar{x}, \sigma^2)$ are not known, where X is the age of a member of a professional society. A random sample $(X_1, \ldots, X_n)$ of X is available. Many estimates of $\bar{x}$ and σ^2 are possible. For example, the sample mean and sample variance can be their estimates, respectively, that is, $\bar{x} \approx \hat{X}, \sigma^2 \approx \hat{V}$. $\bar{x}$ and σ can also be estimated by $\bar{x} \approx X_m$ and $\sigma \approx R/d_n$, where X_m is the ***sample median*** and R is the ***sample range***, defined by

$$X_m = \text{middle value of the sample} = \begin{Bmatrix} X_{i+1} & n \text{ odd} \\ \frac{1}{2}(X_i + X_{i+1}) & n \text{ even} \end{Bmatrix}, \quad X_i \leq X_{i+1}$$

$$R = \max(X_1, \ldots, X_n) - \min(X_1, \ldots, X_n)$$

and $d_n \approx n(n - \frac{1}{2})^{-1/2}$ is a constant. If the sample has the values: 39, 41, 55, 34, 52, 45, 36, then we have

$$\hat{x} = 43.14, \quad \hat{v} = 63.12, \quad x_m = 41, \quad r = 22 \Longrightarrow (r/d_n)^2 = 64.20$$

Note that $\hat{x} \approx x_m$ and $\hat{v} \approx (r/d_n)^2$.

Example 5.6: Hypothesis Testing on Mean of Measurement Error

The measurement error of a device has variance $\sigma^2 = 0.0004$ and an unknown mean $\bar{x}$. Suppose we hypothesize that $\bar{x} = 0$ (i.e., no bias). Should we reject this hypothesis given a sample $(0.01, -0.06, -0.09, 0.04, -0.05, 0.08, -0.03, 0.07)$ of the measurement error? This is the problem of hypothesis testing. Clearly, it would not work by simply comparing the hypothesized value with an estimate of the mean (which cannot be equal in general).

5.5 Parameter Estimation

Many methods are available for parameter estimation. We shall focus on two of them: the maximum likelihood method and the method of moments.

The basic idea of the ***maximum likelihood method*** is the following. If an event occurs in a single observation, then we can reasonably assume that it has a large probability — its likelihood is large. As such, the value of the unknown parameter that is most likely to have produced that event (i.e., that particular sample of data) may be used as the estimate of the parameter.

The ***likelihood function*** $L(x_1, \ldots, x_n; \theta)$ of a parameter θ given a sample $(X_1, \ldots, X_n)$ is the joint PDF of $(X_1, \ldots, X_n)$ pretending that the parameter is known. It is in general a function of the parameter. The maximum likelihood estimate (MLE) of θ is the maximum point (i.e., the peak location — θ value of the peak) of the likelihood function $L(x_1, \ldots, x_n; \theta)$.

Example 5.7: Maximum Likelihood Estimation of Failure Rate

The time X to failure of a system is an exponentially distributed RV with PDF $f(x) = \lambda e^{-\lambda x} u(x)$. However, the failure rate λ is unknown. Given a sample $(X_1, \ldots, X_n)$, we use the maximum likelihood method to estimate λ.

The likelihood function is, since $X_1, \ldots, X_n$ are independent,

$$
\begin{aligned}
L(x_1, \ldots, x_n; \lambda) &= f_{X_1,\ldots,X_n}(x_1, \ldots, x_n | \text{given } \lambda) \\
&= \prod_{i=1}^{n} f_{X_i}(x_i) = \lambda e^{-\lambda x_1} \cdots \lambda e^{-\lambda x_n} u(x_1) \cdots u(x_n) \\
&= \lambda^n e^{-\lambda(x_1 + \cdots + x_n)} u(x_1) \cdots u(x_n) = \lambda^n e^{-\lambda n \hat{x}} \prod_{i=1}^{n} u(x_i)
\end{aligned}
$$

$$
\ln L = n(\ln \lambda - \hat{x}\lambda) + \sum_{i=1}^{n} \ln u(x_i)
$$

where $\hat{x} = \frac{1}{n} \sum_{i=1}^{n} x_i$. Clearly, L and $\ln L$ achieve their maximum at the same value of λ, which is obtained easily by setting $\frac{d}{d\lambda} \ln L = 0$ as

$$
\hat{\lambda} = 1/\hat{X}
$$

That is, the maximum likelihood estimate (MLE) of the failure rate is the reciprocal of the sample mean.

The underlying idea of the ***method of moments*** is that the sample moment should be close to the true moment. Thus, the value of the unknown parameter that makes the moments equal can be used as an estimate of the parameter. Usually, a lower (e.g., the first or second) moment is preferred.

Example 5.8: Method of Moments Estimation of Failure Rate

We now use the method of moments to estimate the failure rate λ of the above example. For this example we will use the first moment. Since the true mean (expected value) of the exponential RV X is equal to $1/\lambda$ (see Example 3.21), by letting

$$1/\lambda = \text{true mean} = \text{sample mean} = \hat{X}$$

we have $\hat{\lambda} = 1/\hat{X}$, which turns out to be equal to the MLE.

Example 5.9: Maximum Likelihood and Method of Moments Estimation

Consider a sample $(X_1, \ldots, X_n)$ of a population X with the following PDF:

$$f(x) = (\theta + 1)x^\theta, \qquad 0 < x < 1, \theta > -1$$

where θ is an unknown parameter to be estimated. The likelihood function is

$$L(x_1, \ldots, x_n; \theta) = (\theta + 1)x_1^\theta \cdots (\theta + 1)x_n^\theta = (\theta + 1)^n (x_1 \cdots x_n)^\theta$$

$$\ln L = n \ln(\theta + 1) + \theta(\ln x_1 + \cdots + \ln x_n)$$

Setting $\frac{d}{d\theta} \ln L = 0$ leads to

$$\frac{n}{\theta + 1} + \sum_{i=1}^{n} \ln x_i = 0$$

Solving this equation yields the maximum likelihood estimate

$$\hat{\theta} = -\left[\frac{n}{\sum_{i=1}^n \ln X_i} + 1\right]$$

Note that

$$\bar{x} = \int_0^1 x(\theta + 1)x^\theta dx = \frac{\theta + 1}{\theta + 2} x^{\theta+2}\Big|_0^1 = \frac{\theta + 1}{\theta + 2}$$

Let $\bar{x} = \hat{x}$. Then the method of moments estimate of θ is $\hat{\theta} = \frac{2\hat{X}-1}{1-\hat{X}}$, which is different from the maximum likelihood estimate.

Desirable Qualities of Estimators

The formula, as a function of the sample suggested by an estimation method, used to approximate an unknown parameter is known as an ***estimator*** of the parameter. As stated before, there are many possible estimators for a particular unknown parameter. A question thus arises: Which one is the best? Unfortunately, the answer to this question is in general not simple because there are many measures of how good an estimator is. Simply put, it depends on what do we mean by the word "best" — that is, "best" in what sense?

A good measure of the quality of an estimator $\hat{\theta}$ of an unknown parameter θ is the ***mean-square error*** (***MSE***), define by

$$\text{MSE}(\hat{\theta}) = E[(\theta - \hat{\theta})^2]$$

Clearly, it is a measure of the difference between the estimator and the parameter. Note that the average error $E[\theta - \hat{\theta}]$ would not work here because a small average error does not guarantee $\hat{\theta}$ and θ are close — it could be the case that positive and negative errors cancel each other. The use of MSE excludes such possibilities. Of course, average absolute error $E[|\theta - \hat{\theta}|]$ could in principle be used instead. It is, however, less tractable mathematically than the MSE.

Another measure of the quality of an estimator $\hat{\theta}$ is the variance of the estimator, defined by

$$\text{var}(\hat{\theta}) = E[(\hat{\theta} - E[\hat{\theta}])^2]$$

It provides a measure of how random the estimator is around its mean.

MSE and variance are two commonly used *quantitative* measures of an estimator. In general, an estimator with a smaller MSE and/or variance is better.

The most commonly used *qualitative* measures of an estimator are *unbiasedness*, *efficiency*, and *consistency*.

An estimator is said to be ***unbiased*** if its estimation error is zero on the average; that is, if

$$E[\theta - \hat{\theta}] = 0$$

The expected error $E[\theta - \hat{\theta}]$ is called the ***bias*** of the estimator. An unbiased estimator has no bias — no estimation error *on the average*. For a given sample (set of data), an unbiased estimator could have a large estimation error. Such error depends on the sample given: for some samples it could be large or positive and for others it could be small or negative. The unbiasedness of an estimator guarantees nothing but that the average of these errors over all possible samples is zero. As such, it is almost necessary for an estimator to be unbiased; otherwise, the estimator is not even right on the average. As explained above, a zero average error is not a guarantee for the estimator to be close to the parameter. It is in most cases just a necessary but not sufficient condition.

An estimator is said to be ***efficient*** if its MSE is the smallest among all possible estimators of the same parameter; that is, if its MSE achieves some lower bound on the MSE for the estimation of the parameter. The most popular such lower bound is the so-called ***Cramer-Rao lower bound***. It states that the MSE of any estimator of a parameter cannot be smaller than the inverse of the ***Fisher information*** I:

$$\text{MSE}(\hat{\theta}) \geq I^{-1}$$

where

$$I = -E\left[\frac{\partial^2}{\partial\theta^2}\ln L(x_1,\ldots,x_n;\theta)\right] = E\left[\left|\frac{\partial}{\partial\theta}\ln L(x_1,\ldots,x_n;\theta)\right|^2\right]$$

and $L(x_1,\ldots,x_n;\theta)$ is the likelihood function of θ. An efficient estimator is the best estimator in the sense of having the smallest estimation error. For an unbiased estimator its MSE is equal to its variance. Thus an unbiased and efficient estimator is guaranteed to be most concentrated around its mean, which is the true but unknown parameter value.

An estimator is said to be ***consistent*** if its estimation error tends to zero as the data (sample) size n increases; that is, if $\hat{\theta} \overset{n\to\infty}{\longrightarrow} \theta$. If the data size is very large, it is very safe to use a consistent estimator because it is guaranteed to be very close to the parameter — arbitrarily close if the data size can be as large as desired. As a result, consistency is a property very desirable for an estimator.

Example 5.10: Comparison of Unbiased Estimators by Their Variances

Given a sample (X_1, X_2) of a population $X \sim \mathcal{N}(\mu, \sigma^2)$ with unknown μ but known σ^2, we have three estimators $\hat{\mu}_1 = \frac{1}{2}X_1 + \frac{1}{2}X_2$, $\hat{\mu}_2 = \frac{2}{5}X_1 + \frac{3}{5}X_2$ and $\hat{\mu}_3 = \frac{1}{3}X_1 + \frac{2}{3}X_2$. All three estimators are unbiased since

$$E[\hat{\mu}_1] = E\left[\frac{1}{2}X_1 + \frac{1}{2}X_2\right] = \mu$$
$$E[\hat{\mu}_2] = E\left[\frac{2}{5}X_1 + \frac{3}{5}X_2\right] = \mu$$
$$E[\hat{\mu}_3] = E\left[\frac{1}{3}X_1 + \frac{2}{3}X_2\right] = \mu$$

Their variances are, respectively

$$\text{var}(\hat{\mu}_1) = \text{var}\left(\frac{1}{2}X_1 + \frac{1}{2}X_2\right) = \frac{1}{4}\sigma^2 + \frac{1}{4}\sigma^2 = \frac{1}{2}\sigma^2$$
$$\text{var}(\hat{\mu}_2) = \text{var}\left(\frac{2}{5}X_1 + \frac{3}{5}X_2\right) = \frac{13}{25}\sigma^2$$
$$\text{var}(\hat{\mu}_3) = \text{var}\left(\frac{1}{3}X_1 + \frac{2}{3}X_2\right) = \frac{5}{9}\sigma^2$$

Clearly, the sample mean estimator $\hat{\mu}_1$ has the smallest variance. In fact, the variance of the estimator decreases when the weights tend to be even.

Some Properties of the Maximum Likelihood Estimator

The maximum likelihood estimator has many nice properties. For example, it is asymptotically unbiased, asymptotically efficient, and asymptotically Gaussian (i.e., it is unbiased, efficient and a Gaussian RV in the limit as the data size approaches infinity). In addition, it is invariant in the following sense. If $g(\cdot)$ is a single-valued function with a unique single-valued inverse function

$g^{-1}(\cdot)$, then the maximum likelihood estimate of $g(\theta)$ is equal to $g(\hat{\theta})$, where $\hat{\theta}$ is the maximum likelihood estimate of θ; that is, $\widehat{g(\theta)} = g(\hat{\theta})$.

In general, the method of moments estimator does not have the many nice properties of the maximum likelihood estimator.

To be more rigorous, among all the solutions of the ***likelihood equation*** $\frac{d}{d\theta} \ln L = 0$, only the one that gives the maximum value of the likelihood function is the maximum likelihood estimator. In view of this, we usually need to check first if the solution is a (local) maximum point by say checking if the second derivative is negative at that point and then use the global (i.e., the largest local) maximum point as the maximum likelihood estimate.

Other Popular Estimation Methods

Another popular estimation method is known as the ***least squares*** (LS) method. Its basic idea is to choose as the estimate the possible value of the true parameter that best fits a model to the data in the sense of having the *least squares* of fitting errors. A common linear data model for θ is

$$X_i = H_i\theta + V_i \qquad i = 1, \dots, n$$

where V_i is the error in the data. Thus, the *least squares estimator* $\hat{\theta}^{\mathrm{LS}}$ of θ is equal to the value of θ that minimizes the following sum of fitting error squares

$$Q = \sum_{i=1}^{n} [X_i - H_i\theta]^2$$

If the parameter θ to be estimated is a random variable, then the two most commonly used estimation methods are the minimum mean-square error and the maximum a posteriori methods.

The basic idea of the ***minimum mean-square error*** (MMSE) estimator is to choose as the estimate the function of the data that gives the smallest expected value of the square of the estimation error. That is, $\hat{\theta}^{\mathrm{MMSE}}$ minimizes the mean of the estimation error square $E[(\theta - \hat{\theta})^2]$. It turns out to be the ***conditional mean*** of θ given the sample $(X_1, \dots, X_n)$:

$$\hat{\theta}^{\mathrm{MMSE}} = E[\theta | X_1, \dots, X_n]$$

which is the mean of the conditional distribution $f(\theta | X_1, \dots, X_n)$ given a sample $(X_1, \dots, X_n)$.

It can also be easily shown that the conditional mean estimator is always unbiased.

The basic idea of the ***maximum a posteriori*** (MAP) estimator is to choose as the estimate the "most probable" value that the random parameter will take, judging from the (particular) sample $(X_1, \dots, X_n)$ drawn as well as prior knowledge. The posterior PDF $f(\theta | X_1, \dots, X_n)$ is a proper measure of how "probable" θ will take on a value given the data (sample) as well as the prior PDF. Hence, the maximum point (i.e., the *mode*) of this posterior PDF can be defined as the sought-after estimate: $\hat{\theta}^{\mathrm{MAP}}$ is the value of θ that maximizes $f(\theta | X_1, \dots, X_n)$.

Note that the maximum likelihood and the maximum a posteriori estimators are the maximum points (i.e., the peak location, also known as ***mode***) of the likelihood function and the posterior PDF of the parameter, respectively.

5.6 Hypothesis Testing

In ***hypothesis testing***, we have one or more hypotheses about the values of some unknown parameters of the population, and we want to decide whether the information contained in the sample supports or rejects the hypotheses.

Consider testing the following single hypothesis on an unknown parameter θ of the population

$$H_0 : \theta = \theta_0$$

where θ_0 is a given constant. Clearly, there are two possible decision errors (mistakes):

- ***Type I error***: We decide that H_0 is false but in fact it is true.
- ***Type II error***: We decide that H_0 is true but in fact it is false.

The type I error probability α is known as the ***significance*** and $(1 - \alpha)$ the ***confidence*** of the test.

Example 5.11: Instrument Calibration — Test on Population Mean

An instrument makes a random measurement error X that is Gaussian distributed with zero mean: $X \sim \mathcal{N}(0, 0.01)$, if it is well calibrated according to the product specification. It has been calibrated some time ago and we want to decide if further calibration is needed by checking if the mean of its measurement error can be accepted as zero. That is, our hypothesis is H_0: $\bar{x} = 0$. We have obtained the following sample of the measurement error:

$$0.1294, -0.0336, 0.1714, 0.2624, 0.0308, 0.1858, 0.2254, -0.0594, -0.044, 0.157$$

Since the sample mean $\hat{X}$ has $\mathcal{N}(\bar{x}, \sigma^2/n) = \mathcal{N}(0, 0.01/10)$ distribution, from Example 3.38, it should be within the interval $(-1.96\sigma/\sqrt{10}, 1.96\sigma/\sqrt{10}) = (-0.062, 0.062)$ with 95% probability if H_0 is true. Based on the above sample, the sample mean $\hat{X} = 0.1025$ is outside the interval. Thus, we should reject H_0 with 95% confidence. After further calibration, we have a sample:

$$-0.040, 0.069, 0.0816, 0.0712, 0.129, 0.0669, 0.1191, -0.1202, -0.002, -0.0157$$

The sample mean $\hat{X} = 0.0359$ is in the interval and thus we should accept H_0 with 95% confidence and no more calibration is needed.

Many hypothesis testing problems can be formulated as the problem of testing a primary hypothesis (known as the ***null hypothesis***)

$$H_0 : \theta - \theta_0$$

against some ***alternative hypothesis***, which may be

$$H_1 : \theta = \theta_1 \qquad \text{or} \qquad H_1 : \theta > \theta_0 \qquad \text{or} \qquad H_1 : \theta < \theta_0 \qquad \text{or} \qquad H_1 : \theta \neq \theta_0$$

It is assumed that one and only one hypothesis is true. The type I and type II error probabilities are then

$$\begin{aligned} \alpha &= P\{\text{type I error}\} = P\{\text{accept } H_1 | H_0 \text{ true}\} = P\{\text{reject } H_0 | H_0 \text{ true}\} \\ \beta &= P\{\text{type II error}\} = P\{\text{accept } H_0 | H_1 \text{ true}\} = P\{\text{reject } H_1 | H_1 \text{ true}\} \end{aligned}$$

Type I and type II errors are often referred to as ***false alarm*** and ***miss*** in engineering.

5.6.1 Neyman-Pearson Hypothesis Testing

Using a sample of a given size, the type I and type II error probabilities of any hypothesis test cannot be reduced simultaneously: reducing one will usually increase the other. The only possible way of reducing both error probabilities simultaneously is to increase the sample size. Due to the success of an optimal theory of binary hypothesis testing developed by Neyman and Pearson, it is customary in practice to control the type I error probability and minimize the type II error probability. As such, the decision error that has a more serious consequence should be assigned the type I error.

Consider the following binary hypothesis testing problem

$$H_0 : \theta = \theta_0 \qquad \text{vs.} \qquad H_1 : \theta = \theta_1$$

and assume we have a sample $X_1, \ldots, X_n$. The celebrated ***Neyman-Pearson lemma*** states that the following ***likelihood ratio test***

- Reject H_1 if $\frac{f(X_1,\ldots,X_n|\theta=\theta_1)}{f(X_1,\ldots,X_n|\theta=\theta_0)} < \lambda$.
- Accept H_1 if $\frac{f(X_1,\ldots,X_n|\theta=\theta_1)}{f(X_1,\ldots,X_n|\theta=\theta_0)} > \lambda$.

is optimal in that it minimizes the type II error probability while maintaining the type I error probability not greater than a given level α. In the test, the likelihood functions $f(x_1, \ldots, x_n|\theta = \theta_0)$ and $f(x_1, \ldots, x_n|\theta = \theta_1)$ are the joint PDF of $X_1, \ldots, X_n$ assuming $\theta = \theta_0$ and $\theta = \theta_1$, respectively; $\frac{f(X_1,\ldots,X_n|\theta=\theta_1)}{f(X_1,\ldots,X_n|\theta=\theta_0)}$ is a number, that is, the ratio of the two likelihood functions evaluated at the sample value $(x_1, \ldots, x_n) = (X_1, \ldots, X_n)$; and threshold λ is determined such that the type I error probability is equal to α.

This theory is elegant but not easy to put to use in some practical situations mainly because the likelihood ratio and the associated threshold are not easy to come by.

5.6.2 Hypothesis Testing Based on Sampling Distributions

In practice, hypothesis testing is more often carried out based on the known distribution of some statistic. Any function of a given sample $(X_1, \ldots, X_n)$ that is not dependent on any unknown parameter is called a ***statistic***. The distributions of statistics are known as ***sampling distributions***.

Test on Mean with Known Variance — The u Test

From Section 5.2, the sample mean $\hat{X} = \frac{1}{n}\sum_{i=1}^{n} X_i$ based on a sample of a population (RV) $X \sim \mathcal{N}(\bar{x}, \sigma^2)$ clearly has the following distribution $\hat{X} \sim \mathcal{N}(\bar{x}, \sigma^2/n)$. This is so because the sample mean has mean and variance equal to $\bar{x}$ and σ^2/n, respectively, and the weighted sum of independent Gaussian RVs is Gaussian distributed.

By the central limit theorem, for a sample of a large size, the sample mean is approximately Gaussian distributed even when the X_i's are not Gaussian distributed since it is the sum of a large number of independent RVs. This can be explained below. The central limit theorem assures us that the standardized RV of the sample mean

$$\tilde{X} = \frac{\hat{X} - E[\hat{X}]}{\sigma/\sqrt{n}}$$

is asymptotically Gaussian distributed as $n \to \infty$ even when X_i's are not Gaussian distributed. As such, $\hat{X}$ is also asymptotically Gaussian distributed as $n \to \infty$ since it is a linear function of $\tilde{X}$.

Consequently, testing of the mean of a population with known variance when it is Gaussian distributed or when the sample size is large can be done based on the fact that the sample mean $\hat{X} \sim \mathcal{N}(\bar{x}, \sigma^2/n)$ in such a case. Specifically, we check if the statistic $\tilde{X}$ falls inside or outside a confidence interval determined by the $\mathcal{N}(\bar{x}, \sigma^2/n)$ distribution. This is illustrated in Examples 5.11 and 5.16.

Test on Mean with Unknown Variance — The t Test

As explained above that the RV $\tilde{X} = \frac{\hat{X} - E[\hat{X}]}{\sigma/\sqrt{n}}$ is Gaussian distributed if the population is Gaussian distributed. Otherwise it is asymptotically Gaussian distributed as $n \to \infty$. However, if the true variance σ^2 of the population is not known and is replaced by the sample variance, then the statistic

$$T = \frac{\hat{X} - E[\hat{X}]}{\sqrt{\hat{V}/n}}$$

has a so-called ***student distribution*** or ***t distribution*** with $n - 1$ degrees of freedom. The t distribution has the following PDF

$$f(t) = \frac{\Gamma[(n+1)/2]}{\sqrt{n\pi}\,\Gamma(n/2)}\left(1 + \frac{t^2}{n}\right)^{-(n+1)/2}$$

where n is called the ***degrees of freedom***. The t distribution with n degrees of freedom is denoted by $t(n)$. The t distribution has a shape similar to but different from that of a Gaussian distribution. It approaches the Gaussian distribution as $n \to \infty$.

The above is a special case of the following. For a standard Gaussian RV X and a chi-square RV Y with unity variance and n degrees of freedom, the following statistic has a t distribution with n degrees of freedom if X and Y are independent:

$$T = \frac{X}{\sqrt{Y/n}} \sim t(n)$$

In the t test, we check to see if the statistic T falls inside or outside of a confidence interval determined by the t distribution. In other words, we replace in the u test the statistic $\tilde{X}$ with T and the $\mathcal{N}(\bar{x}, \sigma^2/n)$ distribution with the t distribution. This is demonstrated in the example below.

Example 5.12: Instrument Calibration — Test on Population Mean with Unknown Variance

Consider again Example 5.11 but assume that the true variance is not known. For the first sample, it was found that the sample variance $\hat{V} = 0.0142$ and the statistic $T = 2.7250$ for the assumed true mean $\bar{x} = 0$. From the t distribution table (not available in this book) of 9 degrees of freedom, we know

$$P\{-2.2622 < T < 2.2622\} = 95\%$$

Since $T = 2.7250$ is outside the two-sided 95% confidence interval $[-2.2622, 2.2622]$, we conclude that calibration is needed for the instrument.

After further calibration, it was found from the second sample that $\hat{V} = 0.0042$ and $T = 1.7483$. Since the new T is inside the two-sided 95% confidence interval $[-2.2622, 2.2622]$, we conclude that no more calibration is needed.

Comparing these results with Example 5.11 we see that the same conclusions were made by the t test even though the variance is assumed not known. However, the margin for $T = 1.7483$ (for the second sample) to be inside the interval $[-2.2622, 2.2622]$ is smaller than that for $\hat{X} = 0.0359$ to be inside $(-0.062, 0.062)$. Thus, the confidence of the above conclusion is lower than in the case of known variance.

We point out that the sample mean and sample variance are independent if the population is Gaussian distributed although they are functions of the same RVs $X_1, \ldots, X_n$.

Chi-Square Test on Variance

It was stated in Section 3.6 that the sum of the squares of n i.i.d. Gaussian RVs is chi-square distributed with n degrees of freedom. Note that in the above the RVs are independent and are thus "free" variables; that is, there are no linear constraints that relate them. As such, if the population X is Gaussian distributed, one might think the statistic (i.e., a scaled sample variance)

$$(n-1)\hat{V} = \sum_{i=1}^{n}(X_i - \hat{X})^2$$

is chi-square distributed with n degrees of freedom because the elements of its sample are independent and Gaussian distributed. In fact, $(n-1)\hat{V}$ is indeed chi-square distributed but the degrees of freedom are $n-1$ rather than n. This is because

$$(X_1 - \hat{X}) + (X_2 - \hat{X}) + \cdots + (X_n - \hat{X}) = X_1 + X_2 + \cdots + X_n - n\hat{X} = 0$$

is a linear constraint and thus only $n-1$ RVs in the n RVs of $(n-1)\hat{V}$ are "free" variables and thus the degrees of freedom are $n-1$.

In view of the above facts, we may perform a ***chi-square test*** to test if $\sigma^2 = \sigma_0^2$ on the value of the unknown variance σ^2 of a Gaussian population $X \sim \mathcal{N}(\bar{x}, \sigma^2)$ given a sample $(X_1, \ldots, X_n)$. If the population mean $\bar{x}$ is unknown, we may check if the scaled sample variance $(n-1)\hat{V}$ is chi-square distributed with $n-1$ degrees of freedom and parameter σ_0^2. If $\bar{x}$ is known, we may check if

$$Q = \sum_{i=1}^{n}(X_i - \bar{x})^2$$

is chi-square distributed with n degrees of freedom and parameter σ_0^2.

Example 5.13: Quality Control by Chi-Square Test

It is known from past experience that the breakdown voltage of a small insulator made by a manufacturer has a variance $\sigma^2 = 100$. We would like to know if the breakdown voltage of a batch of insulators just received has a higher, lower or the same variance as before. For this purpose, we tested twenty samples of the insulators and found their breakdown voltages to be:

$$56.34, 41.34, 47.48, 47.54, 67.03, 57.52, 53.60, 49.71, 47.05, 50.33,$$
$$47.53, 38.52, 52.43, 50.61, 42.68, 50.46, 52.23, 63.84, 43.19, 52.14$$

Suppose we would like to have 95% confidence in our conclusion; that is, the type I error is $\alpha = 5\%$.

The problem as to whether the variance is higher than before is formulated by the following problem of hypothesis testing:

$$H_0 : \sigma^2 = 100 \qquad \text{vs.} \qquad H_1 : \sigma^2 > 100$$

If we know as a fact that the true mean of the breakdown voltage is 50, then $Q = \sum_{i=1}^{20}(X_i - 50)^2$ would be a χ_{20}^2 RV with parameter $\sigma_0^2 = 100$ if H_0 is true, where χ_n^2 stands for a standard chi-square RV with n degrees of freedom. From the companion software P&R (by using "Percentile") or a chi-square table, we know that

$$P\{\chi_{20}^2 > 31.41\} = 5\%$$

that is, a standard χ^2_{20} RV will take on a value within the interval $[0, 31.41]$ with 95% probability. Since for the given sample $Q/100 = 9.41$ falls inside this one-sided lower 95% interval, we cannot reject H_0 and accept H_1; that is, we cannot conclude that the variance is higher than before.

If the mean of the breakdown voltage is unknown, then $(n-1)\hat{V} = \sum_{i=1}^{20}(X_i - \hat{X})^2$ would be a χ^2_{19} RV with parameter $\sigma_0^2 = 100$ if H_0 is true. We found from P&R or a chi-square table that the one-sided lower 95% confidence interval of such a chi-square RV is $[0, 30.14]$. Since for the given sample $(n-1)\hat{V}/100 = 9.35$ falls inside this interval, we cannot conclude that the variance is higher than before.

The problem regarding whether the variance is lower than before is formulated as follows:

$$H_0 : \sigma^2 = 100 \qquad \text{vs.} \qquad H_1 : \sigma^2 < 100$$

Q and $(n-1)\hat{V}$ remain the same as above. The corresponding one-sided *upper* 95% confidence intervals are found to be $[10.85, \infty)$ and $[10.12, \infty)$, respectively. Since for the given sample $Q/100$ and $(n-1)\hat{V}/100$ fall outside the above intervals, respectively, we reject H_0 and accept H_1; that is, we conclude that the variance is lower than before no matter whether the true mean is known or not.

The problem concerning whether the variance is the same as before is formulated as follows:

$$H_0 : \sigma^2 = 100 \qquad \text{vs.} \qquad H_1 : \sigma^2 \neq 100$$

Q and $(n-1)\hat{V}$ remain the same as above. The corresponding two-sided 95% confidence intervals are $[9.59, 34.17]$ and $[8.91, 32.85]$, respectively, which were found from P&R by

$$P\{\chi^2_{20} < 9.59\} = 2.5\% \qquad P\{\chi^2_{20} > 34.17\} = 2.5\%$$
$$P\{\chi^2_{19} < 8.91\} = 2.5\% \qquad P\{\chi^2_{19} > 32.85\} = 2.5\%$$

Since for the given sample $Q/100$ and $(n-1)\hat{V}/100$ fall outside and inside their corresponding intervals, respectively, we conclude that the variance is different from before if we know the mean is really 50, otherwise we cannot conclude that the variance is different than before.

Note that the conclusion could be entirely different whether we know the true mean or not. Also, the conclusion depends largely on our hypotheses. For example, when the true mean is unknown, if we are testing whether the variance is different than before, then we could not conclude that it is different. However, if we are testing whether the variance is lower than before, then we could conclude that it is indeed so.

5.7 Linear Regression and Curve Fitting

Regression techniques are statistical tools that handle the *statistical relation* between two or more variables.

Consider two RVs X and Y. Assume they are related by

$$Y = a + bX + V$$

where $V \sim \mathcal{N}(0, \sigma^2)$ and the coefficients a and b do not depend on X. The problem of ***linear regression*** is to find the estimates $\hat{a}$ and $\hat{b}$ of a and b given a sample (data) $\{(X_1, Y_1), \ldots, (X_n, Y_n)\}$ of the pair of RVs (X, Y).

Geometrically, the sample can be plotted in a scatter diagram, as illustrated in Fig. 5.5. The problem of regression is then that of ***curve fitting*** — find a curve (or a straight line for *linear* regression) that best fits the data points. Assume the fitted straight line is given by

$$\check{Y} = \hat{a} + \hat{b}X$$

It can be shown that $\hat{a} = \hat{y} - \hat{b}\hat{x}$, where $\hat{x}$ and $\hat{y}$ are the values of the sample means $\hat{X}$ and $\hat{Y}$. Thus

$$\check{Y} = \hat{y} + \hat{b}(X - \hat{x}) \tag{5.7}$$

It can be seen that the regression line passes through the centroid $(\hat{x}, \hat{y})$ of the data points in the scatter diagram. It can be shown that least squares and maximum likelihood estimation both lead to

$$\hat{b} = \frac{\sum_{i=1}^{n}(x_i - \hat{x})(y_i - \hat{y})}{\sum_{i=1}^{n}(x_i - \hat{x})^2} \tag{5.8}$$

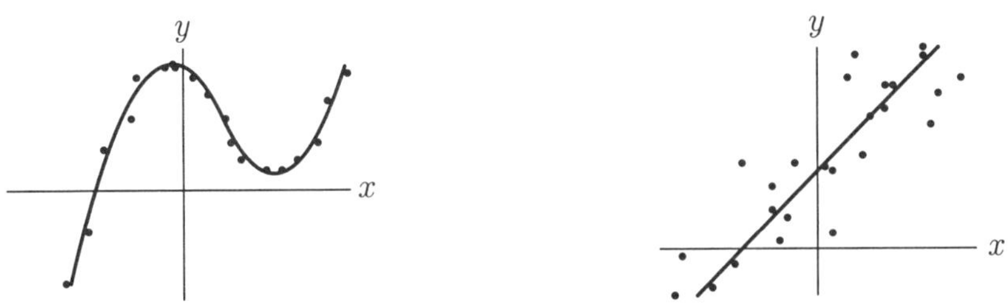

(a) Nonlinear regression (b) Linear regression

Figure 5.5: Curve fitting by linear or nonlinear regression.

Everybody is familiar with causality — the *deterministic* relationship between two phenomena, the cause and effect. In reality, the relationship between variables does not always follow causality. For instance,

- The test score of a student and his/her knowledge of the subject matter of a class are related.
- The humidity and the flashover voltage of an insulator are related.
- The sight of a person is related with his/her age.

Such a non-deterministic relation is known as a *statistical relation* or *correlation.*

Regression analysis is the branch of statistics that studies the statistical relations between two or more variables quantitatively. ***Linear regression*** assumes that the variables are related linearly where the coefficients of linear relation are determined by regression techniques, such as the least squares and maximum likelihood estimation.

(5.7)–(5.8) can be derived as follows using the maximum likelihood estimation. Since $Y = a + bX + V$, where $V \sim \mathcal{N}(0, \sigma^2)$, we have $Y_i = a + bX_i + V_i$ and thus $Y_i \sim \mathcal{N}(a + bX_i, \sigma^2)$. Hence, the likelihood function of $Y_1, \dots, Y_n$ is, since they are independent,

$$\begin{aligned} L &= \prod_{i=1}^{n} f_{Y_i}(y_i) = \prod_{i=1}^{n} \frac{1}{\sqrt{2\pi}\sigma} \exp\left[- \frac{1}{2\sigma^2}(y_i - a - bx_i)^2 \right] \\ &= \frac{1}{(\sqrt{2\pi}\sigma)^n} \exp\left[- \frac{1}{2\sigma^2} \sum_{i=1}^{n} (y_i - a - bx_i)^2 \right] \end{aligned}$$

Clearly, maximizing L is equivalent to minimizing the following (i.e., least squares estimation)

$$Q = \sum_{i=1}^{n} (y_i - a - bx_i)^2$$

Setting to zero the partial derivatives of Q with respect to a and b

$$\begin{aligned} \frac{\partial Q}{\partial a} &= -2 \sum_{i=1}^{n} (y_i - a - bx_i) = 0 \\ \frac{\partial Q}{\partial a} &= -2 \sum_{i=1}^{n} (y_i - a - bx_i) x_i = 0 \end{aligned}$$

yields

$$\begin{aligned} na + n\hat{x}b &= n\hat{y} \\ n\hat{x}a + \sum_{i=1}^{n} x_i^2 b &= \sum_{i=1}^{n} x_i y_i \end{aligned}$$

If the x_i's are not identical, these two equations jointly have a unique solution, given by

$$\begin{aligned} \hat{a} &= \hat{y} - \hat{b}\hat{x} \\ \hat{b} &= \frac{\sum_{i=1}^{n} x_i y_i - n\hat{x}\hat{y}}{\sum_{i=1}^{n} x_i^2 - n\hat{x}^2} = \frac{\sum_{i=1}^{n} (x_i - \hat{x})(y_i - \hat{y})}{\sum_{i=1}^{n} (x_i - \hat{x})^2} \end{aligned}$$

Example 5.14: Linear Regression by P&R

A sample $\{(X_1, Y_1), \ldots, (X_n, Y_n)\}$ of size 50

$$\begin{array}{cc} x_i & y_i \\ 1.9140 & 11.168 \\ 1.1175 & 7.6677 \\ \vdots & \vdots \\ 1.8757 & 11.455 \end{array}$$

of the pair of RVs (X, Y) is given in data file `e5_14.dat` in the companion software P&R. We have

$$\begin{aligned}
\hat{x} &= [1.9140 + 1.1175 + \cdots + 1.8757]/50 = 1.6228 \\
\hat{y} &= [11.168 + 7.6677 + \cdots + 11.455]/50 = 8.6861 \\
\hat{b} &= \frac{[(1.9140)(11.168) + \cdots + (1.8757)(11.455)] - 50(1.6228)(8.6861)}{[1.9140^2 + \cdots + (1.8757)^2] - 50(1.6228)^2} \\
&= 2.9536 \\
\hat{a} &= 8.6861 - (2.9536)(1.6228) = 3.8931
\end{aligned}$$

Consequently, the linear regression of Y on X is given by

$$\check{Y} = 3.8931 + 2.9536X$$

In fact, the above regression equation can be obtained by P&R as follows. Following the procedure described in Example 4.8, the correlation of the data in the file `e5_14.dat` can be obtained. The results are

$$\hat{x} = 1.6228, \quad \hat{y} = 8.6861, \quad \hat{v}_x = 1.616, \quad \hat{v}_y = 15.7545, \quad \rho = 0.94596$$

From the analysis on page 164, we have $\frac{(\check{Y}-\hat{y})}{\sqrt{\hat{v}_y}} = \rho\frac{(X-\hat{x})}{\sqrt{\hat{v}_x}}$. Comparing it with (5.7) yields

$$\hat{b} = \rho\sqrt{\hat{v}_y/\hat{v}_x} = (0.94596)(\sqrt{15.7545})/(\sqrt{1.616}) = 2.9536$$

Consequently, the linear regression of Y on X is given by

$$\check{Y} = \hat{y} + \hat{b}(X - \hat{x}) = 8.6861 + 2.9536(X - 1.6228) = 3.8931 + 2.9536X$$

If $X = 4.294$, the best guess of Y is $3.8931 + (2.9536)(4.294) = 16.5759$.

5.8 Summary and Requirements

The *population* is the entire collection of data. It is a RV. A sample is a subset of the population. The data making up a sample should be independent and identically distributed.

The *sample mean* of a sample $X_1, \ldots, X_n$ is the average value of the sample: $\hat{X} = \frac{1}{n}\sum_{i=1}^{n} X_i$. It is used to approximate the unknown true mean. Its mean is equal to the true mean and its variance is $\sigma_{\hat{x}} = \sigma_x^2/n$. The *sample variance* is given by $\hat{V} = \dfrac{1}{n-1}\sum_{i=1}^{n}(X_i - \hat{X})^2$. It is used to approximate the true variance σ^2. Its expected value is equal to the true variance.

Given the values $x_1, \ldots, x_n$ of a sample of a population, the *empirical PDF* is defined as

$$\bar{f}(x|x_1, \ldots, x_n) = \frac{\text{number of points } x_1, \ldots, x_n \text{ in } [x, x + \Delta x)}{n\Delta x}$$

It can be used to approximate the true PDF of the population.

Statistical inference studies decisions and estimation of unknown parameters of a population based on a its observations — the sample.

Estimation uses the information contained in the sample to provide an approximate value of an unknown parameter that is close to the true value of the parameter. In the *maximum likelihood method*, the value of the parameter that leads to the maximum likelihood of having the particular values of the sample is used as the estimate of the parameter. In the *method of moments*, the value of the parameter that makes the true moment equal to the sample moment is used as the estimate of the parameter.

In *hypothesis testing*, we decide whether the information contained in the sample supports or rejects the hypotheses on certain values of unknown parameters of a population. The decisions are made subject to a certain error probability, called the *significance* of the test.

In *linear regression*, a RV is assumed to be a linear function of another RV. The associated coefficients are estimated from the available data by some estimation methods. Such regression techniques are useful in handling statistical relations between two or more variables.

Basic Requirements

- Know how to calculate the sample mean and sample variance, their expected values, and the variance of the sample mean. Understand what the sample mean and sample variance are used for.
- Know how to compute the empirical PDF and CDF.
- Understand what statistical inference, estimation, and hypothesis testing do, respectively.
- Understand the basic ideas of the maximum likelihood estimation and the method of moments.
- For simple problems, know how to determined the required sample size.

5.9 Additional Examples

5.15 *Minimum sample size.* Given a population of 100 bipolar transistors, one wishes to estimate the mean value of the current gain X with true mean $\bar{x} = 120$ and variance $\sigma_x^2 = 25$. How large a sample size is required to obtain the sample mean $\hat{X}$ with standard deviation not larger than 1% of the true mean?

Solution: Mathematically, the requirement for the sample size is

$$\sigma_{\hat{x}} \leq (1\%)\bar{x} = 0.01 \times 120 = 1.2$$

Note that

$$\sigma_{\hat{x}}^2 = \frac{1}{n}\sigma_x^2 \quad \Longrightarrow \quad n = \frac{\sigma_x^2}{\sigma_{\hat{x}}^2} = \frac{25}{(1.2)^2} = 17.36$$

and that $\sigma_{\hat{x}}^2$ decreases with n. We have $n \geq 18$, that is, at least 18 pieces of data are needed.

5.16* *Teaching performance evaluation.* Suppose we would like to conduct a poll about a professor's teaching performance by giving a single grade on the scale of 100. The professor has taught many students (that is, the population is very large). Assume that the grade given by a student is independent of any other student's grade and has an identical mean equal to the true grade G and an identical standard deviation of 10 points. If we would like to have the resultant grade $\hat{G}$ of this poll be within 5 points of the true grade with probability no smaller than 95%, how many students should be asked?

Solution: Let X_i be the grade given by the ith student being asked. By the assumption, we know that this grade has a standard deviation $\sigma_i = 10$. Suppose that a total of N students is chosen at random, with N to be determined, which is assumed to be large. By the central limit theorem,

$$\sum_{i=1}^{N} \frac{X_i - \bar{x}_i}{\sqrt{\sum_{i=1}^{N} \sigma_i^2}} \overset{N\to\infty}{\longrightarrow} Y \sim \mathcal{N}(0,1)$$

For our problem, $\sqrt{\sum_{i=1}^{N} \sigma_i^2} = \sigma_i \sqrt{\sum_{i=1}^{N}} = 10\sqrt{N}$, $\bar{x}_i = G$, and $\hat{G} = \hat{X} = \frac{1}{N}\sum_{i=1}^{N} X_i$. Since

$$\hat{G} - G = \frac{1}{N}\sum_{i=1}^{N}(X_i - G)$$

we have, for large N,

$$\frac{\sqrt{N}}{10}(\hat{G} - G) = \sum_{i=1}^{N} \frac{X_i - G}{10\sqrt{N}} = \sum_{i=1}^{N} \frac{X_i - \bar{x}_i}{\sqrt{\sum_{i=1}^{N} \sigma_i^2}} \approx Y \sim \mathcal{N}(0,1)$$

Note that "to have the resultant grade $\hat{G}$ of this poll be within 5 points of the true grade with probability no smaller than 95%" in mathematical terms is

$$P\{|\hat{G} - G| \leq 5\} \geq 0.95$$

or equivalently

$$\begin{aligned}0.95 &\le P\{|\hat{G}-G|\le 5\}\\ &= P\left\{\frac{\sqrt{N}}{10}|\hat{G}-G|\le \frac{\sqrt{N}}{10}\times 5\right\}\\ &\approx P\left\{|Y|\le \frac{\sqrt{N}}{10}\times 5\right\}\\ &= P\left\{-0.5\sqrt{N}\le Y\le 0.5\sqrt{N}\right\}\\ &= \Phi\left(0.5\sqrt{N}\right)-\Phi\left(-0.5\sqrt{N}\right)\\ &= 2\Phi\left(0.5\sqrt{N}\right)-1\end{aligned}$$

Thus,

$$\Phi\left(0.5\sqrt{N}\right)\ge \frac{1}{2}(1+0.95)=0.975$$

From Example 3.13, the above is equivalent to

$$0.5\sqrt{N}\ge 1.96 \qquad \Longrightarrow \qquad N\ge 15.37$$

Consequently, at least 16 students should be asked. In this case, we may say the resultant grade $\hat{G}$ has an error of not greater than 5%.

5.17 *MLE of the parameters of a uniform distribution.* Given a sample $(X_1,\ldots,X_n)$ from a population of the following uniform distribution

$$f(x)=\begin{cases}1/(\theta_2-\theta_1) & \theta_1<x\le\theta_2\\ 0 & \text{elsewhere}\end{cases}$$

where $\theta_2>\theta_1$ are unknown, find the method of moments estimate and the maximum likelihood estimate of θ_1 and θ_2.

Solution: The mean and variance of the uniform distribution are

$$\begin{aligned}E[X]&=\frac{1}{2}(\theta_1+\theta_2)\\ \text{var}(X)&=\frac{1}{12}(\theta_2-\theta_1)^2\end{aligned}$$

Hence, let

$$\begin{aligned}E[X]&=\hat{X}\\ \text{var}(X)&=S^2\end{aligned}$$

where $\hat{X}$ is the sample mean and $S^2=\frac{1}{n}\sum_{i=1}^n(X_i-\hat{X})^2$ is the (biased) sample variance. Solving these two equations jointly yields the method of moments estimate

$$\begin{aligned}\theta_1&=\hat{X}-\sqrt{3}S\\ \theta_2&=\hat{X}+\sqrt{3}S\end{aligned}$$

The likelihood function is given by

$$L(x_1, \ldots, x_n; \theta_1, \theta_2) = \prod_{i=1}^{n} f_{X_i}(x_i) = \begin{cases} 1/(\theta_2 - \theta_1)^n & \theta_1 \le x_i \le \theta_2, i = 1, \ldots, n \\ 0 & \text{elsewhere} \end{cases}$$

Clearly, the likelihood equations

$$\frac{\partial \ln L}{\partial \theta_1} = \frac{n}{\theta_2 - \theta_1} = 0$$
$$\frac{\partial \ln L}{\partial \theta_2} = \frac{n}{\theta_2 - \theta_1} = 0$$

do not lead to the maximum likelihood estimate. Note, however, that for $\theta_2 > \theta_1 > 0$, the likelihood function $L(\theta_1, \theta_2)$ increases as either θ_1 increases or θ_2 decreases. On the other hand, since $X_i \sim \mathcal{U}(\theta_1, \theta_2)$, θ_1 must be smaller than every X_i and thus greater than $\min\limits_{1 \le i \le n} X_i$ and θ_2 must be greater than every X_i and thus greater than $\max\limits_{1 \le i \le n} X_i$. Consequently, $L(\theta_1, \theta_2)$ achieves its maximum at $\theta_1 = \min\limits_{1 \le i \le n} X_i, \theta_2 = \max\limits_{1 \le i \le n} X_i$ and thus the maximum likelihood estimate is

$$\hat{\theta}_1 = \min_{1 \le i \le n} X_i$$
$$\hat{\theta}_2 = \max_{1 \le i \le n} X_i$$

5.10 Problems

5.1 *Distribution of a sample.* A population X is Gaussian distributed with mean 2 and variance 4. Find the joint PDF of a sample (X_1, X_2) of size 2 of the population.

5.2 *Distribution of a sample.* A population X is exponentially distributed with parameter $\lambda = 3$. Find the joint PDF of a sample (X_1, X_2) of size 2 of the population.

5.3 *Average age.* Suppose that the age X of a student in a class is a Gaussian RV with $(X - 20) \sim \mathcal{N}(0, 1)$. Given a class of 25 students,

(a) What are the mean and standard deviation of X?
(b) Find the mean $E[\hat{X}]$ and standard deviation $\sigma_{\hat{x}}$ of the average age $\hat{X}$ of this class.
(c) Find the probability that the age of an arbitrary student will be in $(18, 22)$.
(d) Find the probability that the average age of this class will be in $(19.4, 20.6)$.
(e) Find the constant c such that the probability that the average age of this class will be in $(20 - c, 20 + c)$ is 95.4%.

5.4 *Sample mean of random numbers.* A calculator is equipped with a random number generator producing four-digit random numbers that are uniformly distributed over $[0.0000, 0.9999]$. Suppose that the following random numbers are generated: 0.2756, 0.1323, 0.0572, 0.3624, 0.8156, 0.6312, 0.7432, 0.2083, 0.7817.

(a) Find the sample mean.
(b) Find the variance of the sample mean.
(c) To obtain a sample mean whose standard deviation is no greater than 0.01, how large should the sample size be?

5.5 *sample mean and sample variance of Poisson population.* A population X is Poisson distributed with mean 2.4. Find the expected value and variance of the sample mean based on a sample of size 20. Find the expected value of the sample variance.

5.6 *Minimum sample size.* Given a population $X \sim \mathcal{N}(\bar{x}, \sigma^2)$, if it is desirable to have a sample mean $\hat{X}$ whose dispersion from the true mean is smaller than 0.1 with probability 0.954, how large should the sample size be if $\sigma = 0.8$?

5.7 *Distribution of sample mean and sample variance.* Given two independent samples $(X_1, \ldots, X_n)$ and $(Y_1, \ldots, Y_n)$ of two populations $X \sim \mathcal{N}(\bar{x}, \sigma_x^2)$ and $Y \sim \mathcal{N}(\bar{y}, \sigma_y^2)$, respectively, a new sample $(Z_1, \ldots, Z_n)$ is formed, where $Z_i = X_i + Y_i$, $i = 1, 2, \ldots, n$.

(a) Find the mean and variance of the sample mean $\hat{Z} = \frac{1}{n}\sum_{i=1}^n Z_i$.
(b) Find the mean of the sample variance $\hat{V} = \frac{1}{n}\sum_{i=1}^n (Z_i - \hat{Z})^2$.
(c) What is the distribution of $\hat{Z}$?

5.8 *Measurement error.* Suppose that the measurement error of a device is a Gaussian RV with unknown mean and variance. The following data of the measurement error are obtained from a calibration process:

$$0.35, 0.54, -0.48, -0.08, 0.12, 0.45, -0.73, 0.33, -1.11, 0.22,$$
$$1.34, 0.65, 0.46, -0.54, -0.01, -0.05, 0.65, 0.13, -0.65, -0.43$$

(a) Find the sample mean and sample variance of the measurement error given the above sample.
(b) Give an approximate distribution of the measurement error. Give an approximate distribution of the sample mean of the measurement error based on the above sample of size 20. Justify your answer.

5.9 *Defect detection.* We would like to determine if a meter is defective. The reading error is a $\mathcal{N}(0, \sigma_1^2)$ RV if it is not defective. The error would be a $\mathcal{N}(0.5, \sigma_2^2)$ RV should it be defective. Suppose that the following meter reading error data are collected in a calibration process:

$$0.16, -0.43, 1.25, -0.61, -2.42, 0.74, -0.72, 2.33, -0.4, 1.3, 0.04, 0.08, 0.88,$$
$$-0.22, 0.55, -0.51, 0.32, -0.54, -0.64, 0.54, -0.98, -0.40, 0.66, -0.33, 0, 0.52,$$
$$-0.23, 0.33, 0.22, 0.13, 0.34, 0.34, 0.66, -0.23, -0.33, 0.23, -0.12, -0.14, 0.36$$

(a) Find the sample mean and sample variance.

(b) Assuming $\sigma_1 = 0.5$ and $\sigma_2 = 0.6$, is it more probable that the meter is defective?
(c) Assuming σ_1 and σ_2 are unknown, is it more probable that the meter is defective?

5.10 *Recursion of sample mean and sample variance.* Let $\hat{X}_n$ and $\hat{V}_n$ be the sample mean (5.1) and sample variance (5.5) of size n, respectively. Show that

(a) $\hat{X}_{n+1} = \frac{1}{n+1}[n\hat{X}_n + X_{n+1}]$
(b) $\hat{V}_{n+1} = \frac{n-1}{n}\hat{V}_n + \frac{1}{n+1}(X_{n+1} - \hat{X}_n)^2$

Note that these formulas indicate that the sample mean and sample variance can be calculated recursively if a new item is added to the sample.

5.11* *Prediction of election outcome.* Consider a U.S. Senate election in a southern state. Let

$$d = P\{D\} = \text{probability of an arbitrary voter voting for } D$$
$$r = P\{R\} = \text{probability of an arbitrary voter voting for } R$$

It turned out that 2,000,000 people voted. In an exit poll, conducted involving 20,000 voters, candidates D and R received 10,020 and 9,980 votes, respectively. Define a random variable for an arbitrary voter

$$X = \begin{cases} 1 & \text{if the voter votes for } D \\ -1 & \text{if the voter votes for } R \end{cases}$$

and let $Y = (\text{total votes for } D) - (\text{total votes for } R)$.

(a) Assuming a vote is independent of any other vote, determine the expected values and variances of X and Y, respectively, in terms of exact d and r.
(b) Based on the exit poll, use the sample means $\bar{d}$ and $\bar{r}$ of the probabilities d and r, respectively, to determine which candidate (D or R) will win and with how much probability.
(c) Based on the exit poll, predict how many more votes the winner (D or R) will have than the loser (R or D).
(d) Based on the exit poll, determine (approximately) the standard deviation of Y. Give concrete values of the standard deviation.

5.12 *Correlated sample.* Find the mean and variance of the sample mean of a correlated sample: $\{x_1, x_2, x_3\}$, where x_1 and x_2 have correlation coefficient $\rho_{12} = 0.5$, x_3 is independent of x_1 and x_2, $\sigma_{x1}^2 = \sigma_{x2}^2 = \sigma_{x3}^2$.

5.13 *MLE of log-normal distribution parameters.* A sample $(X_1, \ldots, X_n)$ was drawn from a population of the log-normal distribution (3.37) with unknown parameters μ and σ^2. Find the maximum likelihood estimate of μ and σ^2.

5.14 *Estimation of Gaussian variance.* A sample $(X_1, \ldots, X_n)$ is taken from a population $X \sim \mathcal{N}(\mu, \sigma^2)$, where both μ and σ^2 are unknown. Find the maximum likelihood estimate and the method of moments estimate of σ^2. Are these estimators unbiased?

5.15 *Estimation of uniform distribution parameter.* Given a sample $(X_1, \ldots, X_n)$ from a population of the following uniform distribution

$$f(x) = \begin{cases} 1/\theta & 0 < x \leq \theta \\ 0 & \text{elsewhere} \end{cases}$$

where θ is unknown, find the maximum likelihood estimate and the method of moments estimate of θ.

5.16 *u test on average resistance.* The resistance of a resistor manufactured by a manufacturer is Gaussian distributed $\mathcal{N}(\mu, \sigma^2)$ with $\mu = 2000$ and $\sigma = 7$. A new production line was recently put in use. It seems from observations that the tolerance in the resistance (i.e., σ) has not changed but the resistance μ may have a bias. The following sample was then obtained

$$2004, 2003, 1995, 2015, 1997, 1987, 1996, 2011, 2007, 2004, 2012, 1986, 1993, 2005, 1997$$

Can we say that the average resistance has changed with 95% confidence?

5.17 *t test on software bug.* The truncation error in running a software package is uniformly distributed over an interval. It would have zero mean if the software has no bug. A diagnostic program obtained a sample of 30 truncation errors and found the sample mean and sample variance are 7.8×10^{-10} and 2.4×10^{-15}, respectively. From the t distribution table, we know $P\{-2.0452 < t(29) < 2.0452\} = 0.05$. Can we conclude that the software has a bug with 95% confidence?

5.18 *Chi-square test on student readiness.* The test score of a large class of 54 students is Gaussian distributed with an unknown mean and variance σ^2. The prior experience indicates that $\sigma^2 = 120$ if the students prepare the test well; otherwise $\sigma^2 > 120$. The sample mean and sample variance of the mid-term exam scores are equal to 73.6 and 157.8, respectively. Can we conclude that the students did not prepare well for the exam with 95% confidence? What if we need only 90% confidence?

5.19 *Relation between regression parameters and sample statistics.* For two RVs X and Y, their sample means, sample variances and sample correlation coefficient are $\hat{x} = 1.3$, $\hat{y} = 4.67$, $\hat{v}_x = 0.4$, $\hat{v}_y = 2.5$ and $\hat{\rho}_{xy} = 0.87$. Find the linear regression of Y on X (i.e., estimate a and b). If we know that the RV X is equal to 1.4, what is your best guess of the value of Y?

5.20 *Puzzling questions.* Raise two technical questions of your own that you are puzzled by and that are closely related to the material covered in this chapter. These questions should be important, not too simple, and not too involved.

5.11 Computer Exercises

5.1 *Testing of uniform random number generator by sample mean and sample variance.*

(a) Use MATLAB $\mathcal{U}(0, 1)$ random number generator "`rand`" to generate 100,000 random numbers that are uniformly distributed over $(0, 1)$.
(b) Find the sample mean of the random numbers generated.
(c) Find the sample variance of the random numbers generated.
(d) Compare the computed sample mean and sample variance with the true mean and variance.

5.2 *Testing of exponential random numbers by sample mean and sample variance.*

(a) Use MATLAB $\mathcal{U}(0, 1)$ random number generator "`rand`" and the results of Example 3.30 to generate 10,000 random numbers that are exponentially distributed with parameter $\lambda = 4$.
(b) Compute the sample mean and sample variance of the 10,000 random numbers generated.
(c) Compare the computed sample mean and sample variance with the true mean and variance of an exponential RV with parameter $\lambda = 4$.
(d) Compute sample mean and sample variance for an arbitrary 1,000 of the 10,100 random numbers generated and compare them with the true mean and variance of an exponential RV with parameter $\lambda = 4$.

5.3 *Testing of binary random numbers by sample mean and sample variance.*

(a) Use MATLAB $\mathcal{U}(0, 1)$ random number generator "`rand`" to generate 10,000 random numbers $X_1, \ldots, X_{10{,}000}$ with the following point masses, for $i = 1, \ldots, 10000$,

$$P\{X_i = 0\} = 0.35, \quad P\{X_i = 1\} = 0.65$$

(b) Find the (theoretical) mean and variance of X_i.
(c) Compute the sample mean $\hat{x}_{10{,}000}$ and sample variance $\hat{v}_{10{,}000}$ of the 10,000 random numbers generated. Compare them with their theoretical mean and variance.
(d) Compute the sample mean $\hat{x}_{1{,}000}$ and sample variance $\hat{v}_{1{,}000}$ of the first 1,000 of the 10,000 random numbers generated. Compare them with $\hat{x}_{10{,}000}$ and $\hat{v}_{10{,}000}$ and comment on the difference.

5.4 *Testing of ternary random numbers by sample mean and sample variance.*

(a) Use MATLAB $\mathcal{U}(0, 1)$ random number generator "`rand`" to generate 10,000 random numbers $X_1, \ldots, X_{10{,}000}$ with the following point masses, for $i = 1, \ldots, 10000$,

$$P\{X = -1\} = 0.25, \quad P\{X = 0\} = 0.3, \quad P\{X = 1\} = 0.45$$

(b) Find the (theoretical) mean and variance of X_i.

(c) Compute the sample mean $\hat{x}_{10,000}$ and sample variance $\hat{v}_{10,000}$ of the 10,000 random numbers generated. Compare them with their theoretical mean and variance.
(d) Compute the sample mean $\hat{x}_{1,000}$ and sample variance $\hat{v}_{1,000}$ of the first 1,000 of the 10,000 random numbers generated. Compare them with $\hat{x}_{10,000}$ and $\hat{v}_{10,000}$ and comment on the difference.

5.5 *Distribution identification.* The file `m5_5.dat` in the companion software P&R contains a record of data drawn from a population X with a symmetric PDF. Use P&R to do the following.

(a) Compute the sample mean and sample variance.
(b) Compute the empirical PDF.
(c) Identify the distribution of X. Justify your answer.

5.6 *Estimation of Gaussian mean and variance.* The file `m5_6.dat` in the companion software P&R contains a record of data drawn from a population $X \sim \mathcal{N}(\mu, \sigma^2)$, where both μ and σ^2 are unknown. Find the maximum likelihood and the method of moments estimates of μ and σ^2, respectively.

5.7 *Signal detection by u test.* The measurement X of a signal s is corrupted by Gaussian distributed noise $V \sim \mathcal{N}(0, \sigma^2)$, where $\sigma^2 = 1.44$ is known. In other words, $X = s + V$ if the signal is present and $X = V$ if the signal is absent. A set of measurements $(X_1, \ldots, X_{100})$ was made and recorded in file `m5_7.dat` in the companion software P&R. Is the signal present with 95% confidence?

5.8 *Linear regression.* The file `m5_8.dat` in the companion software P&R contains a data record of a pair of RVs (X, Y).

(a) Write a small computer program to find the linear regression of Y on X.
(b) Find the linear regression of Y on X using P&R.
(c) Predict the values of Y given the values of X as 11.5, 32.6 and 22.7, respectively.

5.12 Self-Test Problems

5.1 Answer the following questions briefly

(a) Does the expected value of the sample mean vary with the sample size?
(b) Does the variance of the sample mean vary with the sample size?
(c) Does the expected value of the sample variance vary with the sample size?
(d) What is the primary difference between probability and statistics?
(e) Why should the data making up a sample be independent?
(f) What is the use of empirical distributions?
(g) What is the purpose of estimation?

(h) What is the technical term for the procedure of judging whether an hypothesized value of an unknown parameter should be accepted?
(i) Is it true that linear regression is a good tool for handling cause-and-effect relationships?
(j) If a chi-square RV has 5 degrees of freedom, what does it mean?

5.2 Assume that the test score of a student in a very large class is a Gaussian RV X with unknown mean $\bar{x}$ and variance σ_x^2. Suppose that you have asked 16 students in the class and found their scores to be: 87, 55, 58, 89, 97, 78, 91, 88, 65, 88, 93, 90, 66, 77, 75, 73.

(a) What is your estimate of the average test score?
(b) What is the variance of your estimate of the average test score?
(c) What is your estimate of the variance σ_x^2 of the test score?
(d) Suppose that your score was 92, was your score in the top 10%? Justify your answer.
(e) If a score below 60 corresponds to "fail," what is the probability that one of your fellow students whose score you do not know actually failed the test?

5.3 A sample $(X_1, \ldots, X_n)$ was drawn from a population X of a Poisson distribution with parameter λ. Write down the joint PMF of $X_1, \ldots, X_n$. Find the mean and variance of the sample mean and the mean of the sample variance $\hat{V}$.

5.4 It is known that a population is Poisson distributed with an unknown parameter λ. Given a sample with the following values:

$$3, 2, 3, 2, 1, 2, 3, 3, 3, 2, 0, 3, 2, 0, 4, 2, 5, 4, 4, 0$$

(a) Do sample mean and sample variance have the same expected value for this problem? Find the sample mean and sample variance for this sample.
(b) Write down the likelihood function of λ. Find the maximum likelihood estimate of λ.
(c) Find the method of moments estimates of λ using the first and second moments, respectively.

5.5 The following sample was drawn from a pair of RVs (X, Y)

x_i	0.8	-1.4	-1.1	-0.6	0.1	0.89	1.0	1.4	1.4	0.5
y_i	-0.2	-1.5	-1.2	-0.5	0.05	0.8	1.1	1.3	1.49	0.7

Find the linear regression of Y on X. What is your best guess of the value of Y given the value of $X = 0.77$

5.13 Solutions to Self-Test Problems

5.1 (a) The expected value of the sample mean is equal to the true mean and therefore does not depend on the sample size.

(b) The variance of the sample mean is equal to the true mean over the sample size and thus decreases with sample size.

(c) The expected value of the sample variance is equal to the true variance which does not depend on the sample size.

(d) The primary difference between probability and statistics is that statistics is the science of establishing a probabilistic model of a random phenomenon from the available data while probability assumes the availability of such a model.

(e) Were the data making up a sample not independent, the conclusion of a statistical analysis would depend critically on the choice of the data. As a result, one can arrive at almost any conclusion he or she wants by using different dependent data.

(f) When the true distributions are not known perfectly, empirical distributions are used to approximate the true distributions.

(g) Estimation is to provide a quantity, as a function of the sample, that is approximately equal to the unknown parameter.

(h) It is known as decision or hypothesis testing.

(i) No, linear regression is a tool for handling statistical relations between two or more variables.

(j) It means that the RV can be decomposed as the sum of the squares of 5 *independent* Gaussian RVs.

5.2 (a) Estimated average = sample mean = $\bar{x} = \frac{1}{16}\sum X_i = 79.4$.

(b) Variance of estimated average = variance of sample mean = $\frac{1}{n}\sigma_x^2 \approx \frac{1}{n}\hat{V} = 10.6$.

(c) Estimated σ_x^2 = sample variance = $\frac{1}{n-1}\sum_{i=1}^n (X_i - \bar{x})^2 = \frac{1}{15}\sum_{i=1}^n (X_i - 79.4)^2 = 170$.

(d)

$$\begin{aligned} P\{X \geq 92\} &= 1 - \Phi\left(\frac{92 - \bar{x}}{\sigma_x}\right) \\ &\approx 1 - \Phi\left(\frac{92 - \bar{x}}{\sigma_{\hat{x}}}\right) \\ &= 1 - \Phi\left(\frac{92 - 79.4}{\sqrt{170}}\right) \\ &= 1 - \Phi(0.966) \\ &= 1 - 0.834 = 0.166 \end{aligned}$$

Since $P\{X \geq 92\} \simeq 0.166 > 10\%$, your score was not in the top 10%.

(e)

$$P\{\text{“}F\text{”}\} = P\{X < 60\}$$

$$= P\left\{\frac{X - \bar{x}}{\sigma_x} < \frac{60 - \bar{x}}{\sigma_x}\right\}$$
$$\simeq P\left\{\frac{X - \bar{x}}{\hat{\sigma}_x} < \frac{60 - \bar{x}}{\hat{\sigma}_x}\right\}$$
$$= \Phi\left(\frac{60 - 79.4}{\sqrt{170}}\right)$$
$$= \Phi(-1.49)$$
$$= 6.81\%$$

5.3 Since, for $k = 0, 1, 2, \ldots$,

$$P\{X = x\} = e^{-\lambda}\frac{\lambda^x}{x!}$$

the joint PMF of $X_1, \ldots, X_n$ is

$$\prod_{i=1}^{n}\left(e^{-\lambda}\frac{\lambda^{x_i}}{x_i!}\right) = e^{-n\lambda}\lambda^{\sum_{i=1}^{n} x_i} \Big/ \prod_{i=1}^{n} x_i!$$

Since for Poisson distribution, $E[X] = \lambda$ and $\mathrm{var}(X) = \lambda$, we have

$$E[\hat{X}] = E[X] = \lambda$$
$$\mathrm{var}(\hat{X}) = \mathrm{var}(X)/n = \lambda/n$$
$$E[\hat{V}] = \mathrm{var}(X) = \lambda$$

5.4 (a) For Poisson distribution,

$$\bar{x} = \sigma^2 = \lambda, \quad E[\hat{X}] = \bar{x} = \lambda, \quad E[\hat{V}] = \sigma^2 = \lambda \quad \Longrightarrow \quad E[\hat{X}] = E[\hat{V}]$$

They have the same expected value.

$$\hat{X} = \frac{1}{n}\sum_{i=1}^{n} X_i \quad \Longrightarrow \quad \hat{x} = \frac{1}{20}\sum x_i = 2.4$$

$$\hat{V} = \frac{1}{n-1}\sum_{i=1}^{n}(X_i - \hat{X})^2 \quad \Longrightarrow \quad \hat{v} = \frac{1}{19}\sum_{i=1}^{20}(x_i - \hat{x})^2 = 1.9368$$

(b)

$$L(\lambda) = f_{X_1,\ldots,X_n}(x_1, \ldots, x_n; \lambda) = \prod_{i=1}^{n} f_{X_i}(x_i; \lambda)$$
$$= \prod_{i=1}^{n} e^{-\lambda}\frac{\lambda^{X_i}}{x_i!} = e^{-n\lambda}\lambda^{n\hat{x}} / \prod_{i=1}^{n}(x_i!)$$

For our problem with the given sample,

$$L(\lambda) = e^{-20\lambda}\lambda^{48}/[(2^6)(3!)^6(4!)^3 5!] = 4.161 \times 10^{-22}\lambda^{48}$$
$$\ln L = -n\lambda + n\hat{x}\ln\lambda$$
$$0 \overset{\text{set}}{=} \frac{\partial \ln L}{\partial \lambda} = -n + n\hat{x}/\lambda \Longrightarrow \hat{\lambda}^{\mathrm{ML}} = \hat{X} \overset{\text{this sample}}{=} 2.4$$

(c) Method of moments:

$$\text{Use first moment: } \lambda = \bar{x} \simeq \hat{X} \Longrightarrow \hat{\lambda} = \hat{X} = 2.4$$
$$\text{Use second central moment: } \lambda = \sigma^2 \simeq \hat{V} \Longrightarrow \hat{\lambda} = \hat{V} = 1.9368$$

5.5 Using the following MATLAB routine

```
x=[0.8,-1.4,-1.1,-0.6,0.1,0.89,1.0,1.4,1.4,0.5];
y=[-0.2,-1.5,-1.2,-0.5,0.05,0.8,1.1,1.3,1.49,0.7];
xmean=sum(x)/10
ymean=sum(y)/10
b=sum((x-xmean).*(y-ymean))/sum((x-xmean).^2)
a=ymean-b*xmean
```

we get

$$\hat{x} = 0.2990, \quad \hat{y} = 0.2040, \quad \hat{a} = -0.0877, \quad \hat{b} = 0.9757$$

and thus the linear regression of Y on X is

$$\check{Y} = -0.0877 + 0.9757X$$

The best guess of the value of Y given the value of $X = 0.77$ is

$$\check{Y} = -0.0877 + (0.9757)(0.77) = 0.6636$$

6

RANDOM PROCESSES

Fate, Time, Occasion, Chance, and Change — to these all things are subject.

Percy Bysshe Shelley

The random variable approach is valid only for random problems that are time-invariant. The mathematical tool for time-varying random problems is known as the random processes.

This chapter covers fundamentals of random processes with an emphasis on its first two order time-domain characteristics.

Main Topics

- Concept of Random Processes
- Characterization of Random Processes
- Classification of Random Processes
- Correlation Functions and Their Properties
- Sample Correlation Functions
- Relationship between Two Random Processes
- Gaussian Random Process

6.1 Concept of Random Processes

Many practical, especially engineering, random problems are time dependent. For example, as defined in Chapter 1, a random signal is a random time function. Such time functions are called ***random processes*** in mathematics.

Example 6.1: A Sinusoidal Signal with Binary Random Phase

Consider a sinusoidal signal $X(t) = \cos(t + \phi)$, where ϕ is a RV determined by tossing a coin: $\phi = \left\{ \begin{array}{ll} 0 & \text{head shows} \\ \pi & \text{tail shows} \end{array} \right\}$. Note that

- $X(t)$ is random since its values cannot be determined surely beforehand.
- $X(t)$ is not a RV because a *time function* rather than a *number* is assigned to it given an outcome of the random experiment.
- Sample space $= \{x_1(t), x_2(t)\} = \{\cos t, -\cos t\}$ since $\cos(t + \pi) = -\cos t$.
- For any given time t_1, $X(t_1)$ is a RV with at most two possible values: $\cos t_1$ and $-\cos t_1$.

Example 6.2: Rectangular Pulse Train with Random Pulse Width

Consider a rectangular pulse train with a "period" T, pulse height A, and a random pulse width W that is independent from period to period and is uniformly distributed over the period: $W \sim \mathcal{U}(0, T)$. Note that

- $X(t)$ is random since its values cannot be determined surely beforehand.
- $X(t)$ is not a RV because its value is a time function although for any given time t_1, $X(t_1)$ is a binary RV with two possible values: 0 or A.
- The sample space consists of infinitely many time functions.

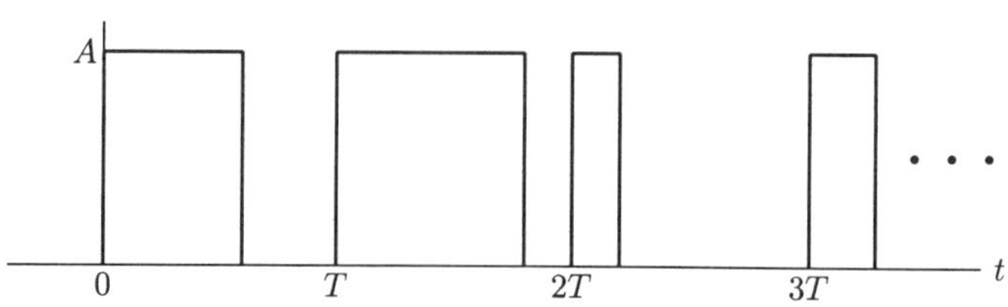

Figure 6.1: A pulse train with a random pulse width.

In general, a ***random process*** or ***stochastic process*** $X(t)$ can be defined as

- a *time-varying random variable* in the sense that the value on which the RV may take is time varying, or
- a *random time function* in the sense that a whole time function, rather than a number, is assigned to each outcome.

Note that

- $X(t)$ at any specific time t_1 is a random variable $X(t_1)$.
- A ***realization*** of $X(t)$ is a time function, called a ***sample path***, ***sample function***, or *member function*, not a number.
- Its sample space (known as ***ensemble***) is a collection of time functions.
- $X(t)$ at distinct time instances could have vastly different distributions.

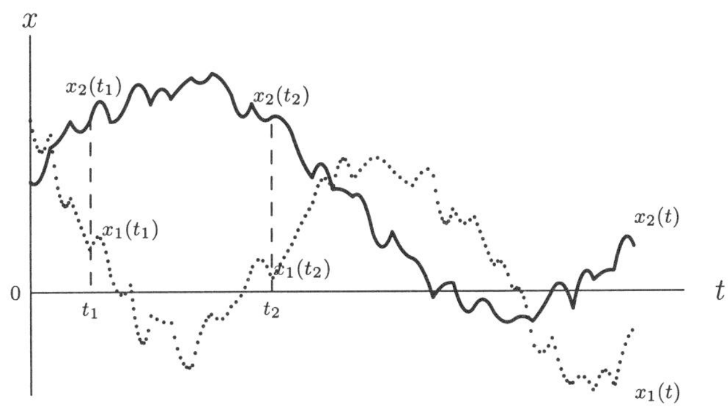

Figure 6.2: Two sample functions of a random process.

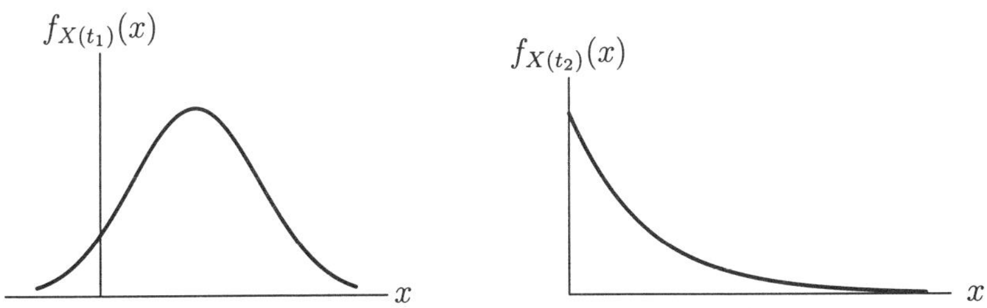

Figure 6.3: Some possible marginal PDFs of a random process.

The RV approach is powerful in handling random phenomena. It will, however, break down if the random phenomena vary with time since a RV does not vary with time. In most engineering applications that involve uncertainty, such as random signals and systems, the random phenomena are not time invariant. Random processes are the right tool for such time-varying random phenomena. They are natural extensions of the concept of RVs.

Recall that a RV assigns a unique *number* x to the outcome ω of a random experiment: $X = X(\omega)$. If to every outcome ω, we assign a unique *time function*, rather than a number, then we have a random process, which may be denoted as $X(t, \omega)$, rather than a RV. Thus, the sample space consists of time functions, called sample functions, rather than numbers.

The term ***stochastic*** is a widely used technical synonym for the word "random."

A random process may be visualized as the process of the random motion of a particle. Its possible motion trajectories are the sample paths (sample functions) of the random process.

In Example 6.1, if a nonzero value of the random process $X(t)$ at a given time is known, then the value of the random process for the entire time horizon is known for sure. More generally, all future values of some random processes can be determined uniquely given the past values. Such processes are sometimes called *deterministic random processes*. Otherwise the random processes are called *non-deterministic*, such as Example 6.2. These terms are somewhat confusing.

A ***random sequence*** or ***discrete-time random process*** $X(n)$ is an (infinite) time sequence of RVs. A random sequence is a special random process that is discrete in time. It is nothing but an infinite sequence of RVs, which is equivalent to an ∞-dimensional random vector. A *finite* sequence of RVs is a degenerated special random process. It may be represented by a finite-dimensional random vector. A random sequence is also known as a ***time series***.

Similarly as in the deterministic case, a random sequence $X[n]$ may be obtained by sampling a (continuous-time) random process $X(t)$ (by a constant sampling rate f); that is $X[n] = X(nT)$, where $T = 1/f$ is the sampling interval. Fig. 6.4 illustrates two sample functions of a random sequence that is the sampled version of the random process of Example 6.1.

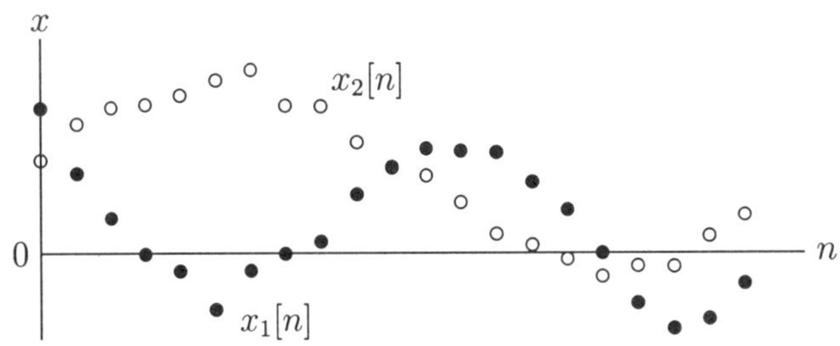

Figure 6.4: Two sample functions of a random sequence.

There are two classes of random pulse trains. The first is random due to the fact that some parameters, e.g., amplitude and/or pulse width, are RVs but time invariant. For this class the amplitude and pulse width are time-invariant from period to period. The amplitude and/or pulse width (or even the period) in the other class are random and time-varying (different for different period). For instance, in Example 6.2 the pulse width is random and time-varying.

Example 6.3: Generation of a Random Pulse Train by P&R

Suppose we want to generate a pulse train similar to the one of Example 6.2 with $T = 2.5$ seconds except that the amplitude A is a time-invariant Gaussian RV: $A \sim \mathcal{N}(3, 0.5^2)$. This can be done easily using the companion software P&R as follows:

S1. Click "RPGenerator" in the main window of P&R.
S2. Click "Pulse Train." The "**Random Pulse Train Generator**" window will appear.
S3. For amplitude choose "Time-Invariant", find and click "Normal" and then enter "3" and "0.25" for the mean and variance. Click "Ok."
S4. For pulse width, choose "Time-Varying", find and click "Uniform" and then enter "0" and "2" as the lower and upper limits, respectively. Click "Ok."
S5. Fill out the remaining parts of the window "**Random Pulse Train Generator**" as shown in Fig. 6.5. Click "Ok." A single sample function of $X(t)$ is then generated and plotted, as shown in Fig. 6.5, and saved to a user-specified data file.

The above steps can be repeated to generate more than one sample function.

Since P&R is a computer software, it can generate only discrete points of $X(t)$. A sampling rate of 10 Hz means that 10 data points per second will be generated and thus 25 points will be generated within each period $T = 2.5$ seconds in this example. Hence, 100 time points means that the data record will have 100 points, covering 4 periods for this example.

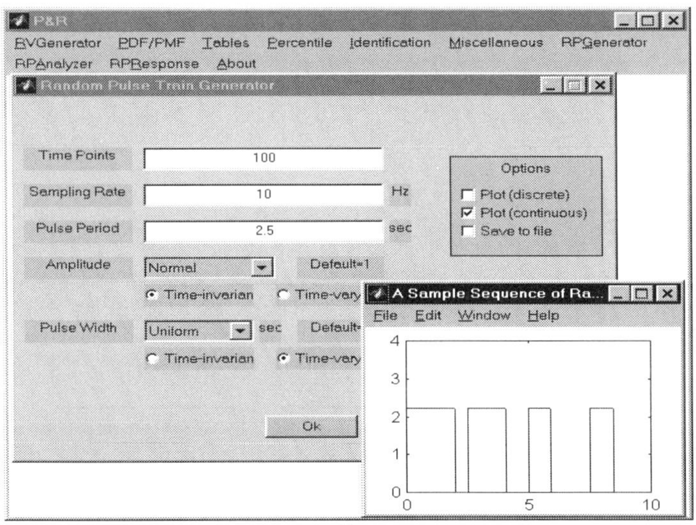

Figure 6.5: Generation of a random pulse train by P&R.

Example 6.4: PDFs of a Random Sinusoidal Signal

A random process $X(t)$ is defined by tossing two dice to determine which of the following three sinusoids to pick:

$$X(t) = \begin{cases} x_1(t) = \cos t & \text{if the sum is 2 or 12} \\ x_2(t) = 2\cos 2t & \text{if the sum is 3 or 7} \\ x_3(t) = 3\cos 3t & \text{otherwise} \end{cases}$$

(a) Determine the probability of each sample path being the right one:

$$\begin{aligned} P\{X(t) = x_1(t)\} &= P\{(1,1)\} + P\{(6,6)\} = \frac{1}{36} + \frac{1}{36} = \frac{1}{18} \\ P\{X(t) = x_2(t)\} &= P\{(1,2),(2,1)\} \\ &\quad + P\{(1,6),(2,5),(3,4),(4,3),(5,2),(6,1)\} \\ &= \frac{2}{36} + \frac{6}{36} = \frac{2}{9} \\ P\{X(t) = x_3(t)\} &= 1 - \frac{1}{18} - \frac{2}{9} = \frac{13}{18} \end{aligned}$$

(b) Determine $f_{X(\pi/4)}(x)$, $f_{X(\pi/2)}(x)$, and $f_{X(\pi/4),X(\pi/2)}(x,y)$:

t	$x_1(t)$	$x_2(t)$	$x_3(t)$
$\pi/4$	0.707	0	−2.12
$\pi/2$	0	−2	0
$P\{X(t) = x_i(t)\}$	1/18	2/9	13/18

$$\begin{aligned} f_{X(\pi/4)}(x) &= \frac{1}{18}\delta(x-0.707) + \frac{2}{9}\delta(x) + \frac{13}{18}\delta(x+2.12) \\ f_{X(\pi/2)}(x) &= \frac{1}{18}\delta(x) + \frac{2}{9}\delta(x+2) + \frac{13}{18}\delta(x) = \frac{7}{9}\delta(x) + \frac{2}{9}\delta(x+2) \\ f_{X(\pi/4),X(\pi/2)}(x,y) &= \underbrace{\frac{1}{18}\delta(x-0.707)\delta(y)}_{X(t)=x_1(t)} + \underbrace{\frac{2}{9}\delta(x)\delta(y+2)}_{X(t)=x_2(t)} \\ &\quad + \underbrace{\frac{13}{18}\delta(x+2.12)\delta(y)}_{X(t)=x_3(t)} \end{aligned}$$

6.2 Characterization of Random Processes

Since $X(t_1)$ is a RV, we have, by (3.43),

$$E[g(X(t_1))] = \begin{cases} \int_{-\infty}^{\infty} g(x) f_{X(t_1)}(x)dx & X(t_1) \text{ is continuous} \\ \sum_i g(x_i) P\{X(t_1) = x_i\} & X(t_1) \text{ is discrete} \end{cases}$$

where $f_{X(t_1)}(x)$ is the PDF of the RV $X(t_1)$. Then, the mean value, mean-square value, and variance of $X(t_1)$ are, by (3.12), (3.11) (3.11),

$$\bar{x}(t_1) = E[X(t_1)] = \int_{-\infty}^{\infty} x f_{X(t_1)}(x)dx \tag{6.1}$$

$$E[X^2(t_1)] = \int_{-\infty}^{\infty} x^2 f_{X(t_1)}(x)dx$$

$$\sigma^2_{x(t_1)} = \int_{-\infty}^{\infty} [x - \bar{x}(t_1)]^2 f_{X(t_1)}(x)dx = E[X^2(t_1)] - [\bar{x}(t_1)]^2 \tag{6.2}$$

Since the above are true for every t_1, t_1 can be replaced with the generic time argument t and thus $\bar{x}(t)$, $E[X^2(t)]$ and $\sigma^2_{x(t)}$ are called the ***mean function***, *mean-square function*, and *variance function*, respectively.

Associated with the values $X(t_1)$ and $X(t_2)$ of $X(t)$ at any two time instants, there is a joint PDF $f_{X(t_1),X(t_2)}(x_1, x_2)$ (i.e., the joint PDF of the two RVs) through which the joint statistics can be determined by (4.20):

$$E[g(X(t_1), X(t_2))] = \int_{-\infty}^{\infty} \int_{-\infty}^{\infty} g(x_1, x_2) f_{X(t_1),X(t_2)}(x_1, x_2)dx_1 dx_2$$

As special cases, we define ***autocorrelation function*** by [see (4.21)]

$$\boxed{R_x(t_1, t_2) = E[X(t_1)X(t_2)]} \tag{6.3}$$

autocovariance function by [see (4.22)]:

$$\boxed{C_x(t_1, t_2) = E\{[X(t_1) - \bar{x}(t_1)][X(t_2) - \bar{x}(t_2)]\}} \tag{6.4}$$

and ***correlation coefficient*** by [see (4.23)]

$$\rho(t_1, t_2) = \frac{C_x(t_1, t_2)}{\sqrt{C_x(t_1, t_1) C_x(t_2, t_2)}} = \frac{C_x(t_1, t_2)}{\sigma_{x(t_1)} \sigma_{x(t_2)}}, \quad -1 \le \rho(t_1, t_2) \le 1 \tag{6.5}$$

Similarly to (4.22) for two RVs, we have

$$\begin{aligned} C_x(t_1, t_2) &= R_x(t_1, t_2) - \bar{x}(t_1)\bar{x}(t_2) \\ \text{autocovariance} &= \text{autocorrelation} - (\text{mean at } t_1) \times (\text{mean at } t_2) \end{aligned} \tag{6.6}$$

Mean, autocorrelation, and autocovariance are often used as the short terms for mean function, autocorrelation function, and autocovariance function, respectively.

Autocorrelation, autocovariance, and correlation coefficient quantify the *coupling of a random process at distinct time instants.* This is the same in principle as the fact that correlation, covariance, and correlation coefficient quantify the coupling between two RVs.

Autocorrelation is used more widely than autocovariance and correlation coefficient for the quantification of coupling of a random process at distinct time instants. This is because many random processes encountered in practice have zero mean and a random process $Y(t)$ with a nonzero mean $\bar{y}(t)$ can always be decomposed as $Y(t) = X(t) + \bar{y}(t)$, where $X(t)$ is a zero-mean random process.

Note the following relations:

$$\begin{aligned} E[X^2(t)] &= R_x(t,t) \\ \sigma^2_{x(t)} &= C_x(t,t) \end{aligned}$$

Example 6.5: First Two Moments of a Sinusoid with Random Binary Phase

Consider again the sinusoidal signal $X(t)$ of Example 6.1. Since $P\{\phi = 0\} = P\{\phi = \pi\} = 1/2$, its mean, mean-square value, variance, autocorrelation, and autocovariance are

$$\begin{aligned}
\bar{x}(t) &= E[X(t)] = E[\cos(t+\phi)] \\
&\overset{(3.43)}{=} P\{\phi = 0\}\cos(t+\phi)|_{\phi=0} + P\{\phi=\pi\}\cos(t+\phi)|_{\phi=\pi} \\
&= \frac{1}{2}\cos t + \frac{1}{2}\cos(t+\pi) = 0 \\
E[X^2(t)] &= E[\cos^2(t+\phi)] = \frac{1}{2}\cos^2 t + \frac{1}{2}\cos^2(t+\pi) = \cos^2 t \\
\sigma^2_{x(t)} &= E[X^2(t)] - [\bar{x}(t)]^2 = \cos^2 t \\
R_x(t_1,t_2) &= E[X(t_1)X(t_2)] = E[\cos(t_1+\phi)\cos(t_2+\phi)] \\
&\overset{?}{=} E\left\{\frac{1}{2}[\cos(t_1-t_2) + \cos(t_1+t_2+2\phi)]\right\} \\
&= \frac{1}{2}\cos(t_1-t_2) + \frac{1}{2}E[\cos(t_1+t_2+2\phi)] \\
&= \frac{1}{2}\cos(t_1-t_2) + \frac{1}{2}\left[\frac{1}{2}\cos(t_1+t_2) + \frac{1}{2}\cos(t_1+t_2+2\pi)\right] \\
&= \frac{1}{2}[\cos(t_1-t_2) + \cos(t_1+t_2)] = \cos t_1 \cdot \cos t_2 \\
C_x(t_1,t_2) &= R_x(t_1,t_2) - \bar{x}(t_1)\bar{x}(t_2) = \cos t_1 \cdot \cos t_2
\end{aligned}$$

It is not easy to find the first two moment functions of the random pulse train of Example 6.2, although the random process looks simple.

Example 6.6: First Two Moments of a Random Sinusoid

Consider again the random process $X(t)$ of Example 6.4.

(a) Determine $\bar{x}(\pi/4)$, $\bar{x}(\pi/2)$, $\sigma^2_{x(\pi/4)}$, and $\sigma^2_{x(\pi/2)}$:

$$
\begin{aligned}
\bar{x}(\pi/4) &= \sum_{i=1}^{3} x_i(\pi/4) P\{X(\pi/4) = x_i(\pi/4)\} \\
&= (0.707)(1/18) + (0)(2/9) + (-2.12)(13/18) = -1.492 \\
\bar{x}(\pi/2) &= (0)(1/18) + (-2)(2/9) + (0)(13/18) = -\frac{4}{9} = -0.4444 \\
E[X^2(\pi/4)] &= \sum_{i=1}^{3} x_i^2(\pi/4) P\{X(\pi/4) = x_i(\pi/4)\} \\
&= (0.707^2)(1/18) + (0^2)(2/9) + (-2.12)^2(13/18) = 3.274 \\
E[X^2(\pi/2)] &= (0^2)(1/18) + (-2)^2(2/9) + 0^2(13/18) = \frac{8}{9} \\
\sigma^2_{x(\pi/4)} &= E[X^2(\pi/4)] - [\bar{x}(\pi/4)]^2 = 3.274 - (-1.492)^2 = 1.048 \\
\sigma^2_{x(\pi/2)} &= E[X^2(\pi/2)] - [\bar{x}(\pi/2)]^2 = 8/9 - (-4/9)^2 = 0.6914
\end{aligned}
$$

(b) Determine $R_x(\pi/4, \pi/2)$ and $C_x(\pi/4, \pi/2)$:

$$
\begin{aligned}
R_x(\pi/4, \pi/2) &= E[X(\pi/4)X(\pi/2)] \\
&= \sum_{i=1}^{3} \sum_{j=1}^{3} x_i(\pi/4) x_j(\pi/2) P\{X(\pi/4) = x_i(\pi/4), \\
&\qquad X(\pi/2) = x_j(\pi/2)\} \\
&\overset{?}{=} \sum_{i=1}^{3} x_i(\pi/4) x_i(\pi/2) P\{X(\pi/4) = x_i(\pi/4), X(\pi/2) = x_i(\pi/2)\} \\
&= 0.707 \cdot 0 \cdot \frac{1}{18} + 0 \cdot (-2) \cdot \frac{2}{9} + (-2.12) \cdot 0 \cdot \frac{13}{18} = 0 \\
C_x(\pi/4, \pi/2) &= R_x(\pi/4, \pi/2) - \bar{x}(\pi/4)\bar{x}(\pi/2) \\
&= 0 - (-1.492)(-0.4444) = -0.6630
\end{aligned}
$$

where $\overset{?}{=}$ follows from the fact that given an outcome of a random event, the *whole* time function is assigned to the random process. Note that $C_x(t_1, t_2)$ as a covariance could be negative and $R_x(\pi/4, \pi/2) = 0$ makes sense because either $X(\pi/2)$ or $X(\pi/4)$ as a RV is equal to zero.

Example 6.7: Mean and Autocorrelation of Random-Phase Sinusoid

Many sinusoids in reality have a random phase. Consider, e.g.,

$$X(t) = A\sin(\omega_0 t + \phi)$$

where $A, \omega_0 > 0$ are known constants, ϕ is a uniform RV over $(-\pi, \pi]$:

$$f(\phi) = \begin{cases} \frac{1}{2\pi} & -\pi < \phi \le \pi \\ 0 & \text{elsewhere} \end{cases}$$

Find the mean and autocorrelation of $X(t)$:

$$\begin{aligned} E[X(t)] &= E[A\sin(\omega_0 t + \phi)] = E[g(\phi)] = \int_{-\infty}^{\infty} A\sin(\omega_0 t + \phi) f(\phi) d\phi \\ &= \frac{A}{2\pi}\int_{-\pi}^{\pi} \sin(\omega_0 t + \phi) d\phi = \frac{A}{2\pi}[-\cos(\omega_0 t + \phi)]\Big|_{\phi=-\pi}^{\pi} = 0 \end{aligned}$$

which does not involve t. This should be the case since a sinusoid with an arbitrary phase has zero average over one period. In fact, this is true if ϕ is any RV with a PDF symmetric about the origin, as can be seen from

$$\int_{-\pi}^{\pi} A\sin(\omega_0 t + \phi) f(\phi) d\phi = 0 \qquad \text{if } f(\phi) \text{ is symmetric about } 0$$

Note that

$$\sin\alpha \sin\beta = \frac{1}{2}[\cos(\alpha - \beta) - \cos(\alpha + \beta)]$$

We have

$$\begin{aligned} R_x(t+\tau, t) &= E[X(t+\tau)X(t)] = E[A^2 \sin(\underbrace{\omega_0 t + \omega_0 \tau + \phi}_{\alpha}) \sin(\underbrace{\omega_0 t + \phi}_{\beta})] \\ &= \frac{A^2}{2}\{E[\cos(\omega_0 \tau)] - E[\cos(2\omega_0 t + \omega_0 \tau + 2\phi)]\} \\ &= \frac{A^2}{2}\left[\cos(\omega_0 \tau) - \int_{-\infty}^{\infty} \cos(2\omega_0 t + \omega_0 \tau + 2\phi) f(\phi) d\phi\right] \\ &= \frac{A^2}{2}\left[\cos(\omega_0 \tau) - \int_{-\pi}^{\pi} \cos(2\omega_0 t + \omega_0 \tau + 2\phi) \cdot \frac{1}{2\pi} d\phi\right] \\ &= \frac{A^2}{2}\cos(\omega_0 \tau) = R(\tau) \quad \text{(depends only on time difference } \tau\text{)} \end{aligned}$$

Note that for this example, $\bar{x}(t)$ does not depend on t and $R_x(t+\tau, t)$ is a function of the *time difference* τ only.

Example 6.8: Generation of a Random Sinusoid by P&R

A random sinusoid (i.e., a sinusoid with random amplitude, phase, and/or frequency) can be generated using the companion software P&R according to the following steps:

S1. Click "RPGenerator" in the main window of P&R.
S2. Click "Sinusoid." The "**Random Sinusoidal Process Generator**" window will appear.
S3. Choose "Constant" for amplitude and frequency and enter "`2`" and "`0.5*pi`" in their corresponding windows, respectively. Click "Ok."
S4. Choose "Uniform" for phase and enter "`-pi`" and "`pi`" as the lower and upper limits, respectively; that is, phase is uniformly distributed over $(-\pi, \pi)$. Click "Ok."
S5. Fill out the remaining parts of the window "**Random Sinusoidal Process Generator**" as shown in Fig. 6.6. Click "Ok." A single sample function of $X(t)$ is then generated and plotted, as shown in Fig. 6.6, and saved to a user-specified data file.

The above steps can be repeated to generate more than one sample function.

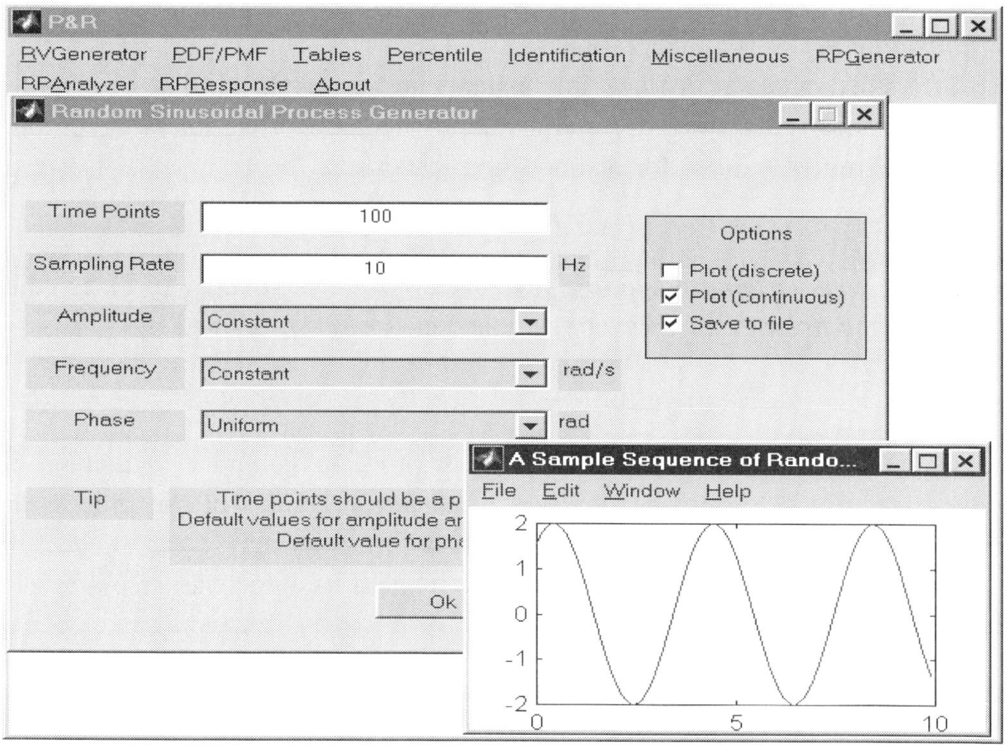

Figure 6.6: Generation of a random sinusoid by P&R.

6.3 Classification of Random Processes

A random process $X(t)$ is said to be

- ***continuous*** if $X(t_1)$ for every fixed t_1 is a continuous RV.
- ***discrete*** if $X(t_1)$ for every fixed t_1 is a discrete RV.
- ***mixed*** if it is neither continuous nor discrete.
- ***strictly stationary*** if none of its marginal and joint PDFs depend on the choice of time origin (this implies that none of its characteristics are time-varying).
- ***(wide-sense) stationary (WSS)*** if neither its mean nor its autocorrelation depends on the choice of time origin; that is, if
  - $E[X(t)]$ does not depend on t, *and*
  - $R_x(t_1, t_2) = R_x(t_1 - t_2) = R_x(\tau)$, $\tau = t_1 - t_2$, for every t_1 and t_2; that is, autocorrelation depends only on the time difference.
- ***nonstationary*** if it is not stationary.
- ***white*** if its values at distinct time instants are uncorrelated; that is, if its autocovariance is always zero (or equivalently, its autocorrelation is the product of the expected values) for distinct time instants:

$$C_x(t_1, t_2) = 0 \qquad [\text{or} \quad R_x(t_1, t_2) = \bar{x}(t_1)\bar{x}(t_2)] \qquad \forall t_1 \neq t_2$$

- ***ergodic*** if its ***ensemble average*** (i.e., mean) $\bar{x}(t)$, defined by (6.1), is equal to the ***time average*** of its every sample function $x(t)$, defined by

$$\bar{x}(\omega) = \lim_{T \to \infty} \frac{1}{2T} \int_{-T}^{T} x(t)dt, \qquad \text{given } \omega \tag{6.7}$$

The concept of stationarity of a random process is similar to that of the steady state of a deterministic process: The *characteristics* of the process are time-invariant even though the process itself is time-varying.

A (wide-sense) stationary process $X(t)$ is white if its autocovariance is always zero for any nonzero τ:

$$C_x(\tau) \triangleq E\{[X(t+\tau) - \bar{x}][X(t) - \bar{x}]\} = 0 \qquad \forall\, \tau \neq 0$$

Its autocorrelation is clearly always zero for any nonzero τ if it has zero mean:

$$R_x(\tau) \triangleq E[X(t+\tau)X(t)] = 0 \qquad \forall \tau \neq 0$$

A random process at a given time, as a RV, is sometimes referred to as its *state*.

If the state of a random process at any given time as a RV is discrete, the process is said to be discrete (or *discrete-valued*, *discrete-state*). If the state is a continuous RV, then the process is continuous (or *continuous-valued*, *continuous-state*). The process in Example 6.2 is discrete because its state has only two possible values, although its ensemble has uncountably many elements. The random process in Example 6.1 is also discrete because its state has at most two possible values, although these values span a continuous range $[-1, 1]$.

The concepts of stationarity, steady state, and time-invariance have much in common. They all refer to the fact that the characteristics of a system, function, or process do not change with time. For a stationary random process, just like the steady state of a deterministic process, the characteristics of the process are *time-invariant* although the process itself is usually time-varying. Likewise, a *nonstationary* process can be thought of as the random process counterpart of the *transient* of a deterministic process. A deterministic function may begin with a transient and eventually settle down to the steady state. Analogously, a nonstationary random process may be ***asymptotically stationary***, meaning that it tends to be stationary as time increases.

It is sometimes extremely hard to check whether a random process is strictly stationary from the definition. In practice, if the primary physical conditions under which the random process is generated do not vary with time, then the random process can be considered (approximately) to be strictly stationary. Many random processes encountered in engineering practice are stationary. For example, the output noise of a radio receiver in steady state can be treated as a stationary random process while the noise is nonstationary during the initial transient (such as right after the receiver is turned on). It can be shown that

$$\text{strict stationarity} \;\underset{\not\Leftarrow}{\Rightarrow}\; \text{wide-sense stationarity}$$

If a process is white, then its values (which are RVs) at distinct time instants are uncorrelated. Whiteness defined this way is also called *wide-sense white* because there are so-called *strictly white* processes. A random process $X(t)$ is ***strictly white*** (or ***independent***) if its values at distinct time instants as RVs are *independent*. White processes are much easier to handle than other random processes. Various physical noises (e.g., thermal noise) are (approximately) white.

The following three commonly used definitions of autocorrelation $R_x(\tau)$:

$$R_x(\tau) = E[X(t+\tau)X(t)], \qquad R_x(\tau) = E[X(t)X(t+\tau)], \qquad R_x(\tau) = E[X(t)X(t-\tau)]$$

are equivalent for a stationary process, although their nonstationary counterparts $R_x(t+\tau, t)$, $R_x(t, t+\tau)$ and $R_x(t, t-\tau)$ are in general all distinct. The first two forms and the last one may result in different *sample autocorrelations* if a finite duration of a random process is used (see Section 6.6). We will not use the second form, albeit most popular, because it is inconsistent with the generally accepted definition for a complex (or vector-valued) process $R_x(t_1, t_2) = E[X(t_1)X^H(t_2)]$:

$$R_x(\tau) = E[X(t+\tau)X^H(t)] = E[X(t)X^H(t-\tau)] = \{E[X(t)X^H(t+\tau)]\}^H \neq E[X(t)X^H(t+\tau)]$$

where H stands for complex conjugate and transposition. Note that the autocorrelation is a matrix for a vector-valued process.

The ergodicity can be interpreted as follows. If a random process is ergodic, every one of its sample functions as time varies visits all possible state (i.e., value) of the process. As such, the entire information of the random process is contained in every single sample function and thus all characteristics of the random process can be determined from any single sample function.

Let ω be an outcome of a random experiment. An ensemble average is obtained by averaging over all $x(t, \omega)$ for a given time instant t and is in general a function of time. A time average is obtained by averaging over all time instants for a given sample function and in general depends on particular outcome ω. Conceptually, we may write

$$\text{ensemble average:} \quad \bar{x}(t) = \sum_{\omega} x(t, \omega) P\{\omega\}, \qquad \text{given } t$$

$$\text{time average:} \quad \bar{x}(\omega) = \lim_{T \to \infty} \frac{1}{2T} \int_{-T}^{T} x(t, \omega) dt, \qquad \text{given } \omega$$

Ergodicity requires that both ensemble average and time average be constants and equal to each other. As a result, for an ergodic random process, its mean (ensemble average) can be obtained from the time average of *any* single realization of the random process. This is extremely important in practice since it is usually the case that only a single sample function of a random process is available. Fortunately, most random processes involved in real-world problems are ergodic. The usefulness of ergodicity is as follows. In most applications, the ensemble average (mean) of a random process cannot be obtained directly. A typical example is the case in which the random process is a signal whose expression or representation is not known. If by some criterion it is known that the process is ergodic, then its mean can be approximated by the finite time average of *any* of its *single* sample functions.

To be more precise, a random process is *ergodic in the mean* if $\bar{x}(t) = \bar{x}(\omega)$, where $\bar{x}(t)$ and $\bar{x}(\omega)$ are the ensemble average (i.e., mean) and time average (for the sample function corresponding to the outcome ω), defined by (6.1) and (6.7), respectively. It is *ergodic in the autocorrelation* if its *ensemble autocorrelation* is equal to the *time autocorrelation* of every one of its sample functions, which are defined by

$$\text{ensemble autocorrelation} = E[X(t+\tau)X(t)]$$

$$\text{time autocorrelation} = \lim_{T \to \infty} \frac{1}{2T} \int_{-T}^{T} x(t+\tau)x(t) dt, \quad \text{for a single sample function}$$

In practice, for any given stationary random we may assume that it is ergodic unless there is a strong indication for the opposite. This is known as the *ergodic hypothesis* or *ergodic principle*. However, if the process is not stationary, then it cannot be ergodic.

The first two moments of an ergodic random process $X(t)$ have clear physical interpretations: The mean $\bar{x}$ is the ***dc component***; the mean-square value is its ***average power***; the variance σ_x^2 is the ***average power*** of its non-dc components; and the standard deviation σ_x is the ***effective value*** of its non-dc components.

Example 6.9: The random process in Example 6.1 is

- discrete since $X(t)$ at every fixed time t has at most two possible values.
- not strictly stationary because its PDF at $t=\pi/2$ and $t=0$ differ:

$$f_{X(\pi/2)}(x)=\delta(x)\neq f_{X(0)}(x)=\frac{1}{2}\delta(x-1)+\frac{1}{2}\delta(x+1)$$

- not WSS because its autocorrelation depends on t as well as τ:

$$R_x(t+\tau,t)\overset{\text{Example 6.5}}{=}\cos(t+\tau)\cos t$$

- not white since its autocovariance $C(t+\tau,t)$ is not always zero for $\tau\neq 0$ (see Example 6.5).
- ergodic in the mean because its time average

$$\bar{x}(\omega)=\lim_{T\to\infty}\frac{1}{2T}\int_{-T}^{T}\cos(t+\phi)dt=\lim_{T\to\infty}\frac{1}{2T}\sin(t+\phi)\Big|_{-T}^{T}=0$$

for every ϕ is equal to its ensemble average $\bar{x}(t)=0$ (see Example 6.5).

Example 6.10: The random process in Example 6.7 is

- continuous since $X(t)$ at every fixed time t is a continuous RV.
- not strictly stationary because it can be shown that $f_{X(\pi/4)}(x)\neq f_{X(\pi/2)}(x)$ but is wide-sense stationary since $\bar{x}(t)$ does not depend on t and $R_x(t+\tau,t)$ is a function of the time difference only.
- not white since its autocorrelation is not always the product of expected values for $\tau\neq 0$:

$$R_x(t+\tau,t)=\frac{A^2}{2}\cos(\omega\tau)\not\equiv\bar{x}(t+\tau)\bar{x}(t)=0$$

- ergodic in both mean and autocorrelation because, for every ϕ,

$$\text{time average } \bar{x}(\omega)=\lim_{T\to\infty}\frac{1}{2T}\int_{-T}^{T}A\sin(\omega_0 t+\phi)dt=0=\text{mean } \bar{x}(t)$$

$$\begin{aligned}\text{time autocorrelation} &= \lim_{T\to\infty}\frac{1}{2T}\int_{-T}^{T}x(t+\tau)x(t)dt\\ &=\lim_{T\to\infty}\frac{1}{2T}\int_{-T}^{T}A^2\sin(\omega_0 t+\omega_0\tau+\phi)\sin(\omega_0 t+\phi)dt\\ &=\frac{1}{2}A^2\lim_{T\to\infty}\frac{1}{2T}\int_{-T}^{T}[\cos(\omega_0\tau)-\cos(2\omega_0 t+2\phi+\omega_0\tau)]dt\\ &=\frac{1}{2}A^2\cos(\omega_0\tau)=\text{ensemble autocorrelation } R(\tau)\end{aligned}$$

6.4 Correlation Functions

- How to measure the ***similarity*** (***correlation***) between two waveforms?
- Given the value of one waveform, how much confidence do we have in estimating the value of the other?
- How much confidence do we have in estimating the value of a waveform at a time given its value at another time?

The correlation function provides a quantitative answer to such questions.

The correlation (coupling, similarity) of the waveforms in Fig. 6.7 are

- Strong (but negative) between $x_2(t)$ and $x_4(t)$.
- Very weak (or non-existent) between $x_3(t)$ and any other waveforms.
- Medium between $x_1(t)$ and $x_2(t)$ (positive).
- Medium between $x_1(t)$ and $x_4(t)$ (negative).

The measure should reflect these facts.

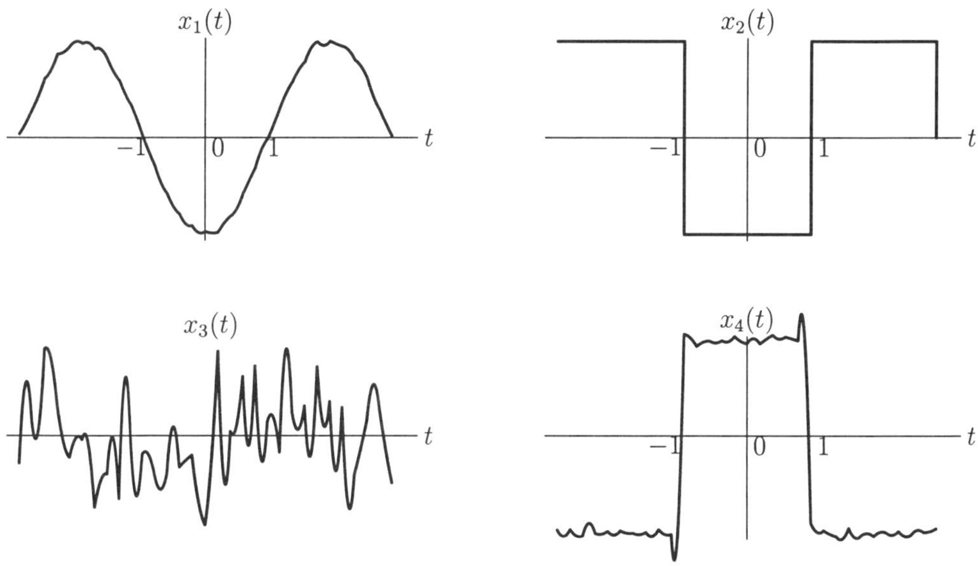

Figure 6.7: Correlation (coupling, similarity) of waveforms.

Correlation of Deterministic Waveforms

- The ***crosscorrelation function*** of two deterministic "power" waveforms (time functions) $x(t)$ and $y(t)$ is defined as

$$r_{xy}(\tau) = \lim_{T\to\infty} \frac{1}{2T} \int_{-T}^{T} x(t+\tau)y(t)dt = \text{time average of } x(t+\tau)y(t) \quad (6.8)$$

  which measures the similarity (coupling, correlation) between the two waveforms (time functions).
- The ***autocorrelation function*** of a deterministic power waveform $x(t)$ is defined as

$$r_x(\tau) = \lim_{T\to\infty} \frac{1}{2T} \int_{-T}^{T} x(t+\tau)x(t)dt = \text{time average of } x(t+\tau)x(t) \quad (6.9)$$

  which measures the similarity (coupling, correlation) between a waveform and its time-shifted version.

A correlation function will have a small/large magnitude if many/few terms in the integral (as a summation) of the product cancel out one another. For example for Fig. 6.7, $r_{x_2x_4}(0)$ has a large magnitude since the product $x_2(t)x_4(t)$ is almost always negative whereas $r_{x_2x_4}(1)$ has a small magnitude since the positive and negative terms of $x_2(t+1)x_4(t)$ cancel out.

Correlation of Random Processes

- The ***crosscorrelation function (CCF)*** of two random processes $X(t)$ and $Y(t)$ is defined as the average crosscorrelation of the two processes:

$$\boxed{R_{xy}(t+\tau, t) \triangleq E[X(t+\tau)Y(t)]} = \text{ensemble average of } X(t+\tau)Y(t) \quad (6.10)$$

- The ***autocorrelation function (ACF)*** of a random process $X(t)$ is defined as the average autocorrelation of the process:

$$\boxed{R_x(t+\tau, t) \triangleq E[X(t+\tau)X(t)]} = \text{ensemble average of } X(t+\tau)X(t) \quad (6.11)$$

These correlation functions are not random. They are the ensemble equivalents of those [i.e., (6.8) and (6.9)] for deterministic waveforms.

The ***instantaneous power*** of a function (or waveform, signal) $x(t)$ is defined as the quantity $|x(t)|^2$. A function $x(t)$ is said to be a (finite) ***energy function*** if it has finite energy over the entire time horizon; that is, if $\int_{-\infty}^{\infty} |x(t)|^2 dt < \infty$. A function $x(t)$ is said to be a (finite) ***power function*** if it has infinite energy but finite average power (but infinite energy) over the entire time horizon; that is, if

$$\lim_{T\to\infty} \frac{1}{2T} \int_{-T}^{T} |x(t)|^2 dt < \infty \qquad \text{but} \qquad \int_{-\infty}^{\infty} |x(t)|^2 dt = \infty$$

Random processes are power functions because their sample functions have an infinite duration and may not even approach zero in the limit.

The correlation functions defined by (6.8) and (6.9) are valid only for power functions. They are always zero for energy functions. For energy functions, the crosscorrelation function is defined by

$$r_{xy}(t+\tau, t) = \int_{-\infty}^{\infty} x(t+\tau)y(t)dt$$

This definition is more commonly used for deterministic functions. It is, however, not valid for power functions since the integral diverges and thus (6.8) and (6.9) should be used as the basis for random processes.

(6.10) can be understood as follows. The ***crosscorrelation*** of two random processes $X(t)$ and $Y(t)$ can be defined as the expected crosscorrelation of $X(t)$ and $Y(t)$, as defined by (6.8) (i.e., crosscorrelation averaged over all sample functions of $X(t)$ and $Y(t)$):

$$\begin{aligned} R_{xy}(t+\tau, t) &= E[r_{xy}(t+\tau, t)] = E\Big[\lim_{T\to\infty} \frac{1}{2T} \int_{-T}^{T} X(t+\tau)Y(t)dt\Big] \\ &= \lim_{T\to\infty} \frac{1}{2T} \int_{-T}^{T} E[X(t+\tau)Y(t)]dt \triangleq \text{time average of } E[X(t+\tau)Y(t)] \end{aligned}$$

If $E[X(t+\tau)Y(t)]$ depends only on τ (e.g., when $X(t), Y(t)$ are jointly wide-sense stationary), then the crosscorrelation is simply equal to $E[X(t+\tau)Y(t)]$, as defined by (6.10). Similarly, the ***autocorrelation*** of a random process $X(t)$ can be interpreted by

$$R_x(t+\tau, t) = E[r_x(t+\tau, t)] = \text{time average of } E[X(t+\tau)X(t)]$$

and if $E[X(t+\tau)X(t)]$ does not depend on t (e.g., when $X(t)$ is wide-sense stationary), then the autocorrelation is simply equal to $E[X(t+\tau)X(t)]$, as defined by (6.11).

To measure the correlation of a random process, it is more convenient to use the ***correlation coefficient***, defined by (6.5).

As for the case of two RVs, autocorrelation, autocovariance, and correlation coefficient of a random process quantifies the degree of *linear* correlation between its values at two time instants. If ρ is positive, then $X(t'+\tau)$ and $X(t')$ are *positively correlated*, which implies that they are likely to have the same sign. If ρ is negative, then $X(t'+\tau)$ and $X(t')$ are *negatively correlated*, which implies that they are likely to have the opposite sign.

For wide-sense stationary random processes,

$$R(\tau) = R(t+\tau, t) = R(t'+\tau, t'), \qquad \forall t, t'$$
$$C(\tau) = C(t+\tau, t) = C(t'+\tau, t'), \qquad \forall t, t'$$
$$\rho(\tau) = \rho(t+\tau, t) = \rho(t'+\tau, t') = \frac{C(\tau)}{C(0)}, \qquad \forall t, t'$$

which depend only on time difference τ.

In general, if a wide-sense stationary random process $X(t)$ has no periodic components, then its correlation between two distinct time instants decreases as the time difference increases; that is, $\tau \uparrow \implies \rho(\tau) \downarrow$. It is thus convenient to define a *correlation time* and the values $X(t'+\tau)$ and $X(t')$ at two time instants are deemed uncorrelated if their time difference τ is larger than the correlation time τ_c for practical purposes. Two commonly used definitions of ***correlation time*** are:

- The correlation time $\tau_{0.05}$ is defined as the value of τ such that the correlation coefficient drops to 0.05 of its peak value 1; that is, $\rho(\tau_{0.05}) = 0.05$.
- The correlation time τ_c is defined as the value of τ such that the area underneath $\rho(\tau)$ is equal to $2\tau_c$; that is,

$$\tau_c = \int_0^\infty \rho(\tau) d\tau \tag{6.12}$$

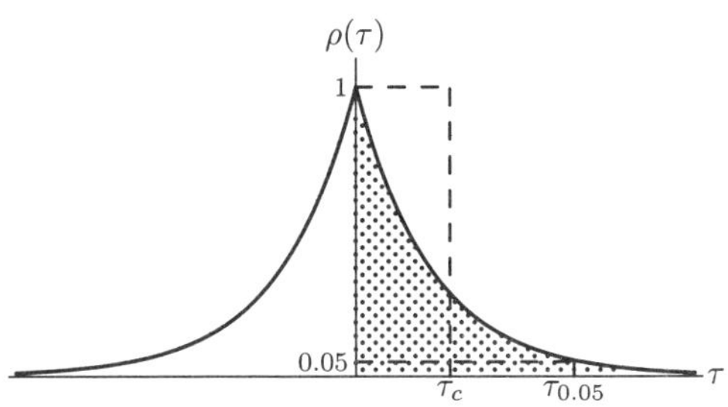

Figure 6.8: Some definitions of correlation time.

It is thus clear that a small correlation time implies a weak correlation between the distinct time instants of the random process. Loosely speaking, in such a case, we may say that the random process varies quickly (in a probabilistic or statistical rather than deterministic sense) since its value at one time can be statistically quite different from that at another time with a small time difference. In this sense, white noise has the quickest possible variation.

Correlation and ***convolution*** of two functions are closely related:

$$r_{xy}(\tau) \triangleq \int_{-\infty}^{\infty} x(t'+\tau)y(t')dt' = \int_{-\infty}^{\infty} x(t)y(t-\tau)dt = \int_{-\infty}^{\infty} x(t)y[-(\tau-t)]dt = x(\tau) * y(-\tau) \tag{6.13}$$

that is, correlation is convolution with one of the functions "flipped" in time.

Example 6.11: A Sinusoid with Random Frequency and Phase

Given $G(t) = A\cos(\omega t + \theta)$, where ω and θ are independent RVs with $\theta \sim \mathcal{U}(0, 2\pi)$ and $\omega \sim \mathcal{U}(\omega_1, \omega_2)$, $\omega_2 \geq \omega_1 \geq 0$, determine the mean $E[G(t)]$, average power $E[G^2(t)]$, and autocorrelation $R_g(\tau)$. Note that

$$\cos(\alpha + \beta) = \cos\alpha\cos\beta - \sin\alpha\sin\beta \tag{6.14}$$

We have

$$\begin{aligned}
E[G(t)] &= E[A\cos(\omega t + \theta)] \overset{(6.14)}{=} AE[\cos\omega t\cos\theta - \sin\omega t\sin\theta] \\
&\overset{?}{=} A\{E[\cos\omega t]\underbrace{E[\cos\theta]}_{=0} - E[\sin\omega t]\underbrace{E[\sin\theta]}_{=0}\} = 0 \\
R_g(\tau) &= E[G(t+\tau)G(t)] = E\left[A\cos(\omega t + \omega\tau + \theta) \cdot A\cos(\omega t + \theta)\right] \\
&= \frac{1}{2}A^2E[\cos\omega\tau + \cos(\underbrace{2\omega t + \omega\tau}_{\alpha} + \underbrace{2\theta}_{\beta})] \\
&= \frac{1}{2}A^2\left\{E[\cos\omega\tau] + E[\cos(2\omega t + \omega\tau)\cos 2\theta - \sin(2\omega t + \omega\tau)\sin 2\theta]\right\} \\
&= \frac{1}{2}A^2\Big\{E[\cos(2\omega t + \omega\tau)]\underbrace{E[\cos 2\theta]}_{=0} - E[\sin(2\omega t + \omega\tau)]\underbrace{E[\sin 2\theta]}_{=0} \\
&\quad + E[\cos\omega\tau]\Big\} \\
&= \frac{1}{2}A^2E[\cos\omega\tau] = \frac{1}{2}A^2\int_{\omega_1}^{\omega_2}\cos\omega\tau \cdot \frac{1}{\omega_2 - \omega_1}d\omega \\
&= \frac{A^2}{2(\omega_2 - \omega_1)} \cdot \frac{1}{\tau} \cdot \sin\omega\tau\Big|_{\omega_1}^{\omega_2} = \frac{A^2}{2\tau(\omega_2 - \omega_1)}[\sin\omega_2\tau - \sin\omega_1\tau] \\
E[G^2(t)] &= R_g(0) = \lim_{\tau\to 0}\frac{A^2}{2(\omega_2 - \omega_1)}\left[\omega_2\frac{\sin\omega_2\tau}{\omega_2\tau} - \omega_1\frac{\sin\omega_1\tau}{\omega_1\tau}\right] \\
&= \frac{A^2}{2(\omega_2 - \omega_1)}(\omega_2 - \omega_1) = \frac{1}{2}A^2
\end{aligned}$$

Alternatively, $E[G^2(t)] = E[A^2\cos^2(\omega t + \theta)] = \frac{A^2}{2}E[1 + \cos 2(\omega t + \theta)] = \frac{A^2}{2}$. Note that in general, if $\theta \sim \mathcal{U}(0, 2\pi)$ and ω is nonrandom or is random but independent of θ, then

$$E[\cos(\omega t + n\theta)] = 0, \qquad n = 0, \pm 1, \pm 2, \ldots \tag{6.15}$$

This makes sense: A sine wave has a zero average over one or more periods.

6.5 Properties of Autocorrelation Functions

Important properties of the autocorrelation function $R_x(\tau)$ of a real-valued *wide-sense stationary* random process $X(t)$ include:

1. It is *even* in the time-shift τ:
$$R_x(-\tau) = E[X(t)X(t+\tau)] = E[X(t+\tau)X(t)] = R_x(\tau) \tag{6.16}$$
which makes sense: $R_x(\tau)$ should be identical for left and right shifts (time delays and time advances).
2. It is bounded by its value at the origin:
$$\boxed{|R_x(\tau)| \le R_x(0)} \tag{6.17}$$
which makes sense: $X(t)$ is most "similar" to itself (without shift). This can be shown as follows:
$$\begin{aligned} 0 \le E\left\{[X(t+\tau) \pm X(t)]^2\right\} &= E\left[X^2(t+\tau) \pm 2X(t+\tau)X(t) + X^2(t)\right] \\ &= R_x(0) \pm 2R_x(\tau) + R_x(0) \end{aligned}$$
3. The above bound $R_x(0)$ equals the mean-square value:
$$\boxed{R_x(0) = E[X(t+0)X(t)] = E[X^2(t)] \stackrel{?}{=} E[X^2] = \text{mean-square value}} \tag{6.18}$$
The mean-square value is called the ***average power*** of the process.
4. If $X(t)$ has a periodic component, then $R_x(\tau)$ will also have a periodic component with the same frequency. For example, assume
$$X(t) = A\cos(\omega_0 t + \theta) + Y(t)$$
where $\theta \sim \mathcal{U}(0, 2\pi)$, and is independent of $Y(t)$. Then
$$\begin{aligned} R_x(\tau) &= E\{[A\cos(\omega_0(t+\tau)+\theta) + Y(t+\tau)][A\cos(\omega_0 t+\theta) + Y(t)]\} \\ &= E[A^2\cos(\omega_0 t+\theta)\cos(\omega_0 t+\omega_0\tau+\theta) + Y(t+\tau)Y(t) \\ &\quad + A\cos(\omega_0 t+\theta)Y(t+\tau) + A\cos(\omega_0 t+\omega_0\tau+\theta)Y(t)] \\ &\stackrel{?}{=} E\left\{A^2\frac{1}{2}\left[\cos\omega_0\tau + \cos(2\omega_0 t - \omega_0\tau + 2\theta)\right]\right\} + R_y(\tau) + 0 + 0 \\ &= \frac{1}{2}A^2\cos\omega_0\tau + R_y(\tau) \end{aligned} \tag{6.19}$$

where $\stackrel{?}{=}$ follows from $\cos\alpha\cos\beta = \frac{1}{2}[\cos(\alpha-\beta) + \cos(\alpha+\beta)]$.

For a deterministic function $x(t)$, its ***instantaneous power*** is defined as $x^2(t)$ and its ***average power*** is defined by

$$p_x = \lim_{T\to\infty}\frac{1}{2T}\int_{-\infty}^{\infty} x^2(t)dt \tag{6.20}$$

Thus, the ***average power*** of a random process $X(t)$ may be defined by

$$P_x = E[p_x] = \lim_{T\to\infty}\frac{1}{2T}\int_{-\infty}^{\infty} E[X^2(t)]dt = \text{time average of } E[X^2(t)] \tag{6.21}$$

Some other properties of the autocorrelation function include:

- If $X(t)$ has a constant (dc) term, then $R_x(\tau)$ will also have a constant (dc) term:

$$\begin{aligned} X(t) = A + Y(t) \quad \Longrightarrow \quad R_x(\tau) &= E\{[A+Y(t+\tau)][A+Y(t)]\} \\ &= A^2 + A[\bar{y}(t+\tau)+\bar{y}(t)] + R_y(\tau) \end{aligned}$$

  This can be seen as a special case of the property 4 above in which the frequency is equal to zero and the phase is not random.
- If $X(t)$ has no periodic component, then $X(t+\tau)$ and $X(t)$ are uncorrelated as $\tau \to \infty$; that is, $C_x(\infty) = 0$ or $R_x(\infty) = (\bar{x})^2$:
- An autocorrelation function cannot have an arbitrary shape. For example, it cannot have jumps, flat tops, or any discontinuity in amplitude.
- For a random process, its autocorrelation function (if exists) is uniquely determined by the joint probability density function. The latter, however, cannot in general be determined uniquely by the former.
- Let

$$\dot{X}(t) \triangleq \frac{d}{dt}X(t)$$

  Then, the autocorrelation function of the derivative of a random process $X(t)$ is the negative second-order derivative of that of $X(t)$:

$$R_{\dot{x}x}(\tau) \triangleq E[\dot{X}(t+\tau)X(t)] = \frac{d}{d\tau}R_x(\tau) \tag{6.22}$$

$$R_{\dot{x}}(\tau) \triangleq E[\dot{X}(t+\tau)\dot{X}(t)] = -\frac{d^2}{d\tau^2}R_x(\tau) \tag{6.23}$$

  This is a special case of (6.37), to be studied later.
- Let $h(t)$ be a deterministic function and let $X(t)$ be wide-sense stationary. Then

$$Y(t) = X(t) + h(t) \Longrightarrow \bar{y}(t) = \bar{x} + h(t), \quad C_y(\tau) = C_x(\tau) \tag{6.24}$$

$$Y(t) = h(t)X(t) \Longrightarrow \bar{y}(t) = \bar{x}h(t), \quad C_y(t+\tau,t) = h(t+\tau)h(t)C_x(\tau) \tag{6.25}$$

  That is, an additive deterministic term of a random process has no effect on its autocovariance and a multiplicative factor acts as a scaling factor. Note that $Y(t)$ is no longer stationary if the additive or multiplicative term $h(t)$ is time-varying.

Example 6.12: Radar Range Determination

In a radar, a signal $X(t)$ is sent out by its transmitter, bounced back by the target, and received by the radar receiver with attenuation and noise.

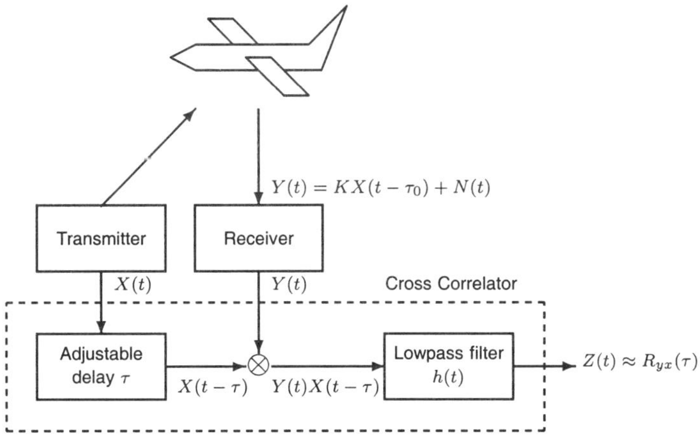

Assume that $X(t)$ and zero-mean noise $N(t)$ are both wide-sense stationary and mutually independent (which is reasonable) and $K > 0$ is a constant, known as attenuation factor. It can been easily shown that

- $Y(t)$ and $X(t)$ are jointly wide-sense stationary (see problem 6.35).
- $R_{yx}(\tau) = KR_x(\tau - \tau_0)$, where τ is known but τ_0 is unknown.

From linear systems theory, we have, since the output of a lowpass filter is approximately equal to its dc term (i.e., average value),

$$\begin{aligned} Z(t) &= [Y(t)X(t-\tau)] * h(t) \approx E[Z(t)] = E\Big[\int_0^t h(\lambda)Y(t-\lambda)\,X(t-\tau-\lambda)d\lambda\Big] \\ &= \int_0^t E\left[Y(t-\lambda)\,X(t-\tau-\lambda)\right] h(\lambda)d\lambda = \int_0^t R_{yx}(\tau)h(\lambda)d\lambda \\ &= R_{yx}(\tau)\underbrace{\int_0^t h(\lambda)d\lambda}_{=1 \text{ by design}} = R_{yx}(\tau) = KR_x(\tau-\tau_0) \end{aligned}$$

Consequently, τ_0 can be determined by adjusting τ such that $Z(t) \approx KR_x(\tau-\tau_0)$ reaches its maximum $KR_x(0)$ (which occurs when $\tau = \tau_0$) and then the distance is determined by:

$$2d = v\tau \qquad \Longrightarrow \qquad \text{distance } d = \frac{v\tau}{2}$$

6.6 Sample Mean and Sample Correlation Functions

It is often the case in practice that mean and correlation functions of random processes cannot be obtained from their definitions because the required density functions or ensemble is not available. They are usually approximated by sample mean and sample correlations.

Consider an *ergodic* random process $X(t)$. Let $x(nT)$ be a sampled (i.e., discrete-time) version of one of its sample functions $x(t)$, where T is the sampling interval. Denote $x(nT)$ by $x[n]$. Assume $x[n]$ has a finite duration from $n = 1$ to $n = N$, which is denoted by $\{x[n]\}_{n=1}^{N}$.

A natural estimate of the mean, known as ***sample mean***, of $X(t)$ based on $\{x[n]\}_{n=1}^{N}$ is

$$\hat{x} = \frac{1}{N}\sum_{i=1}^{N} x[i]$$

Two popular estimators of the autocorrelation function of $X(t)$, known as ***sample autocorrelations***, based on $\{x[n]\}_{n=1}^{N}$ are, for $0 \leq m \ll N$,

$$\text{unbiased:} \quad \hat{R}_x^u[m] \triangleq \hat{R}_x^u(mT) = \frac{1}{N-m}\sum_{i=1}^{N-m} x[i+m]x[i] \tag{6.26}$$

$$\text{biased:} \quad \hat{R}_x[m] \triangleq \hat{R}_x(mT) = \frac{1}{N}\sum_{i=1}^{N-m} x[i+m]x[i] \tag{6.27}$$

The autocorrelation for negative m can be found by $R[-m] = R[m]$.

The biased estimator $\hat{R}_x[m]$ is more widely used than the unbiased estimator $\hat{R}_x^u[m]$ mainly because

- It is usually more accurate in the sense of having a smaller (mean-square) estimation error than $\hat{R}_x^u[m]$.
- It preserves the important property of autocorrelation: $\hat{R}_x[m] \leq \hat{R}_x[0]$, which is not preserved by $\hat{R}_x^u[m]$.

The autocovariance, correlation coefficient of a random process and the crosscorrelation and crosscovariance functions, and correlation coefficient of two ergodic random processes $X(t)$ and $Y(t)$ can also be estimated by their sample quantities, defined similarly on the next page.

$\hat{x}$ and $\hat{R}(\tau)$ are called *sample mean* and *sample autocorrelation* because they are calculated from a single *sample* function. They are also known as ***time average*** and ***time autocorrelation***, respectively.

The estimators $\hat{R}_x^u[m]$ and $\hat{R}_x[m]$ are clearly discrete-time versions of (6.9). Estimator $\hat{R}_x^u[m]$ is *unbiased* because $E(\hat{R}_x^u[m]) = R_x[m]$; that is, $\hat{R}_x^u[m]$ is equal to the true autocorrelation $R_x[m]$ on the average (over all sample functions). $\hat{R}_x[m]$ does not have this desirable property: $E(\hat{R}_x[m]) \neq R_x[m]$, although it is *asymptotically unbiased*: $E(\lim_{N\to\infty} \hat{R}_x^u[m]) = R_x[m]$. In fact, it can be shown easily (see problem 6.36) that

$$E(\hat{R}_x[m]) = \left(1 - \frac{m}{N}\right) R_x[m]$$

For an ergodic process, both of these estimators approach to the true autocorrelation as N increases, which is said to be *consistent* in estimation theory.

If the time difference m is comparable with the data record length N, e.g., $m > 0.2N$, then the estimators $\hat{R}_x^u[m]$ and $\hat{R}_x[m]$ may lead to poor results.

The crosscorrelation of two ergodic random processes $X(t)$ and $Y(t)$ can be estimated based on their discrete-time sample functions $x(n)$ and $y(n)$ with a finite duration from $n = 1$ to $n = N$, for $0 \leq m < N$,

$$\text{unbiased:} \quad \hat{R}_{xy}^u[m] = \frac{1}{N-m} \sum_{i=1}^{N-m} x[i+m]y[i] \tag{6.28}$$

$$\text{biased:} \quad \hat{R}_{xy}[m] = \frac{1}{N} \sum_{i=1}^{N-m} x[i+m]y[i] \tag{6.29}$$

The biased sample autocovariance, crosscovariance, and correlation coefficients are

$$\hat{C}[m] = \frac{1}{N} \sum_{i=1}^{N-m} (x[i+m] - \hat{x})(x[i] - \hat{x})$$

$$\hat{\rho}[m] = \frac{\hat{C}[m]}{\hat{C}[0]}$$

$$\hat{C}_{xy}[m] = \frac{1}{N} \sum_{i=1}^{N-m} (x[i+m] - \hat{x})(y[i] - \hat{y})$$

$$\hat{\rho}_{xy}[m] = \frac{\hat{C}_{xy}[m]}{\sqrt{\hat{C}_x[0]\hat{C}_y[0]}}$$

Note that for sample mean, correlation, and covariance calculated from a finite-duration sample function, in general we have

$$C[m] = R[m] - (\bar{x})^2 \quad \text{but} \quad \hat{C}[m] \neq \hat{R}[m] - (\hat{x})^2$$

$$C_{xy}[m] = R_{xy}[m] - \bar{x}\bar{y} \quad \text{but} \quad \hat{C}_{xy}[m] \neq \hat{R}_{xy}[m] - \hat{x}\hat{y}$$

In practice, sample correlations are widely used even if we do not know whether the random process is ergodic or not. (6.13) can be utilized to implement the calculation of sample correlations.

Example 6.13: Estimation of Mean and Autocorrelation of a Process

The following record (sequence) of 100 data is a discrete-time sample function (with sampling interval $T = 0.1$) of a zero-mean ergodic random process with autocorrelation $R(\tau) = 3e^{-10|\tau|}$: $-2.5770,\ 3.5065,\ -2.3735,\ \cdots, -0.9592$.

The sample mean and correlation functions are computed and compared with the true ones:

$$\hat{x} = \frac{1}{100}\sum_{n=1}^{100} x[n] = \frac{1}{100}[-2.577 + 3.5065 + \cdots - 0.9592] = -0.0458 \approx \bar{x} = 0$$

$$\hat{R}[0] = \frac{1}{100}\sum_{n=1}^{100} x^2[n] = \frac{1}{100}[(-2.577)^2 + \cdots + (-0.9592)^2] = 2.845 \approx R(0) = 3$$

$$\hat{C}[0] = \frac{1}{100}\sum_{n=1}^{100} (x[n] - \hat{x})^2 = \frac{1}{100}[(-2.5312)^2 + \cdots + (-0.9134)^2] = 2.843$$

$$\hat{R}[1] = \frac{1}{100}\sum_{n=1}^{100-1} x[n+1]x[n] = 0.7868 \approx R(T) = R(0.1) = 3e^{-1} = 1.1036$$

$$\hat{C}[1] = \frac{1}{100}\sum_{n=1}^{100-1} (x[n+1] - \hat{x})(x[n] - \hat{x}) = 0.7871 \approx C(T) = R(T) = 1.1036$$

$$\hat{\rho}[1] = \frac{\hat{C}[1]}{\hat{C}[0]} = 0.2769 \approx \rho(T) = \frac{C(T)}{C(0)} = 0.3679$$

$$\hat{R}[2] = \frac{1}{100}\sum_{n=1}^{100-2} x[n+2]x[n] = 0.3319 \approx R(2T) = R(0.2) = 3e^{-2} = 0.406$$

Note that $\hat{R}[m] \approx \hat{C}[m]$ although in principle $R[m] = C[m]$ for this example.

The true and sample autocorrelations, computed by P&R, are shown in Fig. 6.9. Better accuracy is achievable if the data record is longer.

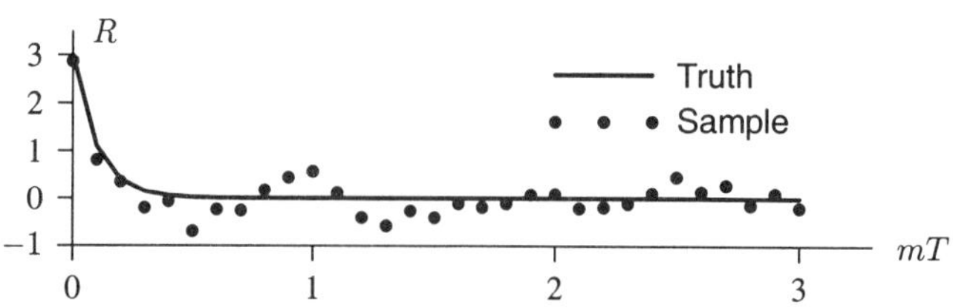

Figure 6.9: Comparison of true and sample autocorrelation functions.

Example 6.14: Correlation Coefficient of a Random Sinusoid

Consider a random process $X(t) = 2\sin(0.5\pi t + \phi)$, where $\phi \sim \mathcal{U}(-\pi, \pi)$.

Fig. 6.10 (a) and (b) illustrate two sample functions of $X(t)$, generated by the companion software P&R. Plots (c) and (d) show the true and 100-point sample correlation coefficient functions (with 10 Hz sampling rate) of $X(t)$, computed also by P&R. Note the difference between the two coefficient functions. The difference, arising from the fact that only 100 points were used in the latter, increases with the time difference τ. Plots (e) and (f) give the 1000-point sample correlation coefficient (with 10 Hz sampling rate), also computed by P&R. Clearly, the accuracy of the sample correlation coefficients for the first 100 points increases greatly using 1000 points, rather than 100 points.

This example demonstrates that the computed sample correlations are accurate only for small time difference compared with the data record length.

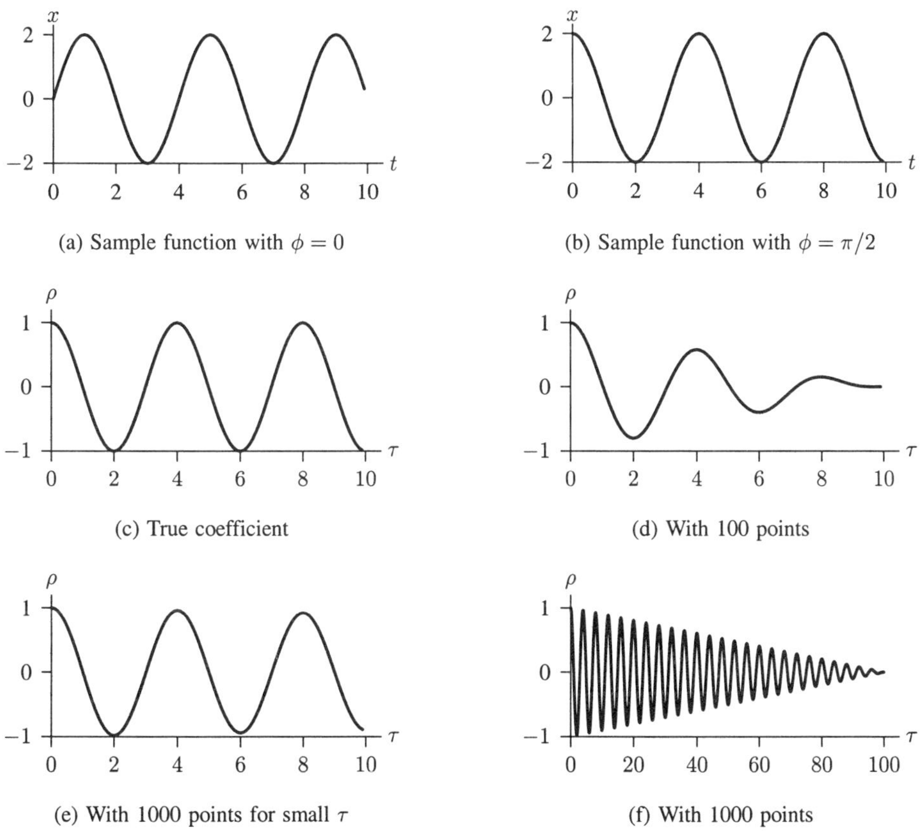

(a) Sample function with $\phi = 0$

(b) Sample function with $\phi = \pi/2$

(c) True coefficient

(d) With 100 points

(e) With 1000 points for small τ

(f) With 1000 points

Figure 6.10: Sample correlation coefficients of a random sinusoid.

Example 6.15: Computation of Sample Autocorrelations by P&R

Example 6.14 can be computed using the companion software P&R. First, generate a 1000-point discrete sample function (with a sampling rate $f_s = 10$ Hz) of the random sinusoid $X(t)$, as in Example 6.8, and store it in `e6_15.dat`. Then the correlation coefficient of $X(t)$ can be computed using P&R with the following steps:

S1. Click "RPAnalyzer" in the main window of P&R.

S2. Click "Time Domain." You will be prompted to enter the name of a data file. Choose "data" subdirectory, enter the data file name "`e6_15.dat`" and click "Ok."

S3. The window "**Time-Domain Analyzer**" will appear, with the computed sample mean and sample variance. Set the "Max time difference" to `100`. Click "Corr-Coeff", as shown in Fig. 6.11. Then the correlation coefficients $\rho(i/f_s)$ are computed, plotted and saved to a user-specified data file for the first 101 points ($i = 0, 1, \ldots, 100$), which correspond to $\tau \leq 10$ since sampling rate is 10 Hz. Fig. 6.10(e) was generated this way.

S4. Fig. 6.10(f) can be generated by repeating the above steps with "Max time difference" setting to 1000.

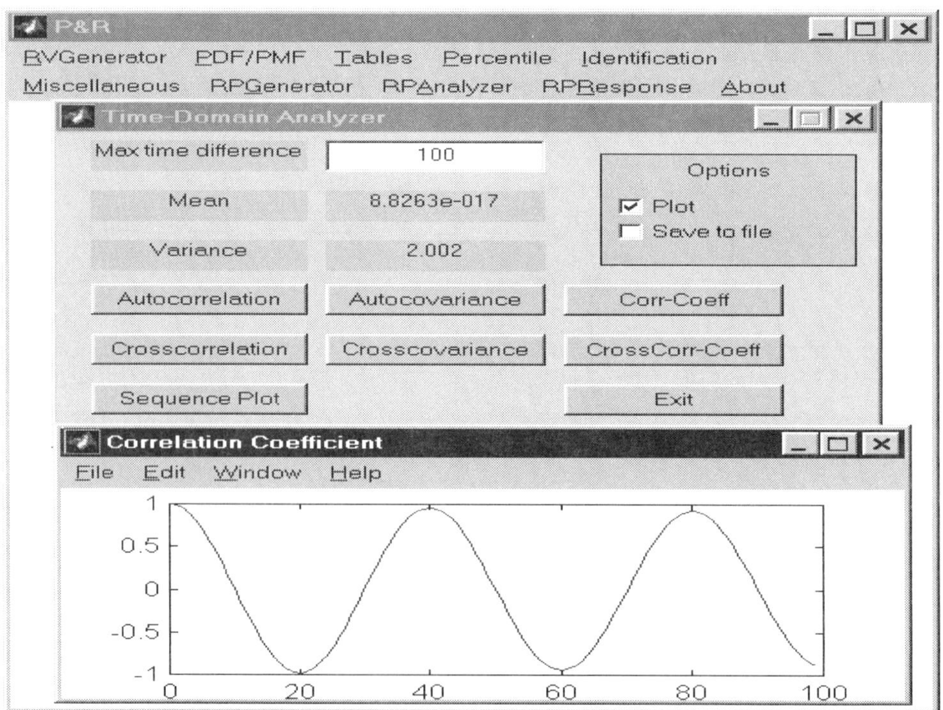

Figure 6.11: Computation of correlation functions by P&R.

6.7 Relationship Between Two Random Processes

Two random processes $X(t)$ and $Y(t)$ are said to be

- ***jointly (wide-sense) stationary*** (***JWSS***) if
  - $X(t)$ and $Y(t)$ are both wide-sense stationary, *and*
  - $R_{xy}(t+\tau,t)$ and $R_{yx}(t+\tau,t)$ depend only on the time difference τ; that is,

$$R_{xy}(t+\tau,t) = R_{xy}(\tau)$$
$$R_{yx}(t+\tau,t) = R_{yx}(\tau)$$

- ***uncorrelated*** if their crosscorrelation is equal to the product of their mean functions for all time t_1 and t_2:

$$R_{xy}(t_1,t_2) = \bar{x}(t_1)\bar{y}(t_2), \qquad \forall t_1, t_2 \tag{6.30}$$

  or equivalently, their crosscovariance is zero for all time t_1 and t_2:

$$C_{xy}(t_1,t_2) \triangleq E\{[X(t_1) - \bar{x}(t_1)][Y(t_2) - \bar{y}(t_2)]\} = 0, \qquad \forall t_1, t_2 \tag{6.31}$$

- ***orthogonal*** if their crosscorrelation is zero for all time t_1 and t_2:

$$R_{xy}(t_1,t_2) = 0, \qquad \forall t_1, t_2$$

- ***independent*** if every set of RVs $X(t_1),\ldots,X(t_n)$ is independent of every set of RVs $Y(t_1'),\ldots,Y(t_m')$.

  Clearly, if $X(t)$ and $Y(t)$ are jointly wide-sense stationary, then

$$X(t) \text{ and } Y(t) \text{ are uncorrelated} \iff R_{xy}(\tau) = \bar{x}\bar{y} \tag{6.32}$$

(6.31) and (6.30) are clearly equivalent in view of (6.6). (6.30) follows clearly from (4.26) since $X(t_1)$ and $Y(t_2)$ are simply two RVs.

Two RVs are said to be orthogonal if their correlation is zero. This is because two RVs can be viewed as two vectors in a linear space and they are perpendicular if their correlation is zero. The orthogonality of two random processes is a generalization of this concept.

Example 6.16: Correlation of Weighted Sums of Random Processes

Given two JWSS processes $X(t)$ and $Y(t)$, let

$$U(t) = aX(t) + bY(t) \qquad V(t) = cX(t) + dY(t)$$

where a, b, c, d are real constants. Determine if $U(t)$ and $V(t)$ are JWSS.

(a) First check if $U(t)$ and $V(t)$ are wide-sense stationary: Note first
$X(t)$ and $Y(t)$ are JWSS $\implies$ $X(t)$ and $Y(t)$ are both WSS

$$\begin{aligned}
E[U(t)] &= E[aX(t) + bY(t)] \stackrel{?}{=} a\bar{x} + b\bar{y} \triangleq \bar{u} \quad \text{(time invariant)}\\
E[V(t)] &= E[cX(t) + dY(t)] = c\bar{x} + d\bar{y} \triangleq \bar{v} \quad \text{(time invariant)}\\
R_u(t,t') &= E[U(t)U(t')] = E\{[aX(t) + bY(t)][aX(t') + bY(t')]\}\\
&= E[a^2X(t)X(t') + abX(t)Y(t') + abY(t)X(t') + b^2Y(t)Y(t')]\\
&= a^2R_x(t,t') + abR_{xy}(t,t') + abR_{yx}(t,t') + b^2R_y(t,t')
\end{aligned}$$

Thus, since $X(t)$ and $Y(t)$ are jointly stationary,

$$R_u(t+\tau,t) = a^2R_x(\tau) + abR_{xy}(\tau) + abR_{yx}(\tau) + b^2R_y(\tau) \triangleq R_u(\tau)$$

and by symmetry,

$$R_v(t+\tau,t) = c^2R_x(\tau) + cdR_{xy}(\tau) + cdR_{yx}(\tau) + d^2R_y(\tau) \triangleq R_v(\tau)$$

which are functions of τ only. Hence, $U(t)$ and $V(t)$ are both wide-sense stationary.

(b) Check the other two conditions for JWSS of $U(t)$ and $V(t)$:

$$\begin{aligned}
R_{uv}(t',t) &= E[U(t')V(t)] = E\{[aX(t') + bY(t')][cX(t) + dY(t)]\}\\
&= E[acX(t')X(t) + adX(t')Y(t) + bcY(t')X(t) + bdY(t')Y(t)]\\
R_{uv}(t+\tau,t) &\stackrel{?}{=} acR_x(\tau) + adR_{xy}(\tau) + bcR_{yx}(\tau) + bdR_y(\tau) \triangleq R_{uv}(\tau)\\
R_{vu}(t+\tau,t) &= caR_x(\tau) + cbR_{xy}(\tau) + daR_{yx}(\tau) + dbR_y(\tau) \triangleq R_{vu}(\tau)
\end{aligned}$$

Thus, $U(t)$ and $V(t)$ are jointly wide-sense stationary. Note that

- Every pair of the four processes $X(t)$, $Y(t)$, $U(t)$ and $V(t)$ is JWSS.
- If $X(t), Y(t)$ are zero-mean and uncorrelated, $U(t), V(t)$ are zero-mean also but not necessarily uncorrelated: $R_{uv}(\tau) = acR_x(\tau) + bdR_y(\tau) \neq 0$.

6.8 Properties of Crosscorrelation Functions

Important properties of the crosscorrelation function of two *jointly* wide-sense stationary random processes $X(t)$ and $Y(t)$ include

1. $R_{xy}(\tau)$ and $R_{yx}(\tau)$ are mirror images of each other:

$$R_{xy}(\tau) = E[X(t+\tau)Y(t)] = E[Y(t)X(t+\tau)] = R_{yx}(-\tau) \tag{6.33}$$

as depicted in Fig. 6.12. This makes sense: Shifting $Y(t)$ in one direction is equivalent to shifting $X(t)$ in the other direction as far as crosscorrelation is concerned. A special case is

$$R_{xy}(0) = R_{yx}(0) \tag{6.34}$$

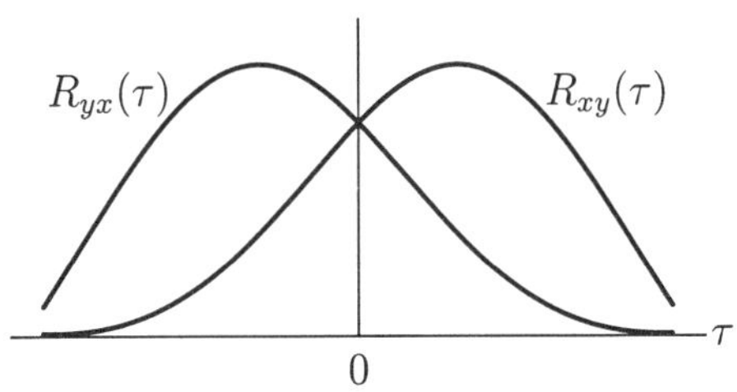

Figure 6.12: Mirror image relation between $R_{xy}(\tau)$ and $R_{yx}(\tau)$.

2. The crosscorrelation is bounded by the geometric and arithmetic averages of $R_x(0)$ and $R_y(0)$:

$$|R_{xy}(\tau)| \le \underbrace{[R_x(0)R_y(0)]^{1/2}}_{\text{geometric average}} \le \underbrace{\frac{1}{2}[R_x(0) + R_y(0)]}_{\text{arithmetic average}} \tag{6.35}$$

The first inequality implies that the crosscorrelation coefficient is bounded by 1 in magnitude: $|\rho_{xy}(\tau)| \le 1$.

3. Let $\dot{X}(t) \triangleq \frac{d}{dt}X(t)$. Then

$$R_{\dot{x}y}(\tau) \triangleq E[\dot{X}(t+\tau)Y(t)] = \frac{d}{d\tau}R_{xy}(\tau) \tag{6.36}$$

$$R_{\dot{x}\dot{y}}(\tau) \triangleq E[\dot{X}(t+\tau)\dot{Y}(t)] = -\frac{d^2}{d\tau^2}R_{xy}(\tau) \tag{6.37}$$

The second inequality in (6.35) follows from the fact that for positive numbers, geometric average is never larger than arithmetic average. The first inequality can be derived easily as in problem 6.33 following the same procedure as (4.25). Alternatively, it can be shown as follows, which demonstrates the use of Schwarz inequality. Since

$$\begin{aligned} R_{xy}(\tau) &= E[X(t+\tau)Y(t)], \qquad & R_x(0) &= E[X^2] \\ R_{yx}(\tau) &= E[Y(t+\tau)X(t)], \qquad & R_y(0) &= E[Y^2] \end{aligned}$$

the first inequality follows clearly from the following ***Schwarz inequality***:

$$\{E[X(t)Y(t')]\}^2 \le E[X^2(t)]E[Y^2(t')], \qquad \forall t, t' \tag{6.38}$$

where the equality holds if and only if $Y(t') = aX(t)$ for some real-valued scalar a. The Schwarz inequality can be shown as follows. Since the average of a positive quantity must be positive, $E[(\alpha X - Y)^2] = 0$ if and only if $Y = \alpha X$; that is, the only solution to the equation $E[(\alpha X - Y)^2] = 0$ is $Y = \alpha X$. On the other hand, the equation

$$E[(\alpha X - Y)^2] = E[X^2]\alpha^2 - 2E[XY]\alpha + E[Y^2] = 0$$

has its solution(s) given by

$$\alpha = \frac{2E[XY] \pm \sqrt{4(E[XY])^2 - 4E[X^2]E[Y^2]}}{2E[X^2]} \tag{6.39}$$

It has a unique solution if and only if $(E[XY])^2 = E[X^2]E[Y^2]$ and thus $(E[XY])^2 = E[X^2]E[Y^2]$ if and only if $Y = \alpha X$. In addition, if $Y \neq \alpha X$, then $E[(\alpha X - Y)^2] = 0$ has no (real-valued) solution, which corresponds to, according to (6.39),

$$(E[XY])^2 < E[X^2]E[Y^2]$$

In this case, $E[(\alpha X - Y)^2] > 0$. In summary, the Schwarz inequality holds because otherwise $E[(\alpha X - Y)^2] = 0$ would have two distinct (real-valued) solutions, which is impossible.

(6.36) follows from below:

$$\begin{aligned} \frac{d}{d\tau}R_{xy}(\tau) &= \lim_{\epsilon \to 0} \frac{R_{xy}(\tau+\epsilon) - R_{xy}(\tau)}{\epsilon} = \lim_{\epsilon \to 0} \frac{E[X(t+\tau+\epsilon)Y(t)] - E[X(t+\tau)Y(t)]}{\epsilon} \\ &= E\left[\lim_{\epsilon \to 0} \frac{X(t+\tau+\epsilon)Y(t) - X(t+\tau)Y(t)}{\epsilon}\right] \\ &= E\left[\lim_{\epsilon \to 0} \frac{X(t+\tau+\epsilon) - X(t+\tau)}{\epsilon} Y(t)\right] = E[\dot{X}(t+\tau)Y(t)] = R_{\dot{x}y}(\tau) \end{aligned}$$

and similarly for (6.37). More generally, let $X^{(n)}(t) = \frac{d^n X(t)}{dt^n}$. Then

$$R_{x^{(n)}y^{(m)}}(\tau) = (-1)^m \frac{d^{n+m} R_{xy}(\tau)}{d\tau^{n+m}} \tag{6.40}$$

If $X(t)$ and $Y(t)$ are orthogonal, then

$$\begin{aligned} R_{xy}(\tau) &= R_{yx}(\tau) = 0 \\ R_{x+y}(\tau) &= R_x(\tau) + R_y(\tau) \end{aligned}$$

Example 6.17: Computation of Sample Crosscorrelation by P&R

Consider two random processes $X(t) = \sin(\omega_0 t + \phi)$ and $Y(t) = \cos(\omega_0 t + \phi)$, where ω_0 is a nonrandom constant and $\phi \sim \mathcal{U}(-\pi, \pi)$. Sample functions of $X(t)$ and $Y(t)$ can be generated by the companion software P&R following Example 6.8. Suppose that two 500-point discrete sample functions (with sampling rate $f_s = 10$ Hz) of $X(t)$ and $Y(t)$ are stored in the data files `e6_17x.dat` and `e6_17y.dat`. Then the crosscorrelation coefficient $\rho_{xy}(i/f_s), i = 0, 1, \ldots, 50$ of $X(t)$ and $Y(t)$ can be computed using P&R with the following steps:

S1. Click "RPAnalyzer" in the main window of P&R.
S2. Click "Time Domain." You will be prompted to enter the name of a data file. Choose "data" and enter the data file name "`e6_17x.dat`" and click "Ok."
S3. The window "**Time-Domain Analyzer**" will appear, with the computed sample mean and sample variance. Set "Max time difference" to `50`. Click "CrossCorr-Coeff", as shown in Fig. 6.13.
S4. You will be prompted for another data file name. Choose "data" subdirectory, enter the data file name "`e6_17y.dat`" and click "Ok." Then the crosscorrelation coefficients $\rho(i/f_s)$ are computed, plotted and saved to a user-specified data file for the first 51 points ($i = 0, 1, \ldots, 50$), which correspond to $\tau \leq 10$ since sampling rate is 10 Hz.

Note that the result of Fig. 6.13 agrees with the theoretical results of self-test problem 6.6.

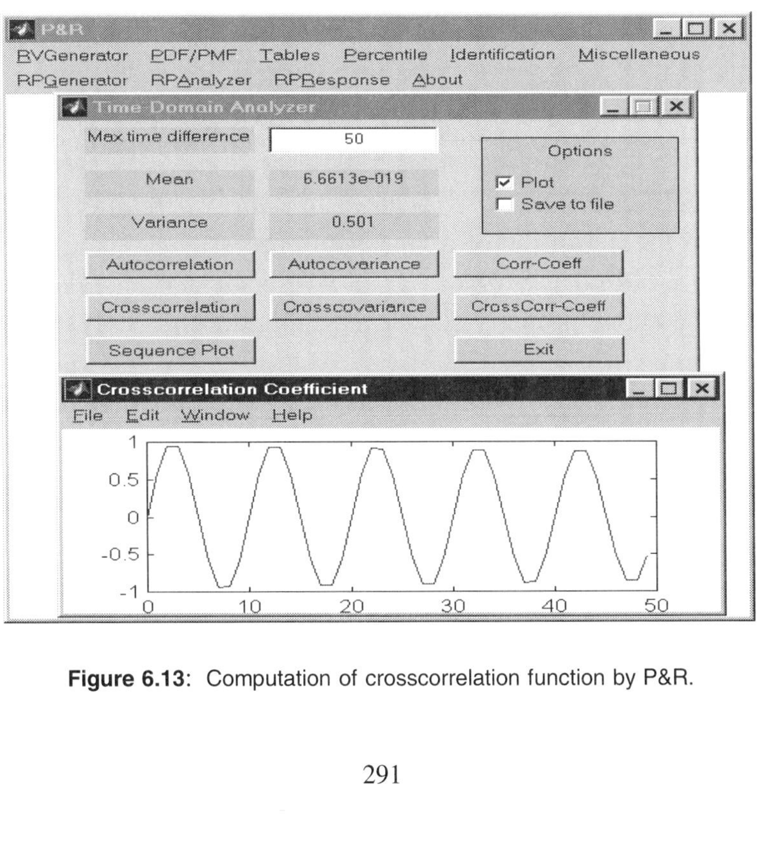

Figure 6.13: Computation of crosscorrelation function by P&R.

6.9 Gaussian Random Processes

Similarly as for the RV case, a number of random processes are frequently encountered in the theory and application and thus are given names. The most important of them is the Gaussian process.

A ***Gaussian process*** is such a random process $X(t)$ that, for any n and any set of time instants $t_1, t_2, \ldots, t_n$, the RVs $X(t_1), X(t_2), \ldots, X(t_n)$ are *jointly Gaussian*; that is, they have the following joint PDF:

$$f_{X(t_1),X(t_2),\ldots,X(t_n)}(x_1, x_2, \ldots, x_n) = \frac{1}{\sqrt{|2\pi C_{\mathbf{x}}|}} \exp\left[-\frac{1}{2}(\mathbf{x} - \bar{\mathbf{x}})' C_{\mathbf{x}}^{-1} (\mathbf{x} - \bar{\mathbf{x}})\right]$$

where $\mathbf{x} = [x_1, x_2, \ldots, x_n]'$, $\bar{\mathbf{x}} = [\bar{x}_1, \bar{x}_2, \ldots, \bar{x}_n]'$ is its mean, and $C_{\mathbf{x}} = \text{cov}(\mathbf{X})$ is the covariance of the RVs $X(t_1), X(t_2), \ldots, X(t_n)$, defined by (4.30) and is given explicitly as

$$C_{\mathbf{x}} = \begin{bmatrix} C_{11} & C_{12} & \cdots & C_{1n} \\ C_{21} & C_{22} & \cdots & C_{2n} \\ \vdots & \vdots & \ddots & \vdots \\ C_{n1} & C_{n2} & \cdots & C_{nn} \end{bmatrix}$$

where $C_{ij} = \text{cov}[X(t_i), X(t_j)] = E\{[X(t_i) - \bar{X}(t_i)][X(t_j) - \bar{X}(t_j)]\}$.

Gaussian processes have many nice and distinctive properties, e.g.,

1. *It is completely determined by its mean and autocorrelation functions.* This follows from the definition of a Gaussian process.
2. *It is strictly stationary if and only if it is wide-sense stationary.* This is clear from the above property since everything depends only on the mean and autocorrelation.
3. *It is independent (strictly white) if and only if it is uncorrelated (wide-sense white).*
4. *Every linear function of a Gaussian process is a Gaussian process.* This follows from the fact that a family of RVs are Gaussian if and only if every linear function of them is Gaussian.
5. *The response of a linear system to a Gaussian process is again a Gaussian process.*

Gaussian processes are especially useful in modeling noise processes.

For a Gaussian process $X(t)$, its mean conditioned on its values at some time instants are linear in those values; that is, $E[X(t)|X(t_1),\ldots,X(t_n)]$ is linear in $X(t_1),\ldots,X(t_n)$ for any positive integer n and any set $t_1,\ldots,t_n$. This has important consequence in estimation and filtering theory.

The popularity of the Gaussian process is a result of its mathematical tractability, due to its nice properties, and the central limit theorem.

Two random processes $X(t)$ and $Y(t)$ are said to be ***jointly Gaussian*** if the RVs $X(t_1)$, $X(t_2),\ldots,X(t_n)$ and $Y(t'_1)$, $Y(t'_2),\ldots,Y(t'_m)$ are *jointly Gaussian* for any positive integers n, m and any two sets of time instants t_1, $t_2,\ldots,t_n$ and t'_1, $t'_2,\ldots,t'_m$.

Example 6.18: Generation of a Gaussian Process

A Gaussian process $X(t)$ with mean $\bar{x}(t)$ and autocorrelation function $R_x(\tau) = ae^{-b|\tau|}$, $a, b > 0$, can be generated as follows. Let

$$Y(t) = X(t) - \bar{x}(t)$$

Then, by (6.24), $Y(t)$ has zero mean and autocorrelation function

$$R_y(\tau) = R_x(\tau) = ae^{-b|\tau|}$$

A sequence of n RVs $Y(T), Y(2T),\ldots,Y(nT)$ can thus be obtained by generating an n-dimensional jointly Gaussian random vector $\mathbf{Y} = [Y(T), Y(2T),\ldots,Y(nT)]'$. Finally, a discrete-time sequence of the Gaussian process $X(t)$ at $t = iT, i = 1, 2,\ldots,n$, is generated by adding $\bar{x}(iT)$ to the ith term of the sequence $Y(T), Y(2T),\ldots,Y(nT)$:

$$X(iT) = Y(iT) + \bar{x}(iT)$$

The required n-dimensional jointly Gaussian random vector $\mathbf{Y}$ can be generated by an extension of Computer Exercise 4.5 as follows.

S1. Generate a vector $\mathbf{Z}$ consists of n independent standard Gaussian RVs (i.e., with zero mean and unity variance) using, say, MATLAB command $\texttt{randn}(n, 1)$.

S2. $\mathbf{Y} = C^{1/2}\mathbf{Z}$ is the sought-after random vector, where $C^{1/2}$ is the so-called standard deviation matrix, which is the positive (semi)definite square-root matrix of the covariance matrix

$$C = a\begin{bmatrix} 1 & e^{-bT} & \cdots & e^{-(n-1)bT} \\ e^{-bT} & 1 & \cdots & e^{-(n-2)bT} \\ \vdots & \vdots & \ddots & \vdots \\ e^{-(n-1)bT} & e^{-(n-2)bT} & \cdots & 1 \end{bmatrix}$$

Note that $C^{1/2}$ can be obtained using MATLAB command $\texttt{sqrtm}(C)$.

An alternative method of generating Gaussian processes is studied in Chapter 8.

Example 6.19: Generation of a Gaussian Process by P&R

The companion software P&R can be used to generate a Gaussian process $X(t)$. Assume $X(t)$ has mean $\bar{x}(t) = 2e^{-t^2}\cos t$ and autocorrelation $R_x(\tau) = 6e^{-12|\tau|}$. Since P&R is a computer software, it can only generate discrete points of $X(t)$. Let us assume we want to generate 200 such points with a sampling rate of 10 Hz. This can be done by the following steps:

S1. Click "RPGenerator" in the main window of P&R. Click "Gaussian."

S2. Fill out the window "**Gaussian Process Generator**" as shown in Fig. 6.14. Click "Ok." The Gaussian process is then generated.

The Gaussian process generated is saved to a user-specified data file and plotted as shown in Fig. 6.14. Note that

- The input to "Mean" must be a valid MATLAB expression. For instance in this example, "`2*exp(-t.^2).*cos(t)`" cannot be replaced by something like "`2exp(-t2)cost`", which would not work. if $x = [x_1, x_2]'$ and $y = [y_1, y_2]'$, then `x.*y` and `x.^2` yield $[x_1y_1, x_2y_2]'$ and $[x_1^2, x_2^2]'$, respectively. The dot operations `.*` and `.^` are necessary here since `t` is a vector (of dimension 200). See MATLAB instruction.
- Gaussian white process can also be generated by choosing "White" on the window.

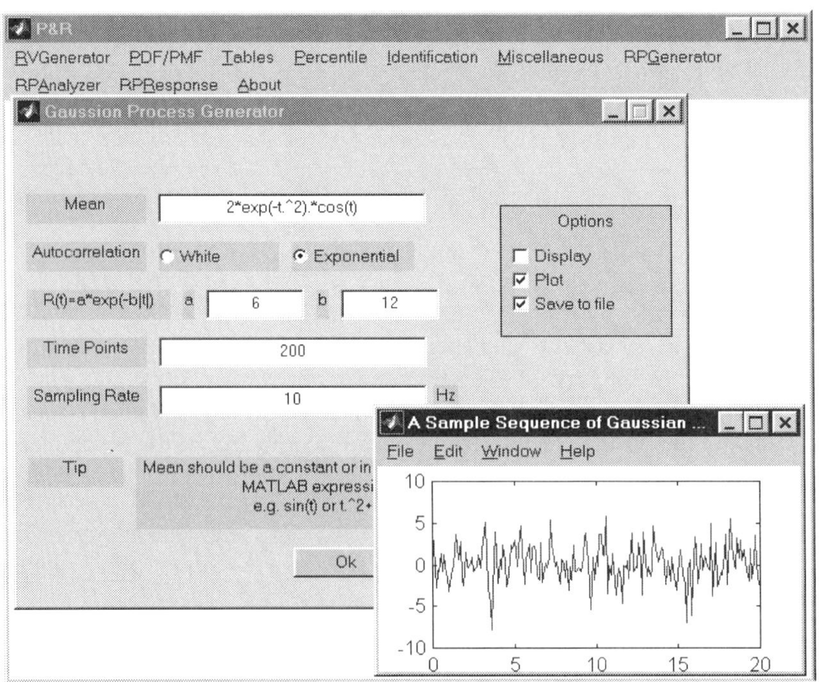

Figure 6.14: Generation of a Gaussian process by P&R.

6.10 Summary and Requirements

A *random process* is a time-varying RV or a random time function. For a fixed time, it is a RV; given the outcome of the random experiment, a time function, known as *sample function*, rather than a number is assigned to it.

The definitions of *mean* $\bar{x}(t)$, *autocorrelation* $R_x(t_1, t_2)$, *autocovariance* $C_x(t_1, t_2)$, and *correlation coefficient* $\rho(t_1, t_2)$ of a random process $X(t)$ follow from the corresponding definitions for RVs $X(t_1)$ and $X(t_2)$:

$$\begin{aligned} \bar{x}(t_1) - E[X(t_1)] &= \int_{-\infty}^{\infty} x f_{X(t_1)}(x)dx \\ R_x(t_1, t_2) &= E[X(t_1)X(t_2)] \\ C_x(t_1, t_2) &= E\{[X(t_1) - \bar{x}(t_1)][X(t_2) - \bar{x}(t_2)]\} \\ \rho(t_1, t_2) &= \frac{C_x(t_1, t_2)}{\sigma_x(t_1)\sigma_x(t_2)}, \qquad -1 \le \rho(t_1, t_2) \le 1 \end{aligned}$$

A *stationary* random process has a constant mean and its autocorrelation depends only on the time difference; that is,

$$\begin{aligned} \bar{x}(t) &= \bar{x} \\ R_x(t_1, t_2) &= R_x(t_1 - t_2) = R_x(\tau), \qquad \tau = t_1 - t_2, \forall t_1, t_2 \end{aligned}$$

A random process is *white* if its values at distinct time instants are uncorrelated, or equivalently $C_x(t_1, t_2) = 0$ or $R_x(t_1, t_2) = \bar{x}(t_1)\bar{x}(t_2)$ for every $t_1 \neq t_2$. The ensemble average (i.e., mean) of an *ergodic* random process is equal to the time average of any of its sample functions.

The *crosscorrelation* of two random processes $X(t)$ and $Y(t)$ is a measure of the similarity or coupling between $X(t)$ and $Y(t)$:

$$R_{xy}(t + \tau, t) \triangleq E[X(t + \tau)Y(t)]$$

A large $|\rho_{xy}(t + \tau, t)|$ implies that $Y(t_1)$ can be predicted largely from $X(t_1 + \tau)$ and vice versa.

The autocorrelation of a stationary random process has many nice properties: It is even in the time difference τ; it is bounded in magnitude by its value at the origin, which is equal to the mean-square value (i.e., average power); it has a periodic component if the random process has a periodic component.

Two random processes are *jointly stationary* if they are both stationary and their crosscorrelations depend only on the time difference. They are *uncorrelated* if their crosscovariance is always zero. They are *orthogonal* if their crosscorrelation is always zero. They are *independent* if the RVs of one random process at arbitrary time instants are independent of the RVs of the other random process at arbitrary time instants. Crosscorrelation of jointly stationary random processes are mirror images of each other: $R_{xy}(\tau) = R_{yx}(-\tau)$. They are also bounded in magnitude by the geometric average and arithmetic average of the autocorrelations of the two processes.

Basic Requirements

- Know how to find mean, correlation, and covariance functions.
- Comprehend the concepts of a random process and its stationarity and whiteness.
- Know how to compute sample correlation functions by your own and by the companion software P&R.
- Know how to determine if a random process is stationary, white, or ergodic.
- Be familiar with the properties of autocorrelation and crosscorrelation functions.
- Understand the concept of uncorrelatedness and independence of two random processes.
- Know how to find the marginal and joint PDFs of a random process for simple problems.

The emphasis of the chapter is on general concept of a random process and its autocorrelation and stationarity.

6.11 Additional Examples

6.20 *Sinusoid with uniform amplitude and phase.* Find the autocorrelation function $R_x(t+\tau, t)$ of the continuous random process $X(t) = A\cos(\omega_0 t + \phi)$, where ω_0 is a known constant, A is a RV uniformly distributed over $(0, 1)$, and ϕ is a RV uniformly distributed over $(0, 2\pi)$. Assume that A and ϕ are independent.

Solution: Since A and ϕ are independent, A^2 and $[\cos\omega_0\tau + \cos(2\omega_0 t + \omega_0\tau + 2\phi)]$ are uncorrelated. Thus, we have

$$\begin{aligned}
R_x(t+\tau, t) &= E\left[A\cos(\omega_0 t + \omega_0\tau + \phi)\cdot A\cos(\omega_0 t + \phi)\right] \\
&= \frac{1}{2}E\{A^2[\cos\omega_0\tau + \cos(2\omega_0 t + \omega_0\tau + 2\phi)]\} \\
&= \frac{1}{2}E[A^2]E[\cos\omega_0\tau + \cos(2\omega_0 t + \omega_0\tau + 2\phi)] \\
&= \frac{1}{2}\{\sigma_A^2 + [E(A)]^2\}\{E(\cos\omega_0\tau) + E[\cos(2\omega_0 t + \omega_0\tau + 2\phi)]\} \\
&= \frac{1}{2}\left[\frac{(1-0)^2}{12} + \left(\frac{1+0}{2}\right)^2\right](\cos\omega_0\tau + 0) \\
&= \frac{1}{6}\cos\omega_0\tau = R(\tau)
\end{aligned}$$

6.21 *Sinusoid with uniform amplitude, frequency, and phase.* Find the autocorrelation function $R_x(t+\tau, t)$ of the continuous random process $X(t) = A\cos(\omega t + \phi)$, where ω, A, and ϕ are mutually independent RVs uniformly distributed over $(1, 2)$, $(0, 1)$, and $(0, \pi/2)$, respectively.

Solution: Since A, ω, and ϕ are independent, $f(A)$, $g(\omega)$ and $h(\theta)$ are uncorrelated for any f, g and h. Thus, we have

$$R_x(t+\tau, t) = E\left[A\cos(\omega t + \omega\tau + \theta)\cdot A\cos(\omega t + \theta)\right]$$

$$
\begin{aligned}
&= \frac{1}{2}E\{A^2[\cos\omega\tau + \cos(2\omega t + \omega\tau + 2\theta)]\} \\
&= \frac{1}{2}E[A^2]E[\cos\omega\tau + \cos(2\omega t + \omega\tau)\cos 2\theta - \sin(2\omega t + \omega\tau)\sin 2\theta] \\
&= \frac{1}{2}\{\sigma_A^2 + [E(A)]^2\}\{E(\cos\omega\tau) + E[\cos(2t+\tau)\omega]E[\cos 2\theta] \\
&\quad - E[\sin(2t+\tau)\omega]E[\sin 2\theta]\} \\
&= \frac{1}{2}\left[\frac{(1-0)^2}{12} + \left(\frac{1+0}{2}\right)^2\right]\left\{\int_1^2 \frac{\cos\omega\tau}{2-1}d\omega\right. \\
&\quad \left. + \int_1^2 \frac{\cos(2t+\tau)\omega}{2-1}d\omega \int_0^{\pi/2}\frac{\cos 2\theta}{\pi/2}d\theta - \int_1^2 \frac{\sin(2t+\tau)\omega}{2-1}d\omega \int_0^{\pi/2}\frac{\sin 2\theta}{\pi/2}d\theta\right\} \\
&= \frac{\sin 2\tau - \sin\tau}{6\tau} + \frac{\sin 2(2t+\tau) - \sin(2t+\tau)}{2t+\tau}\frac{\sin\pi - \sin 0}{6\pi} \\
&\quad - \frac{\cos 2(2t+\tau) - \cos 2(t+\tau)}{2t+\tau}\frac{\cos\pi - \cos 0}{6\pi} \\
&= \frac{\sin 2\tau - \sin\tau}{6\tau} + \frac{\cos 2(2t+\tau) - \cos 2(t+\tau)}{3\pi(2t+\tau)}
\end{aligned}
$$

Note that the random process $X(t)$ is not (wide-sense) stationary since $R_x(t+\tau, t)$ depends on t as well as τ.

6.22 *Secondary voltage of a transformer with random tap position.* The primary voltage of a transformer is a deterministic waveform $x(t) = 10\cos(120\pi t + 60°)$ and the secondary voltage of a transformer is a random process $Y(t)$ having the sample space $\{y_1(t), y_2(t)\}$, where $y_i(t) = A_i\cos(120\pi t + 60°)$ with $A_1 = 1100\sqrt{2}$ and $A_2 = 1000\sqrt{2}$. $Y(t) = y_1(t)$ or $Y(t) = y_2(t)$ depends on whether the transformer is in tap position 1 or 2, respectively, which is random with equal probability.

(a) Find the probabilities that $Y(t) = y_1(t)$ and $Y(t) = y_2(t)$, respectively.
(b) Find the mean function $\bar{y}(t)$ and mean $\bar{y}(0)$ of $Y(t)$.
(c) Find the autocorrelation $R_y(t+\tau, t)$ of $Y(t)$.
(d) Is $Y(t)$ wide-sense stationary? Justify your answer.

Solution:

(a) Since the tap position of the transformer is random with equal probability, we have

$$P\{Y(t) = y_1(t)\} = P\{Y(t) = y_2(t)\} = \frac{1}{2}$$

(b) The mean function $\bar{y}(t)$ and mean $\bar{y}(0)$ are

$$\bar{y}(t) = E[Y(t)] = \sum_{i=1}^{2} A_i\cos(120\pi t + 60°)P\{Y(t) = y_i(t)\}$$

$$
\begin{aligned}
&= \cos(120\pi t + 60^\circ)\left[1100\sqrt{2}P\{Y(t) = y_1(t)\} + 1000\sqrt{2}P\{Y(t) = y_2(t)\}\right] \\
&= \frac{1}{2}(1100\sqrt{2} + 1000\sqrt{2})\cos(120\pi t + 60^\circ) \\
&= 1050\sqrt{2}\cos(120\pi t + 60^\circ) \\
\bar{y}(0) = \bar{y}(t)|_{t=0} &= 1050\sqrt{2}\cos 60^\circ = 525\sqrt{2} = 742.46
\end{aligned}
$$

(c) The autocorrelation function $R_y(t+\tau, t)$ is

$$
\begin{aligned}
R_y(t+\tau, t) &= E[Y(t+\tau)Y(t)] \\
&= E[A\cos(120\pi(t+\tau) + 60^\circ)A\cos(120\pi t + 60^\circ)] \\
&= E[A^2]\cos(120\pi(t+\tau) + 60^\circ)\cos(120\pi t + 60^\circ) \\
&= \cos(120\pi t + 60^\circ)\cos(120\pi(t+\tau) + 60^\circ)\sum_{i=1}^{2} A_i^2 P\{Y(t) = y_i(t)\} \\
&= \cos(120\pi t + 60^\circ)\cos(120\pi(t+\tau) + 60^\circ)\frac{1}{2}\left[(1100\sqrt{2})^2 + (1000\sqrt{2})^2\right] \\
&= 2210000\cos(120\pi t + 60^\circ)\cos(120\pi(t+\tau) + 60^\circ)
\end{aligned}
$$

(d) Since $\bar{y}(t)$ is actually a function of time (not time invariant), $Y(t)$ is not wide-sense stationary.

6.23 *Stationarity and ergodicity of a RV as a random process.* Determine if random process $Y(t) = aX$ is stationary or ergodic, where $a \neq 0$ is a constant and X is a RV with nonzero variance.

Solution: Since

$$
\begin{aligned}
E[Y(t)] &= aE[X] = \text{constant} \\
R_y(t+\tau, t) &= E[Y(t+\tau)Y(t)] = E[aXaX] = a^2E[X^2] = \text{constant}
\end{aligned}
$$

$Y(t)$ is wide-sense stationary. This makes perfect sense since the characteristics of $Y(t)$ do not change with respect to time. Note that the time average of an arbitrary sample function of $Y(t)$ is

$$
\lim_{T\to\infty}\frac{1}{2T}\int_{-T}^{T} aX dt = aX
$$

which depends on the value of x and thus is not always equal to the ensemble average $\bar{x}$. Consequently, the process is not ergodic (in the mean), which makes sense since the mean of $Y(t)$ cannot be obtained by an arbitrary sample function of $Y(t)$.

6.24 *Sum of random process and RV with increasing variance.* Consider a wide-sense stationary process $X(t)$ with zero mean and autocorrelation $R_x(\tau)$. Let $Y(t) = X(t) + At$, where A is a zero-mean random variable with unity variance, which is independent of $X(t)$.

(a) Find the mean $E[Y(t)]$ and autocorrelation $R_y(t+\tau, t)$ of $Y(t)$.

(b) Is $Y(t)$ wide-sense stationary?
(c) Find the crosscorrelation $R_{xy}(t+\tau, t)$ between $X(t+\tau)$ and $Y(t)$.
(d) Are $X(t)$ and $Y(t)$ jointly wide-sense stationary?

Solution:

(a)

$$
\begin{aligned}
R_y(t+\tau, t) &= E\{[X(t+\tau) + A(t+\tau)][X(t) + At]\} \\
&= E[X(t)X(t+\tau) + A^2 t(t+\tau) + X(t)A(t+\tau) + AtX(t+\tau)] \\
&= R_x(\tau) + E[A^2]t(t+\tau) + \bar{x}(t)\bar{A}(t+\tau) + \bar{A}tx(t+\tau) \\
&= R_x(\tau) + t(t+\tau)
\end{aligned}
$$

(b) Since $R_y(t+\tau, t)$ depends on t as well as τ, $Y(t)$ is not wide-sense stationary.
(c)

$$
\begin{aligned}
R_{xy}(t+\tau, t) &= E\{X(t+\tau)[X(t) + At]\} \\
&= E[X(t)X(t+\tau) + X(t+\tau)At] \\
&= R_x(\tau) + \bar{x}(t)\bar{A}t = R_x(\tau)
\end{aligned}
$$

(d) Since $Y(t)$ is not wide-sense stationary, $X(t)$ and $Y(t)$ are not jointly wide-sense stationary.

6.25 *Average power of various components.* A stationary random process $X(t)$ has autocorrelation $R(\tau) = 20 + 5\cos(2\tau) + 10e^{-4|\tau|}$. Find (a) the average power of its dc component, ac (i.e., periodic) components, non-dc components, and non-periodic components; (b) average power of $X(t)$; and (c) the variance of $X(t)$.

Solution: $R(\tau)$ can be decomposed into periodic and non-periodic components as

$$
R(\tau) = R_1(\tau) + R_2(\tau) = 5\cos(2\tau) + 20 + 10e^{-4|\tau|}
$$

where the periodic and non-periodic components are, respectively,

$$
\begin{aligned}
R_1(\tau) &= 5\cos(2\tau) = \frac{1}{2}A^2\cos(2\tau) \\
R_2(\tau) &= 20 + 10e^{-4|\tau|}
\end{aligned}
$$

By property 4 of autocorrelation function, in particular (6.19), the periodic (ac) component of $X(t)$ is $A\cos(2t+\theta)$. It has zero mean. Its average power is $R_1(0) = \frac{1}{2}A^2 = 5$. The average power of the non-periodic components is $R_2(0) = 20 + 10 = 30$. From the property of autocorrelation function described on page 280, for the non-periodic component, $\lim_{\tau\to\infty} R_2(\tau) = (\bar{x})^2$. Thus, the dc component is equal to

$$
\bar{x} = \pm\sqrt{\lim_{\tau\to\infty} R_2(\tau)} = \pm\sqrt{\lim_{\tau\to\infty}(20 + 10e^{-4|\tau|})} = \pm 4.472
$$

It has the power $\lim_{\tau\to\infty} R_2(\tau) = (\bar{x})^2$=20. The average power of $X(t)$ is

$$E[X^2] = R(0) = 5 + 20 + 10 = 35$$

The variance is

$$\sigma^2 = E[X^2] - (\bar{x})^2 = R(0) - (\bar{x})^2 = 35 - 20 = 15$$

which is the average power of the non-dc components. Note that $10e^{-4|\tau|}$ is a non-dc and non-ac component. It has the average power of 10.

6.26* *First-order polynomial in time with random coefficient.* Given a random process $X(t) = Y + Zt$, where Y and Z are two standard Gaussian RVs, independent of each other, find

(a) the mean and variance of $X(t)$
(b) the marginal PDF and CDF of $X(t)$
(c) the joint CDF of $X(t)$: $F_{X(t),X(t')}(x_1, x_2)$

Solution: For a fixed t, $X(t)$ is a RV.

(a)

$$\left.\begin{array}{r} Y \sim \mathcal{N}(0,1) \\ Z \sim \mathcal{N}(0,1) \Longrightarrow Zt \sim \mathcal{N}(0,t^2) \\ Y, Z \text{ and thus } Y, Zt \text{ are independent} \end{array}\right\} \overset{(4.34)}{\Longrightarrow} \left\{\begin{array}{l} X(t) = Y + Zt \\ X(t) \sim \mathcal{N}(0, 1+t^2) \end{array}\right.$$

Thus for a fixed t, $X(t)$ is a zero-mean Gaussian RV with variance $1 + t^2$.

(b) The PDF of $X(t)$ at a fixed time t is, from part (a),

$$f_{X(t)}(x) = \frac{1}{\sqrt{2\pi(1+t^2)}} e^{-\frac{x^2}{2(1+t^2)}}$$

The CDF of $X(t)$ at a fixed time t is thus

$$F_{X(t)}(x) = \int_{-\infty}^{x} \frac{1}{\sqrt{2\pi(1+t^2)}} e^{-\frac{x^2}{2(1+t^2)}} dx$$

(c) Note that

$$\begin{bmatrix} X(t) \\ X(t') \end{bmatrix} = \begin{bmatrix} 1 & t \\ 1 & t' \end{bmatrix} \begin{bmatrix} Y \\ Z \end{bmatrix}$$

and Y and Z are independent Gaussian RVs. Since $X(t)$ and $X(t')$ are weighted sum of Y and Z, they are jointly Gaussian with zero mean and covariance

$$\begin{aligned} \text{cov}[X(t), X(t')] &= E[(X(t) - \bar{x}(t))(X(t') - \bar{x}(t'))] \\ &= E[(Y + Zt)(Y + Zt')] \\ &= E\left[Y^2 + Z^2tt' + YZt' + YZt\right] \\ &= E[Y^2] + E[Z^2]tt' + E[Y]E[Z](t' + t) \\ &= 1 + tt' \end{aligned}$$

6.12 Problems

6.1 *PDF of a discrete random process.* For the random process $X(t)$ of Example 6.1, find

(a) $f_{X(\pi/4)}(x), f_{X(\pi/2)}(x)$, and $f_{X(\pi)}(x)$
(b) $f_{X(\pi/4),X(\pi/2)}(x_1, x_2)$ and $f_{X(\pi/4),X(\pi)}(x_1, x_2)$
(c) $\bar{x}(\pi/4)$, $\bar{x}(\pi)$, and $\bar{x}(t)$ in terms of t
(d) $R_x(\pi/2, \pi)$ and $C_x(t+\tau, t)$ for every t and τ

6.2 *PDF and CDF of a discrete random process.* Given a random process $X(t) = A \sin t$, where A is a discrete RV with point masses

$$P\{A=1\} = 0.4, \qquad P\{A=2\} = 0.3, \qquad P\{A=3\} = 0.3$$

(a) Find the marginal CDF $F_{X(t)}(x)$ for all t.
(b) Find the joint CDF $F_{X(\pi/4),X(0)}(x_1, x_2)$.

6.3 *PDF of a simple random process.* Find the mean, autocorrelation, variance, and autocovariance of the random process $X(t) = tU$, where $U \sim \mathcal{U}(0,1)$.

6.4 *Mean and autocorrelation.* Given a random process $X(t) = Ae^{-at}$, where a is a nonrandom constant and $A \sim \mathcal{U}(0,1)$, find the mean and autocorrelation of $X(t)$.

6.5 *Mean and autocorrelation.* Given a random process $Z(t) = X \sin 2t + Y \cos t + t$, where X and Y are independent RVs with $\bar{x} = 1, \sigma_x^2 = 2, \bar{y} = 2, \sigma_y^2 = 4$, find the mean and autocorrelation of $Z(t)$.

6.6 *Mean and autocorrelation.* Given a random process $Z(t) = Xt + Yt^2$, where X and Y are independent RVs with $\bar{x} = 1, \sigma_x^2 = 1, \bar{y} = 2, \sigma_y^2 = 4$, find the mean and autocorrelation of $Z(t)$.

6.7 *From autocorrelation to mean and variance.* A random process $X(t)$ has autocorrelation $R(\tau) = 1 + 4\cos(7\tau) + 6e^{-3|\tau|}$. Find the mean, mean-square value, and variance of $X(t)$.

6.8 *From autocorrelation to mean and variance.* Find the mean and variance of a stationary random process with autocorrelation function $R(\tau) = 7 + \dfrac{2}{1+9\tau^2}$.

6.9 *Sinusoid with random amplitude.* Consider a sine wave $X(t) = A \sin \omega_0 t$ with random amplitude, where ω_0 is known constant and A is a standard Gaussian RV.

(a) Find the PDF at $t = 0$ and $t = \pi/2\omega_0$ of $X(t)$.
(b) Find the mean, autocorrelation and autocovariance of $X(t)$.

6.10 *Sinusoid with random phase.* Consider the random process $X(t) = A\cos(\omega_0 t + \theta)$, where A and ω_0 are real constants and $\theta \sim \mathcal{U}(0, 2\pi)$.

(a) Find the mean function $E[X(t)]$ and the autocorrelation function $R_x(t+\tau, t)$. Is $X(t)$ wide-sense stationary? Why?
(b) Assume now $\theta \sim \mathcal{U}(0, \pi/2)$. Determine the mean-square value $E[X^2(t)]$. Is $E[X^2(t)]$ deterministic? time dependent? Is the new random process $X(t)$ wide-sense stationary? Why? If $E[X^2(t)]$ is time varying, what is its time average?

6.11 *Sinusoid with random phase.* Consider a random process $X(t) = A\cos(\omega_0 t + \phi)$, where A and ω_0 are nonrandom positive constants and ϕ is a RV uniformly distributed over $(0, \pi)$, i.e., $\phi \sim \mathcal{U}(0, \pi)$.

(a) Find the mean function $\bar{x}(t)$ of $X(t)$.
(b) Find the autocorrelation $R_x(t+\tau, t)$ and autocovariance $C_x(t+\tau, t)$ of $X(t)$.
(c) Is $X(t)$ discrete/continuous/mixed? wide-sense stationary? or white? Justify your answer.

6.12 *Sinusoid with random phase.* Given a random process $X(t) = A\cos(\omega_0 t + n\phi)$ with $\phi \sim \mathcal{U}(0, 2\pi/m)$ and A and ω_0 are nonrandom constants, where m and n are two arbitrary but fixed positive integers, find the mean and autocorrelation function of $X(t)$. Your results should be valid for all positive integers n and m.

6.13 *Mean and autocorrelation.* Consider a random process $X(t) = Ae^{2t}\sin 4t$, where A is a RV with mean 1 and variance 2.

(a) Find the mean and autocorrelation function of $Y(t) = \dot{X}(t)$.
(b) Find the mean and autocorrelation function of $Z(t) = \int_0^t X(\tau)d\tau$.

6.14 *Stationarity of two simple random processes.* Determine the wide-sense stationarity of random processes $Y(t) = aX$ and $Z(t) = tX$, where X is a RV and a is constant.

6.15 *Classification of a random process.* For the random process $X(t)$ in Example 6.4,

(a) Is $X(t)$ continuous, discrete, or mixed?
(b) Is $X(t)$ strictly stationary? Is it wide-sense stationary?
(c) Is $X(t)$ white?
(d) Is $X(t)$ ergodic?

6.16 *Stationarity check.* Suppose that random processes $X_i(t)$, $i = 1, \ldots, 5$, have time-invariant mean functions and the following autocorrelation functions, respectively. Determine if they are wide-sense stationary.

(a) $R_{x_1}(t_1, t_2) = \sin 2t_1 \cos t_2 - \cos 2t_1 \sin t_2$.
(b) $R_{x_2}(t_1, t_2) = e^{t_2}e^{-t_1}$.
(c) $R_{x_3}(t_1, t_2) = e^{-t_2^2}e^{t_1^2}$.
(d) $R_{x_4}(t_1, t_2) = e^{|t_1 - t_2|}E[X_4(t_1)]$.

6.17 *Stationarity of product of random processes.* Show that the product $Z(t) = X(t)Y(t)$ of two independent and stationary random processes is also stationary.

6.18 *Stationarity of sine and cosine waves.* Consider two independent, stationary, and zero-mean random processes $X(t)$ and $Y(t)$ with a common autocorrelation $R(\tau)$. Let

$$U(t) = X(t)\cos\omega_0 t$$
$$V(t) = Y(t)\sin\omega_0 t$$
$$W(t) = U(t) + V(t)$$

where ω_0 is a positive constant. Determine if $U(t), V(t), W(t)$ are stationary, respectively.

6.19 *Stationarity of sinusoid with random amplitude and phase.* Given a random process $X(t) = A\cos(120\pi t + \phi)$, where $A \sim \mathcal{N}(0,1)$ is independent of $\phi \sim \mathcal{U}(-\pi, \pi)$, determine if $X(t)$ is wide-sense stationary.

6.20 *Ergodicity of sum of random in-phase and quadrature components.* Consider random process $X(t) = Y\cos\omega_0 t + Z\sin\omega_0 t$, where ω_0 is a nonrandom constant, Y and Z are zero-mean RVs.

(a) Determine if $X(t)$ is ergodic in the mean.
(b) Determine if $X(t)$ is ergodic in the mean-square value.

6.21 *Sum of stationary and deterministic processes.* Let $X(t)$ be a wide-sense stationary random process with mean $\bar{x}$ and autocorrelation $R_x(\tau)$. Let $Y(t) = X(t) + g(t)$, where $g(t)$ is a deterministic function.

(a) Find the mean, autocovariance, and autocorrelation of $Y(t)$.
(b) Is $Y(t)$ wide-sense stationary?

6.22 *Power of signal and noise.* For the random process $X(t)$ of Example 6.7. Let $Y(t) = X(t) + N(t)$, where $N(t)$ is a random noise process with average power $P_n = E[N^2(t)]$ and is uncorrelated with $X(t)$.

(a) Find the average power of $X(t)$.
(b) Find the average power of $Y(t)$.
(c) If $N(t) = B\cos(\omega_1 t + \theta)$, where B and ω_1 are known and $\theta \sim \mathcal{U}(-\pi, \pi)$, what are the mean and average power of $N(t)$?

6.23 *Sum of random in-phase and quadrature components.* Consider the random process $X(t) = Y\sin\omega_0 t + Z\cos\omega_0 t$, where ω_0 is a constant; Y and Z are uncorrelated RVs with zero-mean and a common variance σ^2.

(a) Find the mean function of $X(t)$.
(b) Find the autocorrelation function of $X(t)$.

(c) Is $X(t)$ wide-sense stationary?
(d) Is $X(t)$ strictly stationary?

6.24 *Autocorrelation of derivative.* Find the autocorrelation function of the derivative $\dot{X}(t)$ of the random process $X(t)$ with autocorrelation $R_x(\tau) = 30e^{-\tau^2/2}\cos(20\tau)$.

6.25* *Mixed random process determined by differential equation.* A mixed random process $X(t)$ at any time takes on a value

- equal to the maximum and minimum values $\pm x_m$ with equal probability 0.2
- equal to zero with probability 0.4
- uniformly distributed over $[-x_m, x_m]$ with the remaining probability mass 0.2

Assume that the random process satisfies the following differential equation

$$\dot{X}(t) = -aX(t) + N(t)$$

where $N(t)$ is a stationary zero-mean white noise process with autocorrelation function

$$R_n(\tau) = E[N(t+\tau)N(t)] = 2a\sigma^2\delta(\tau)$$

(a) Find the marginal probability density function.
(b) Find the mean function of $X(t)$.
(c) Find the mean square function of $X(t)$.
(d) Find the autocorrelation function of $X(t)$.
(e) Is $X(t)$ wide-sense stationary?
(f) Is $X(t)$ strictly stationary?

6.26 *Properties of autocorrelation.* Determine which of the functions in Fig. 6.15 can/cannot be considered autocorrelation functions. Justify your answer.

6.27 *Properties of autocorrelation.* Determine which of the functions in Fig. 6.16 can/cannot be considered autocorrelation functions. Justify your answer.

6.28 *Sinusoids with random phase.* Consider a random process $X(t) = A\cos(\omega_0 t + \theta)$, where A and ω_0 are constants, $\theta \sim \mathcal{U}(0, 2\pi)$. Let $Y(t) = X^2(t)$.

(a) Find the mean function of $Y(t)$.
(b) Find the autocorrelation function of $Y(t)$.
(c) Find the crosscorrelation function of $X(t)$ and $Y(t)$.
(d) Are $X(t)$ and $Y(t)$ wide-sense stationary?
(e) Are $X(t)$ and $Y(t)$ jointly wide-sense stationary? Are they orthogonal?

6.29 *Stationarity of random frequency components.* Consider two independent and zero-mean RVs X and Y with a common variance σ^2. Let

$$U(t) = X\cos(\omega_0 t) + Y\sin(\omega_0 t)$$
$$V(t) = X\cos(2\omega_0 t) + Y\sin(2\omega_0 t)$$

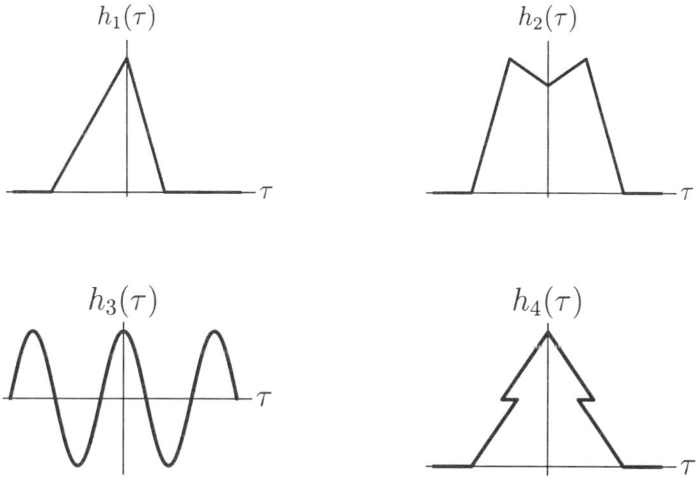

Figure 6.15: Candidate functions as autocorrelation functions.

where ω_0 is a positive constant. Determine if $U(t)$ and $V(t)$ are jointly stationary.

6.30 *Stationarity of two random processes.* A zero-mean wide-sense stationary random process $X(t)$ has an autocorrelation function $R_x(\tau)$. Define a new process $Z(t) = X(t) + Yt$, where Y is a zero-mean unity-variance RV that is independent of $X(t)$.

(a) Determine the autocorrelation function $R_z(t+\tau, t)$. Is $Z(t)$ wide-sense stationary? Why?
(b) Determine the cross-correlation function $R_{xz}(t+\tau, t)$. Are $X(t)$ and $Z(t)$ jointly wide-sense stationary? Why?

6.31 *Orthogonality of sine and cosine functions.* Consider random processes $X(t)$, $Y(t)$ and $Z(t)$, defined by

$$X(t) = \sin(\omega_0 t + \phi)$$
$$Y(t) = \cos(\omega_0 t + \theta)$$
$$Z(t) = \sin(\omega_0 t + \varphi)$$

where ω_0 is nonrandom, $\phi \sim \mathcal{U}(0, 2\pi)$, $\theta \sim \mathcal{U}(0, \pi)$ and $\varphi \sim \mathcal{U}(0, \pi/2)$ are independent.

(a) Find the mean functions of $X(t), Y(t)$ and $Z(t)$.
(b) Find the crosscorrelations of $X(t), Y(t)$ and $Z(t)$ and the average cross powers $E[X(t)Y(t)]$, $E[X(t)Z(t)]$ and $E[Y(t)Z(t)]$.
(c) Are random processes $X(t), Y(t)$ and $Z(t)$ orthogonal?

6.32 *Correlation of sum and difference.* Two uncorrelated zero-mean random processes $X(t)$ and $Y(t)$ have autocorrelation functions $R_x(\tau) = 2\cos\tau$ and $R_y(\tau) = 3e^{-|\tau|}$. Find

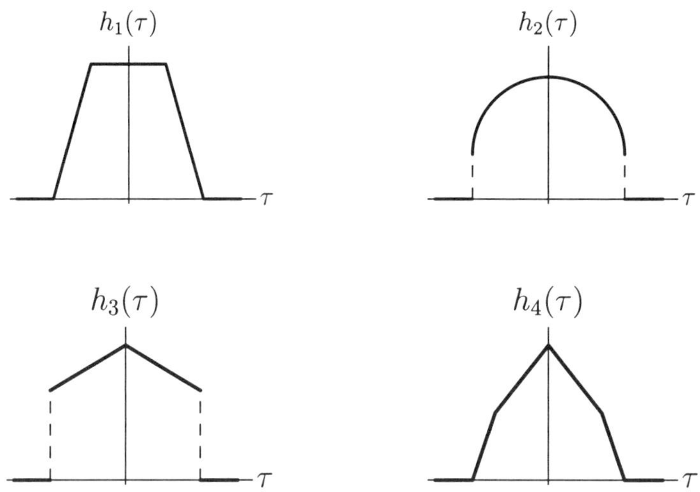

Figure 6.16: Candidate functions as autocorrelation functions.

(a) the autocorrelation of $U(t) = X(t) + Y(t)$
(b) the autocorrelation of $V(t) = X(t) - Y(t)$
(c) the crosscorrelation of $U(t)$ and $V(t)$

6.33 *Correlation of random processes proportional to two RVs.* Given two random processes $X(t) = Zt, Y(t) = Ze^{-t}$, where Z is a RV, find $R_x(\tau)$, $R_y(\tau)$, and $R_{xy}(\tau)$.

6.34 *Properties of crosscorrelation.* For stationary random processes $X(t)$ and $Y(t)$, show the first inequality in (6.35) using the following inequality

$$E\left[\left(\frac{X(t)}{\sqrt{R_x(0)}} \pm \frac{Y(t)}{\sqrt{R_y(0)}}\right)^2\right] \geq 0$$

6.35 *Stationarity of signal plus noise.* For Example 6.12, show that $X(t)$ and $Y(t)$ are jointly wide-sense stationary.

6.36 *Expected values of sample autocorrelations.* Find the expected values of the unbiased and biased sample autocorrelations (6.26) and (6.27), respectively, in terms of the true autocorrelation $R(\tau)$.

6.37 *Joint PDF of Gaussian process.* Given a stationary Gaussian random process $X(t)$ with mean 5 and autocorrelation $R(\tau) = 3.4e^{-4.6|\tau|}$, find the joint PDF of $X(t_1)$ and $X(t_1+3)$.

6.38 *Puzzling questions.* Raise two technical questions of your own that you are puzzled by and that are closely related to the material covered in this chapter. These questions should be important, not too simple, and not too involved.

6.13 Computer Exercises

6.1 *Generation and testing of a random sinusoid.*

(a) Write a MATLAB routine `random_sin.m` to generate 100 discrete-time sequences $x(nT)$ that are sampled (i.e., discrete-time) versions of the sample functions of the random process $X(t)$ of Example 6.7 for time period from zero to 1 second, with $A = 1$, $\omega_0 = 10\pi$ and $T = 0.02$ second (i.e., sampling rate is 50 Hz). Plot three of these sequences in one plot.
(b) Find the average value of the 100 sequences at $t = nT = 0.4$ second. Justify your result.

6.2 *Generation and testing of a random pulse train.*

(a) Write a MATLAB routine `random_pulse_w.m` to generate 100 discrete-time sequences $x[n]$ that are the sampled versions (with 10 Hz sampling rate) of the sample functions of the random process $X(t)$ of Example 6.2 for time period from zero to 5 seconds, with $A = 1$ and the period $T = 1$ second. Plot two of these sequences in two separate plots.
(b) Find the average value of the 100 sequences at $t = 1.3$ seconds. Justify your result.
(c) Use MATLAB routine `hist` to plot the histogram of the 100 sequences at $t = 1.3$ seconds using 10 bins. Justify your result.
(d) Use the companion software P&R to generate one sequence and plot the sequence as in part (a).

6.3 *Generation and testing of a random pulse train.* Consider a pulse train $X(t)$ with an amplitude A that is time-invariant and uniformly distributed over $(0, 1)$; that is, $A \sim \mathcal{U}(0, 1)$. The pulse has a constant width that is 0.5 second and the period is 1 second.

(a) Write a MATLAB routine `random_pulse_a.m` to generate 100 discrete-time sequences $x[n]$ that are sampled versions (with 10 Hz sampling rate) of the sample functions of the random pulse train $X(t)$ for time period from zero to 5 seconds. Plot two of these sequences in two separate plots.
(b) Find the average value of the 100 sequences at $t = 1.4$ seconds. Justify your result.
(c) Use MATLAB command `hist` to plot the histogram of the 100 sequences at $t = 1.4$ seconds using 10 bins. Justify your result.
(d) Use the companion software P&R to generate one sequence and plot the sequence as in part (a).

6.4 *Estimation of autocorrelation from a sample function.* The data file `m6_4.dat` in the companion software P&R contains a sequence. It is a sampled version of a sample function of an ergodic random process $X(t)$ with sampling interval T; that is, the random sequence is $X(T)$, $X(2T)$, $\ldots$, $X(nT)$.

(a) Write a MATLAB *function subroutine* `autocorr.m` to find the autocorrelation $R_x(mT)$ of $X(t)$ for an arbitrary m and a random sequence $X(T)$, $X(2T)$, ... , $X(nT)$ with a finite duration (n points). Justify your solution.
(b) Use routine `autocorr.m` to find $R_x(T)$, $R_x(3T)$ and $R_x(-2T)$ using the data file.
(c) Compare your results with those obtained by using P&R.

6.5 *Ergodicity check.* The data file `m6_5.dat` in the companion software P&R contains [a sequence that is a sampled version of] a sample function of the random process $G(t)$ of Example 6.11 with $\omega_1 = 5\pi, \omega_2 = 16\pi$. Use MATLAB to determine if $G(t)$ is ergodic using this data file. Justify your answer. Compare your results with those obtained by using P&R.

6.6 *Estimation of crosscorrelation from sample functions.* The data files `m6_6x.dat` and `m6_6y.dat` in the companion software P&R contains two sequences $x[n]$ and $y[n]$, respectively. They are the sampled versions of two sample functions of two random processes $X(t)$ and $Y(t)$, respectively, with sampling interval T.

(a) Write a MATLAB *function subroutine* `crosscorr_coeff.m` to estimate the cross-correlation coefficient $\rho_{xy}(kT)$ for an arbitrary m using random sequences $X(T)$, $X(2T)$, ... , $X(nT)$ and $Y(T)$, $Y(2T)$, ... , $Y(mT)$. Justify your solution.
(b) Use routine `crosscorr_coeff.m` to find $\rho_{xy}(T)$ and $\rho_{yx}(3T)$ using the data file.
(c) Compare your results with those obtained by using P&R.

6.7 *Correlation coefficient of a random sinusoid.* Consider a random sinusoidal signal $X(t) = A\sin(2\pi t + \phi)$, where A and ϕ are two independent RVs, with $A \sim \mathcal{U}(0, 2)$ and $\phi \sim \mathcal{U}(0, 2\pi)$.

(a) Find the theoretical correlation coefficient of $X(t)$.
(b) Use the companion software P&R to generate a 100-point sampled version of a sample function of $X(t)$ with 10 Hz sampling rate.
(c) Use P&R to compute the correlation coefficient of $X(t)$ based on the 100-point discrete-time sample function of $X(t)$ obtained in (b).
(d) Compare results of (a) and (c) and make comments.
(e) Repeat (b), (c), and (d) with 100 points replaced by 1000 points.

6.8 *Generation and testing of a Gaussian process.*

(a) Write a MATLAB routine `gauss_proc.m` to generate 100 discrete-time sequences $x[n]$ that are sampled versions (with 10 Hz sampling rate) of the sample functions of a Gaussian process with zero mean and autocorrelation $R_x(\tau) = e^{-15|\tau|}$ for time period from zero to one second. Plot three of these sequences in one plot.
(b) Find the average value of the 100 sequences at $t = 0.6$ second. Justify your result.
(c) Find the average autocorrelation over the 100 sequences for $\tau = 0.1$ second. Justify your result.

(d) Use MATLAB command `hist` to plot the histogram of the 100 sequences at $t = 0.6$ second using 8 bins. Justify your result.
(e) Use the companion software P&R to generate one sequence, plot the sequence as in part (a), and find the autocorrelation over the 100 sequences for $\tau < 0.1$ second. Compare the results with those in parts (a) and (c).

6.9 *Sample mean of a colored sequence of finite length.* Use the companion software P&R to do the following.

(a) Generate two discrete-time (with 10 Hz sampling rate and 100 time-points) sample functions of a zero-mean Gaussian process $X(t)$ with $R(\tau) = 4e^{-0.3|\tau|}$. Compute the sample means of the two sample functions.
(b) Generate two discrete-time (with 10 Hz sampling rate and 100 time-points) sample functions of a zero-mean Gaussian process $Y(t)$ with $R(\tau) = 4e^{-12|\tau|}$. Compute the sample means of the two sample functions.
(c) Generate two discrete-time (with 10 Hz sampling rate and 500 time-points) sample functions of the zero-mean Gaussian processes $X(t)$ and $Y(t)$ of parts (a) and (b), respectively. Compute the sample means of the two sample functions.
(d) Compare and discuss the results obtained in (a), (b), and (c).

6.14 Self-Test Problems

6.1 Answer the following questions briefly.

(a) Are sample functions of a random process deterministic or random?
(b) What is the probability $P\{X(t_1) = x_1(t), X(t_2) = x_2(t)\}$, where $x_1(t)$ and $x_2(t)$ are two sample functions of $X(t)$?
(c) A random process has a time-varying mean function and its autocorrelation depends only on the time difference. Is it stationary?
(d) If the values $X(t_1)$ and $X(t_2)$ of a random process $X(t)$ at two arbitrary and distinct time instants are uncorrelated, what is this random process called?
(e) Given an arbitrary sample function of a random process, is it true that its average over time is equal to the mean of the random process?
(f) Can a random process having different means at distinct times be ergodic?
(g) Is it true that the mean of a sinusoid with a random phase uniformly distributed over an interval is always equal to zero?
(h) Is it true that $R_x(\tau) \leq R_x(0)$ and $R_{xy}(\tau) \leq R_{xy}(0)$? Why or why not?
(i) Is it true that if $X(t_1)$ and $Y(t_1)$ are uncorrelated then $X(t)$ and $Y(t)$ are uncorrelated?

6.2 *Sinusoid with binary phase.* For Example 6.1, determine

(a) $f_{X(\pi/2)}(x)$, $f_{X(\pi)}(x)$, and $f_{X(\pi/2),X(\pi)}(x, y)$.
(b) $\bar{x}(\pi/2)$ and $\sigma^2_{x(\pi/2)}$.

(c) $R_x(\pi/4, \pi/2)$.

6.3 *Sinusoid with ternary amplitude.* Given a random process $X(t) = A\cos t$, where A is a discrete RV with the following probability masses

$$P\{A = 0\} = 1/2, \qquad P\{A = 1\} = 2/5, \qquad P\{A = -1\} = 1/10$$

(a) Find the mean function of $X(t)$.
(b) Find the autocorrelation function of $X(t)$.
(c) Is $X(t)$ wide-sense stationary? Is it ergodic?
(d) Find the marginal PDF and CDF of $X(t)$ for all t.

6.4 *Sinusoid with random phase.* Given a random process $X(t) = A\cos(\omega_0 t + \frac{1}{n}\phi)$ with $\phi \sim \mathcal{U}(0, 2m\pi)$ and A and ω_0 are nonrandom constants, where m and n are two arbitrary but fixed positive integers, find the mean and autocorrelation function of $X(t)$. Your results should be valid for all positive integers n and m.

6.5 *Product of random and deterministic processes.* Let $X(t)$ be a wide-sense stationary random process with mean $\bar{x}$ and autocorrelation $R_x(\tau)$ and $Y(t) = X(t)g(t)$, where $g(t)$ is a deterministic function.

(a) Find the mean, autocovariance, and autocorrelation of $Y(t)$.
(b) Is $Y(t)$ wide-sense stationary?

6.6 *Orthogonality of sine and cosine functions.* Consider two random processes $X(t)$ and $Y(t)$, defined by

$$X(t) = \sin(\omega_0 t + \phi)$$
$$Y(t) = \cos(\omega_0 t + \phi)$$

where ω_0 is nonrandom and $\phi \sim \mathcal{U}(-\pi, \pi)$.

(a) Find the average cross power $E[X(t)Y(t)]$.
(b) Are random variables $X(t_1)$ and $Y(t_1)$ orthogonal for a fixed t_1?
(c) Are random processes $X(t)$ and $Y(t)$ orthogonal?

6.15 Solutions to Self-Test Problems

6.1 (a) Deterministic because a sample function of a random process is the function given an outcome of the random experiment.
(b) The probability is equal to zero (an impossible event) because given an outcome of a random event, the *whole* time function is assigned to the random process. In other words, it is not allowed to change the time function halfway.
(c) No, the mean function of a stationary random process must be constant.

(d) It is said to be a white random process.
(e) It is true if and only if the random process is ergodic.
(f) No, ergodicity requires that the time average (which is a number) be equal to the mean. This would be impossible if there are two values of the mean.
(g) No. The mean of a sinusoid $\sin(\omega_0 t + \phi)$ with a random phase $\phi \sim \mathcal{U}(\theta_0, \theta_0 + 2\pi)$ (i.e., uniformly distributed over an interval of length 2π) is always zero. The same is true if $\phi \sim \mathcal{U}(\theta_0, \theta_0 + 2\pi/n)$. For other cases, it is better to find the mean by the formula. See problem 6.12 and self-test problem 6.4.
(h) It is always true that $R_x(\tau) \le R_x(0)$ because a random process is "most similar" to itself (i.e., without time shift $\tau = 0$). For crosscorrelation, however, it is *not* always true that $R_{xy}(\tau) \le R_{xy}(0)$. For example, $X(t)$ and $Y(t) = X(t+3)$ are "most similar" when $\tau = 3$; that is, $R_{xy}(\tau) \le R_{xy}(3)$ because $R_{xy}(3) = E[X(t)Y(t-3)] = E[X^2(t)]$.
(i) No, $X(t)$ and $Y(t)$ are uncorrelated if and only if $X(t_1)$ and $Y(t_2)$ are uncorrelated for every t_1 and t_2; that is, for the same time and for different time.

6.2 (a) Since

t	$x_1(t)$	$x_2(t)$
$\pi/4$	0.707	−0.707
$\pi/2$	0	0
π	−1	1
$P\{X(t) = x_i(t)\}$	1/2	1/2

we have

$$f_{X(\pi/2)}(x) = \frac{1}{2}\delta(x) + \frac{1}{2}\delta(x) = \delta(x)$$
$$f_{X(\pi)}(x) = \frac{1}{2}\delta(x+1) + \frac{1}{2}\delta(x-1)$$
$$f_{X(\pi/2)X(\pi)}(x,y) = \underbrace{\frac{1}{2}\delta(x)\delta(y+1)}_{X(t)=x_1(t)} + \underbrace{\frac{1}{2}\delta(x)\delta(y-1)}_{X(t)=x_2(t)}$$

(b)

$$\bar{x}(\pi/2) = \sum_{i=1}^{2} x_i(\pi/2)P\{X(\pi/2) = x_i(\pi/2)\} = 0 \cdot \frac{1}{2} + 0 \cdot \frac{1}{2} = 0$$
$$E[X^2(\pi/2)] = \sum_{i=1}^{2} x_i^2(\pi/2)P\{X(\pi/2) = x_i(\pi/2)\} = 0^2 \cdot \frac{1}{2} + 0^2 \cdot \frac{1}{2} = 0$$
$$\sigma^2_{x(\pi/2)} = E[X^2(\pi/2)] - [\bar{x}(\pi/2)]^2 = 0 - 0 = 0$$

Actually, we know $X(\pi/2) = 0$ for sure, which agrees with $\sigma^2_{x(\pi/2)} = 0$.

(c)

$$\begin{aligned}
R_x(\pi/4, \pi/2) &= E[X(\pi/4)X(\pi/2)] \\
&= \sum_{i=1}^{2}\sum_{j=1}^{2} x_i(\pi/4)x_j(\pi/2)P\{X(\pi/4) = x_i(\pi/4), X(\pi/2) = x_j(\pi/2)\} \\
&= \sum_{i=1}^{2} x_i(\pi/4)x_i(\pi/2)P\{X(\pi/4) = x_i(\pi/4), X(\pi/2) = x_i(\pi/2)\} \\
&= \sum_{i=1}^{2} x_i(\pi/4)x_i(\pi/2)P\{X(\pi/4) = x_i(\pi/4)\} \\
&= 0.707 \cdot 0 \cdot \frac{1}{2} + (-0.707) \cdot 0 \cdot \frac{1}{2} \\
&= 0
\end{aligned}$$

where the double summation is reduced to single summation since $X(t)$ cannot be *both* $x_1(t)$ and $x_2(t)$. In other words, $X(t) = x_1(t)$ and $X(t') = x_2(t')$ cannot be both true. Note that $R_x(\pi/4, \pi/2) = 0$ agrees with the fact that $X(\pi/2) = 0$ for sure, since $X(\pi/2) = 0$ implies that it is not similar to (correlated with) any other value. For this reason, the probability should better be replaced by $P\{X(t) = x_i(t)\}$.

6.3 (a) $E[X(t)] = (2/5)\cos t - (1/10)\cos t = 0.3\cos t$.

(b) Since $E[A^2] = (0)(2/5) + (1^2)(2/5) + (-1)^2(1/10) = 0.5$,

$$R(t+\tau, t) = E[A\cos(t+\tau)A\cos t] = E[A^2]\cos(t+\tau)\cos t = 0.5\cos(t+\tau)\cos t$$

(c) Since $E[X(t)]$ depends on time, $X(t)$ is not stationary and thus not ergodic, either.

(d) The marginal PDF and CDF of $X(t)$ are

$$\begin{aligned}
f_{X(t)}(x) &= \frac{1}{2}\delta(x) + \frac{2}{5}\delta(x - \cos t) + \frac{1}{10}\delta(x + \cos t) \\
F_{X(t)}(x) &= \frac{1}{2}u(x) + \frac{2}{5}u(x - \cos t) + \frac{1}{10}u(x + \cos t)
\end{aligned}$$

6.4 (a)

$$\begin{aligned}
\bar{x}(t) &= E[X(t)] \\
&= \int_0^{2m\pi} A\cos(\omega_0 t + \phi/n)\frac{1}{2m\pi}d\phi \\
&= A\frac{n}{2m\pi}\sin(\omega_0 t + \frac{\phi}{n})\Big|_0^{2m\pi} \\
&= A\frac{n}{2m\pi}\left[\sin\left(\omega_0 t + \frac{2m\pi}{n}\right) - \sin(\omega_0 t)\right] \\
&= \begin{cases} A\frac{n}{2m\pi}[\sin(\omega_0 t + \frac{2m\pi}{n}) - \sin(\omega_0 t)] & \frac{m}{n} \neq \text{integer} \\ 0 & \frac{m}{n} = \text{integer} \end{cases}
\end{aligned}$$

(b)

$$R_x(t+\tau,t) = E\Big[A^2\cos\Big(\omega_0 t+\omega_0\tau+\frac{\phi}{n}\Big)\cos\Big(\omega_0 t+\frac{\phi}{n}\Big)\Big]$$
$$= \frac{A^2}{2}E\Big[\cos(\omega_0\tau)+\cos\Big(2\omega_0 t+\omega_0\tau+\frac{2\phi}{n}\Big)\Big]$$
$$= \frac{A^2}{2}\Big[\cos(\omega_0\tau)+\int_0^{2m\pi}\cos\Big(2\omega_0 t+\omega_0\tau+\frac{2\phi}{n}\Big)\frac{1}{2m\pi}d\phi\Big]$$
$$= \frac{A^2}{2}\Big[\cos(\omega_0\tau)+\frac{n/2}{2m\pi}\sin\Big(2\omega_0 t+\omega_0\tau+\frac{2\phi}{n}\Big)\Big|_0^{2m\pi}\Big]$$
$$= \frac{A^2}{2}\Big[\cos(\omega_0\tau)+\frac{n}{4m\pi}\Big(\sin(2\omega_0 t+\omega_0\tau+\frac{4m\pi}{n})-\sin(2\omega_0 t+\omega_0\tau)\Big)\Big]$$
$$= \begin{cases} \frac{A^2}{2}\cos(\omega_0\tau) & \frac{2m}{n} = \text{integer} \\ \frac{A^2}{2}\Big[\cos(\omega_0\tau)+\frac{n}{4m\pi}\Big(\sin[\omega_0(2t+\tau)+\frac{4m\pi}{n}]-\sin\omega_0(2t+\tau)\Big)\Big] & \frac{2m}{n} \neq \text{integer}\end{cases}$$

6.5 (a)

$$E[Y(t)] = E[X(t)g(t)] = E[X(t)]g(t) = \bar{x}g(t)$$
$$\begin{aligned} C_y(t+\tau,t) &= E[(X(t+\tau)g(t+\tau)-\bar{x}g(t+\tau))(X(t)g(t)-\bar{x}g(t))] \\ &= E[(X(t+\tau)-\bar{x})g(t+\tau)(X(t)-\bar{x})g(t)] \\ &= E[(X(t+\tau)-\bar{x})(X(t)-\bar{x})]g(t+\tau)g(t) \\ &= C_x(\tau)g(t+\tau)g(t) \end{aligned}$$
$$R_y(t+\tau,t) = C_y(t+\tau,t)+\bar{y}(t+\tau)\bar{y}(t) = [C_x(t)+(\bar{x})^2]g(t+\tau)g(t)$$

(b) Clearly, $Y(t)$ is not wide-sense stationary (even though $X(t)$ is stationary) since $C_y(t+\tau,t)$ depends on t as well as τ.

6.6 (a)

$$E[X(t)Y(t)] = E[\cos(\omega_0 t+\phi)\sin(\omega_0 t+\phi)] = \frac{1}{2}\int_{-\pi}^{\pi}\sin(2\omega_0 t+2\phi)\frac{1}{2\pi}d\phi = 0$$

(b) From part (a), $E[X(t_1)Y(t_1)] = 0 \implies X(t_1)\perp Y(t_1), \quad \forall t_1$

(c) Since

$$\begin{aligned} E[X(t_1)Y(t_2)] &= E\left[\sin(\omega_0 t_1+\phi)\cos(\omega_0 t_2+\phi)\right] \\ &= \frac{1}{2}\{E[\sin\omega_0(t_1-t_2)]+\underbrace{E[\sin(\omega_0 t_1+\omega_0 t_2+2\phi)]}_{=0}\} \\ &= \frac{1}{2}\sin[\omega_0(t_1-t_2)] \\ &= \frac{1}{2}\sin(\omega_0\tau) \end{aligned}$$

is not always equal to zero, $X(t)$ and $Y(t)$ are not orthogonal.

7

POWER SPECTRAL DENSITY

The "spectre" of power spectra is a vivid presence rather than a ghostly apparition in engineering and science.

Author

Frequency-domain methods, such as Fourier transform, are useful in science and engineering. This chapter covers the characterization of random processes in frequency domain.

Main Topics

- Concept of Power Spectral Density
- Properties of Power Spectral Density
- White Noise
- Power Spectrum Estimation
- Cross-Power Spectrum and Its Properties

7.1 Concept of Power Spectral Density

Frequency-domain tools, such as Fourier series and Fourier transform, are widely used in engineering.

The ***Fourier transform*** of a function (signal) $f(t)$ is defined by

$$F(\omega) = \int_{-\infty}^{\infty} f(t)e^{-j\omega t}dt = \mathcal{F}[f(t)] \tag{7.1}$$

and is called the (*amplitude*) ***spectrum*** (or ***spectral density***) of $f(t)$. It describes the distribution of its relative amplitude strength with respect to frequency. $F(\omega)$ and $f(t)$ are a Fourier transform pair. They contain the same information. $f(t)$ can be recovered from $F(\omega)$ by inverse Fourier transform:

$$f(t) = \mathcal{F}^{-1}[F(\omega)] = \frac{1}{2\pi}\int_{-\infty}^{\infty} F(\omega)e^{j\omega t}d\omega \tag{7.2}$$

Fourier transform has many nice properties, listed in Table 7.4. For signal and system analysis, the most important one is the convolution property that time-domain convolution amounts to frequency-domain multiplication.

Example 7.1: Fourier Transform of a Pulse

The Fourier transform of $f_1(t) = \left\{ \begin{array}{ll} A & 0 \le t \le T \\ 0 & \text{elsewhere} \end{array} \right\}$ is

$$F_1(\omega) = AT\frac{\sin(\omega T/2)}{\omega T/2}e^{-j\omega T/2} \tag{7.3}$$

The most prevalent frequency component of $f_1(t)$ is dc, which makes sense. There are no $\pm 2\pi/T$, $\pm 4\pi/T$, ... components in the pulse.

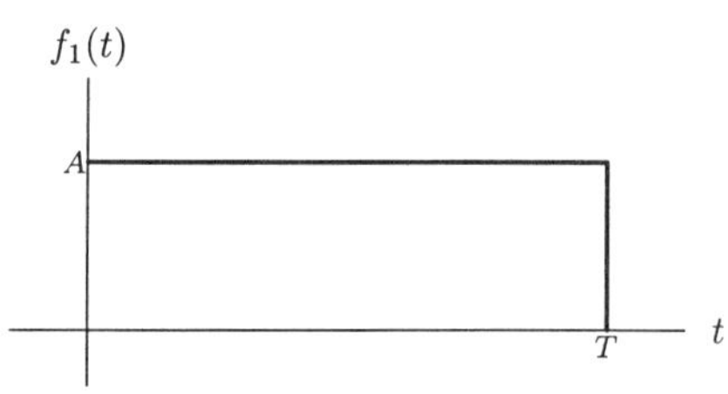

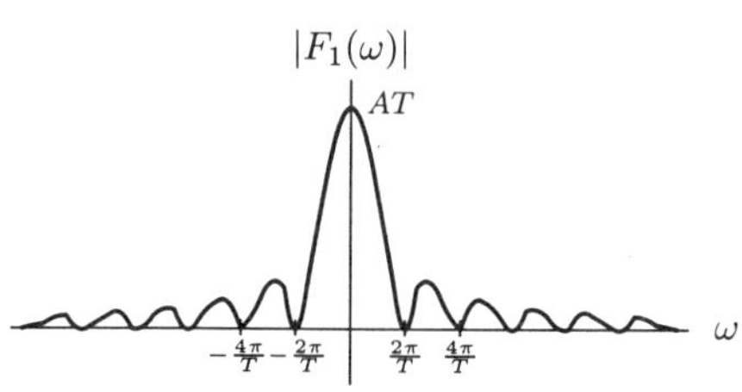

As we have learned, Fourier series, Fourier transform, and Laplace transform are powerful tools for electric circuits, signal and system analysis, solving differential equations, etc. They simplify greatly certain problems that may be complex in time domain.

The English word "spectrum" came from the Latin word "spectre."

Fourier series

$$f(t) = \sum_{n=-\infty}^{\infty} \alpha_n e^{jn\omega_0 t} = \sum_{n=-\infty}^{\infty} F(n\omega_0) e^{jn\omega_0 t}$$

is valid only for a *periodic* function (waveform) with a period $T = 2\pi/\omega_0$. It gives distribution of the amplitudes of *discrete* (fundamental and harmonic) frequency components. An aperiodic function, however, does not have a Fourier series expansion. A possible solution is to use Fourier transform $f(t) = \int_{-\infty}^{\infty} F(\omega) e^{j\omega t} \frac{d\omega}{2\pi}$. Comparing with the Fourier series coefficients $\alpha_n = F(n\omega_0), -\infty < n < \infty$, it can be seen that the Fourier transform $F(\omega)$ provides a description of the distribution of the amplitudes over a *continuous* frequency spectrum. Note, however, the difference between Fourier series and Fourier transform: $\alpha_n \neq 0$ implies that $f(t)$ has a sinusoidal component with frequency $n\omega_0$ (i.e., nth harmonic), while $F(\omega_1) \neq 0$ does not imply that $f(t)$ has a sinusoidal component with frequency ω_1. It simply means that there is an *integral* part in $f(t)$ that has frequency ω_1. For instance, consider Example 7.1. Although $|F_1(\omega)|$ has the peak at $\omega = 0$, the signal $f_1(t)$ actually has no dc term since its average over the entire time horizon is zero.

With the help of impulse (delta) functions, the (extended) Fourier transform is valid for periodic as well as aperiodic functions because the discrete frequency components can be represented by the delta functions. As a result, Fourier transform covers Fourier series as a special case.

(7.3) can be obtained as follows

$$\begin{aligned}
F_1(\omega) &= \int_0^T A e^{-j\omega t} dt \\
&= A \frac{1}{-j\omega} e^{-j\omega t} \Big|_0^T \\
&= A \frac{1 - e^{-j\omega T}}{j\omega} \\
&= 2A e^{-j\omega T/2} \frac{e^{j\omega T/2} - e^{-j\omega T/2}}{2j\omega} \\
&= \frac{2}{\omega} A e^{-j\omega T/2} \sin(\omega T/2)
\end{aligned}$$

Note that the height of the main lobe in $F_1(\omega)$ is proportional to A and T, which makes good sense: As A or T increases, the pulse has more significant "dc term."

Example 7.2: Fourier Transform of a Symmetric Pulse

The Fourier transform of

$$f_2(t) = \begin{cases} A & |t| \leq T/2 \\ 0 & \text{elsewhere} \end{cases}$$

is

$$F_2(\omega) = AT\frac{\sin(\omega T/2)}{\omega T/2}$$

Note that $f_2(t)$ is $f_1(t)$ time shifted by $-T/2$. Clearly, the difference between $F_1(\omega)$ of Example 7.1 and $F_2(\omega)$ of this example makes sense from the time-shifting property of Fourier transform (see Table 7.4) — a shift in time corresponds to a phase difference only (i.e., the amplitude does not change):

$$f_2(t) = f_1(t - T/2) \qquad \Longleftrightarrow \qquad \begin{cases} |F_2(\omega)| = |F_1(\omega)| \\ \angle F_2(\omega) = \angle F_1(\omega) + \omega T/2 \end{cases}$$

A prism performs a Fourier transform *naturally*. This is not surprising because everybody knows that if white light enters into a prism, what comes out is a complete spectrum of colored light, ranging from red to violet, as shown below.

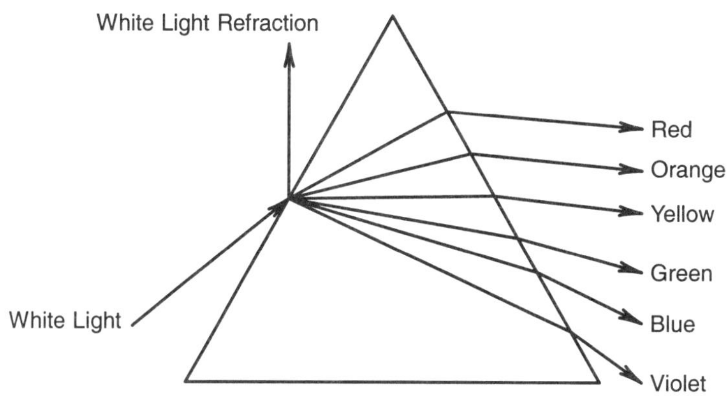

Figure 7.1: Fourier transform of white light by a prism.

Not every function has a Fourier transform. The following conditions are sufficient (but not necessary) for the existence of the Fourier transform of a function $f(t)$:

- *Dirichlet conditions*: $f(t)$ has *bounded variation* meaning that it is bounded with at most a finite number of maxima and minima and a finite number of discontinuities in any finite period;
- $f(t)$ is *absolute integrable*; that is, $\int_{-\infty}^{\infty} |f(t)|dt < \infty$.

Consider naturally the extension of Fourier transform to a random process. A direct extension to a random process $X(t)$ itself would not work: not only would its Fourier transform $\mathcal{F}[X(t)]$ be random and clumsy to use, but also it would exist only if the Fourier transforms of all its sample functions exist, which is almost never the case. A solution is to consider instead the (nonrandom) average power of $X(t)$, of which the Fourier transform usually exists.

The *average power* of a deterministic function $x(t)$ is defined by

$$p_x \triangleq \lim_{T\to\infty} \frac{1}{2T} \int_{-T}^{T} x^2(t)dt \tag{7.4}$$

which can be viewed as an abstraction of average power dissipation of the current $i(t)$ through (or voltage $v(t)$ across) a 1Ω resistor:

$$\text{power} \triangleq \lim_{T\to\infty} \frac{1}{2T} \int_{-T}^{T} Ri^2(t)dt = \lim_{T\to\infty} \frac{1}{2T} \int_{-T}^{T} \frac{v^2(t)}{R}dt$$

The ***average power*** of a random process $X(t)$ is similarly defined by

$$P_x \triangleq E\Big[\lim_{T\to\infty} \frac{1}{2T} \int_{-T}^{T} X^2(t)dt\Big] = \lim_{T\to\infty} \frac{1}{2T} \int_{-T}^{T} E\left[X^2(t)\right] dt \tag{7.5}$$

which is the *time average of the mean-square value*. For stationary $X(t)$, the *average power is the time-invariant mean-square value*:

$$P_x = \lim_{T\to\infty} \frac{1}{2T} \int_{-T}^{T} R_x(0)dt = R_x(0) = E(X^2) \tag{7.6}$$

The ***power spectral density*** (***PSD***) or ***power spectrum*** of a random process $X(t)$ is defined as the Fourier transform of its average power. If $X(t)$ is stationary, it turns out to be (see next page) the Fourier transform of its autocorrelation:

$$\boxed{S_x(\omega) = \mathcal{F}[R_x(\tau)] \triangleq \int_{-\infty}^{\infty} R_x(\tau)e^{-j\omega\tau}d\tau} \tag{7.7}$$

Power spectrum is a measure of how average power is distributed with respect to frequency. $S_x(\omega)$ and $R_x(\tau)$ are a Fourier transform pair:

$$\boxed{R_x(\tau) = \mathcal{F}^{-1}[S_x(\omega)] = \frac{1}{2\pi} \int_{-\infty}^{\infty} S_x(\omega)e^{j\omega\tau}d\omega} \tag{7.8}$$

Average power equals the integral of power spectrum over all frequencies:

$$P_x = R_x(\tau)\Big|_{\tau=0} = \int_{-\infty}^{\infty} S_x(\omega)e^{j\omega\tau}\frac{d\omega}{2\pi}\Big|_{\tau=0} = \frac{1}{2\pi}\int_{-\infty}^{\infty} S_x(\omega)d\omega \tag{7.9}$$

A random process is actually a collection of sample functions in which exactly which one is picked depends on the outcome of the underlying random experiment. As such, the Fourier transform of a random process would be the weighted sum of the Fourier transforms of these sample functions, where the weights are the probabilities of the corresponding outcomes. Consequently, the Fourier transform of a random process would not exist provided at least one of the sample functions does not have a Fourier transform.

At first glance, since Laplace transform exists for a larger class of functions than Fourier transform, the existence problem of the Fourier transform of a random process may be avoided or alleviated by using the (two-sided) Laplace transform. It turns out, however, that the corresponding inverse transform may suffer from a similar existence problem.

In plain words, (7.5) is

$$\text{average power} = \text{time average of mean-square value}$$

It was stated in Chapter 6 that the mean of an ergodic random process is its ***dc component***; the mean-square value is its ***average power***; the variance is the ***average power*** of its non-dc component; and the standard deviation is the ***effective value*** of its non-dc component. This can be explained as follows. Since the mean of an ergodic process is its constant part, it is the dc component, which has power $\bar{x}^2$; its average power is by (7.6) the mean-square value; its variance $\sigma^2 = E[X^2] - \bar{x}^2 =$ average power – dc power and is thus the power of its non-dc component; since the effective value of a signal is the square root of its power, the standard deviation is the effective value of its non-dc component.

From circuit analysis, $Ri^2(t)$ and $v^2(t)/R$ are the instantaneous powers consumed in a resistor with R resistance. For an abstract signal or function $x(t)$, it is thus reasonable to define the quantity $|x(t)|^2$ as its ***instantaneous power***. Accordingly, the ***total energy*** of a (nonrandom) function $x(t)$ with a finite nonzero duration $[-T, T]$ is defined by

$$E_x = \int_{-\infty}^{\infty} |x(t)|^2 dt = \int_{-T}^{T} |x(t)|^2 dt$$

Consequently, it is natural to define the average power by (7.4). By the *Parseval theorem* (also known as the *Rayleigh energy theorem*) for Fourier transforms (see Table 7.4),

$$E_x = \int_{-\infty}^{\infty} |x(t)|^2 dt = \int_{-\infty}^{\infty} |F_x(\omega)|^2 \frac{d\omega}{2\pi} \tag{7.10}$$

where $F_x(\omega)$ is the Fourier transform of $x(t)$, we may view $|F_x(\omega)|^2$ as the ***energy spectral density*** and thus $\dfrac{|F_x(\omega)|^2}{2T}$ as the *average* power spectral density. With this understanding, (7.10) simply means that the energy in time domain and in frequency domain are equal. To be valid for functions with an infinite duration, we may thus define the ***power spectral density*** via limit by

$$s_x(\omega) = \lim_{T\to\infty} \frac{|F_x(\omega)|^2}{2T} \tag{7.11}$$

Since the time-correlation property (see Table 7.4) of Fourier transforms states that the magnitude square of the Fourier transform of a time function is the Fourier transform of the autocorrelation of the function, this definition is equivalent to the following

$$\begin{aligned} s_x(\omega) &= \lim_{T\to\infty} \frac{1}{2T} \mathcal{F}\left[\int_{-T}^{T} x(t+\tau)x(t)dt \right] \\ &= \mathcal{F}\left[\lim_{T\to\infty} \frac{1}{2T} \int_{-T}^{T} x(t+\tau)x(t)dt \right] \\ &= \mathcal{F}\{A\langle x(t+\tau)x(t)\rangle\} \end{aligned}$$

where $A\langle g(t)\rangle \triangleq \lim_{T\to\infty} \frac{1}{2T} \int_{-T}^{T} g(t)dt$ is the ***time average*** of $g(t)$.

For a random process $X(t)$, the above can be thought to be valid for each sample function and its Fourier transform is random, depending on which sample function is the true one. As a result, the (average) power spectral density may be defined by

$$\begin{aligned} S_x(\omega) &= E\left[\lim_{T\to\infty} \frac{|F_X(\omega)|^2}{2T} \right] \\ &= \lim_{T\to\infty} \frac{E[|F_X(\omega)|^2]}{2T} \qquad (7.12) \\ &= \mathcal{F}\{A\langle E[x(t+\tau)x(t)]\rangle\} = \mathcal{F}\{A\langle R_x(t+\tau,t)\rangle\} \\ &= \mathcal{F}\left[\lim_{T\to\infty} \frac{1}{2T} \int_{-T}^{T} R_x(t+\tau,t)dt \right] \qquad (7.13) \end{aligned}$$

That is, it turns out that ***power spectrum*** *is the Fourier transform of the time average of the autocorrelation function*:

$$\begin{aligned} S_x(\omega) &= \mathcal{F}[\text{time average of } R_x(t+\tau,t)] \\ \text{time average of } R_x(t+\tau,t) &= \mathcal{F}^{-1}[S_x(\omega)] \end{aligned}$$

Clearly, this relation, known as ***Wiener-Khinchine theorem***, is more general than the one given by (7.7) because it is valid for nonstationary as well as stationary processes. It clearly reduces to (7.7) for wide-sense stationary processes. From the above discussion, it is clear that $S_x(\omega)$ thus defined deserves the name "*power spectral density*." Similarly, $|F_X(\omega)|^2$ can be justified as the ***energy spectral density***.

In the above, $F_X(\omega)$ is the Fourier transform of the truncated version of $X(t)$ with a finite duration $[-T, T]$, which usually exists. Note that the Fourier transform of the untruncated $X(t)$ usually does not exist since its existence requires the existence of Fourier transforms of all the sample functions of $X(t)$. That is why we take Fourier transform before taking limit and not the opposite.

Unlike Fourier transform of a deterministic function, *power spectrum carries no phase (time shift) information* about the random process because "power" does not depend on time shift (phase).

Stronger results, e.g., (7.17) of the next section, than (7.9) hold for the relation between the power spectrum and the average power.

Example 7.3: Power of a Signal with Random Phase and DC Bias

Determine the power spectrum $S_x(\omega)$ and average power P_x of the random process $X(t) = A + B\cos(\omega_0 t + \theta)$, where $\theta \sim \mathcal{U}(0, 2\pi)$, A, B, are independent RVs, and $\omega_0 > 0$ is a constant. Note that

$$\begin{aligned} R_x(t+\tau, t) &= E\{[A + B\cos(\omega_0 t + \omega_0\tau + \theta)][A + B\cos(\omega_0 t + \theta)]\} \\ &= \overline{A^2} + \bar{A}\bar{B}\{\underbrace{E[\cos(\omega_0 t + \theta)]}_{=0} + \underbrace{E[\cos(\omega_0 t + \omega_0\tau + \theta)]}_{=0}\} \\ &\quad + \overline{B^2}E[\cos(\omega_0 t + \theta)\cos(\omega_0 t + \omega_0\tau + \theta)] \\ &= \overline{A^2} + \frac{1}{2}\overline{B^2}E[\cos(\omega_0\tau) + \cos(2\omega_0 t + \omega_0\tau + 2\theta)] \\ &= \overline{A^2} + \frac{1}{2}\overline{B^2}\cos(\omega_0\tau) = R_x(\tau) \end{aligned} \tag{7.14}$$

We have, from Example 6.11,

$$\begin{aligned} P_x &= R_x(0) = \overline{A^2} + \frac{1}{2}\overline{B^2} = \text{dc power} + \text{ac power} \\ S_x(\omega) &= \mathcal{F}[R_x(\tau)] = \overline{A^2}\mathcal{F}[1] + \frac{1}{2}\overline{B^2}\mathcal{F}[\cos(\omega_0\tau)] \\ &\overset{\text{Table 7.5}}{=} 2\pi\overline{A^2}\delta(\omega) + \frac{1}{2}\pi\overline{B^2}[\delta(\omega - \omega_0) + \delta(\omega + \omega_0)] \end{aligned}$$

Alternatively,

$$P_x = \frac{1}{2\pi}\int_{-\infty}^{\infty} S_x(\omega)d\omega \overset{(3.15)}{=} \frac{1}{2\pi}[2\pi\overline{A^2} + \frac{\pi}{2}\overline{B^2} + \frac{\pi}{2}\overline{B^2}] = \overline{A^2} + \frac{1}{2}\overline{B^2}$$

Note that a dc term in $X(t)$ corresponds to a $\delta(\omega)$ term in $S(\omega)$ and a sinusoid with $\omega = \omega_0$ in $X(t)$ corresponds to a pair of $\delta(\omega - \omega_0)$ and $\delta(\omega + \omega_0)$.

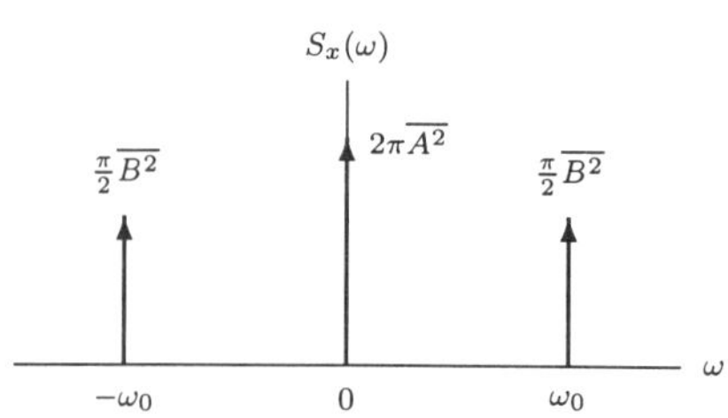

7.2 Properties of Power Spectral Density

Since $S(\omega)$ is a Fourier transform, it has all properties of Fourier transforms, such as those listed in Table 7.4. Furthermore,

1. $S(\omega)$ is *real-valued* and *nonnegative*: $\boxed{S(\omega) \geq 0}$, which can be seen intuitively from (7.12) since the expected value of a nonnegative quantity is nonnegative. This is consistent with that $S(\omega)$ is a density function.
2. $S_x(\omega)$ is *even* in ω:

$$\begin{aligned} S(-\omega) &= \int_{-\infty}^{\infty} R(\tau)e^{-j(-\omega)\tau}d\tau \overset{t=-\tau}{=} \int_{\infty}^{-\infty} R(-t)e^{-j\omega t}d(-t) \\ &\overset{(6.16)}{=} \int_{-\infty}^{\infty} R(t)e^{-j\omega t}dt = S(\omega) \end{aligned} \tag{7.15}$$

3. *Average power* is equal to the integral of $S(\omega)$ [i.e., area underneath $S(\omega)$]:

$$\boxed{P_x \overset{(7.9)}{=} \int_{-\infty}^{\infty} S_x(\omega)\frac{d\omega}{2\pi} \overset{(7.15)}{=} 2\int_{0}^{\infty} S_x(\omega)\frac{d\omega}{2\pi}} \tag{7.16}$$

Average power in the frequency band $[\omega_1, \omega_2]$ for $\omega_2 \geq \omega_1 \geq 0$ is given by

$$\boxed{P_{x[\omega_1,\omega_2]} = \frac{1}{\pi}\int_{\omega_1}^{\omega_2} S_x(\omega)d\omega = 2\int_{\omega_1}^{\omega_2} S_x(\omega)\frac{d\omega}{2\pi}} \tag{7.17}$$

If frequency f instead of angular frequency ω is used, then P_x is the area under $S(\omega)$ over $-\infty < f < \infty$ and $P_{x[f_1,f_2]}$ is the area under $S(\omega)$ over $[-f_2, -f_1]$ and $[f_1, f_2]$ (or *two times* the area under $S(\omega)$ over $[f_1, f_2]$). In (7.16)–(7.17), negative frequency spectrum has been folded into the positive frequency spectrum by the even symmetry of $S(\omega)$ since negative frequency has no direct physical meaning.

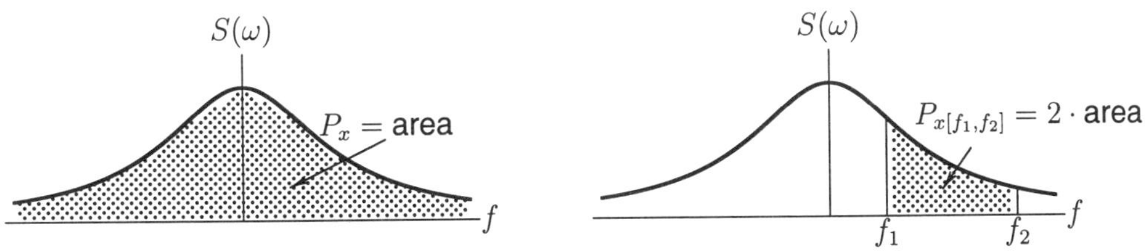

Figure 7.2: Relation between average power and power spectrum.

Property 2 is valid only if $X(t)$ is real-valued because it relies on $R_x(-\tau) = R_x(\tau)$. Thus,

$$S_x(\omega) = 2\int_0^{\infty} R_x(\tau)\cos(\omega\tau)d\tau$$

It follows easily from the fact that for a real-valued random process, $R_x(\tau) = R_x(-\tau)$ and $e^{-j\omega\tau} = \cos(\omega\tau) - j\sin(\omega\tau)$. For complex-valued random processes, it is easy to check that $R_x(-\tau) \neq R_x(\tau)$ but $R_x(-\tau) = R_x^*(\tau)$.

(7.17) will be shown in Example 8.6. Caution should be taken when using it when $S(\omega)$ contains a $\delta(\omega)$. (In this case, only half of the impulse should be counted in the integral.)

Table 7.1: Additional properties of autocorrelation and power spectrum.

Random Process	*Autocorrelation*	*Power Spectrum*
$X(t)$	$R_x(\tau)$	$S_x(\omega)$
$\alpha X(t)$	$\lvert\alpha\rvert^2 R_x(\tau)$	$\lvert\alpha\rvert^2 S_x(\omega)$
$X(t)+b$ with $E[X(t)]=0$	$R_x(\tau)+\lvert b\rvert^2$	$S_x(\omega)+2\pi\lvert b\rvert^2\delta(\omega)$
$X_1(t)+X_2(t)$ with $X_1(t)\perp X_2(t)$	$R_{X_1}(\tau)+R_{X_2}(\tau)$	$S_{X_1}(\omega)+S_{X_2}(\omega)$
$\dfrac{d^n X(t)}{dt^n}$	$(-1)^n\dfrac{d^{2n}R_x(\tau)}{d\tau^{2n}}$	$\omega^{2n}S_x(\omega)$
$X(t)e^{j\omega_0 t}$ with nonrandom ω_0	$R_x(\tau)e^{j\omega_0\tau}$	$S_x(\omega-\omega_0)$
$X(t)\cos(\omega_0 t+\theta)$ with $\theta\sim\mathcal{U}(0,2\pi)$	$R_x(\tau)\cos(\omega_0\tau)$	$\frac{1}{2}[S_x(\omega+\omega_0)+S_x(\omega-\omega_0)]$

Some additional properties are given in Table 7.1. They are interpreted as follows.

- The first property is consistent with the interpretation that autocorrelation and power spectrum are quantities associated directly with power (squared value) rather than amplitude of the process.
- The second property indicates that a constant term in the process is a dc ($\omega = 0$) term.
- The third property indicates that if $X_1(t)$ and $X_2(t)$ are orthogonal, the power of their sum is simply the sum of their powers. Note that uncorrelatedness or even independence is not sufficient here because the power of their sum would not be the sum of their powers if they both have a dc component.
- The differentiation property can be obtained by a repeated application of (6.23) and the differentiation property of Fourier transforms. As a special case, the power spectrum of the derivative $\dot{X}(t) = dX(t)/dt$ (if it exists) is ω^2 times that of $X(t)$:

$$S_{\dot{x}}(\omega) = \omega^2 S_x(\omega) \tag{7.18}$$

  This follows from (6.23) or (6.37) with $Y(t) = X(t)$ and the differentiation property of the Fourier transform.
- The second last property implies that we can treat $e^{j\omega_0 t}$ as a phase-shift factor, exactly the same as in systems applications and signal processing.
- The last property, known as the *modulation property*, is very useful in communications.

Table 7.2: Some popular autocorrelations and power spectra.

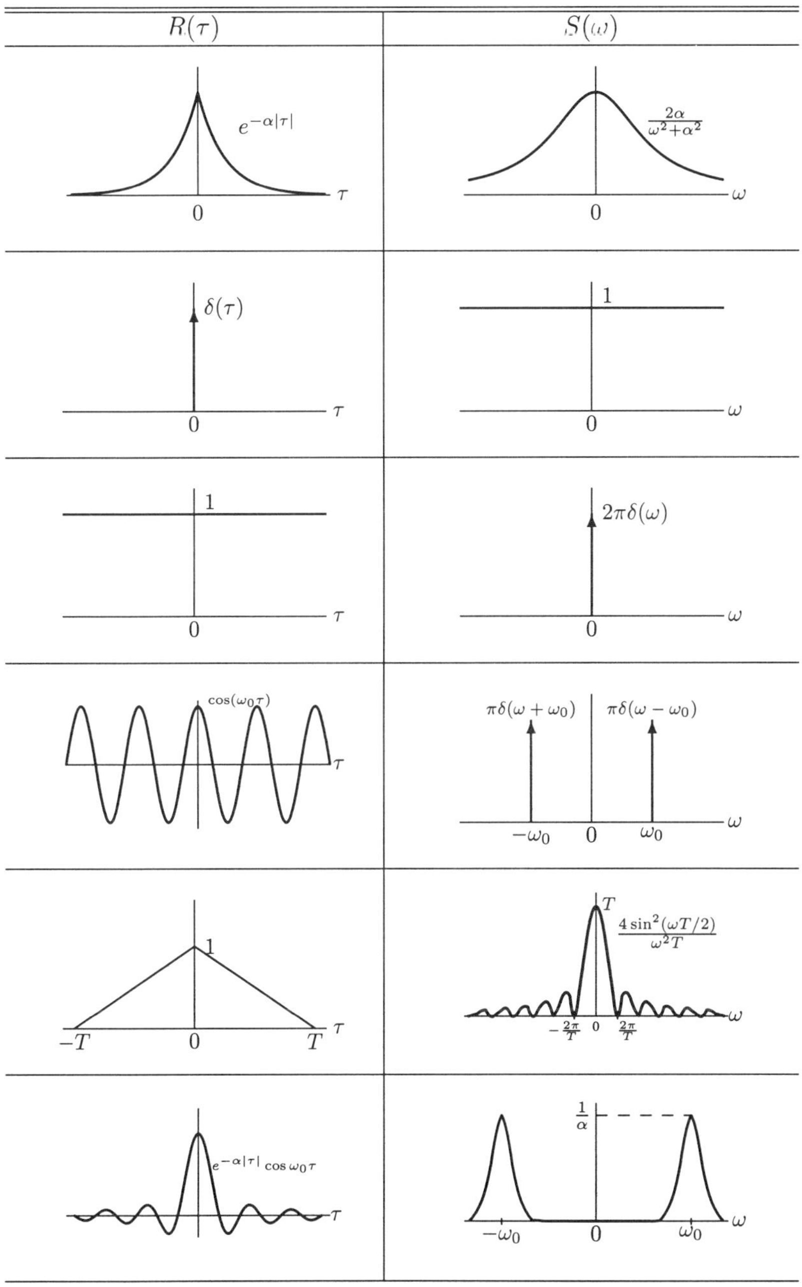

Example 7.4: Determination of Various Power from Power Spectrum

A wide-sense stationary process $X(t)$ has power spectrum $S(\omega) = \frac{4}{\omega^2+3}$.

(a) Find the average power of $X(t)$:

$$\begin{aligned} P_x &= R(0) = \mathcal{F}^{-1}\Big[\frac{4}{\omega^2+3}\Big]\Big|_{\tau=0} = \frac{4}{2\sqrt{3}}\mathcal{F}^{-1}\Big[\frac{(2)(\sqrt{3})}{\omega^2+(\sqrt{3})^2}\Big]\Big|_{\tau=0} \\ &\overset{\text{Table 7.5}}{=} \frac{2}{\sqrt{3}}e^{-\sqrt{3}|\tau|}\Big|_{\tau=0} = \frac{2}{\sqrt{3}} \end{aligned}$$

(b) Find the average power over frequency band $0 < \omega < 3$: Since $\frac{d}{dx}\tan^{-1} x = \frac{1}{x^2+1}$, we have

$$\begin{aligned} P_{x(0,3)} &\overset{(7.17)}{=} \frac{1}{\pi}\int_0^3 \frac{4}{\omega^2+3}d\omega = \frac{4}{\sqrt{3}\pi}\int_0^3 \frac{1}{(\omega/\sqrt{3})^2+1}d(\omega/\sqrt{3}) \\ &= \frac{4}{\sqrt{3}\pi}\tan^{-1}(\omega/\sqrt{3})\Big|_0^3 = \frac{4}{\sqrt{3}\pi}\frac{\pi}{3} = \frac{4}{3\sqrt{3}} \end{aligned}$$

Note that in (7.17) frequency band $\omega_1 < \omega < \omega_2$ actually means the frequency bands $\omega_1 < |\omega| < \omega_2$ (because negative frequency has no direct physical meaning and $S(-\omega) = S(\omega)$). Thus, in the above, frequency band $0 < \omega < 3$ actually means the two-sided frequency band $-3 < \omega < 3$.

(c) Find the average power of its non-dc components: Since the average power of non-dc components is the variance $\sigma_x^2 = E[X^2] - \bar{x}^2 = P_x - \bar{x}^2$, we need only to find $\bar{x}$. Assume $\bar{x} = b$ and let $Y(t)$ be the zero-mean part of $X(t)$; that is, $X(t) = Y(t) + b$. Then

$$\begin{aligned} R_x(\tau) &= E\{[Y(t+\tau)+b][Y(t)+b]\} \\ &= E[Y(t+\tau)Y(t)] + E[Y(t+\tau)]b + bE[Y(t)] + b^2 = R_y(\tau) + b^2 \end{aligned}$$

This leads to $S_x(\omega) = S_y(\omega) + 2\pi b^2\delta(\omega)$. Thus, if there should be a constant (dc) term in $X(t)$, its power spectrum would have a delta function at $\omega = 0$. More generally, *a random process $X(t)$ has a sinusoidal component with frequency ω_0 if and only if its power spectrum has a pair of delta functions at $\omega = \pm\omega_0$* (see Example 7.3). Since $S_x(\omega)$ of the current problem does not have a delta function at $\omega = 0$, $\bar{x} = b = 0$ and finally, the variance $\sigma_x^2 = P_x - \bar{x}^2 = P_x = 2/\sqrt{3}$, although $S(\omega)|_{\omega=0} \neq 0$.

7.3 White Noise

A *zero-mean* noise process $N(t)$ is said to be ***white noise*** if it has no coupling between distinct time instants, or equivalently, its power spectrum is constant:

$$\boxed{R_n(\tau) = S_0\delta(\tau) \qquad \Longleftrightarrow \qquad S_n(\omega) = S_0} \tag{7.19}$$

Figure 7.3: Autocorrelation and power spectrum of white noise.

where S_0 is called its ***intensity***. This is analogous to white light, which has a spectrum constant over all visible light frequencies. White noise is a popular *mathematical model* of physical (e.g., thermal) noises. It is

- *completely unpredictable* because its future is uncorrelated with (or independent of) its past and present — worst case
- *easiest to handle* because its value at one time is uncorrelated with (or independent of) any other time — easiest case
- physically *unrealizable* because it has infinite power: $P_n = R_n(0) = \infty$

A noise process that is not white is called ***colored noise***. A typical example of colored noise is one with the following exponential autocorrelation

$$R_n(\tau) = S_0 e^{-\alpha|\tau|} \qquad \Longleftrightarrow \qquad S_n(\omega) = \frac{2\alpha S_0}{\alpha^2 + \omega^2} \tag{7.20}$$

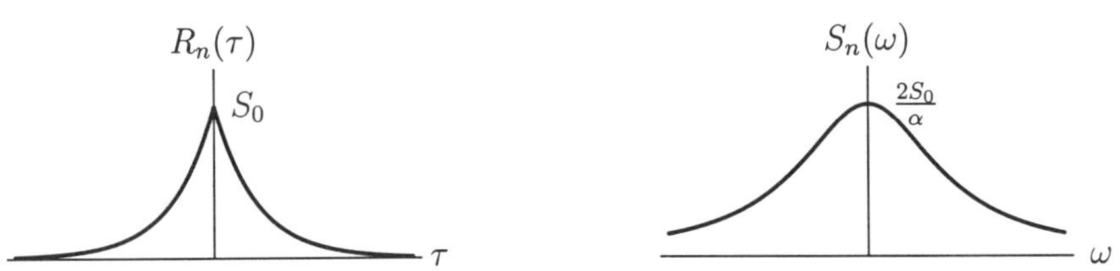

Figure 7.4: Autocorrelation and power spectrum of a colored noise process.

Thermal noise has a power spectrum that is constant up to very high frequency and then decreases. A more realistic mathematical model for such zero-mean noise is the so-called ***bandlimited white noise***. Its *lowpass* version has the following power spectral density

$$S_n(\omega) = \begin{cases} S_0 & |\omega| \leq W = 2\pi B \\ 0 & \text{elsewhere} \end{cases} \tag{7.21}$$

where B (or W) is called its ***bandwidth*** in Hz (or in radians per second); S_0 its intensity. The corresponding autocorrelation function can be shown to be (see problem 7.17), by inverse Fourier transform

$$R_n(\tau) = \mathcal{F}^{-1}[S_n(\omega)] = \frac{S_0 W}{\pi} \frac{\sin(W\tau)}{W\tau} \tag{7.22}$$

In this case

$$P_n = \left\{ \begin{array}{ll} R_n(0) = \lim\limits_{\tau \to 0} \dfrac{S_0 W}{\pi} \dfrac{\sin(W\tau)}{W\tau} & \text{or} \\ \dfrac{1}{\pi} \displaystyle\int_0^{\infty} S_n(\omega) d\omega = \dfrac{1}{\pi} \displaystyle\int_0^{W} S_n(\omega) d\omega & \end{array} \right\} = \frac{S_0 W}{\pi} \tag{7.23}$$

which has finite power and is thus physically "realizable."

It is clear that bandlimited white noise is colored but it becomes white noise as bandwidth goes to infinity.

Sometimes it is desirable to consider a white noise process that is *independent*, rather than just uncorrelated, across time. Such noise is known as ***strictly white noise*** or ***strict-sense white noise***.

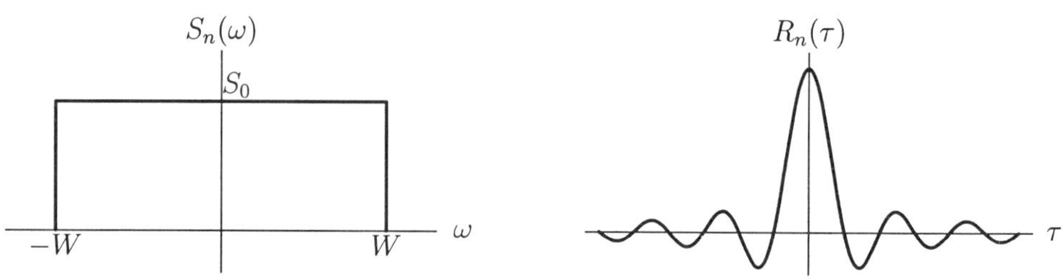

Figure 7.5: Power spectrum and autocorrelation of lowpass white noise.

White noise with $R(\tau) = S_0\delta(\tau)$ in (7.19) clearly means that there is no coupling between any two distinct time instants. It has a correlation coefficient

$$\rho(\tau) = \frac{R(\tau)}{R(0)} = \begin{cases} 1 & \tau = 0 \\ 0 & \tau \neq 0 \end{cases}$$

α in (7.20) is sometimes called the *correlation time constant.*

Bandlimited white noise can also be *bandpass*. Its power spectrum is

$$S_n(\omega) = \begin{cases} S_0 & |\omega \pm \omega_c| < W/2 \\ 0 & \text{elsewhere} \end{cases} \tag{7.24}$$

and its autocorrelation is (see problem 7.18 for a proof)

$$R_n(\tau) = \frac{S_0 W}{\pi} \frac{\sin(W\tau/2)}{W\tau/2} \cos(\omega_c \tau) \tag{7.25}$$

which has a bandwidth of $B = \frac{W}{2\pi}$ and power given by (7.23).

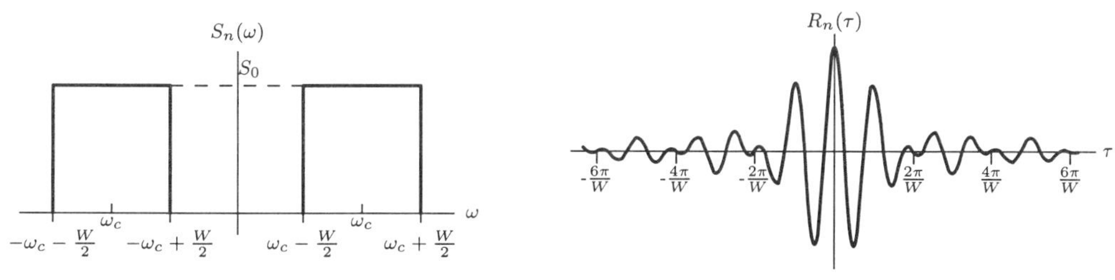

Figure 7.6: Power spectrum and autocorrelation of bandpass white noise.

In fact, any random process with a power spectrum that has a flat top or sharp corner is not really physically realizable. This is the case for bandlimited white noise. Unlike white noise, however, it can be implemented within an arbitrary accuracy.

Note the factors W and $W/2$ in (7.21) and (7.24), respectively. As a result, although lowpass (LP) is usually equivalent to bandpass (BP) with $\omega_c = 0$, for bandlimited white noise, $S_n^{LP}(\omega) \neq S_n^{BP}(\omega)|_{\omega_c=0}$. Instead, we have

$$S_n^{LP}(\omega) = S_n^{BP}(\omega)|_{\omega_c=W/2}, \qquad R_n^{LP}(\tau) = R_n^{BP}(\tau)|_{\omega_c=W/2}$$

The correlation coefficients of lowpass and bandpass white noise are, respectively,

$$\rho_n^{LP}(\tau) = \frac{C_n(\tau)}{C_n(0)} = \frac{R_n(\tau)}{R_n(0)} \overset{(7.22)}{=} \frac{\sin(W\tau)}{W\tau} \tag{7.26}$$

$$\rho_n^{BP}(\tau) = \frac{C_n(\tau)}{C_n(0)} = \frac{R_n(\tau)}{R_n(0)} \overset{(7.25)}{=} \frac{\sin(W\tau/2)}{W\tau/2} \cos(\omega_c\tau) \tag{7.27}$$

The definition of white noise based on (7.19) is valid only for *zero-mean* noise processes. More generally, white noise is defined as a noise process whose *autocovariance* is a delta function

$$C_n(\tau) = S_0\delta(\tau) \quad \Longleftrightarrow \quad S_n(\omega) = S_0 + 2\pi[E(N)]^2\delta(\omega)$$

where $E(N)$ is the time-invariant mean of the noise $N(t)$.

Example 7.5: Power Spectrum of Nonzero-Mean White Noise

Find the power spectrum and autocorrelation of a white noise with mean 10V and a power spectrum of its non-dc component equal to $100V^2$/Hz.

First, it should be identified that $E(N) = 10$, $S_0 = 100$ and thus $C_n(\tau) = S_0\delta(\tau) = 100\delta(\tau)$. Thus

$$R_n(\tau) = C_n(\tau) + [E(N)]^2 = 100\delta(\tau) + 100$$

Taking Fourier transform yields

$$S_n(\omega) = S_0 + 2\pi[E(N)]^2\delta(\omega) = 100 + 200\pi\delta(\omega)$$

Although white noise is not physically realizable, its study is very meaningful for engineering practice. One of the best examples to demonstrate that proper abstraction of real-world problems is important is that of the historical evolution of the number systems.

Positive integers (*natural numbers*) are the first abstraction necessary for counting. This was learned in our early childhood. The well-known mathematician Kronecker had a famous saying "Natural numbers were created by God and all the other numbers were created by man."[1] *Negative integers* and zero were introduced only because we need to perform subtraction (in other words, only for the convenience of subtraction). Decimal points, fractions, and in general, *rational numbers*[2], were introduced for the convenience of division. *Irrational numbers* were introduced due to the need of such mathematical operations as taking square root (e.g., $\sqrt{2}$ is an irrational number). The operation on rational and irrational numbers led to *real numbers*. The introduction of *complex numbers* had not been generally accepted until the one-to-one correspondence was found between these numbers and the points on the plane in Cartesian coordinates. In the modern era, numbers have various extensions. One direction is the *vector*, *matrix*, and *array* of an arbitrary dimension. Another direction is the *cardinal number* of an infinite set[3].

[1] Die ganzen zahlen hat Gott gemacht, alles andere ist Menschenwerk.

[2] A rational number is one that can be written in the form of n/m, where m and n are integers. Otherwise, it is an irrational number.

[3] Simply put, the cardinal number describes how large a (finite or infinite) set is, or more specific, how many elements a set has. Different infinite sets could have different numbers of elements. For example, the set of positive integers has a smaller cardinal number than the set of real numbers. It has, however, the same cardinal number as the set of rational numbers has, which is not easily acceptable by people new to this theory.

Example 7.6: Signal Detection

We want to decide if there is a signal $S(t) = A\sin(\omega_0\pi t+\theta)$ in the noisy random process $X(t)$, where $\theta \sim \mathcal{U}(0, 2\pi)$. In other words, there are two possibilities:

$$\begin{array}{ll} X(t) = S(t) + N(t) & \text{if the signal is present} \\ X(t) = N(t) & \text{if the signal is absent} \end{array}$$

where $N(t)$ is white noise with power spectrum S_0, and we want to determine which one is true.

$X(t)$ would have the following autocorrelation and power spectrum were the signal present

$$R_x(\tau) = R_s(\tau) + R_n(\tau) = \frac{1}{2}A^2\cos(\omega_0\tau) + S_0\delta(\tau)$$
$$S_x(\tau) = S_s(\tau) + S_n(\tau) = \frac{\pi}{2}A^2[\delta(\omega-\omega_0) + \delta(\omega+\omega_0)] + S_0$$

or the following autocorrelation and power spectrum were the signal absent:

$$R_x(\tau) = S_0\delta(\tau)$$
$$S_x(\tau) = S_0$$

Consequently, if a delta function (i.e., high peak) in the power spectrum or a sinusoid in the autocorrelation of $X(t)$ is found, then we can declare that there is a sinusoidal signal in $X(t)$, as for the case shown in Fig. 7.7 where the signal is observed to have a frequency around $1/8$ of the sampling frequency f_s.

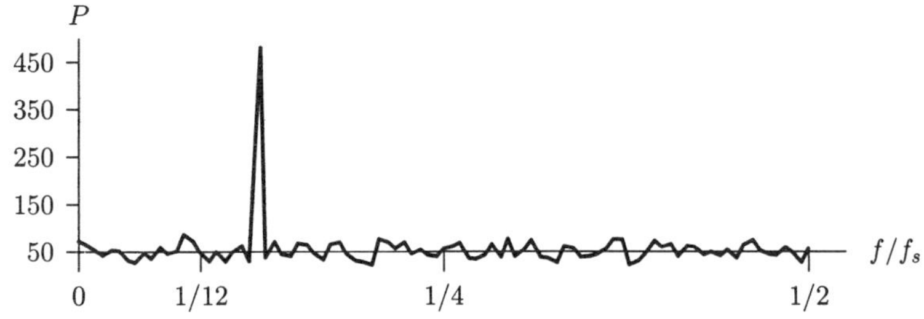

Figure 7.7: Power spectrum of a signal imbedded in noise.

Example 7.7: Power of a Signal Amplitude-Modulated by Noise

Consider a random process $Y(t) = N(t)\cos(\omega_0 t + \theta)$, where $\theta \sim \mathcal{U}(0, 2\pi)$; $N(t)$ is a bandlimited white noise with intensity S_0 and bandwidth W; $N(t)$ is independent of θ; and $\omega_0 > 0$ is a known constant. Find $S_y(\omega)$ and P_y.

Solution:

$$\begin{aligned}
E[Y(t)] &= E[N(t)\cos(\omega_0 t + \theta)] \overset{\text{independence}}{=} E[N(t)]E[\cos(\omega_0 t + \theta)] \overset{(6.15)}{=} 0 \\
R_y(t+\tau, t) &= E[N(t+\tau)\cos(\omega_0 t + \omega_0 \tau + \theta) N(t)\cos(\omega_0 t + \theta)] \\
&= E[N(t+\tau)N(t)] \cdot \frac{1}{2} E[\cos(\omega_0 \tau) + \cos(2\omega_0 t + \omega_0 \tau + 2\theta)] \\
&= \frac{1}{2} R_n(\tau)\cos(\omega_0 \tau) = R_y(\tau) \qquad (7.28)
\end{aligned}$$

Thus, $Y(t)$ is wide-sense stationary. Now

$$\begin{aligned}
P_y &= R_y(0) = \frac{1}{2} R_n(0)\cos 0 = \frac{1}{2} P_n \overset{(7.23)}{=} \frac{S_0 W}{2\pi} \\
S_y(\omega) &= \mathcal{F}[R_y(\tau)] = \frac{1}{2}\mathcal{F}[R_n(\tau)\cos(\omega_0 \tau)] \\
&\overset{\text{Table 7.4}}{=} \frac{1}{4}[S_n(\omega - \omega_0) + S_n(\omega + \omega_0)] \qquad (7.29)
\end{aligned}$$

where $S_n(\omega)$ is the power spectrum of $N(t)$ with intensity S_0 and bandwidth W. (7.29) follows from the frequency shifting property of Fourier transform (Table 7.4).

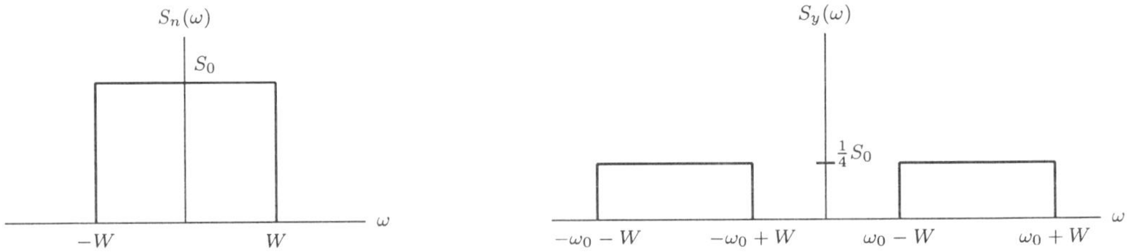

Figure 7.8: Power spectrum of amplitude modulation by noise.

7.4 Power Spectrum Estimation

An important issue in practice is how to estimate the power spectrum of a stationary random process, given a finite-length data record of the process.

There are two classes of methods for power spectrum estimation: parametric and nonparametric. A *parametric method* estimate the power spectrum of a random process based on an assumed model for the process and is thus also known as a *model-based* method. Since the model is assumed, the estimation of the power spectrum thus reduces to the estimation of the parameters of the model, hence the name. If the model is a good fit to the process, the corresponding parametric method can usually yield better estimates than the nonparametric methods, where no assumptions are made with respect to the random process. The parametric methods for power spectrum estimation is beyond the scope of this book.

Unfortunately, there is no simple and accurate method for power spectrum estimation that is versatile. In fact, power spectrum estimation is in a virtual "Tower of Babel" of methods.

One of the most commonly used nonparametric estimators of power spectrum of a random process $X[n]$ is the so-called ***periodogram***:

$$\hat{S}_x(\omega) = \frac{1}{N}|X_N(\omega)|^2 = \frac{1}{N}\left|\sum_{n=1}^{N} x[n]e^{-j\omega n}\right|^2 \tag{7.30}$$

where $x[n]$ is a discrete-time sample function of the random process; $X_N(\omega)$ is its *discrete Fourier transform* (DFT); and N is the length of the data record. A straightforward implementation of the periodogram requires, for each frequency ω, the calculation of the DFT of $x[n]$, which is usually done by the so-called *fast Fourier transform* (FFT) — computationally efficient implementations of the DFT.

The periodogram can be justified by (7.11): It is clear that (7.30) is the discrete-time counterpart of (7.11), which is clearly related to the true power spectrum, defined by (7.13).

Although widely used, the periodogram is in general not a good spectral estimator: It is biased in the sense that its average value is not necessarily equal to the true power spectrum; and as the length of the data record increases, its variance does not decrease although its mean approaches to the true power spectrum. To increase its accuracy, an average over a number of periodograms, each computed on the basis of a segment of the data record, is often used.

MATLAB signal processing toolbox has a routine `psd` that calculates the power spectrum of a signal (sequence), albeit not accurately. This routine is based on the so-called Welch's averaging modified periodogram method. It divides the signal sequence into a number of possibly overlapping segments and calculate the periodogram for each segment. The average of these periodograms is then taken as the estimate of the power spectrum. A so-called Hanning window may be used to smooth out any end effects caused by the abrupt transition between adjacent segments prior to computing the periodogram.

The power spectrum is estimated in the companion software P&R based on the Welch's method of averaging periodogram, similarly as in the MATLAB function `psd`. However, two additional windows are available: rectangular and Hamming. In addition, two choices are available for the spectrum type: smoothed and unsmoothed. These will be described below in Example 7.8. To use P&R, each record of the signal of which the PSD is to be estimated should be stored in a data file.

Another general method of estimating power spectrum is to compute the sample autocorrelation function first and then compute its discrete Fourier transform.

Example 7.8: Estimation of Power Spectrum of a Random Pulse Train by P&R

Given a data record, its power spectrum can be easily estimated by the companion software P&R.

Suppose that a data record of a 1024-point (with 100 Hz sampling rate) pulse train with a random amplitude is stored in the data file `e7_8.dat`. The pulse train has a "period" of 0.1 second, a pulse width of 0.05 second and an amplitude that is time-varying and uniformly distributed over $(0, 4)$. It can be generated by the companion software P&R following the steps shown in Example 6.3.

Then the power spectrum of this pulse train can be estimated according to the following simple steps:

S1. Click "RPAnalyzer" on the main window of P&R.
S2. Click "Frequency Domain." You will be prompted to enter the name of a data file.
S3. Choose "data" subdirectory and enter the data file name "`e7_8.dat`" and click "Ok."
S4. Fill out the window "**Power Spectrum Estimator**" as shown in Fig. 7.9 and click "Ok." Then a MATLAB figure window will appear and the positive part of the power spectrum of the data stored in the data file is plotted.

Please note the following:

- The segment length is better to be in the power of 2 for the convenience of FFT. It should not exceed the length of the data record. A shorter segment length will result in a smoother PSD curve, at the cost of losing frequency resolution. The highest frequency resolution is achieved by setting the segment length to that of the data record. See Computer Exercise 7.6.
- Overlapped N is the number of overlapping samples between two adjacent segments used in the Welch's method, described before. It is better to be in the power of 2 and should not exceed the segment length. A smoother PSD curve is obtained if this number is set to a larger value.
- The sampling rate f_s is used only in two places: (a) to determine the fundamental interval, as described below; and (b) to scale the magnitude of the power spectrum curve in the case of white noise so as to ensure the correct value for a discrete-time signal. As a result, you need to enter the correct f_s only for white noise. In other case you may enter an arbitrarily large sampling rate if you do not know the true sampling rate.
- The maximum frequency for the PSD plot can be set to any positive number not larger than one half of the sampling rate f_s. This is because all the higher frequencies of a discrete-time signal can be folded into the *fundamental interval* $[-f_s/2, f_s/2]$ in the sense that they are

identical to (i.e., indistinguishable from) the corresponding frequencies in the fundamental interval. You may of course choose a maximum frequency for plot smaller than $f_s/2$ to zoom in the lower frequency part. The left and right bottom plots in Fig. 7.9 correspond to the choices of 50 Hz and 20 Hz for the maximum frequency for plot, respectively.

- The rectangular window corresponds to the division of the data record into segments without any windowing technique to smooth out the end effects due to the abrupt transition between adjacent segments.
- The smoothed spectrum type uses an FFT of a sequence of length that is approximately four times that of the data record. The sequence is expanded by zero-padding, a commonly used technique in FFT processing for interpolating the values of the computed spectrum at more frequencies so that smoother curves are resulted. The unsmoothed version does not utilize such technique.

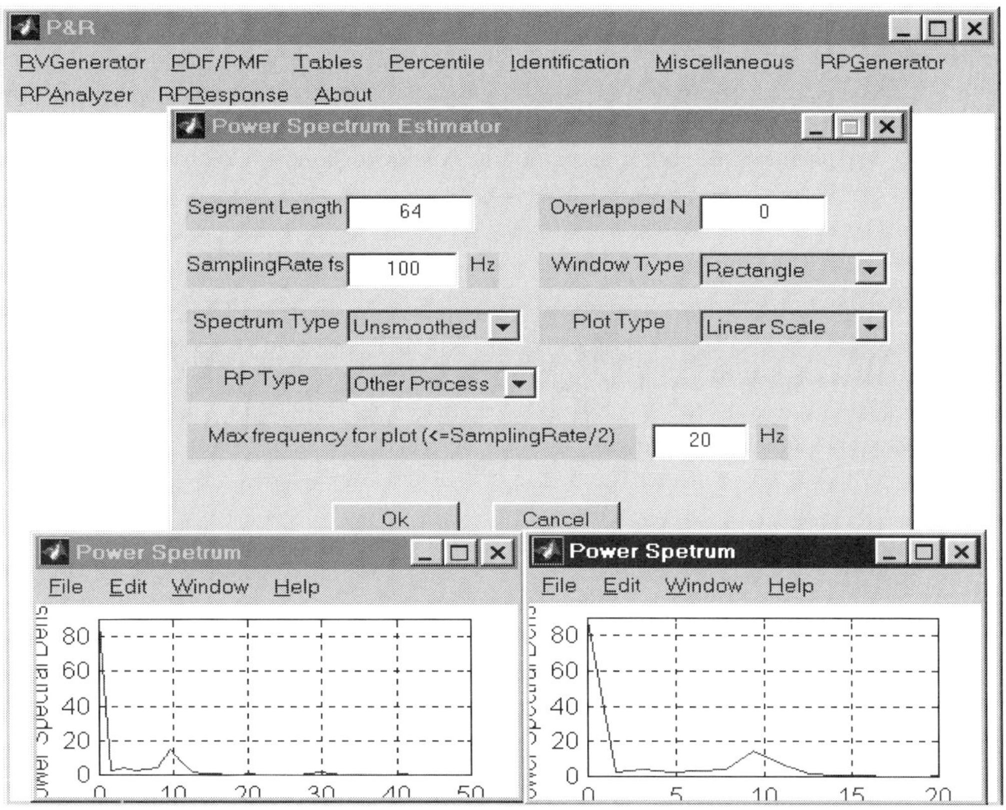

Figure 7.9: Estimating power spectrum of a random pulse train by P&R.

Example 7.9: Estimation of Power Spectrum of White Noise

Fig. 7.10 illustrates a 512-point (with 2000 Hz sampling rate) sample function of a zero-mean Gaussian white noise sequence, with a variance of 4, generated by the companion software P&R following similar steps as illustrated in Example 6.19.

Fig. 7.11 gives the estimated power spectrum of the white noise, obtained by P&R following similar steps as the previous example. Note that the estimated power spectrum is not a constant although the theoretical one is. Since the sampling frequency is 2000 Hz, the fundamental interval for the sampled random sequence is $(-1000, 1000)$ Hz. In other words, by the sampling theorem, the highest frequency component that can be recovered without distortion (known as aliasing) is 1000 Hz. Recall that the average power, which is 4 in this example, is equal to the area underneath the power spectrum curve. For a random sequence, that is $2000C = 4$. Thus, $C = 2 \times 10^{-3}$ is the theoretical value (constant) of the power spectrum of the sampled white noise. The estimated power spectrum of Fig. 7.11 agrees with this analysis. This is the case only when "White Sequence" is selected and the correct sampling rate is entered, along with a rectangular window, for an unsmoothed spectrum. If a window of another type, such as Hanning window, were used, the magnitude of the PSD would change greatly and the shape might also change. If the smoothed spectrum were sought, the magnitude might change slightly. If "Other Processes" were selected, the shape of the PSD curve would be correct but the magnitude of the plot would correspond to the spectral density of the continuous-time white noise, rather than the sampled one.

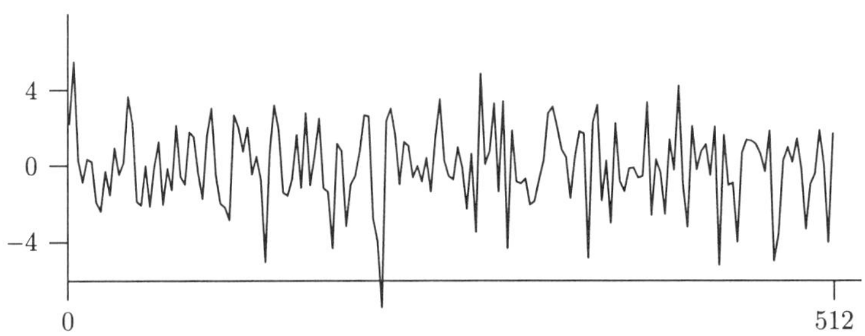

Figure 7.10: A sample function of a white sequence.

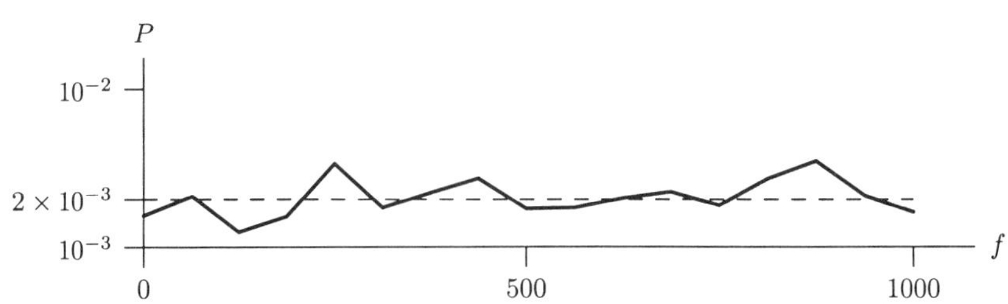

Figure 7.11: The estimated power spectrum of the white sequence.

Example 7.10: Estimation of Power Spectrum of a Sinusoid

A 180-point discrete-time (with sampling rate $f_s = 720$ Hz) sample function of the sinusoid of Example 6.7 with amplitude $A = 2\sqrt{2}$, frequency $\omega_0 = 120\pi$ and a random phase uniformly distributed over $(-\pi, \pi)$ is given. The autocorrelation of this random sinusoid is (see Example 6.7) a single cosine function $R(\tau) = \frac{1}{2}A^2\cos(\omega_0\tau)$. Its Fourier transform (i.e., power spectrum) is theoretically a delta function located at ω_0:

$$S(\omega) = \frac{\pi}{2}A^2[\delta(\omega - \omega_0) + \delta(\omega + \omega_0)]$$

However, for a sequence $x[n], n = 0, 1, \ldots, L-1$ of finite length L ($L \leq 180$ in this example is the *segment* length), its power spectrum turns out to be

$$S(\omega) = \frac{\pi}{2}A^2\left[\frac{\sin[(\omega-\omega_0)L/2]}{\sin[(\omega-\omega_0)/2]}e^{-j(\omega-\omega_0)(L-1)/2} + \frac{\sin[(\omega+\omega_0)L/2]}{\sin[(\omega+\omega_0)/2]}e^{-j(\omega+\omega_0)(L-1)/2}\right]$$

The magnitude plot of the positive frequency part is thus

$$\frac{\pi}{2}A^2\frac{|\sin[(\omega-\omega_0)L/2]|}{|\sin[(\omega-\omega_0)/2]|}$$

Consequently, the above should be divided by 2π if the power spectrum is in terms of frequency f in Hz, rather than angular frequency ω in radians per second. This leads to $A^2L/4$ per Hz as the height of the peak in the power spectrum. Fig. 7.12 shows the estimated power spectrum with a peak height of 360, which was obtained by the companion software P&R with the segment length set to the data record length of 180 to yield the maximum frequency resolution, in agreement with the above analysis of peak height: $A^2L/4 = (8)(180)/4 = 360$/Hz. The peak would be sharper if more data points were used.

Note that the location of the peak times the sampling rate is equal to the frequency of the sinusoid: $(1/12)(720) = 60$ Hz.

For the dc term $\bar{x}$ of a discrete-time signal of length L, its power spectrum is $S(\omega) = 2\pi(\bar{x})^2\sin(\omega L/2)/\sin(\omega/2)$. Consequently, the corresponding peak height per Hz is A^2L instead of $A^2L/4$.

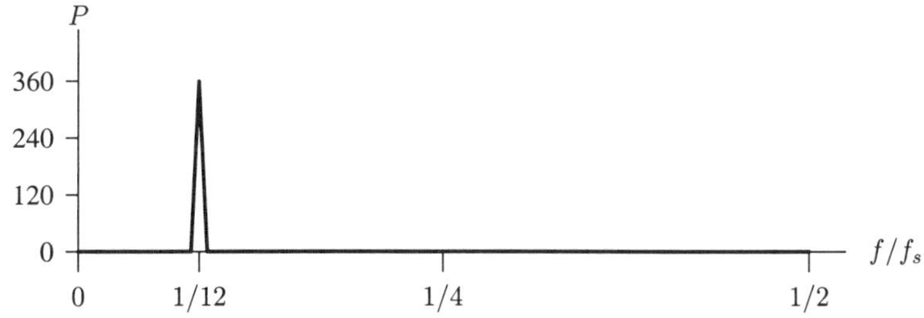

Figure 7.12: The estimated power spectrum of a random sinusoid.

7.5 Cross Power Spectrum

Given jointly wide-sense stationary random processes $X(t)$ and $Y(t)$, their sum $Z(t) = X(t) + Y(t)$ has the following autocorrelation and power spectrum

$$\begin{aligned} R_z(\tau) &= E\{[X(t+\tau) + Y(t+\tau)][X(t) + Y(t)]\} \\ &= R_x(\tau) + R_y(\tau) + R_{xy}(\tau) + R_{yx}(\tau) \end{aligned} \tag{7.31}$$

$$\begin{aligned} S_z(\omega) &= \mathcal{F}[R_x(\tau) + R_y(\tau) + R_{xy}(\tau) + R_{yx}(\tau)] \\ &= S_x(\omega) + S_y(\omega) + \mathcal{F}[R_{xy}(\tau)] + \mathcal{F}[R_{yx}(\tau)] \end{aligned} \tag{7.32}$$

Or

$$S_z(\omega) = S_x(\omega) + S_y(\omega) + S_{xy}(\omega) + S_{yx}(\omega) \tag{7.33}$$

if we define the ***cross-power spectral density*** as the Fourier transform of the crosscorrelation

$$S_{xy}(\omega) = \mathcal{F}[R_{xy}(\tau)] = \int_{-\infty}^{\infty} R_{xy}(\tau)e^{-j\omega\tau}\,d\tau \tag{7.34}$$

Likewise for $S_{yx}(\omega)$.

Unlike the (auto-)power spectrum, the cross-power spectrum need not be real, nonnegative, or an even function of ω. They have, however, the following properties for real-valued random processes:

1. Let * denote complex conjugate. Then

$$S_{yx}(\omega) = S_{xy}(\omega)^* = S_{xy}(-\omega) \tag{7.35}$$

2. Their real parts $\text{Re}[S_{yx}(\omega)] = \text{Re}[S_{xy}(\omega)]$ are even in ω.
3. Their imaginary parts $\text{Im}[S_{yx}(\omega)] = -\text{Im}[S_{xy}(\omega)]$ are odd in ω.
4. If $X(t)$ and $Y(t)$ are uncorrelated and have constant means, then

$$S_{xy}(\omega) = S_{yx}(\omega) = 2\pi E[X]E[Y]\delta(\omega) \tag{7.36}$$

This reduces to, if $X(t)$ and $Y(t)$ are orthogonal (see Section 6.7),

$$S_{xy}(\omega) = S_{yx}(\omega) = 0 \tag{7.37}$$

$$S_z(\omega) = S_x(\omega) + S_y(\omega) \tag{7.38}$$

$$S_{xz}(\omega) = S_x(\omega) \tag{7.39}$$

where $Z(t) = X(t) + Y(t)$.

The cross-power spectrum $S_{xy}(\omega)$ does not have a physical interpretation as simple and clear as the power spectrum $S_x(\omega)$.

The average ***cross power*** of two nonrandom waveforms $x(t)$ and $y(t)$ can be defined as

$$p_{xy} = \lim_{T\to\infty} \frac{1}{2T}\int_{-T}^{T} x(t)y(t)dt = r_{xy}(0)$$

where $r_{xy}(\tau)$ is the crosscorrelation function defined by (6.8). For random processes $X(t)$ and $Y(t)$, the expected (average) cross power can be defined by

$$\begin{aligned} P_{xy} &= E[p_{xy}] \\ &= E\left[\lim_{T\to\infty} \frac{1}{2T}\int_{-T}^{T} X(t)Y(t)dt\right] \\ &= E\left[\lim_{T\to\infty} \frac{1}{2T}\frac{1}{2\pi}\int_{-\infty}^{\infty} F_x(\omega)F_y(\omega)^* d\omega\right] && (7.40) \\ &= \frac{1}{2\pi}\int_{-\infty}^{\infty} \lim_{T\to\infty} \frac{1}{2T} E[F_x(\omega)F_y(\omega)^*]d\omega && (7.41) \end{aligned}$$

where $F_x(\omega)$ and $F_y(\omega)$ are the Fourier transforms of the truncated version of $X(t)$ and $Y(t)$ that have a finite duration $[-T, T]$, and (7.40) follows from the following *Parseval theorem* for two real-valued functions $x(t)$ and $y(t)$, both with a finite duration $[-T, T]$,

$$\int_{-T}^{T} x(t)y(t)dt = \frac{1}{2\pi}\int_{-\infty}^{\infty} F_x(\omega)F_y(\omega)^* d\omega$$

From the above and following a discussion similar to the one given on page 321, $F_x(\omega)F_y(\omega)^*$ can be called *cross-energy spectrum* and the *cross-power spectrum* $S_{xy}(\omega)$ can be defined as

$$\begin{aligned} S_{xy}(\omega) &= \lim_{T\to\infty} \frac{E[F_x(\omega)F_y(\omega)^*]}{2T} \\ &= E\left[\lim_{T\to\infty} \frac{F_x(\omega)F_y(\omega)^*}{2T}\right] \\ &= E\left[\lim_{T\to\infty} \frac{1}{2T}\int_{-T}^{T} X(t+\tau)Y(t)dt\right] && (7.42) \\ &= E[\mathcal{F}\{A\langle x(t+\tau)y(t)\rangle\}] \\ &= \mathcal{F}\{A\langle E[x(t+\tau)y(t)]\rangle\} \\ &= \mathcal{F}\{A\langle R_{xy}(t+\tau, t)\rangle\} && (7.43) \end{aligned}$$

where $A\langle g(t)\rangle$ is the time average of $g(t)$ and (7.42) follows from the correlation property of Fourier transforms (Table 7.4), which implies that the product of the above Fourier transforms is the Fourier transform of the crosscorrelation of the two functions. Consequently, ***cross-power spectrum*** *can be defined as the Fourier transform of the time average of the crosscorrelation function:*

$$\begin{aligned} S_{xy}(\omega) &= \mathcal{F}[\text{time average of } R_{xy}(t+\tau, t)] \\ \text{time average of } R_{xy}(t+\tau, t) &= \mathcal{F}^{-1}[S_{xy}(\omega)] \end{aligned}$$

Clearly, this relation is more general than the one given by (7.34) because it is valid for nonstationary as well as jointly stationary processes. It clearly reduces to (7.34) for jointly stationary processes. It is thus clear that $S_{xy}(\omega)$ so defined deserves the name "cross-power spectral density."

From the above discussion, it is clear that we have the following relationship

$$P_{xy} = \frac{1}{2\pi}\int_{-\infty}^{\infty} S_{xy}(\omega)d\omega \tag{7.44}$$

$$P_{yx} = \frac{1}{2\pi}\int_{-\infty}^{\infty} S_{yx}(\omega)d\omega \tag{7.45}$$

Property 1 follows from the following directly

$$\begin{aligned}
\int_{-\infty}^{\infty} R_{yx}(\tau)e^{-j\omega\tau}d\tau &= \int_{-\infty}^{\infty} E[Y(t+\tau)X(t)]e^{-j\omega\tau}d\tau \\
&= \int_{-\infty}^{\infty} E[X(t)Y(t+\tau)]e^{-j\omega\tau}d\tau \\
&= \int_{-\infty}^{\infty} R_{xy}(-\tau)e^{-j\omega\tau}d\tau \\
&= \int_{-\infty}^{\infty} R_{xy}(\tau)e^{j\omega\tau}d\tau = S_{xy}(-\omega) \\
&= \left[\int_{-\infty}^{\infty} R_{xy}(\tau)e^{-j\omega\tau}d\tau\right]^* \\
&= S_{xy}(\omega)^*
\end{aligned}$$

Properties 2 and 3 follow from the following facts directly:

$$\begin{aligned}
&S_{yx}(\omega) = S_{xy}(\omega)^* \implies \text{Re}[S_{yx}(\omega)] = \text{Re}[S_{xy}(\omega)], \quad \text{Im}[S_{yx}(\omega)] = -\text{Im}[S_{xy}(\omega)^*] \\
&\text{Re}[S_{xy}(\omega)] = \text{Re}[S_{xy}(\omega)^*] = \text{Re}[S_{xy}(-\omega)] \implies \text{they are even in } \omega \\
&\text{Im}[S_{xy}(\omega)] = -\text{Im}[S_{xy}(\omega)^*] = -\text{Im}[S_{xy}(-\omega)] \implies \text{they are odd in } \omega
\end{aligned}$$

(7.37) follows from (7.36) since $\bar{x}\bar{y} \overset{\text{uncorrelated}}{=} E[XY] \overset{\text{orthogonal}}{=} 0$. (7.38) follows from (7.34) and (7.37) directly. It indicates that $P_z = P_x + P_y$, that is, if $X(t)$ and $Y(t)$ are orthogonal, then the power of their sum is equal to the sum of their powers. (7.39) makes sense since $XZ = X(X+Y) = XX + XY$ and $X(t)$ and $Y(t)$ are orthogonal.

It can also be shown

$$\left|\int_a^b S_{xy}(\omega)d\omega\right|^2 \leq \int_a^b S_x(\omega)d\omega \int_a^b S_y(\omega)d\omega \tag{7.46}$$

$$|S_{xy}(\omega)|^2 \leq S_x(\omega)S_y(\omega) \tag{7.47}$$

$$S_{x\dot{x}}(\omega) = j\omega S_x(\omega) \tag{7.48}$$

(7.46) will be shown in Chapter 8 using (6.35) with $\tau = 0$. (7.47) can be shown using Schwarz inequality. The derivation of (7.48) is left as an exercise (problem 7.24).

Example 7.11: Power Spectrum of Sum of Two Random Processes

Determine the autocorrelation and power spectrum of $Z(t) = X(t) + Y(t)$, where

$$X(t) = A + B\cos(\omega_0 t + \theta)$$
$$Y(t) = N(t)\cos(\omega_0 t + \theta)$$

$\omega_0 > 0$ is a constant; A, B, and θ are independent RVs with $\theta \sim \mathcal{U}(0, 2\pi)$; and $N(t)$, independent of A, B, and θ, is a bandlimited white noise with a constant (nonzero) mean. Clearly, from Example 7.3,

$$\begin{aligned}
R_{xy}(\tau) &= E\{[A + B\cos(\omega_0 t + \omega_0\tau + \theta)]N(t)\cos(\omega_0 t + \theta)\} \\
&= E[AN(t)\cos(\omega_0 t + \theta)] \\
&\quad + E[B\cos(\omega_0 t + \omega_0\tau + \theta)N(t)\cos(\omega_0 t + \theta)] \\
&= \bar{A}\bar{N}\underbrace{E[\cos(\omega_0 t + \theta)]}_{=0 \text{ by } (6.15)} + \bar{B}\bar{N}\frac{E[\cos(\omega_0\tau) + \cos(2\omega_0 t + \omega_0\tau + 2\theta)]}{2} \\
&= \frac{1}{2}\bar{B}\bar{N}\cos(\omega_0\tau)
\end{aligned}$$

$$R_{yx}(\tau) = R_{xy}(-\tau) = \frac{1}{2}\bar{B}\bar{N}\cos(\omega_0\tau)$$

$$S_{xy}(\omega) = \mathcal{F}[R_{xy}(\tau)] \overset{\text{Table 7.5}}{=} \frac{\pi}{2}\bar{B}\bar{N}[\delta(\omega - \omega_0) + \delta(\omega + \omega_0)]$$

Hence, from Examples 7.3 and 7.7,

$$\begin{aligned}
R_z(\tau) &\overset{(7.31)}{=} R_x(\tau) + R_y(\tau) + R_{xy}(\tau) + R_{yx}(\tau) \\
&\underset{(7.28)}{\overset{(7.14)}{=}} \overline{A^2} + \frac{1}{2}\overline{B^2}\cos\omega_0\tau + \frac{1}{2}R_n(\tau)\cos(\omega_0\tau) + \bar{B}\bar{N}\cos(\omega_0\tau) \\
S_z(\omega) &= \mathcal{F}[R_z(\tau)] \\
&\overset{\text{Table 7.4}}{=} 2\pi\overline{A^2}\delta(\omega) + \frac{1}{2}\pi\overline{B^2}[\delta(\omega - \omega_0) + \delta(\omega + \omega_0)] \\
&\quad + \frac{1}{4}[S_n(\omega - \omega_0) + S_n(\omega + \omega_0)] + \pi\bar{B}\bar{N}[\delta(\omega - \omega_0) + \delta(\omega + \omega_0)] \\
&= 2\pi\overline{A^2}\delta(\omega) + \pi\left(\frac{1}{2}\overline{B^2} + \bar{B}\bar{N}\right)[\delta(\omega - \omega_0) + \delta(\omega + \omega_0)] \\
&\quad + \frac{1}{4}[S_n(\omega - \omega_0) + S_n(\omega + \omega_0)]
\end{aligned}$$

7.6 Power Spectrum in Laplace Domain

It is sometimes, e.g., in systems applications, more desirable to use Laplace transform than Fourier transform. Thus, we may define the ***power spectrum*** as the two-sided Laplace transform of the autocorrelation function:

$$S_x(s) \triangleq \mathcal{L}_2[R_x(\tau)] \triangleq \int_{-\infty}^{\infty} R_x(\tau)e^{-s\tau}d\tau = \int_{-\infty}^{\infty} R_x(\tau)e^{-j\omega\tau}d\tau\Big|_{\omega=-js} \tag{7.49}$$

The more popular one-sided Laplace transform is applicable only if $R(\tau) = 0, \forall\tau < 0$, which is never true (except for white noise) since $R(-\tau) = R(\tau)$. Clearly

$$S_x(s) = S_x(\omega)|_{\omega=-js}, \qquad S_x(\omega) = S_x(s)|_{s=j\omega} \tag{7.50}$$

For a random process with the following (proper) *rational* power spectrum

$$S_x(s) = \frac{c(s)c(-s)}{d(s)d(-s)}$$

its average power is given by

$$P_x = R_x(0) = \mathcal{L}^{-1}[S_x(s)]\Big|_{s=0} = \frac{1}{2\pi j}\int_{-j\infty}^{j\infty} \frac{c(s)c(-s)}{d(s)d(-s)}ds \triangleq I_n \tag{7.51}$$

where $c(s)$ and $d(s)$ are two polynomials in s, given by

$$c(s) = c_{n-1}s^{n-1} + \cdots + c_1 s + c_0, \qquad d(s) = d_n s^n + \cdots + d_1 s + d_0$$

with $d(s)$ having a higher degree than $c(s)$ and no zero coefficients, the following table is very useful in determining the power of $X(t)$.

Table 7.3: Some useful integrals I_n (average powers).

$$I_1 = \frac{c_0^2}{2d_0d_1} \qquad I_2 = \frac{c_1^2d_0 + c_0^2d_2}{2d_0d_1d_2}$$

$$I_3 = \frac{c_2^2d_0d_1 + (c_1^2 - 2c_0c_2)d_0d_3 + c_0^2d_2d_3}{2d_0d_3(d_1d_2 - d_0d_3)}$$

$$I_4 = \frac{c_3^2d_0(d_1d_2 - d_0d_3) + d_0d_1d_4(c_2^2 - 2c_1c_3) + d_0d_3d_4(c_1^2 - 2c_0c_2) + c_0^2d_4(d_2d_3 - d_1d_4)}{2d_0d_4(d_1d_2d_3 - d_0d_3^2 - d_1^2d_4)}$$

Example 7.12: Determination of Power from Power Spectrum

Determine the average power of a random process $X(t)$ with power spectrum:

$$S_x(\omega) = \frac{\omega^2}{\omega^4 + 52\omega^2 + 576}$$

The following is not a good way to obtain the power for such a problem,

$$P_x = \frac{1}{2\pi}\int_{-\infty}^{\infty} S_x(\omega)d\omega = \frac{1}{2\pi}\int_{\infty}^{\infty} \frac{\omega^2}{\omega^4 + 52\omega^2 + 576}d\omega$$

because it is difficult to evaluate the integral involved. A better way is,

$$R_x(\tau) = \mathcal{F}^{-1}[S_x(\omega)] = \mathcal{F}^{-1}\Big[\frac{\omega^2}{(\omega^2+36)(\omega^2+16)}\Big] = \mathcal{F}^{-1}\Big[\frac{1.8}{\omega^2+6^2} - \frac{0.8}{\omega^2+4^2}\Big]$$

$$= \mathcal{F}^{-1}\Big[0.15\frac{2(6)}{\omega^2+6^2} - 0.1\frac{2(4)}{\omega^2+4^2}\Big] = 0.15e^{-6|\tau|} - 0.1e^{-4|\tau|}$$

and thus, $P_x = R_x(0) = 0.05$. Note that $R_x(\tau)$ is obtained as a by-product. Alternatively, we can express $S_x(s)$ as

$$S_x(s) = \frac{-s^2}{s^4 - 52s^2 + 576} = \frac{s(-s)}{(s+6)(s+4)(s-6)(s-4)} = \frac{c(s)c(-s)}{d(s)d(-s)}$$

and then use Table 7.3 to determine the average power:

$$P_x = R_x(0) = \frac{1}{2\pi j}\int_{-j\infty}^{j\infty} S_x(s)ds = \frac{1}{2\pi j}\int_{-j\infty}^{j\infty} \frac{-s^2}{s^4 - 52s^2 + 576}ds$$

Since

$$\begin{aligned} c(s) &= c_{n-1}s^{n-1} + \cdots + c_1 s + c_0 = s, \quad \Longrightarrow \quad c_0 = 0, c_1 = 1 \\ d(s) &= d_n s^n + \cdots + d_1 s + d_0 = (s+6)(s+4) = s^2 + 10s + 24, \\ &\Longrightarrow d_0 = 24, d_1 = 10, d_2 = 1 \end{aligned}$$

from Table 7.3, we have

$$P_x = \frac{1}{2\pi j}\int_{-j\infty}^{j\infty} \frac{s(-s)}{(s^2+10s+24)(s^2-10s+24)}ds = I_2$$

$$= \frac{c_1^2 d_0 + c_0^2 d_2}{2d_0 d_1 d_2} = \frac{1^2(24) + 0^2(1)}{2(24)(10)(1)} = 0.05$$

7.7 Appendix: Some Facts About Fourier Transforms

Some important properties of Fourier transforms are listed in Table 7.4. Some most frequently used Fourier transform pairs are listed in Table 7.5.

Table 7.4: Salient properties of Fourier transforms.

Description	$f(t) = \mathcal{F}^{-1}[F(\omega)]$	$F(\omega) = \mathcal{F}[f(t)]$	*Notes*
Definition	$\frac{1}{2\pi}\int_{-\infty}^{\infty} F(\omega)e^{j\omega\tau}d\omega$	$\int_{-\infty}^{\infty} f(t)e^{-j\omega\tau}d\tau$	
Shifting	$f(t-t_0)$	$F(\omega)e^{-j\omega t_0}$	t_0: real constant
	$f(t)e^{j\omega_0 t}$	$F(\omega-\omega_0)$	ω_0: real constant
Modulation	$f(t)\cos(\omega_0 t)$	$\frac{1}{2}[F(\omega-\omega_0)+F(\omega+\omega_0)]$	ω_0: real constant
	$f(t)e^{j\omega_0 t}$	$F(\omega-\omega_0)$	ω_0: real constant
Scaling (dilation)	$f(\alpha t)$	$\frac{1}{\lvert\alpha\rvert}F(\omega/\alpha)$	$\alpha \neq 0$: real constant
Duality	$F(t)$	$2\pi f(-\omega)$	
Conjugation	$f(t)^*$	$F^*(-\omega)$	
	$f(-t)^*$	$F(\omega)^*$	
Differentiation	$\frac{d^n f(t)}{dt^n}$	$(j\omega)^n F(\omega)$	
	$(-jt)^n f(t)$	$\frac{d^n F(\omega)}{d\omega^n}$	
Integration	$\int_{-\infty}^{t} f(x)dx$	$\pi F(0)\delta(\omega) + \frac{F(\omega)}{j\omega}$	
	$\pi f(0)\delta(t) - \frac{f(t)}{jt}$	$\int_{-\infty}^{\omega} F(x)dx$	
Linearity	$\sum_i \alpha_i f_i(t)$	$\sum_i \alpha_i F_i(\omega)$	α_i: complex constants
			$F_i(\omega) = \mathcal{F}[f_i(t)], \quad \forall i$
Convolution	$f_1(t) * f_2(t)$	$F_1(\omega)F_2(\omega)$	$F_i(\omega) = \mathcal{F}[f_i(t)], i = 1, 2$
	$2\pi f_1(t)f_2(t)$	$F_1(\omega) * F_2(\omega)$	$F_i(\omega) = \mathcal{F}[f_i(t)], i = 1, 2$
Correlation	$f_1(t) \circ f_2(t)$	$F_1(\omega)F_2(\omega)^*$	$F_i(\omega) = \mathcal{F}[f_i(t)], i = 1, 2$
	$2\pi f_1(t)f_2(t)^*$	$F_1(\omega) \circ F_2(\omega)$	$F_i(\omega) = \mathcal{F}[f_i(t)], i = 1, 2$
Parseval theorem	$\int_{-\infty}^{\infty} f_1(t)f_2(t)^*dt =$	$\frac{1}{2\pi}\int_{-\infty}^{\infty} F_1(\omega)F_2(\omega)^*d\omega$	$F_i(\omega) = \mathcal{F}[f_i(t)], i = 1, 2$
	$\int_{-\infty}^{\infty} \lvert f(t)\rvert^2 dt =$	$\frac{1}{2\pi}\int_{-\infty}^{\infty} \lvert F(\omega)\rvert^2 d\omega$	

Convolution is defined as $f_1(y) * f_2(y) \triangleq \int_{-\infty}^{\infty} f_1(x)f_2(y-x)dx$

Correlation is defined as $f_1(y) \circ f_2(y) \triangleq \int_{-\infty}^{\infty} f_1(x)f_2(x-y)dx$

Table 7.5: Some frequently used Fourier transform pairs.

$f(t)$	$F(\omega)$	*Notes*
α	$2\pi\alpha\delta(\omega)$	α: constant
$\delta(t)$	1	
$u(t)$	$\pi\delta(\omega) + \dfrac{1}{j\omega}$	
$\delta(t - t_0)$	$e^{-j\omega t_0}$	t_0: real constant
$\cos(\omega_0 t)$	$\pi[\delta(\omega - \omega_0) + \delta(\omega + \omega_0)]$	
$\sin(\omega_0 t)$	$-j\pi[\delta(\omega - \omega_0) - \delta(\omega + \omega_0)]$	
$e^{j\omega_0 t}$	$2\pi\delta(\omega - \omega_0)$	
$\dfrac{1}{2}\delta(t) - \dfrac{1}{j2\pi t}$	$u(\omega)$	
$e^{-\alpha\|t\|}$	$\dfrac{2\alpha}{\omega^2 + \alpha^2}$	$\alpha > 0$
$u(t)e^{-\alpha t}$	$\dfrac{1}{j\omega + \alpha}$	$\alpha > 0$
$u(-t)e^{\alpha t}$	$\dfrac{1}{j\omega - \alpha}$	$\alpha > 0$

7.8 Summary and Requirements

Power spectrum of a random process is the Fourier transform of the autocorrelation of the process, given by

$$S_x(\omega) = \mathcal{F}[R_x(\tau)] \triangleq \int_{-\infty}^{\infty} R_x(\tau)e^{-j\omega\tau}d\tau$$

It is the correct frequency description of a random process because Fourier transform of the random process itself does not exist. It describes the distribution of the average power of the process with respect to frequency. It, however, does not contain any information of the time shift or phase since it is the "*power*" spectrum and is always real and nonnegative (not a complex quantity and thus has no phase). It is an even function of frequency. The autocorrelation function can be obtained from the power spectrum by inverse Fourier transform.

The *average power* of a random process is the integral of the power spectrum over all frequencies:

$$P_x = R_x(0) = \frac{1}{2\pi}\int_{-\infty}^{\infty} S_x(\omega)d\omega$$

The average power in the frequency band $[\omega_1, \omega_2]$ for $\omega_2 \geq \omega_1 \geq 0$ is given by

$$P_{x[\omega_1,\omega_2]} = \frac{1}{\pi}\int_{\omega_1}^{\omega_2} S_x(\omega)d\omega$$

White noise is a zero-mean noise process whose power spectrum is constant over the entire frequency range. This amounts to having its autocorrelation equal to a delta function. The values of a white noise process at distinct time instants are uncorrelated. This makes it easy to handle and thus white noise is a popular random process model for many engineering applications. If a noise process is not white, then it is called *colored noise*. If the power spectrum of a random process is constant over a finite frequency band and zero elsewhere, then it is bandlimited white.

Cross-power spectrum is the Fourier transform of the crosscorrelation function. It is introduced for convenience to handle the coupling between random processes. It has some nice properties but it is not an even function of frequency and could be complex-valued.

Power spectrum can also be defined using two-sided Laplace transform.

Basic Requirements

- Comprehend the concept of power spectrum.
- Know how to find the power spectrum of a random process and how to find the autocorrelation and average power from the power spectrum.
- Be familiar with the properties of power spectrum.
- Be familiar with the concept of white noise and its power spectrum and autocorrelation.
- Be skillful in the various methods of finding average power.
- Be familiar with the power spectrum, average power, and autocorrelation of sinusoids with random phase, frequency, and/or amplitude.
- Understand the concept of cross-power spectrum and know how to find it.

The emphasis of the chapter is on the concept of power spectrum, its relation with autocorrelation, and white noise.

7.9 Additional Examples

7.13 *Average power of a sinusoid.* Consider a waveform $X(t) = A\cos(\omega_0 t + \phi)$. Find the *average* power of $X(t)$ if

(a) nothing in $X(t)$ is random
(b) $\phi \sim \mathcal{U}(-\pi, \pi)$ and nothing else in $X(t)$ is random
(c) $\omega_0 \sim \mathcal{U}(100, 200)$ and nothing else in $X(t)$ is random
(d) $A \sim \mathcal{N}(10, 1)$ and nothing else in $X(t)$ is random
(e) Compare and discuss the results obtained in the above parts.

Solution:

(a)

$$P_x \overset{(7.4)}{=} \lim_{T\to\infty} \frac{1}{2T} \int_{-T}^{T} x^2(t)dt$$

$$
\begin{aligned}
&= \lim_{T\to\infty} \frac{1}{2T} \int_{-T}^{T} \frac{1}{2} A^2[1 + \cos 2(\omega_0 t + \theta)]dt \\
&= \frac{1}{2}A^2 + \lim_{T\to\infty} \frac{1}{2T} \frac{\sin 2(\omega_0 T + \theta) - \sin 2(-\omega_0 T + \theta)}{2\omega_0} \\
&= \frac{1}{2}A^2
\end{aligned}
$$

(b) From Example 6.7, $X(t) = A\sin(\omega_0 t + \phi)$ with $\phi \sim \mathcal{U}(-\pi, \pi)$ has autocorrelation $R(\tau) = \frac{A^2}{2}\cos(\omega_0 \tau)$ and thus average power $P_x = A^2/2$. Since $A\sin(\omega_0 t + \phi)$ is just a time-shifted version of $A\cos(\omega_0 t + \phi)$, they should have the same average power $P_x = A^2/2$.

(c) Following Example 6.11, we have

$$
\begin{aligned}
R_x(t+\tau, t) &= E[A\cos(\omega_0 t + \omega_0 \tau + \phi) \cdot A\cos(\omega_0 t + \phi)] \\
&= \frac{1}{2}A^2 E[\cos\omega_0\tau + \cos(2\omega_0 t + \omega_0\tau + 2\phi)] \\
&= \frac{1}{2}A^2 \int_{100}^{200} [\cos(\omega_0\tau) + \cos(2\omega_0 t + \omega_0\tau + 2\phi)] \frac{1}{200-100} d\omega_0 \\
&= \frac{A^2}{200}\left[\frac{1}{\tau}\sin\omega_0\tau + \frac{1}{2t+\tau}\sin[(2t+\tau)\omega_0 + 2\phi]\right]_{\omega_0=100}^{200} \\
&= \frac{A^2}{200}\left[\frac{\sin 200\tau - \sin 100\tau}{\tau}\right. \\
&\quad \left. + \frac{\sin[200(2t+\tau) + 2\phi] - \sin[100(2t+\tau) + 2\phi]}{2t+\tau}\right] \\
E[X(t)^2] &= \lim_{\tau\to 0} R_x(\tau) \\
&= \frac{A^2}{200}\left\{\lim_{\tau\to 0}\left[200\frac{\sin 200\tau}{200\tau} - 100\frac{\sin 100\tau}{100\tau}\right]\right. \\
&\quad \left. + \frac{\sin(400t + 2\phi) - \sin(200t + 2\phi)}{2t}\right\} \\
&= \frac{A^2}{200}\left[100 + \frac{\sin(400t + 2\phi) - \sin(200t + 2\phi)}{2t}\right]
\end{aligned}
$$

Note that the mean-square value is time varying. The average power is

$$
\begin{aligned}
P_x &\overset{(7.5)}{=} \lim_{T\to\infty} \frac{1}{2T} \int_{-T}^{T} E[X(t)^2]dt \\
&= \lim_{T\to\infty} \frac{1}{2T} \int_{-T}^{T} \frac{A^2}{200}\left[100 + \frac{\sin(400t + 2\phi) - \sin(200t + 2\phi)}{2t}\right] dt \\
&= \frac{A^2}{2}
\end{aligned}
$$

(d)

$$
R_x(t+\tau, t) = E[A\cos(\omega_0 t + \omega_0\tau + \phi) A\cos(\omega_0 t + \phi)]
$$

$$
\begin{aligned}
&= E[A^2]\cos(\omega_0 t + \omega_0\tau + \phi)\cos(\omega_0 t + \phi) \\
&= [\sigma_A^2 + (E[A])^2]\cos(\omega_0 t + \omega_0\tau + \phi)\cos(\omega_0 t + \phi) \\
&= [1 + 10^2]\cos(\omega_0 t + \phi)\cos(\omega_0 t + \omega_0\tau + \phi) \\
&= 101\cos(\omega_0 t + \phi)\cos(\omega_0 t + \omega_0\tau + \phi)
\end{aligned}
$$

Thus, the mean-square value is given by

$$
E[X(t)^2] = R_x(t,t) = 101[\cos(\omega_0 t + \phi)]^2 = \frac{101}{2}[1 + \cos 2(\omega_0 t + \phi)]
$$

which is time varying, and thus the average power is

$$
\begin{aligned}
P_x &= \lim_{T\to\infty}\frac{1}{2T}\int_{-T}^{T} E[X(t)^2]dt \\
&= \lim_{T\to\infty}\frac{1}{2T}\int_{-T}^{T}\frac{101}{2}[1 + \cos 2(\omega_0 t + \phi)]dt = \frac{101}{2} \\
&= 50.5
\end{aligned}
$$

(e) The expected instantaneous power (i.e., the mean-square value) depends on what is random and the corresponding distribution. The average power is always a time-invariant constant, which is equal to $A^2/2$ if A is not random. Note the similarity between the average power of a random process and that of a deterministic function.

7.14 *Find frequency, mean and power from PSD.* Given the following power spectrum of a random process $X(t)$ with a nonrandom dc term

$$
S_x(\omega) = 8\pi\delta(\omega) + 1.2\pi\delta(\omega - 2.3\pi) + 1.2\pi\delta(\omega + 2.3\pi) + 4\pi\delta(\omega - 5) + 4\pi\delta(\omega + 5)
$$

(a) What frequencies are included in $X(t)$? How large are these frequency components in magnitude?
(b) Find the mean value and average power of $X(t)$.
(c) Find the variance of $X(t)$.

Solution:

(a)

$$
S_x(\omega) = 8\pi\underbrace{\delta(\omega)}_{\text{dc}} + 1.2\pi\underbrace{[\delta(\omega - 2.3\pi) + \delta(\omega + 2.3\pi)]}_{\omega_1 = 2.3\pi \implies f = 1.15} + 4\pi\underbrace{[\delta(\omega - 5) + \delta(\omega + 5)]}_{\omega_2 = 5 \implies f = 0.796}
$$

$$
R_x(\tau) \overset{\text{Table 7.5}}{=} 4 + 1.2\cos(2.3\pi\tau) + 4\cos(5\tau)
$$

Thus,

$$
\text{dc term in } R(\tau) = 4 \implies \text{dc power} = 4 \implies X(t) \text{ has a dc term} = 2
$$

$$\left.\begin{aligned} f_1 &= 1.15 \text{ has a power of } \tfrac{1}{2}A^2 = 1.2 \Longrightarrow A = 1.55 \\ f_2 &= 0.796 \text{ has a power of } \tfrac{1}{2}A^2 = 4 \Longrightarrow A = 2.83 \end{aligned}\right\} \text{relative to dc term, which is 1}$$

(b) mean value = dc value = 4.
Average power = mean-square value $= R_x(0) = 4 + 1.2 + 4 = 9.2$.

(c) $\text{var}(x) =$ mean-square value $-$ (mean value)$^2 = 9.2 - 4 = 5.2$.

7.15 *Find autocorrelation from PSD.* A random process has the power spectrum $S(\omega) = \dfrac{\omega^2+4}{\omega^4+8\omega^2+7}$. Find its autocorrelation function.

Solution: The autocorrelation function is given by the inverse Fourier transform

$$\begin{aligned} R(\tau) &= \mathcal{F}^{-1}[S(\omega)] \\ &= \mathcal{F}^{-1}\left[\frac{\omega^2+4}{\omega^4+8\omega^2+7}\right] \\ &= \mathcal{F}^{-1}\left[\frac{\omega^2+4}{(\omega^2+1)(\omega^2+7)}\right] \\ &= \mathcal{F}^{-1}\left[\frac{1/2}{\omega^2+1} + \frac{1/2}{\omega^2+7}\right] \\ &= \mathcal{F}^{-1}\left[\frac{1}{4}\frac{2\cdot 1}{\omega^2+1^2} + \frac{1}{4\sqrt{7}}\frac{2\cdot\sqrt{7}}{\omega^2+(\sqrt{7})^2}\right] \\ &= \frac{1}{4}e^{-|\tau|} + \frac{1}{4\sqrt{7}}e^{-\sqrt{7}|\tau|} \end{aligned}$$

7.16 *Find PSD from autocorrelation.* A random process has autocorrelation $R(\tau) = 7e^{-5|\tau|} + 2.4e^{-2|\tau|}\cos(3\pi\tau) + 3\cos(4\pi\tau)$. Find its power spectrum.

Solution: By linearity of Fourier transform,

$$\begin{aligned} S(\omega) &= \mathcal{F}[R(\tau)] \\ &= 7\mathcal{F}[e^{-5|\tau|}] + 2.4\mathcal{F}[e^{-2|\tau|}\cos(3\pi\tau)] + 3\mathcal{F}[\cos(4\pi\tau)] \\ &= 7\frac{(2)(5)}{\omega^2+5^2} + (2.4)(1/2)\left[\frac{(2)(2)}{(\omega-3\pi)^2+2^2} + \frac{(2)(2)}{(\omega+3\pi)^2+2^2}\right] \\ &\quad + 3\pi[\delta(\omega-4\pi)+\delta(\omega+4\pi)] \\ &= \frac{70}{\omega^2+25} + \frac{4.8}{(\omega-3\pi)^2+4} + \frac{4.8}{(\omega+3\pi)^2+4} + 3\pi[\delta(\omega-4\pi)+\delta(\omega+4\pi)] \end{aligned}$$

where use has been made of the modulation property of Fourier transforms:

$$\mathcal{F}[f(t)\cos(\omega_0 t)] = \frac{1}{2}[F(\omega-\omega_0) + F(\omega+\omega_0)]$$

7.17 *Requirements to be a PSD.* Determine which of the following functions are power spectrum, which are not:

(a) $S_1(\omega) = e^{-(\omega-2)^2}$

(b) $S_2(\omega) = \dfrac{\omega^2}{1 + 2\omega^2 + j\omega^4}$

(c) $S_3(\omega) = \dfrac{\omega^2 + 5}{(\omega^2 + 1)(\omega^4 + 3)}$

(d) $S_4(\omega) = \dfrac{\cos 2\omega}{1 + 2\omega^2 + \omega^4}$

(e) $S_5(\omega) = \dfrac{\omega^2 + 3}{\omega^4 + \omega^2 - 2}$

Justify your answer.

Solution:

(a) $S_1(\omega) = e^{-(\omega-2)^2}$ is not a power spectrum because it is not an even function of ω (although it is symmetric about $\omega = 2$).

(b) $S_2(\omega) = \dfrac{\omega^2}{1 + 2\omega^2 + j\omega^4}$ is not a power spectrum because it is not real-valued.

(c) $S_3(\omega) = \dfrac{\omega^2 + 5}{(\omega^2 + 1)(\omega^4 + 3)}$ is a power spectrum.

(d) $S_4(\omega) = \dfrac{\cos 2\omega}{1 + 2\omega^2 + \omega^4}$ is not a power spectrum because it is not always nonnegative.

(e) $S_5(\omega) = \dfrac{\omega^2 + 3}{\omega^4 + \omega^2 - 2} = \dfrac{\omega^2 + 3}{(\omega^2 - 1)(\omega^2 + 2)}$ is not a power spectrum because it is not always nonnegative.

7.18 *Sum of time-shifted versions.* Suppose that a wide-sense stationary random process $X(t)$ has power spectrum $S_x(\omega)$. Find the power spectrum $S_y(\omega)$ of $Y(t) = X(t) + X(t - t_0)$ in terms of $S_x(\omega)$, where t_0 is a nonrandom constant.

Solution:

$$\begin{aligned}
R_y(\tau) &= E[Y(t+\tau)Y(t)] \\
&= E\{[X(t+\tau) + X(t - t_0 + \tau)][X(t) + X(t - t_0)]\} \\
&= E[X(t)X(t+\tau) + X(t)X(t - t_0 + \tau) + X(t - t_0)X(t+\tau) \\
&\quad + X(t - t_0)X(t - t_0 + \tau)] \\
&= R_x(\tau) + R_x(\tau + t_0) + R_x(\tau - t_0) + R_x(\tau) \\
&= 2R_x(\tau) + R_x(\tau + t_0) + R_x(\tau - t_0) \\
S_y(\omega) &= \mathcal{F}[R_y(\tau)] \\
&= \mathcal{F}[2R_x(\tau)] + \mathcal{F}[R_x(\tau + t_0)] + \mathcal{F}[R_x(\tau - t_0)] \\
&= 2S_x(\omega) + e^{j\omega t_0} S_x(\omega) + e^{-j\omega t_0} S_x(\omega)
\end{aligned}$$

$$= 2S_x(\omega) + (e^{j\omega t_0} + e^{-j\omega t_0})S_x(\omega)$$
$$= 2S_x(\omega)[1 + \cos \omega t_0]$$

where use has been made of the time-shifting property of Fourier transforms.

7.10 Problems

7.1 *Average power of a sinusoid.* Consider a waveform $X(t) = A\cos(\omega_0 t + \phi)$. Find the average power of $X(t)$ if

(a) $\phi \sim \mathcal{U}(\theta_0, \theta_0 + 2\pi)$, where θ_0 is a given constant, and nothing else in $X(t)$ is random
(b) $\phi \sim \mathcal{U}(\theta_0, \theta_0 + \frac{1}{3}\pi)$, where θ_0 is a given constant, and nothing else in $X(t)$ is random
(c) A is a RV, with mean $\bar{A}$ and variance σ_A^2, and nothing else in $X(t)$ is random
(d) Compare and discuss the results obtained in the above parts.

7.2 *Power spectrum of a sinusoid.* Consider a waveform $X(t) = A\cos(\omega_0 t + \phi)$. Find the power spectrum of $X(t)$ if

(a) $\phi \sim \mathcal{U}(\theta_0, \theta_0 + 2\pi)$, where θ_0 is a given constant, and nothing else in $X(t)$ is random
(b) $\phi \sim \mathcal{U}(\theta_0, \theta_0 + \frac{1}{3}\pi)$, where θ_0 is a given constant, and nothing else in $X(t)$ is random
(c) A is a RV, with mean $\bar{A}$ and variance σ_A^2, and nothing else in $X(t)$ is random
(d) Compare and discuss the results obtained in the above parts.

7.3 *Average power by two definitions.* Consider a random process $X(t) = \cos(\omega_0 t + \phi)$, where $\phi \sim U(0, \pi/2)$.

(a) Find its average power using (7.5).
(b) Find its average power using (7.16), where $S(\omega)$ is obtained by

$$S(\omega) = E\left[\lim_{T\to\infty} \frac{|F_X(\omega)|^2}{2T}\right]$$

(c) Are these two methods lead to the same results?

7.4 *Find power spectrum from autocorrelation.* A random process has autocorrelation $R(\tau) = a + be^{-c|\tau|}[\cos(2\tau)]^2$, where a, b, c are real-valued constants. Find its power spectrum.

7.5 *RV as a random process.* A random signal $X(t)$ has only a dc component A, where A is a zero-mean (time-invariant) random variable with unity variance. Find the autocorrelation and power spectrum of $X(t)$.

7.6 *Effect of scaling.* Given a stationary random process $X(t)$ and its autocorrelation function $R_x(\tau)$ and power spectrum $S_x(\omega)$, find the autocorrelation and power spectrum of $Y(t) = \alpha X(t)$, where α is a nonrandom real constant.

7.7 *Effect of adding a constant.* Given a zero-mean stationary random process $X(t)$ and its autocorrelation function $R_x(\tau)$ and power spectrum $S_x(\omega)$, find the autocorrelation and power spectrum of $Y(t) = X(t) + a$, where a is a nonzero nonrandom real constant.

7.8 *Find power spectrum from autocorrelation.* Given a wide-sense stationary random process $X(t)$, find its power spectrum if its autocorrelation is (a) $R(\tau) = \sigma^2 e^{-2b|\tau|}, b > 0$; (b) $R(\tau) = \sigma^2 \cos(\omega_0 \tau)$.

7.9 *Power spectrum of a triangular autocorrelation.* Find the power spectrum of the random process with the following triangular autocorrelation function

$$R(\tau) = \begin{cases} 1 - \frac{|\tau|}{T} & |\tau| < T \\ 0 & |\tau| \geq T \end{cases}$$

where T is a positive number.

7.10 *Power spectrum of frequency shifting.* Given a stationary random process $X(t)$ and its autocorrelation function $R_x(\tau)$ and power spectrum $S_x(\omega)$, find the autocorrelation and power spectrum of $Y(t) = X(t)e^{j\omega_0 t}$, where ω_0 is a nonrandom real constant.

7.11 *Power spectrum of derivative.* Show that (7.18) holds.

7.12 *Requirements to be a power spectrum.* Can the following functions be power spectra of some random processes? Justify your answer. (a) $\dfrac{\omega}{1+\omega^2}$; (b) $\cos(3\omega + \pi/4)$; (c) $e^{-|\omega|}$; (d) $e^{-|\omega+1|}$; (e) $\dfrac{2}{(\omega-1)^2+3} + \dfrac{2}{(\omega+1)^2+3}$.

7.13 *Requirements to be a power spectrum.* Determine which of the following functions are power spectra and which are not. Justify your answer.

$$S_1(\omega) = e^{-(\omega+1)} \qquad S_2(\omega) = \frac{1}{\sqrt{1-\omega^2}}$$

$$S_3(\omega) = \frac{\omega^2+5}{(\omega^2+2)(\omega^4-3)} \qquad S_4(\omega) = \frac{\omega^2+6}{1+\omega^2-2\omega^4}$$

$$S_5(\omega) = \frac{\omega^2+1}{(\omega^2+2)^2} \qquad S_6(\omega) = \frac{\omega^2+1}{1+\omega^2+4\omega^4} - \delta(\omega)$$

$$S_7(\omega) = \frac{|\sin\omega|}{1+3\omega^2+\omega^4} \qquad S_8(\omega) = \frac{j\omega^2}{1+2\omega^2+\omega^4}$$

$$S_9(\omega) = e^{-|\omega|} + \delta(\omega)$$

7.14 *Find power from power spectrum.* Determine the average power of the stationary random process $X(t)$ with the power spectrum $S(\omega) = \dfrac{4(\omega^2+1)}{(\omega^2+16)(\omega^2+4)}$.

7.15 *Find power from power spectrum.* Find the average power of the random process with the following power spectrum

$$S(\omega) = \begin{cases} 1 - \frac{|\omega|}{4\pi} & |\omega| < 4\pi \\ 0 & |\omega| \geq 4\pi \end{cases}$$

7.16 *Find power from power spectrum.* Two stationary random processes $X_1(t)$ and $X_2(t)$ have the following power spectra

$$S_{x_1}(\omega) = \frac{\omega^2 + 1}{\omega^4 + 5\omega^2 + 6}$$

$$S_{x_2}(\omega) = 10\pi\delta(\omega) + \frac{12}{9 + \omega^2} + \frac{6}{9 + (\omega - 4)^2} + \frac{6}{9 + (\omega + 4)^2}$$

Find their average powers, respectively.

7.17 *Autocorrelation of lowpass white noise.* Given a bandlimited noise process with power spectrum given by (7.21), show that its autocorrelation function is given by (7.22).

7.18 *Autocorrelation of bandpass white noise.* Given a bandlimited noise process with power spectrum given by (7.24), show that its autocorrelation function is given by (7.25).

7.19 *Cross-power spectrum of orthogonal processes.* Show that if real-valued random processes $X(t)$ and $Y(t)$ are orthogonal (i.e., if $E[X(t)Y(t')] = 0, \forall t, t'$), then

$$S_{xy}(\omega) = S_{yx}(\omega) = 0$$

7.20 *Cross-power spectrum of uncorrelated processes.* Show that if real-valued random processes $X(t)$ and $Y(t)$ are uncorrelated and have constant means, then

$$S_{xy}(\omega) = S_{yx}(\omega) = 2\pi E[X]E[Y]\delta(\omega)$$

7.21 *Sum of two random processes.* Given stationary random processes $X_1(t)$ and $X_2(t)$, along with their autocorrelation functions $R_{x_1}(\tau)$ and $R_{x_2}(\tau)$, and power spectra, $S_{x_1}(\omega)$ and $S_{x_2}(\omega)$, find the autocorrelation and power spectrum of the sum $X(t) = X_1(t) + X_2(t)$ if

(a) $X_1(t)$ and $X_2(t)$ are orthogonal; that is, $E[X_1(t)X_2(t')] = 0, \forall t, t'$
(b) $X_1(t)$ and $X_2(t)$ are uncorrelated; that is, $E[X_1(t)X_2(t')] = E[X_1]E[X_2], \forall t, t'$
(c) $X_1(t)$ and $X_2(t)$ are independent; that is, $X_1(t_1), X_1(t_2), \ldots, X_1(t_n)$ and $X_2(t_1')$, $X_2(t_2'), \ldots, X_2(t_m')$ are mutually independent for every n and m. Assume that they have time-invariant but in general nonzero mean functions.

Justify your answers.

7.22 *Sum of sine and cosine waveforms.* For two random processes $X(t) = 2\cos(\omega_0 t + \phi)$ and $Y(t) = 5\sin(\omega_0 t + \phi) + 10$, where ω_0 is a nonrandom positive number and $\phi \sim \mathcal{U}(\frac{\pi}{4}, \frac{9\pi}{4})$, form a new random process $Z(t) = X(t) + Y(t)$. Find

(a) average power of $X(t)$, $Y(t)$, and $Z(t)$
(b) power spectra of $X(t)$, $Y(t)$, and $Z(t)$

7.23 *Cross-power spectrum of weighted sums of processes.* Find the cross-power spectrum $S_{uv}(\omega)$ of random processes

$$U(t) = aX(t) + bY(t)$$
$$V(t) = cX(t) + dY(t)$$

where $X(t)$ and $Y(t)$ are jointly wide-sense stationary with power spectra $S_x(\omega), S_y(\omega)$, $S_{xy}(\omega)$.

7.24 *Cross-power spectrum of random process and its derivative.* Show that the cross-power spectrum of a stationary random process $X(t)$ with power spectrum $S_x(\omega)$ and its derivative $\dot{X}(t)$ is given by

$$S_{x\dot{x}}(\omega) = j\omega S_x(\omega)$$

7.25 *Sum of independent processes.* Given two mutually independent wide-sense stationary random processes $X(t)$ and $Y(t)$ with power spectra $S_x(\omega), S_y(\omega)$ and nonzero means $\bar{x}, \bar{y}$, respectively, find the power spectrum $S_z(\omega)$ of

$$Z(t) = X(t) + Y(t)$$

7.26 *Power spectrum of weighted sum of processes.* Consider a weighted sum of stationary and orthogonal random processes $X_i(t)$, given by

$$X(t) = \sum_{i=1}^{n} a_i X_i(t)$$

where a_i are real-valued constants. Show that the power spectrum of $X(t)$ is given by

$$S_x(\omega) = \sum_{i=1}^{n} a_i^2 S_{x_i}(\omega)$$

where $S_{x_i}(\omega)$ is the power spectrum of $X_i(t)$.

7.27 *Puzzling questions.* Raise two technical questions of your own that you are puzzled by and that are closely related to the material covered in this chapter. These questions should be important, not too simple, and not too involved.

7.11 Computer Exercises

7.1 *PSD of a random sinusoid*

(a) Generate a sequence of 150 data points by sampling (with 15 Hz sampling rate) a sample function of the random process $X(t) = 4\sin(5.3\pi t + \phi)$, where $\phi \sim \mathcal{U}(0, 2\pi)$.
(b) Use MATLAB routine `psd` to plot the power spectral density of the sequence in (a).
(c) Repeat (a) and (b) using the companion software P&R and compare with results of (a) and (b).

7.2 *PSD of a pulse train of random amplitude.* Consider a pulse train with period $T = 1$ second, pulse width $W = 0.4$ second and a random amplitude A that is uniformly distributed over $(0, 4)$. The pulse train has the same amplitude over all periods. Use the companion software P&R to do the following.

(a) Generate a discrete-time (with 10 Hz sampling rate and 256 time points) sample function of this pulse train. Plot it.
(b) Estimate the mean, autocorrelation, and power spectrum of the random pulse train. Plot the autocorrelation and power spectrum.

7.3 *PSD of a pulse train of random amplitude.* Consider a pulse train with period $T = 1$ second, pulse width $W = 0.4$ second and a random amplitude A that is uniformly distributed over $(0, 4)$. The amplitude of different periods are independent. Use the companion software P&R to do the following

(a) Generate a discrete-time (with 10 Hz sampling rate and 256 time points) sample function of this pulse train. Plot it.
(b) Estimate the mean, autocorrelation and power spectrum of the random pulse train. Plot the autocorrelation and power spectrum.

7.4 *PSD of a Gaussian white noise.* Use the companion software P&R to do the following.

(a) Generate a discrete-time (with 8 Hz sampling rate and 500 time points) sample function of a zero-mean Gaussian white noise process. Plot it.
(b) Estimate the mean, autocorrelation, and power spectrum of the Gaussian noise process generated in (a). Plot the autocorrelation (the first 20 points) and power spectrum.
(c) Repeat (a) and (b) with 5000 time points rather than 500 points.
(d) Comment on the accuracy of your results.

7.5 *PSD estimation.* The data files `m7_5x.dat` and `m7_5y.dat` in the companion software P&R contain two data records that are discrete-time (with 1000 Hz sampling rate) sample functions of two zero-mean ergodic random processes $X(t)$ and $Y(t)$, respectively. Use P&R to calculate and plot the power spectra of $X(t)$ and $Y(t)$, respectively. Are $X(t)$ and $Y(t)$ white?

7.6 *Resolution of power spectrum.*

(a) Use the companion software P&R to generate two discrete-time (with 10 Hz sampling rate and 100 time points) sample functions of two random sinusoids $X(t) = 2.6\sin(4.2\pi t + \phi)$ and $Y(t) = 2.1\sin(4.4\pi t + \theta)$, respectively, where $\phi \sim \mathcal{U}(0, 4\pi)$ and $\theta \sim \mathcal{U}(-\pi, \pi)$ are independent. Save the functions as data files `m7_6x.dat` and `m7_6y.dat`, respectively.

(b) Let $Z(t) = X(t)+Y(t)$. Use the following MATLAB routine to read in the two data records from the two data files, add the corresponding terms to yield a data record for $Z(t)$ and save it as `m7_6z.dat`:

```
cd data
load m7_6x.dat
load m7_6y.dat
z=m7_6x+m7_6y;
save m7_6z.dat z -ascii
cd ..
```

(c) Use P&R to estimate and plot the power spectrum of $Z(t)$.

(d) Repeat (a), (b), and (c) using 10 Hz sampling rate and 1000 time points. Adjust sequence length.

(e) Discuss the frequency resolutions of the two power spectra of $Z(t)$ based on your results.

7.7 *Estimating power spectrum of colored noise.*

(a) Use the companion software P&R to generate a 256-point (with 1000 Hz sampling rate) sample function of a zero-mean Gaussian colored noise with an exponential autocorrelation $R(\tau) = 3e^{-0.4|\tau|}$.

(b) Use the companion software P&R to estimate the power spectrum of the Gaussian colored noise sequence. Set the segment length to 256.

(c) Repeat part (b) with the segment length set to 32.

(d) Find the theoretical power spectrum of the colored noise in question.

(e) Compare your results in parts (b), (c), and (d).

7.12 Self-Test Problems

7.1 Answer the following questions briefly.

(a) Why don't we define spectrum of a random process as the Fourier transform of the process directly?

(b) Does the power spectrum of a random process contain the information of time shift of the process?

(c) Can the power spectrum of a random process be negative? Can the cross-power spectrum of two random processes be negative?

(d) Is the power spectrum the Fourier transform of the average power of a random process?
(e) What does a pair of delta functions at $\omega = \pm\omega_1$ in the power spectrum indicate?
(f) How do you understand the existence of the factor $1/2\pi$ in (7.9)?
(g) Is it possible to have "white" noise with autocorrelation function $R(\tau) = S_0\delta(\tau - t_0)$ for some $t_0 \neq 0$?
(h) Are the values of bandlimited white noise uncorrelated at distinct time instants?
(i) Is it true that a lowpass white noise and a bandpass white noise with the same intensity and bandwidth have the same average power?

7.2 Consider a waveform $X(t) = A\cos(\omega_0 t + \phi)$. Find the average power of $X(t)$ if

(a) $\phi \sim \mathcal{U}(\theta_0, \theta_0 + 2\pi)$, where θ_0 is a given constant, and nothing else in $X(t)$ is random
(b) $\phi \sim \mathcal{U}(\theta_0, \theta_0 + \frac{1}{2}\pi)$, where θ_0 is a given constant, and nothing else in $X(t)$ is random
(c) $\omega_0 \sim \mathcal{U}(\omega_1, \omega_2)$, where $\omega_2 > \omega_1 > 0$ are given, and nothing else in $X(t)$ is random. Note that the following is bounded:

$$\lim_{T\to\infty} \int_{-T}^{T} \frac{\sin(2\omega t + 2\phi)}{2t} dt$$

(d) A is a RV with mean $\bar{A}$ and variance σ_A^2, and nothing else in $X(t)$ is random
(e) $\phi \sim \mathcal{U}(\theta_0, \theta_0 + 2\pi)$ with θ_0 given; $\omega_0 \sim \mathcal{U}(\omega_1, \omega_2)$ with $\omega_2 > \omega_1 > 0$ given; A is a RV with mean $\bar{A}$ and variance σ_A^2; and ϕ, ω_0, A are mutually independent.
(f) Compare and discuss the results obtained in the above parts.

7.3 Given a random process $Y(t) = X(t)\cos\omega_0 t$, where ω_0 is constant and $X(t)$ is stationary random process with power spectrum $S_x(\omega)$, find the power spectrum of $Y(t)$.

7.4 Find the autocorrelation and average power of the random process with power spectrum $S(\omega) = \dfrac{3\omega^2 + 4}{2\omega^4 + 6\omega^2 + 4}$.

7.5 Find the autocorrelation and power spectrum of random process $X(t) = Y\sin\omega_0 t + Z\cos\omega_0 t$, where Y and Z are independent zero-mean RVs with the same variance σ^2.

7.13 Solutions to Self-Test Problems

7.1 (a) Because the required Fourier transform does not exist.
(b) No, power spectrum is the spectrum of "power" only. Two purely time-shifted versions of the same random process would have the power spectrum.
(c) The power spectrum of a random process *cannot* be negative. The cross-power spectrum of two random processes *can* be negative or even complex.
(d) No, the power spectrum is not the Fourier transform of the average power, but the Fourier transform of the autocorrelation function. This is similar to the case where the "amplitude" spectrum of a deterministic function is not the Fourier transform of the amplitude itself but the Fourier transform of the whole function.

(e) It indicates that the random process has a component of frequency ω_1.

(f) Because the integral is over the range of radian frequency ω rather than frequency f. The factor $1/2\pi$ would disappear were the integral over f.

(g) It is impossible to have a random process with $R(\tau) = S_0\delta(\tau - t_0)$ for some $t_0 \neq 0$ because this autocorrelation does not satisfy the condition $R(\tau) \leq R(0)$.

(h) No, they are actually correlated but the correlation is weak if the time difference is not small.

(i) Yes, since the area under the power spectral densities are equal.

7.2 Average power is given by

$$\begin{aligned} P_x &= E[X(t)] \\ &= E[A^2 \cos^2(\omega_0 t + \phi)] \overset{\text{if } A \text{ not random}}{=} \frac{A^2}{2} E[1 + \cos(2\omega_0 t + 2\phi)] \\ &= \frac{A^2}{2} + \frac{A^2}{2} E[\cos(2\omega_0 t + 2\phi)] \end{aligned}$$

(a)

$$P_x = \frac{A^2}{2} + \frac{A^2}{2} \int_{\theta_0}^{\theta_0 + 2\pi} \frac{1}{2\pi} \cos(2\omega_0 t + 2\phi) d\phi = \frac{A^2}{2}$$

(b)

$$\begin{aligned} E[X^2(t)] &= \frac{A^2}{2} + \frac{A^2}{2} \int_{\theta_0}^{\theta_0 + \frac{\pi}{2}} \frac{1}{\pi/2} \cos(2\omega_0 t + 2\phi) d\phi \\ &= \frac{A^2}{2} + \frac{A^2}{2\pi} \sin(2\omega_0 t + 2\phi) \Big|_{\theta_0}^{\theta_0 + \frac{\pi}{2}} \\ &= \frac{A^2}{2} - \frac{A^2}{\pi} \sin(2\omega_0 t + 2\theta_0) \end{aligned}$$

which is time variant. Thus the average power is the time average:

$$P_x = A\langle E[X^2(t)]\rangle = \lim_{T\to\infty} \frac{1}{2T} \int_{-T}^{T} \left[\frac{A^2}{2} - \frac{A^2}{\pi} \sin(2\omega_0 t + 2\theta_0)\right] dt = \frac{A^2}{2}$$

(c)

$$\begin{aligned} E[X^2(t)] &= \frac{A^2}{2} + \frac{A^2}{2} \int_{\omega_1}^{\omega_2} \frac{1}{\omega_2 - \omega_1} \cos(2\omega_0 t + 2\phi) d\phi \\ &= \frac{A^2}{2} + \frac{A^2}{\omega_2 - \omega_1} \frac{1}{2t} \sin(2\omega_0 t + 2\phi) \Big|_{\omega_1}^{\omega_2} \\ &= \frac{A^2}{2} - \frac{A^2}{4t} \frac{\sin(2\omega_2 t + 2\phi) - \sin(2\omega_1 t + 2\phi)}{\omega_2 - \omega_1} \end{aligned}$$

which is also time variant. Thus the average power is the time average:

$$\begin{aligned} P_x &= A\langle E[X^2(t)]\rangle \\ &= \lim_{T\to\infty}\frac{1}{2T}\int_{-T}^{T}\left[\frac{A^2}{2}-\frac{A^2}{4t}\frac{\sin(2\omega_2 t+2\phi)-\sin(2\omega_1 t+2\phi)}{\omega_2-\omega_1}\right]dt \\ &= \frac{A^2}{2} \end{aligned}$$

where

$$\lim_{T\to\infty}\frac{1}{2T}\int_{-T}^{T}\frac{A^2}{4t}\frac{\sin(2\omega_2 t+2\phi)-\sin(2\omega_1 t+2\phi)}{\omega_2-\omega_1}dt=0$$

follows from the fact that

$$\lim_{T\to\infty}\int_{-T}^{T}\frac{\sin(2\omega t+2\phi)}{2t}dt$$

is bounded.

(d)

$$P_x = E[A^2\cos^2(\omega_0 t+\phi)] = E[A^2]\cos^2(\omega_0 t+\phi) = [\bar{A}^2+\sigma_A^2]\cos^2(\omega_0 t+\phi)$$

(e)

$$\begin{aligned} P_x &= E[A^2\cos^2(\omega_0 t+\phi)] \\ &= E[A^2]E[\frac{1}{2}(1+\cos(2\omega_0 t+2\phi))] \\ &= [(\bar{A})^2+\sigma_A^2]\frac{1}{2}E[1+\cos 2\omega_0 t\cos 2\phi-\sin 2\omega_0 t\sin 2\phi] \\ &= \frac{1}{2}[(\bar{A})^2+\sigma_A^2][1+E[\cos 2\omega_0 t]\underbrace{E[\cos 2\phi]}_{=0}-E[\sin 2\omega_0 t]\underbrace{E[\sin 2\phi]}_{=0}] \\ &= \frac{1}{2}[(\bar{A})^2+\sigma_A^2] \end{aligned}$$

(f) The average power of $A\cos(\omega_0 t+\phi)$ is equal to one half of the mean-square value of A if the phase ϕ is not random or is uniformly distributed.

7.3 The autocorrelation function of $Y(t)$ is

$$\begin{aligned} R_y(t+\tau,t) &= E[Y(t+\tau)Y(t)] \\ &= E[X(t+\tau)\cos\omega_0(t+\tau)X(t)\cos\omega_0 t] \\ &= E[X(t+\tau)X(t)]\cos\omega_0 t\cos\omega_0(t+\tau) \\ &= R_x(\tau)\cos\omega_0 t\cos\omega_0(t+\tau) \end{aligned}$$

Clearly, $Y(t)$ is nonstationary since the above is a function of t as well as τ. The time average of the above is

$$\begin{aligned}
A\langle R_y(t+\tau,t)\rangle &\overset{(7.5)}{=} A\langle R_x(\tau)\cos\omega_0 t\cos\omega_0(t+\tau)\rangle \\
&= \frac{1}{2}R_x(\tau)A\langle\cos\omega_0\tau+\cos(2\omega_0 t+\omega_0\tau)\rangle \\
&= \frac{1}{2}R_x(\tau)\cos\omega_0\tau+\frac{1}{2}R_x(\tau)\lim_{T\to\infty}\frac{1}{2T}\int_{-T}^{T}\cos(2\omega_0 t+\omega_0\tau)dt \\
&= \frac{1}{2}R_x(\tau)\cos\omega_0\tau
\end{aligned}$$

Thus the power spectrum is

$$\begin{aligned}
S_y(\omega) &= \mathcal{F}[A\langle R_y(t+\tau,t)\rangle] \\
&= \mathcal{F}[\frac{1}{2}R_x(\tau)\cos\omega_0\tau] \\
&\overset{\text{Example 7.7}}{=} \frac{1}{4}[S_x(\omega-\omega_0)+S_x(\omega+\omega_0)]
\end{aligned}$$

7.4

$$\begin{aligned}
R(\tau) &= \mathcal{F}^{-1}[S(\omega)] \\
&= \mathcal{F}^{-1}\left[\frac{3\omega^2+4}{2\omega^4+6\omega^2+4}\right] \\
&= \mathcal{F}^{-1}\left[\frac{3\omega^2+4}{(2\omega^2+2)(\omega^2+2)}\right] \\
&= \mathcal{F}^{-1}\left[\frac{1}{2\omega^2+2}+\frac{1}{\omega^2+2}\right] \\
&= \mathcal{F}^{-1}\left[\frac{1}{2}\frac{1}{\omega^2+1}+\frac{1}{\omega^2+2}\right] \\
&= \mathcal{F}^{-1}\left[\frac{1}{4}\frac{2\cdot 1}{\omega^2+1^2}\right]+\mathcal{F}^{-1}\left[\frac{1}{2\sqrt{2}}\frac{2\cdot\sqrt{2}}{\omega^2+(\sqrt{2})^2}\right] \\
&= \frac{1}{4}e^{-|\tau|}+\frac{1}{2\sqrt{2}}e^{-\sqrt{2}|\tau|}
\end{aligned}$$

7.5

$$\begin{aligned}
R(\tau) &= E[X(t+\tau)X(t)] \\
&= E\{[Y\sin(\omega_0(t+\tau))+Z\cos(\omega_0(t+\tau))][Y\sin\omega_0 t+Z\cos\omega_0 t]\} \\
&= E[Y^2]\sin\omega_0 t\sin(\omega_0(t+\tau))+E[Z^2]\cos\omega_0 t\cos(\omega_0(t+\tau)) \\
&\quad +E[YZ][\sin\omega_0 t\cos(\omega_0(t+\tau))+\cos\omega_0 t\sin(\omega_0(t+\tau))] \\
&= \sigma^2[\sin\omega_0 t\sin(\omega_0(t+\tau))+\cos\omega_0 t\cos(\omega_0(t+\tau))]+\bar{y}\bar{z}\sin(2\omega_0 t+\omega_0\tau) \\
&= \sigma^2\cos(\omega_0\tau) \\
S(\omega) &= \mathcal{F}[R(\tau)] = \sigma^2\pi[\delta(\omega-\omega_0)+\delta(\omega+\omega_0)]
\end{aligned}$$

8

LINEAR SYSTEMS WITH RANDOM INPUTS

Man is a species that invents its own responses. It is out of this unique ability to invent, to improvise, his responses that cultures are born.

Ashley Montagu

What we have studied in the previous chapters is applied to the analysis of linear systems with random input in this chapter.

Main Topics

- Deterministic Linear Systems
- Time-Domain Analysis
- Frequency-Domain Analysis
- Linear Systems with White Noise Input
- Equivalent Noise Bandwidth

8.1 Deterministic Linear Systems

From the input-output viewpoint, a ***linear system*** $y(t) = L[x(t)]$ is one of which the response (output) $y(t)$ is related to the input $x(t)$ *linearly*; that is, with the following ***linearity property***:

$$\alpha_1 y_1(t) + \alpha_2 y_2(t) = L[\alpha_1 x_1(t) + \alpha_2 x_2(t)] \qquad \forall \alpha_1, \alpha_2, x_1(t), x_2(t) \tag{8.1}$$

where $y_1(t) = L[x_1(t)]$ and $y_2(t) = L[x_2(t)]$ are the responses to input $x_1(t)$ and $x_2(t)$, respectively. In other words, *the weighted sum of inputs of a linear system results in the same weighted sum of the corresponding outputs.*

Example 8.1: Differentiator as a Linear System

A differentiator (whose output is the derivative of its input) is a linear system because $y(t) = \frac{d}{dt}x(t)$ implies

$$\alpha_1 y_1(t) + \alpha_2 y_2(t) = \frac{d}{dt}[\alpha_1 x_1(t) + \alpha_2 x_2(t)] \qquad \forall \alpha_1, \alpha_2, x_1(t), x_2(t)$$

A linear system can be represented by its ***impulse response*** function $h(t, t_0) = L[\delta(t - t_0)]$, which is the response of the system to a delta function $\delta(t - t_0)$ (i.e., unit impulse applied at time t_0) as the input. In general, it depends on the impulse application time t_0 as well as the current time t.

A linear system is said to be ***time-invariant*** if and only if

- all the coefficients of the linear differential (or difference) equation(s) that describes the system are time invariant; or equivalently,
- its response to the impulse input $\delta(t - t_0)$ applied at any time t_0 is $h(t - t_0)$, meaning that it depends only on the time difference, rather than the two time instants. In other words, a time shift in the input will result only in the same time shift in the output: $y(t - t_0) = L[x(t - t_0)]$.

Powerful methods are available for ***linear time- invariant*** (***LTI***) systems.

Time invariance of a system, steady state of a deterministic time function, and stationarity of a random process are similar concepts. They all refer to the fact that the characteristics of a system, function, or process do not change with time.

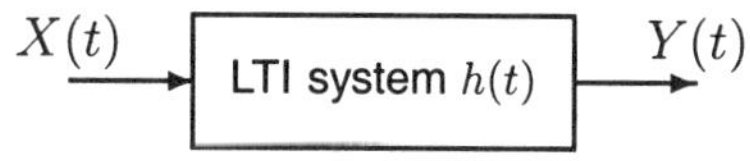

Figure 8.1: A linear time-invariant system.

In general, a linear or nonlinear system is said to be

- ***causal*** if its response (effect) does not come before its input (cause)
- ***stable*** if every bounded input results in bounded response (output)

An LTI system can be represented in terms of its

- impulse response function $h(t)$ in time domain
- ***frequency response function*** $H(\omega)$ in frequency domain as the Fourier transform of $h(t)$ or ***transfer function*** $H(s)$ as the Laplace transform of $h(t)$
- *state variables* in state space

An LTI system is

- ***causal*** if and only if $h(t) = 0, \ \forall t < 0$, which makes sense since $h(t)$ is the response to input $\delta(t)$ at $t = 0$ and $h(t') \neq 0$ for some $t' < 0$ implies a response prior to the input
- ***stable*** if and only if its impulse response function is absolutely integrable

$$\int_{-\infty}^{\infty} |h(t)| dt < \infty$$

or equivalently, all poles of its transfer function $H(s)$ are in the negative (left) half plane (excluding the imaginary axis for bounded input bounded output (BIBO) stability)

A noncausal system has the response (effect) that comes before the input (cause) and is thus not ***physically realizable***. The study of noncausal systems is, however, meaningful in both theory and practice. An unstable system cannot be used in most applications since its output may grow unbounded even for nice bounded input.

In this chapter, we shall study LTI systems with random process as input. In these cases, both response (output) and state of the system are random.

The response of a linear system has two components: response to input and response to the system's nonzero initial condition. We are concerned only with the response to input.

From both input-output and state-space viewpoints, in addition to the input-output linearity property (8.1), a linear system has to satisfy two more conditions:

- *state linearity*: the state of the system is linear with respect to the initial state
- *response additivity*: the responses to zero-input and zero-state are additive: $y(t) = y_{zi}(t) + y_{zs}(t)$, where $y_{zi}(t)$ and $y_{zs}(t)$ are zero-input and zero-state responses, respectively

In general, a linear or nonlinear system is said to be ***time-invariant*** if its characteristics are invariant with respect to time; or more specifically, if

- all the coefficients of the differential (or difference) equation(s) that describes the system are time invariant; or equivalently,
- a shift in time in the input will result *only* in the same time shift in the output.

Both linear systems and time-invariant systems are abstraction of physical systems. There is hardly any physical system that is either perfectly linear over the entire (state or input-output) space, due to e.g., saturation. It is also hard to find a system that is time invariant over an infinite period of time, due to e.g., aging. However, a great many of physical systems may be approximated quite accurately by the models of linear time-invariant systems over the space and time of most interest.

In essence, input-output description of a system looks the system from outside; whereas the state-space description is an internal representation of the system.

The stability used here is the so-called bounded input bounded output (BIBO) stability. It is an external property of the system, meaning that it is a property observed from outside of the system. There is another popular concept of stability, the *Lyapunov* (asymptotical) *stability*, which has a closer relation with the state-space representation of the system. It is an internal property of the system under the assumption of no external excitation. Mathematically speaking, the input-output description of a system can be viewed as a mapping from the input to the output. As such, the BIBO stability is equivalent to that this mapping is bounded. In contrast, the Lyapunov stability states that the state as a function of time is bounded for any small initial perturbation. In other words, if a system is Lyapunov stable, then its state will stay in a small neighborhood of the equilibrium point for any small initial disturbance, provided that no external excitation is applied. Although different, these two concepts of stability are closely related. For example, for a large class of time-invariant systems,

$$\text{Lyapunov asymptotical stability} \Longleftrightarrow \text{BIBO stability}$$

Note that the system itself considered in this chapter is deterministic. Its output and state are random only because its input is random.

It is also assumed throughout this chapter that the system has zero initial conditions. Or more rigorously, the system is relaxed at $t = -\infty$. A system is ***relaxed*** at $t = t_0$ if and only if its output is determined solely and uniquely by its input on and after t_0.

Example 8.2: An RC Circuit as a Linear System

Consider the RC circuit shown in Fig. 8.2. Treat the voltages $x(t)$ and $y(t)$ as the input and output. Let $a = 1/RC$. Clearly, it has the frequency response

$$H(\omega) = \frac{Y(\omega)}{X(\omega)} = \frac{\frac{1}{j\omega C} I(\omega)}{(R + \frac{1}{j\omega C}) I(\omega)} = \frac{1/(RC)}{j\omega + 1/(RC)} = \frac{a}{j\omega + a}$$

Thus, its impulse response is, denoting by $u(t)$ the unit step function,

$$h(t) = \mathcal{F}^{-1}[H(\omega)] = \mathcal{F}^{-1}\left[\frac{a}{j\omega + a}\right] \overset{\text{Table 7.5}}{=} ae^{-at}u(t)$$

Note that this circuit is

- a *linear system* since the following is true from the circuit theory:

$$\alpha_1 y_1(t) + \alpha_2 y_2(t) = L[\alpha_1 x_1(t) + \alpha_2 x_2(t)] \qquad \forall \alpha_1, \alpha_2, x_1(t), x_2(t)$$

- *time-invariant* provided RC is a constant
- *causal* since $h(t) = 0$ for all $t < 0$. This must be the case since a system is causal if and only if it is physically realizable and the circuit is a physical system in the real world.
- *stable* since $H(s)$ has only poles in the left half plane or

$$\int_{-\infty}^{\infty} |h(t)| dt = \int_{-\infty}^{\infty} |ae^{-at}u(t)| dt = \int_0^{\infty} ae^{-at} dt = -e^{-at}\Big|_0^{\infty} = 1 < \infty$$

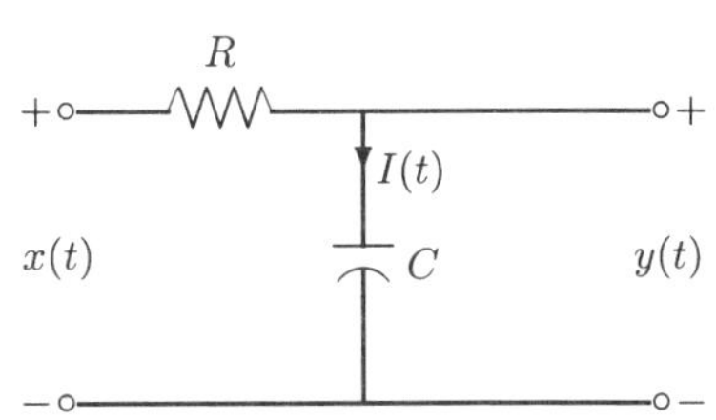

Figure 8.2: An RC circuit.

8.2 Time-Domain Analysis

Recall that the output $y(t)$ of an LTI system is the convolution of its input $x(t)$ and impulse response function $h(t)$:

$$y(t) = x(t) * h(t) = h(t) * x(t) \tag{8.2}$$

where the ***convolution*** of two functions is defined by

$$f(t) * g(t) = g(t) * f(t) = \int_{-\infty}^{\infty} g(\lambda) f(t-\lambda) d\lambda = \int_{-\infty}^{\infty} f(\lambda) g(t-\lambda) d\lambda$$

Since this holds for every corresponding pair of sample functions of the random processes $X(t)$ as input and $Y(t)$ as output, we may write

$$\boxed{Y(t) = X(t) * h(t)} \tag{8.3}$$

The ***mean*** *of the output* is the convolution of the input mean and the impulse response

$$\boxed{E[Y(t)] = E[X(t)] * h(t)} \tag{8.4}$$

or

$$\text{average response} = \text{response to average input}$$

which should be the case: (8.2) holds for every corresponding pair of sample functions of $X(t)$ and $Y(t)$ and thus should be valid for their means. In particular, if the input $X(t)$ has a time-invariant mean, so does the output $Y(t)$:

$$\begin{aligned} E[Y(t)] &= \bar{y} = \bar{x} \textstyle\int_{-\infty}^{\infty} h(t) dt = \bar{x} H(0) \\ \text{output dc component} &= (\text{input dc component}) \times (\text{system's dc gain}) \end{aligned} \tag{8.5}$$

where $H(0)$ is the ***dc gain*** of the system, given by

$$H(0) = H(\omega)\big|_{\omega=0} = \int_{-\infty}^{\infty} h(t) e^{-j\omega t} dt \Big|_{\omega=0} = \int_{-\infty}^{\infty} h(t) dt \tag{8.6}$$

The ***crosscorrelation*** of the output $Y(t)$ and a *stationary* input $X(t)$ is

$$\begin{aligned} R_{xy}(t+\tau, t) &= E[X(t+\tau)Y(t)] = E\left[X(t+\tau) \int_{-\infty}^{\infty} h(\lambda) X(t-\lambda) d\lambda\right] \\ &= \int_{-\infty}^{\infty} h(\lambda) E[X(t+\tau) X(t-\lambda)] d\lambda = \int_{-\infty}^{\infty} h(\lambda) R_x(\tau+\lambda) d\lambda \\ &\overset{t=-\lambda}{=} \int_{-\infty}^{\infty} h(-t) R_x(\tau - t) dt \overset{g(t)=h(-t)}{=} \int_{-\infty}^{\infty} g(t) R_x(\tau - t) dt \\ &= R_x(\tau) * g(\tau) = R_x(\tau) * h(-\tau) \end{aligned}$$

Thus

$$R_{xy}(\tau) = R_x(\tau) * h(-\tau) = \int_{-\infty}^{\infty} h(-t)R_x(\tau - t)dt$$
$$= \int_{-\infty}^{\infty} h(t)R_x(\tau + t)dt \tag{8.7}$$

$$R_{yx}(\tau) = R_{xy}(-\tau) = R_x(-\tau) * h(\tau) \overset{(6.16)}{=} R_x(\tau) * h(\tau) \tag{8.8}$$

They are usually nonzero since $Y(t)$ is the response to $X(t)$ and thus is correlated with $X(t)$.

The ***autocorrelation*** *of the response* $Y(t)$ to an input $X(t)$ is

$$R_y(t+\tau, t) = E[Y(t+\tau)Y(t)] = E\Big[\Big(\int_{-\infty}^{\infty} h(\lambda)X(t+\tau-\lambda)d\lambda\Big)Y(t)\Big]$$
$$= \int_{-\infty}^{\infty} h(\lambda)E[X(t+\tau-\lambda)Y(t)]d\lambda$$
$$= \int_{-\infty}^{\infty} h(\lambda)R_{xy}(t+\tau-\lambda, t)d\lambda$$

For a (wide-sense) stationary input, it becomes

$$R_y(\tau) = \int_{-\infty}^{\infty} h(\lambda)R_{xy}(\tau - \lambda)d\lambda = R_{xy}(\tau) * h(\tau) \tag{8.9}$$

Similarly,

$$R_y(\tau) = R_{yx}(\tau) * h(-\tau) \tag{8.10}$$

Consequently, the autocorrelations of the output and stationary input of an LTI system have the following relationship, substituting (8.7) into (8.9),

$$R_y(\tau) = \int_{-\infty}^{\infty}\int_{-\infty}^{\infty} h(t_1)h(t_2)R_x(\tau - t_1 + t_2)dt_1dt_2$$

or, by (8.8) and (8.10),

$$\boxed{R_y(\tau) = R_x(\tau) * h(\tau) * h(-\tau)} \tag{8.11}$$

The above indicates that for an LTI system,

$$\text{input is WSS} \quad \Longrightarrow \quad \text{output and input are jointly WSS} \tag{8.12}$$

This makes sense: Since a time-invariant system and a stationary process both have time-invariant characteristics, it is expected that the response of such a system to such an input would also have time-invariant characteristics.

The ***average power*** *of the response* $Y(t)$ to a stationary input is

$$P_y = E[|Y(t)|^2] = R_y(0)$$

(8.4) can also be shown as follows:

$$E[Y(t)] = E\left[\int_{-\infty}^{\infty} h(\lambda)X(t-\lambda)d\lambda\right] = \int_{-\infty}^{\infty} h(\lambda)E[X(t-\lambda)]d\lambda = h(t) * E[X(t)]$$
$$= E\left[\int_{-\infty}^{\infty} h(t-\lambda)X(\lambda)d\lambda\right] = \int_{-\infty}^{\infty} h(t-\lambda)E[X(\lambda)]d\lambda = E[X(t)] * h(t)$$

In the above and on several occasions, we have interchanged the order of integration and expectation. This is not always legitimate. In general, for a random process $X(t)$ and a nonrandom function $f(t)$, such order interchange is valid, that is

$$E\left[\int_a^b X(t)f(t)dt\right] = \int_a^b E[X(t)]f(t)dt$$

if $X(t)$ is bounded on the interval $[a, b]$, meaning that none of the sample functions of $X(t)$ tend to ∞ on any point on $[a, b]$ (which is true for all physical processes), and

$$\int_a^b E[|X(t)|]|f(t)|dt < \infty$$

where a and b could be infinite. These conditions are satisfied in almost all practical situations, including our cases.

(8.5) indicates that the average output will be zero if either the input has zero mean or the system has zero dc gain (i.e., the system's passband does not include $\omega = 0$).

For a *causal* LTI system, since $h(t) = 0, \forall t < 0$, its convolution with a random or nonrandom time function $g(t)$ can be simplified to

$$h(t) * g(t) = \int_0^{\infty} h(\lambda)g(t-\lambda)d\lambda = \int_{-\infty}^{t} g(\lambda)h(t-\lambda)d\lambda$$

where in our case, $g(t) = x(t), X(t), E[X(t)], h(-t),$ or $R(\tau)$. For example,

$$R_y(\tau) = \int_0^{\infty}\int_0^{\infty} h(t_1)h(t_2)R_x(\tau - t_1 + t_2)dt_1dt_2$$
$$P_y = E[|Y(t)|^2] = R_y(0) = \int_0^{\infty}\int_0^{\infty} h(t_1)h(t_2)R_x(t_2 - t_1)dt_1dt_2$$
$$R_{yx}(\tau) = \int_0^{\infty} h(t)R_x(-\tau + t)dt$$

The derivation of (8.10) is left as an exercise (problem 8.2).

In general, the response of a linear or nonlinear time-invariant system to a strictly stationary random input is always strictly stationary. This can be easily understood from the explanation following (8.12). Rigorously speaking, (8.12) is true only when the system was relaxed at $t = -\infty$ and input applied at $t = -\infty$. In other words, the response of a linear system, relaxed at $t = -\infty$, to a wide-sense stationary input, applied at $t = -\infty$, is also wide-sense stationary. However, the response of a linear stable system, relaxed at $t \leq 0$, to a wide-sense stationary input, applied at $t = 0$, is in general not wide-sense stationary but *asymptotically wide-sense stationary* (i.e., as time increases, it tends to be wide-sense stationary) due to the possible transient caused by the sudden application of input at $t = 0$.

Example 8.3: System Identification by Input-Output Crosscorrelation

A popular method of identifying (measuring the impulse response $h(t)$ of) an unknown LTI system is using the following system:

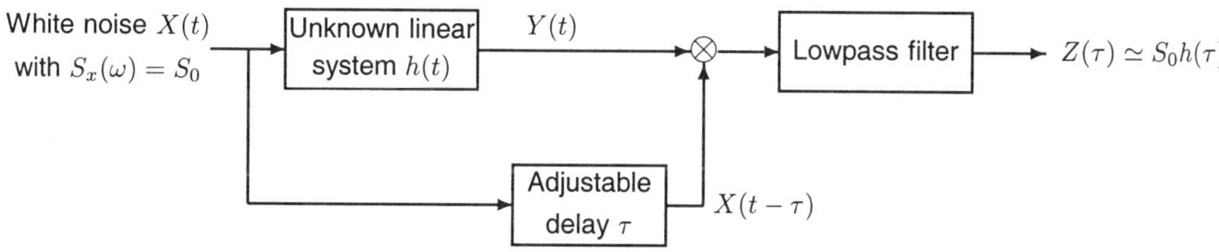

Figure 8.3: Input-output crosscorrelation approach to system identification.

Since the lowpass filter operates in such a way that its output is approximately the time average of its input, if $X(t)$ is ergodic white noise with autocorrelation $R_x(\tau) = S_0\delta(\tau)$, then we have, for $\tau \geq 0$,

$$\begin{aligned} Z(\tau) &\approx A\langle Y(t)X(t-\tau)\rangle && (A\langle\cdot\rangle \text{ stands for time average}) \\ &= E[Y(t)X(t-\tau)] && (\text{by ergodicity of } X(t)) \\ &= R_{yx}(\tau) = \int_{-\infty}^{\infty} h(t)R_x(\tau-t)dt = \int_{-\infty}^{\infty} h(t)S_0\delta(\tau-t)dt = S_0 h(\tau) \end{aligned}$$

Note that $Z(\tau)$ depends only on τ and thus if τ is fixed, then $Z(\tau)$ is (nearly) constant. Consequently, the impulse response can be obtained from $Z(\tau)$ by varying τ: $h(\tau) = Z(\tau)/S_0$, where the proportionality factor S_0, i.e., the power spectral intensity of the input noise, is known.

The following remarks are in order.

- Intuitively, identifying an unknown system by measuring its input-output relationship makes perfect sense. What is somewhat surprising is that measuring the crosscorrelation alone is sufficient.
- In practice, the input $X(t)$ need not be white provided that its power spectrum is nearly flat over an interval (called *bandwidth*, to be studied later) that is large compared with that of the unknown system.
- An advantage of this ***white-noise approach*** to ***system identification*** is that it is applicable in real time (while the system is in normal operation) by superimposing a small random input on the normal input.

Example 8.4: Time-Domain Analysis of RC Circuit with White Noise Input

Consider the RC circuit of Fig. 8.2 on p. 365. Let $a = 1/RC$. Suppose now that the input $x(t)$ is a stationary white noise process $X_1(t)$ with zero mean and autocorrelation $R_{x_1}(\tau) = S_0\delta(\tau)$. Recall that it has the impulse response function

$$h(t) = ae^{-at}u(t)$$

The autocorrelation of the output of this LTI system is given by, from (8.11)

$$\begin{aligned}
R_{y_1}(\tau) &= S_0\delta(\tau) * h(\tau) * h(-\tau) \\
&= S_0\left[\int_{-\infty}^{\infty} h(t)\delta(\tau - t)dt\right] * h(-\tau) \\
&= S_0[h(\tau) * h(-\tau)] \qquad (8.13) \\
&= S_0\int_{-\infty}^{\infty} h(-t)h(\tau - t)dt \\
&= S_0\int_{-\infty}^{\infty} ae^{at}u(-t)ae^{-a(\tau-t)}u(\tau - t)dt \\
&= a^2S_0e^{-a\tau}\int_{-\infty}^{0} e^{2at}u(\tau - t)dt \\
&= \begin{cases} a^2S_0e^{-a\tau}\int_{-\infty}^{0} e^{2at}dt & \tau \geq 0 \\ a^2S_0e^{-a\tau}\int_{-\infty}^{\tau} e^{2at}dt & \tau < 0 \end{cases} \\
&= \begin{cases} \frac{1}{2}aS_0e^{-a\tau} & \tau \geq 0 \\ \frac{a}{2}S_0e^{a\tau} & \tau < 0 \end{cases} \\
&= \frac{a}{2}S_0e^{-a|\tau|} = \frac{S_0}{2RC}e^{-|\tau|/RC} \qquad (8.14)
\end{aligned}$$

and the average power of the output of this LTI system is given by

$$P_{y_1} = E[|Y_1(t)|^2] = R_{y_1}(0) = \frac{aS_0}{2}$$

Note that the calculation for $\tau < 0$ is actually redundant in view of the symmetry property of the autocorrelation: $R(-\tau) = R(\tau)$. It, however, provides a check of the result obtained.

It is interesting to note that the time constant $a^{-1} = RC$ of the RC circuit is also the time constant for $R_{y_1}(\tau)$ to die out.

Example 8.5: Time-Domain Analysis of RC Circuit with Colored Noise Input

If the input to the previous RC circuit is colored noise $X_2(t)$ with autocorrelation $R_{x_2}(\tau) = S_2 e^{-b|\tau|}$, where $b \neq a, b \geq 0$, the autocorrelation of the output is shown on the next page to be, for $\tau \geq 0$,

$$\begin{aligned} R_{y_2}(\tau) &= R_{x_2}(\tau) * h(\tau) * h(-\tau) \\ &= \frac{aS_2}{a^2 - b^2}\left[ae^{-b\tau} - be^{-a\tau}\right] \end{aligned} \tag{8.15}$$

Thus, by symmetry $R(-\tau) = R(\tau)$, we have

$$R_{y_2}(\tau) = \frac{aS_2}{a^2 - b^2}\left[ae^{-b|\tau|} - be^{-a|\tau|}\right] \tag{8.16}$$

Note that if $S_2 = \frac{1}{2}bS_0$ then $R_{y_2}(\tau)$ tends to $R_{y_1}(\tau)$ of the previous example as b increases; that is,

$$\begin{aligned} R_{y_2}(\tau)|_{S_2=\frac{1}{2}bS_0} &= \frac{\frac{1}{2}baS_0}{a^2 - b^2}\left[ae^{-b|\tau|} - be^{-a|\tau|}\right] \\ &\xrightarrow{b\to\infty} \frac{1}{2}aS_0 e^{-a|\tau|} = R_{y_1}(\tau) \end{aligned}$$

This makes sense since the colored noise input $X_2(t)$ tends to be white (i.e., tends to have the same autocorrelation as the white noise $X_1(t)$) under this condition:

$$R_{x_2}(\tau)|_{S_2=\frac{1}{2}bS_0} = \frac{1}{2}bS_0 e^{-b|\tau|} \xrightarrow{b\to\infty} S_0\delta(\tau) = R_{x_1}(\tau)$$

which follows from

$$\begin{aligned} \mathcal{F}\left[\lim_{b\to\infty}\frac{1}{2}be^{-b|\tau|}\right] &= \lim_{b\to\infty}\mathcal{F}\left[\frac{1}{2}be^{-b|\tau|}\right] \overset{\text{Table 7.5}}{=} \lim_{b\to\infty}\frac{b^2}{\omega^2 + b^2} \\ &= 1 = \mathcal{F}[\delta(\tau)] \end{aligned}$$

A simpler solution to this problem is using a frequency-domain method, studied in the next section.

The results of these two examples will be used repeatedly later.

If the white noise in Example 8.4 has a nonzero mean $\bar{x}$, then the autocorrelation function calculated in Example 8.4 should be replaced by its corresponding autocovariance function. The mean of the output is given by

$$E[Y_1(t)] = \bar{x}H(0) = \bar{x}$$

which makes perfect sense: "The dc output voltage is equal to the dc input voltage" since the capacitor should be treated as open circuit as far as dc voltage is concerned.

The corresponding autocorrelation function is given by

$$R_{y_1} = R_{y_1} + (\bar{x})^2 = \frac{a}{2}S_0 e^{-a|\tau|} + (\bar{x})^2$$

(8.15) can be derived as follows:

$$\begin{aligned}
R_{y_2}(\tau) &= R_{x_2}(\tau) * h(\tau) * h(-\tau) \\
&= \int_{-\infty}^{\infty}\int_{-\infty}^{\infty} h(t_1)h(t_2)R_{x_2}(\tau - t_1 + t_2)dt_1 dt_2 \\
&= \int_{-\infty}^{\infty}\int_{-\infty}^{\infty} ae^{-at_1}u(t_1)ae^{-at_2}u(t_2)S_2 e^{-b|\tau - t_1 + t_2|}dt_1 dt_2 \\
&= a^2 S_2 \int_0^{\infty} e^{-at_2}\left[\int_0^{\tau+t_2} e^{-b(\tau - t_1 + t_2)}e^{-at_1}dt_1 + \int_{\tau+t_2}^{\infty} e^{b(\tau - t_1 + t_2)}e^{-at_1}dt_1\right]dt_2 \\
&= a^2 S_2 \int_0^{\infty} e^{-at_2}\left[e^{-b(\tau+t_2)}\int_0^{\tau+t_2} e^{-(a-b)t_1}dt_1 + e^{b(\tau+t_2)}\int_{\tau+t_2}^{\infty} e^{-(a+b)t_1}dt_1\right]dt_2 \\
&= a^2 S_2 \int_0^{\infty} e^{-at_2}\left[\frac{e^{-b(\tau+t_2)}}{a-b}\left(1 - e^{-(a-b)(\tau+t_2)}\right) + \frac{e^{b(\tau+t_2)}}{a+b}e^{-(a+b)(\tau+t_2)}\right]dt_2 \\
&= a^2 S_2 \int_0^{\infty} \left[\frac{1}{a-b}e^{-b\tau}e^{-(a+b)t_2} + \left(\frac{1}{a+b} - \frac{1}{a-b}\right)e^{-a\tau}e^{-2at_2}\right]dt_2 \\
&= \frac{a^2 S_2}{a^2 - b^2}\left[-e^{-b\tau}e^{-(a+b)t_2} + \frac{b}{a}e^{-a\tau}e^{-2at_2}\right]_0^{\infty} \\
&= \frac{aS_2}{a^2 - b^2}\left[ae^{-b\tau} - be^{-a\tau}\right]
\end{aligned}$$

For Example 8.5, the crosscorrelations can be found to be

$$\begin{aligned}
R_{x_1 y_1}(\tau) &= R_{x_1}(\tau) * h(-\tau) \\
&= \int_{-\infty}^{\infty} h(-t)R_{x_1}(\tau - t)dt \\
&= \int_{-\infty}^{\infty} h(-t)S_0\delta(\tau - t)dt \\
&= S_0 h(-\tau) \\
&= aS_0 e^{a\tau}u(-\tau) \\
R_{x_2 y_2}(\tau) &= R_{x_2}(\tau) * h(-\tau) \\
&= \int_{-\infty}^{\infty} h(-t)R_{x_2}(\tau - t)dt
\end{aligned}$$

$$
\begin{aligned}
&= \int_{-\infty}^{\infty} ae^{at}u(-t)S_2e^{-b|\tau-t|}dt \\
&= \int_{-\infty}^{0} ae^{at}S_2e^{-b|\tau-t|}dt \\
&= aS_2\left[\int_{-\infty}^{\min(0,\tau)} e^{at}e^{-b(\tau-t)}dt + \int_{\min(0,\tau)}^{0} e^{at}e^{b(\tau-t)}dt\right] \\
&= aS_2\left[\frac{e^{-b\tau}}{a+b}e^{(a+b)t}\Big|_{-\infty}^{\min(0,\tau)} + \frac{e^{b\tau}}{a-b}e^{(a-b)t}\Big|_{\min(0,\tau)}^{0}\right] \\
&= aS_2\left[\frac{1}{a+b}e^{-b\tau}e^{(a+b)\min(0,\tau)} + \frac{1}{a-b}e^{b\tau}\left(1-e^{(a-b)\min(0,\tau)}\right)\right] \\
&= \begin{cases} \dfrac{aS_2}{a+b}e^{-b\tau} & \tau > 0 \\ \dfrac{aS_2}{a^2-b^2}\left[(a+b)e^{b\tau}-2be^{a\tau}\right] & \tau \le 0 \end{cases}
\end{aligned}
$$

The other crosscorrelations $R_{y_1x_1}(\tau)$ and $R_{y_2x_2}(\tau)$ follow from symmetry $R_{xy}(-\tau) = R_{yx}(\tau)$. Note that for $\tau > 0$, $R_{x_1y_1}(\tau) = 0$ but $R_{x_2y_2}(\tau) \neq 0$. This makes sense: the current output $Y(t)$ of a causal system depends only on the past and current $X(t)$. If $X(t)$ is white noise, its future value $X(t+\tau)$ (for $\tau > 0$) is uncorrelated with its past and current values and thus uncorrelated with the current output $Y(t)$. Hence $R_{xy}(\tau) = E[X(t+\tau)Y(t)] = 0$. If $X(t)$ is colored noise, however, $X(t+\tau)$ (for $\tau > 0$) is correlated with its past and current values and thus correlated with the current output $Y(t)$. Hence $R_{xy}(\tau) = E[X(t+\tau)Y(t)] \neq 0$.

Note that

$$
\lim_{b\to\infty}\left[R_{x_2y_2}(\tau)\Big|_{S_2=\frac{1}{2}bS_0}\right] = \left\{\begin{array}{ll} 0 & \tau > 0 \\ aS_0e^{a\tau} & \tau \le 0 \end{array}\right\} = aS_0e^{a\tau}u(-\tau) = R_{x_1y_1}(\tau)
$$

which shows again that the colored noise tends to be white when $S_2 = \frac{1}{2}bS_0$ and $b \to \infty$.

(8.16) can also be obtained from the crosscorrelation calculated above using (8.9). For $\tau \ge 0$,

$$
\begin{aligned}
R_{y_2}(\tau) &= R_{x_2y_2}(\tau) * h(\tau) \\
&= \int_{-\infty}^{\infty} h(\tau-t)R_{x_2y_2}(t)dt \\
&= \int_{-\infty}^{\infty} R_{x_2y_2}(t)ae^{-a(\tau-t)}u(\tau-t)dt \\
&= ae^{-a\tau}\left[\int_{-\infty}^{0}\frac{aS_2}{a^2-b^2}\left((a+b)e^{bt}-2be^{at}\right)e^{at}dt + \int_0^{\tau}\frac{aS_2}{a+b}e^{-bt}e^{at}dt\right] \\
&= \frac{a^2S_2}{a+b}e^{-a\tau}\left(\frac{1}{a-b}\left[e^{(a+b)t}-\frac{b}{a}e^{2at}\right]_{-\infty}^{0} + \frac{1}{a-b}e^{(a-b)t}\Big|_0^{\tau}\right) \\
&= \frac{aS_2}{a^2-b^2}\left[ae^{-b\tau}-be^{-a\tau}\right]
\end{aligned}
$$

Then, (8.16) follows from the symmetry $R(-\tau) = R(\tau)$.

8.3 Frequency-Domain Analysis

For a nonrandom input $x(t)$, by the convolution property of Fourier (or Laplace) transforms, we have

$$y(t) = h(t) * x(t) \Longleftrightarrow Y(\omega) = H(\omega)X(\omega) \Longleftrightarrow Y(s) = H(s)X(s)$$

where $H(\omega)$ and $H(s)$ are the ***frequency response*** and ***transfer function*** of the system, respectively. This property is the most important reason for the popularity of Fourier (Laplace) transforms in signals and systems analysis.

For a random process, the corresponding frequency representation is its power spectral density. The ***power spectral density*** (***PSD***) of the output $Y(t)$ is related to that of a stationary input $X(t)$ by

$$\boxed{S_y(\omega) = |H(\omega)|^2 S_x(\omega)} \tag{8.17}$$

where $|H(\omega)|^2$ is called ***power gain*** (or ***power transfer function***) of the system. (8.17) is one of the most important results in frequency domain. It states that

output power spectrum = (power gain) $\times$ (input power spectrum)

It follows from (8.11) and the convolution and conjugation properties of Fourier transforms:

$$\begin{array}{ccccccc} R_y(\tau) & = & R_x(\tau) & * & h(\tau) & * & h(-\tau) \\ \updownarrow & & \updownarrow & & \updownarrow & & \updownarrow \\ S_y(\omega) & = & S_x(\omega) & \cdot & H(\omega) & \cdot & H(\omega)^* \end{array}$$

(8.17) implies that *a system may be viewed as a filter* that allows/disallows *selectively* certain frequency components of the input process to pass. Note that *$S(\omega)$ and $|H(\omega)|^2$ are associated with the average power of the process. They carry no information about the phase*: A process $Y(t)$ and any of its purely time-shifted version $Y(t-\tau)$ have the same power spectrum $S_y(\omega)$.

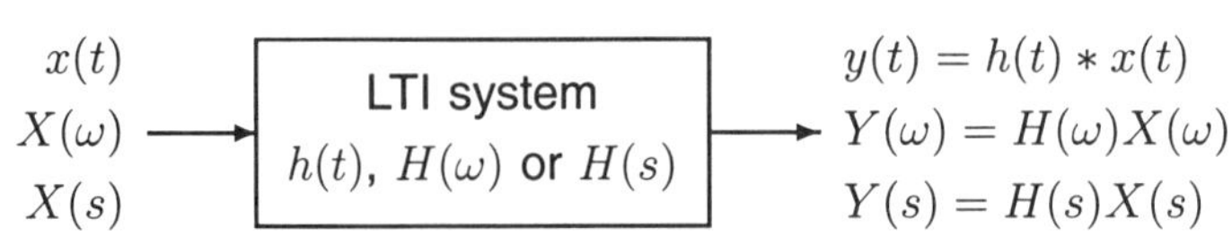

Figure 8.4: Input-output relationship of an LTI system with deterministic input.

Since autocorrelation and power spectrum are a Fourier transform pair,

$$\boxed{R_y(\tau) = \mathcal{F}^{-1}[|H(\omega)|^2 S_x(\omega)]} \tag{8.18}$$

the ***power** of the response* $Y(t)$ to a stationary input $X(t)$ is then

$$\boxed{P_y = R_y(0) = \frac{1}{2\pi}\int_{-\infty}^{\infty} S_y(\omega)d\omega = \frac{1}{2\pi}\int_{-\infty}^{\infty} |H(\omega)|^2 S_x(\omega)d\omega} \tag{8.19}$$

Similar to (8.17), the input-output cross-power spectra follow from (8.7)–(8.8):

$$\begin{array}{ccccc} R_{xy}(\tau) & = & R_x(\tau) & * & h(-\tau) \\ \updownarrow & & \updownarrow & & \updownarrow \\ S_{xy}(\omega) & = & S_x(\omega) & \cdot & H(\omega)^* \end{array} \qquad \begin{array}{ccccc} R_{yx}(\tau) & = & R_x(\tau) & * & h(\tau) \\ \updownarrow & & \updownarrow & & \updownarrow \\ S_{yx}(\omega) & = & S_x(\omega) & \cdot & H(\omega) \end{array} \tag{8.20}$$

It also follows from (8.17) and (8.20) or from (8.11) and (8.7) that

$$S_y(\omega) = S_{xy}(\omega)H(\omega) = S_{yx}(\omega)H(\omega)^*$$

Example 8.6: Output Power of an Ideal Bandpass System

(7.17) can now be shown as follows. Consider that a random process $X(t)$ is applied to an ideal LTI bandpass system with frequency response

$$H(\omega) = \begin{cases} 1 & 0 < \omega_1 \le |\omega| \le \omega_2 \\ 0 & \text{elsewhere} \end{cases}$$

This system (filter) allows all frequency components of $X(t)$ in the band $[\omega_1, \omega_2]$ to pass without distortion and nothing else to pass. Thus,

$$\begin{aligned} P_{x[\omega_1,\omega_2]} = P_y &= \frac{1}{2\pi}\int_{-\infty}^{\infty} S_y(\omega)d\omega = \frac{1}{2\pi}\int_{-\infty}^{\infty} |H(\omega)|^2 S_x(\omega)d\omega \\ &= \frac{1}{\pi}\int_0^{\infty} |H(\omega)|^2 S_x(\omega)d\omega = \frac{1}{\pi}\int_{\omega_1}^{\omega_2} S_x(\omega)d\omega \end{aligned}$$

Since the above is valid for any $\omega_1 \le \omega_2$, we have

$$S_x(\omega) \approx \lim_{\omega_1 \to \omega_2} \frac{1}{\omega_2 - \omega_1}\frac{1}{\pi}\int_{\omega_1}^{\omega_2} S_x(\omega)d\omega = \lim_{\omega_1 \to \omega_2} \frac{P_{x[\omega_1,\omega_2]}}{\omega_2 - \omega_1} \ge 0$$

Thus, property 1 of the power spectrum on p. 323 is also shown.

(8.17) can also be derived directly:

$$\begin{aligned}
S_y(\omega) &= \mathcal{F}[R_y(\tau)] = \int_{-\infty}^{\infty} R_y(\tau)e^{-j\omega\tau}d\tau \\
&= \int_{-\infty}^{\infty}\left[\int_{-\infty}^{\infty}\int_{-\infty}^{\infty} h(t_1)h(t_2)R_x(\tau+t_2-t_1)dt_1dt_2\right]e^{-j\omega\tau}d\tau \\
&= \int_{-\infty}^{\infty}\int_{-\infty}^{\infty}\int_{-\infty}^{\infty} h(t_1)h(t_2)R_x(t)e^{-j\omega(t-t_2+t_1)}dtdt_1dt_2 \\
&= \underbrace{\int_{-\infty}^{\infty} h(t_1)e^{-j\omega t_1}dt_1}_{H(\omega)} \underbrace{\int_{-\infty}^{\infty} h(t_2)e^{j\omega t_2}dt_2}_{H(\omega)^*} \underbrace{\int_{-\infty}^{\infty} R_y(t)e^{-j\omega t}dt}_{S_x(\omega)} \\
&= H(\omega)H(\omega)^*S_x(\omega) = |H(\omega)|^2S_x(\omega)
\end{aligned}$$

(8.20) can also be derived directly:

$$\begin{aligned}
S_{yx}(\omega) &= \mathcal{F}[R_{yx}(\tau)] \\
&= \int_{-\infty}^{\infty}\left[\int_{-\infty}^{\infty} h(\lambda)R_x(-\tau+\lambda)d\lambda\right]e^{-j\omega\tau}d\tau \\
&= \int_{-\infty}^{\infty}\int_{-\infty}^{\infty} h(\lambda)R_x(-t)e^{-j\omega(t+\lambda)}dtd\lambda \\
&= \underbrace{\int_{-\infty}^{\infty} h(\lambda)e^{-j\omega\lambda}d\lambda}_{H(\omega)} \underbrace{\int_{-\infty}^{\infty} R_x(t)e^{-j\omega t}dt}_{S_x(\omega)} \\
&= S_x(\omega)H(\omega) \\
S_{xy}(\omega) &= S_{yx}(-\omega) = S_x(-\omega)H(-\omega) = S_x(\omega)H(\omega)^*
\end{aligned}$$

If (two-sided) Laplace transforms, instead of Fourier transforms, are desirable and used, then the power spectrum of the output is given by

$$S_y(s) = S_y(\omega)|_{\omega=-js} = S_x(j\omega)H(j\omega)H(-j\omega)|_{\omega=-js} = S_x(s)H(s)H(-s)$$

that is,

$$S_y(s) = H(s)H(-s)S_x(s)$$

where $H(s)H(-s)$ is the ***power gain*** or ***power transfer function*** now.

Similarly, the cross-power spectra are given by

$$\begin{aligned}
S_{yx}(s) &= S_{yx}(\omega)|_{\omega=-js} = S_x(j\omega)H(j\omega)|_{\omega=-js} = S_x(s)H(s) \\
S_{xy}(s) &= S_x(s)H(-s)
\end{aligned}$$

The ***power** of the response* $Y(t)$ to a stationary input $X(t)$ is

$$\begin{aligned}
P_y &= \mathcal{L}^{-1}[S_y(s)]|_{\tau=0} = \mathcal{L}^{-1}[S_x(s)H(s)H(-s)]|_{\tau=0} \\
&= \frac{1}{2\pi j}\int_{-j\infty}^{j\infty} S_x(s)H(s)H(-s)ds
\end{aligned}$$

Example 8.7: Frequency-Domain Analysis of the RC Circuit

Consider again Examples 8.4 and 8.5 for the RC circuit. The power gain of the system is

$$|H(\omega)|^2 = \left|\frac{a}{j\omega + a}\right|^2 = \frac{|a|^2}{|j\omega + a|^2} = \frac{a^2}{(a + j\omega)(a - j\omega)} = \frac{a^2}{\omega^2 + a^2}$$

The power spectra of the white noise and colored noise inputs are

$$S_{x_1}(\omega) = S_0$$
$$S_{x_2}(\omega) = \mathcal{F}[S_2 e^{-b|\tau|}] \overset{\text{Table 7.5}}{=} \frac{2bS_2}{\omega^2 + b^2}$$

The power spectra of the corresponding outputs are

$$S_{y_1}(\omega) = |H(\omega)|^2 S_{x_1}(\omega) = \frac{a^2 S_0}{\omega^2 + a^2} \tag{8.21}$$

$$S_{y_2}(\omega) = |H(\omega)|^2 S_{x_2}(\omega) = \frac{a^2}{\omega^2 + a^2}\frac{2bS_2}{\omega^2 + b^2} \tag{8.22}$$

The autocorrelations of the corresponding outputs are

$$\begin{aligned}
R_{y_1}(\tau) &= \mathcal{F}^{-1}[S_{y_1}(\omega)] = \mathcal{F}^{-1}\Big[\frac{a^2 S_0}{\omega^2 + a^2}\Big] \overset{\text{Table 7.5}}{=} \frac{a}{2} S_0 e^{-a|\tau|} \\
R_{y_2}(\tau) &= \mathcal{F}^{-1}[S_{y_2}(\omega)] = \mathcal{F}^{-1}\Big[\frac{a^2}{\omega^2 + a^2}\frac{2bS_2}{\omega^2 + b^2}\Big] \\
&= \mathcal{F}^{-1}\Big[\frac{2a^2 bS_2}{a^2 - b^2}\Big(\frac{1}{\omega^2 + b^2} - \frac{1}{\omega^2 + a^2}\Big)\Big] \\
&\overset{\text{Table 7.5}}{=} \frac{aS_2}{a^2 - b^2}\left[ae^{-b|\tau|} - be^{-a|\tau|}\right]
\end{aligned}$$

The average powers of the corresponding outputs are

$$P_{y_1} = R_{y_1}(0) = \frac{a}{2} S_0, \qquad P_{y_2} = R_{y_2}(0) = \frac{aS_2}{a + b}$$

Compared with Examples 8.4 and 8.5, frequency-domain analysis is clearly much simpler than the time-domain method. This is true, in general, when the system's frequency response function and input power spectrum are both rational (i.e., ratios of polynomials).

Example 8.8: Frequency-Domain Analysis of an RL Circuit

Consider the following circuit excited by zero-mean white noise $X(t)$ with $S_x(\omega) = S_0$. Let $a = R/L$. From the circuit, we have

$$H(\omega) = \frac{Y(\omega)}{X(\omega)} = \frac{\alpha R I(\omega)}{(j\omega L + R)I(\omega)} = \frac{\alpha R}{j\omega L + R} = \frac{\alpha a}{j\omega + a}$$

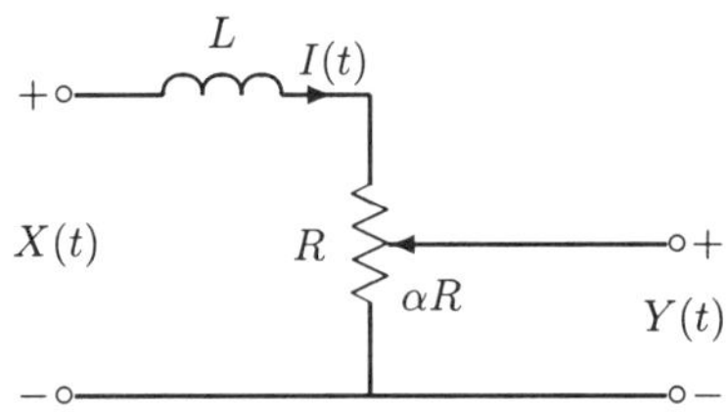

(a) Determine $S_y(\omega)$ and $S_{yx}(\omega)$:

$$S_y(\omega) = S_x(\omega)|H(\omega)|^2 = S_0 \cdot \left|\frac{\alpha a}{j\omega + a}\right|^2 = \frac{S_0 \alpha^2 a^2}{\omega^2 + a^2}$$

$$S_{yx}(\omega) = S_x(\omega)H(\omega) = \frac{\alpha a S_0}{j\omega + a}$$

(b) Determine $R_y(\tau)$ and $R_{yx}(\tau)$:

$$R_y(\tau) = \mathcal{F}^{-1}[S_y(\omega)] = \mathcal{F}^{-1}\left[\frac{1}{2}\alpha^2 a S_0 \frac{2a}{\omega^2 + a^2}\right] \overset{\text{Table 7.5}}{=} \frac{1}{2}\alpha^2 a S_0 e^{-a|\tau|}$$

$$R_{yx}(\tau) = \mathcal{F}^{-1}[S_{yx}(\omega)] = \mathcal{F}^{-1}\left[\frac{\alpha a S_0}{j\omega + a}\right] \overset{\text{Table 7.5}}{=} \alpha a S_0 e^{-a\tau} u(\tau)$$

where the unit step function guarantees that $Y(t)$ is uncorrelated with future $X(t)$, which makes sense since $X(t)$ is white noise.

(c) Determine P_y, $\bar{y}$, and σ_y^2:

$$P_y = R_y(0) = \frac{1}{2}\alpha^2 a S_0, \qquad \bar{y} = \bar{x}H(0) = \alpha\bar{x} = 0$$

$$\sigma_y^2 = E[Y^2] - (\bar{y})^2 = P_y - (\bar{y})^2 = \frac{1}{2}\alpha^2 a S_0 - \alpha^2(\bar{x})^2 = \alpha^2\left(\frac{1}{2}a S_0 - (\bar{x})^2\right)$$

($\bar{y} = \alpha\bar{x}$) makes sense: output dc component = $\alpha \times$ (input dc component).

Example 8.9: Probabilistic Analysis of RL Circuit by P&R

Consider the RL circuit of the last example above with $R = 2\Omega$ and $L = 100mH$. The input is white noise $X(t)$ with $S(\omega) = 3\pi$. We demonstrate how the companion software P&R can be used to analyze this circuit.

Suppose that the data record in file `e8_9x.dat` is a discrete-time sample function of the white noise input. Assume $\alpha = 0.5$. Note that $a = R/L = 20$ and $\alpha a = 10$. Then the response of the system, and autocorrelations and power spectra of the input and output can be obtained by P&R according to the following steps:

S1. Click "RPResponse" in the main window of P&R.
S2. Click "Rational System." You will be prompted to enter the name of a data file.
S3. Choose "data" subdirectory, enter the data file name "`e8_9x.dat`" and click "Ok."
S4. Fill out the window "**Response Generator for Rational System**" as shown in Fig. 8.5 and click "Ok." You will be prompted for a data file name since "Save to file" box is checked. Choose "data" subdirectory and enter the data file name "`e8_9y.dat`." Click "Ok." Then a MATLAB figure window will appear with a plot of the output $y(t)$ (see Fig. 8.6 (a)).

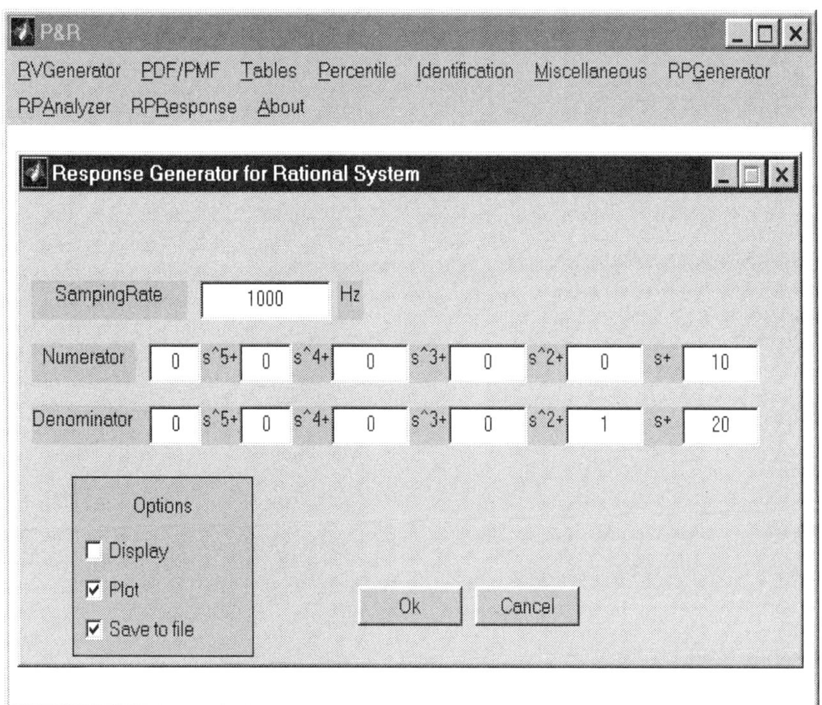

Figure 8.5: Generating response of a rational system by P&R.

S5. Follow the steps in Example 6.15 to get two autocorrelation plots for the input and output contained in `e8_9x.dat` and `e8_9y.dat`, respectively. The plots for the small τ are shown in Fig. 8.6 (c) and (d).

S6. Follow the steps in Example 6.17 to get an crosscorrelation plot for the output and input (see Fig. 8.6 (b)).

S7. Follow the steps in Example 7.8 to get two power spectrum plots for the input and output (see Fig. 8.6 (e) and (f)).

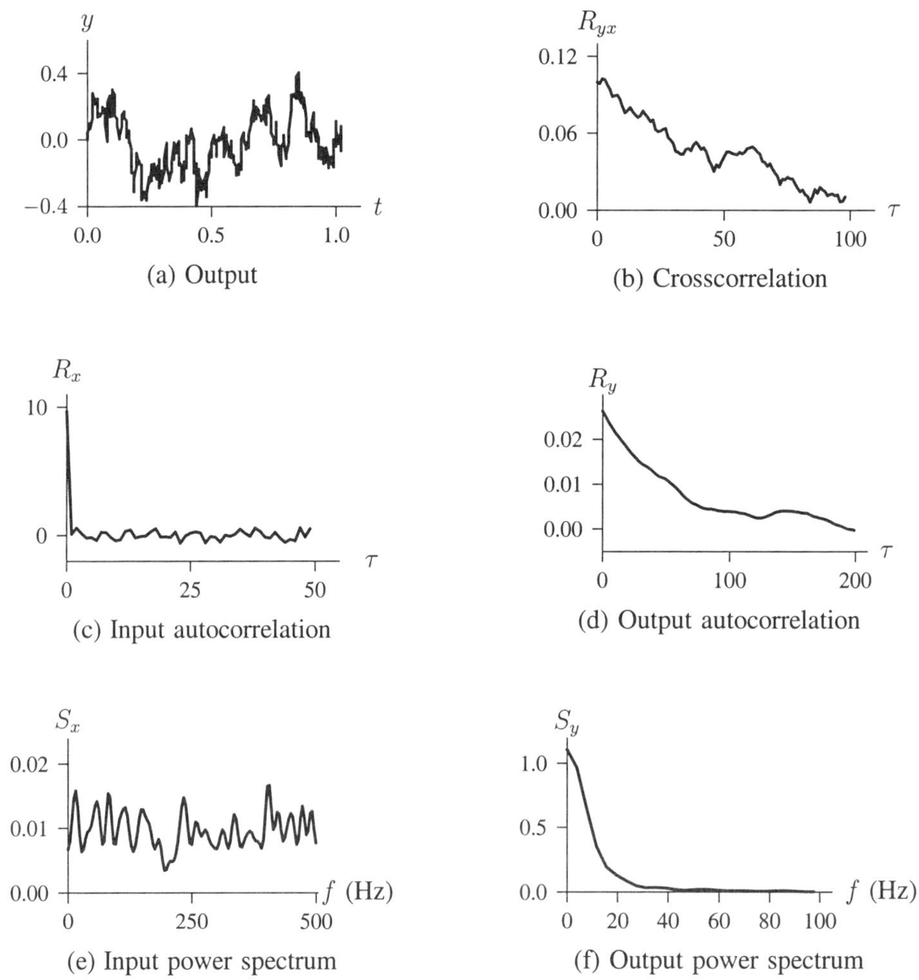

Figure 8.6: Correlation and power spectra of the input and output of the rational system.

8.4 Summary of Input-Output Relationships

For a linear time-invariant system, the relationships between its input and output in terms of some "amplitude" functions are summarized in Fig. 8.7, where x_i and y_i are a pair of sample functions of the input $X(t)$ and output $Y(t)$; X_i and Y_i are the corresponding Fourier (Laplace) transforms; the use of notations $\bar{x}$ and $\bar{y}$ implicitly assumes that $X(t)$ has a time-invariant mean.

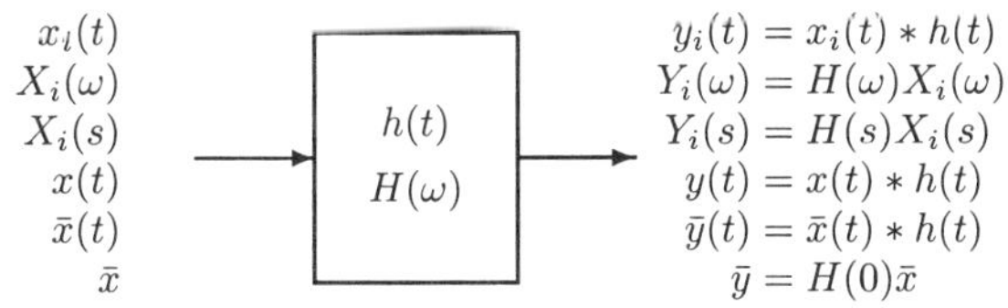

Figure 8.7: "Amplitude" relationships between input and output of an LTI system.

The relationships between its stationary input and output in terms of some "power" (quadratic) functions are summarized in Fig. 8.8. For these relationships, the LTI system should be replaced with the cascade of itself and its time-reversed (adjoint) system.

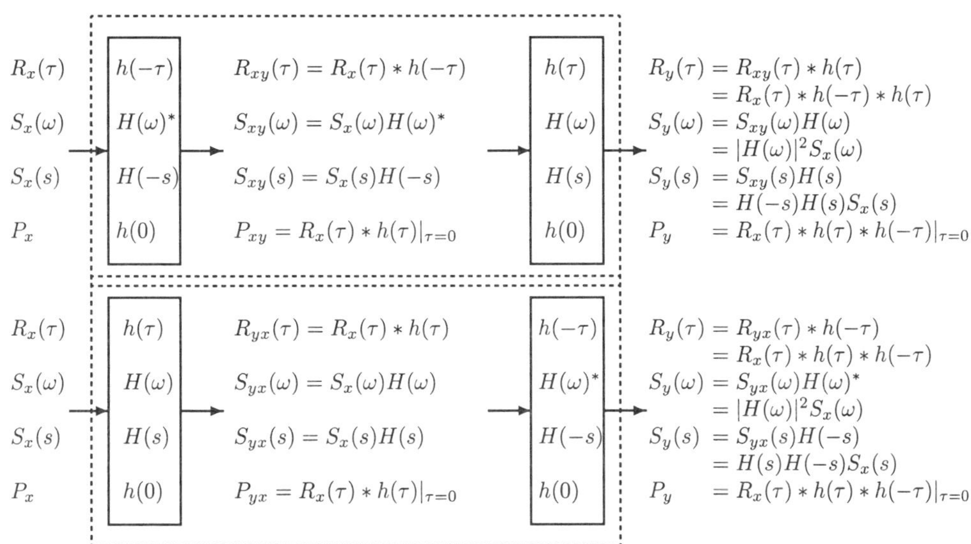

Figure 8.8: "Power" relationships between input and output of an LTI system.

In engineering practice, frequency-domain methods are usually simpler and thus more preferable. Frequency-domain analysis is usually simpler than time-domain analysis in the cases where the frequency response (or transfer) function of the system and the input power spectrum are both rational (a ratio of polynomials). This advantage is greater for a more complex system. If, however, the impulse response and the autocorrelation function of the input have simple nonzero form over a finite time interval, then time-domain methods are usually simpler.

Clearly, for "amplitude" relationships, the output is the convolution in time domain (or multiplication in frequency domain) of the input and the system's impulse response function (or frequency response, transfer function). Fig. 8.7 is valid for arbitrary (nonstationary as well as stationary) input, except for the last line, which holds only for input with a time-invariant mean.

For "power" relationships, the output is the *two-fold* convolution in time domain (or double multiplications in frequency domain) of the input and the system's impulse response (or frequency response, transfer) function. Or equivalently, the output is the convolution in time domain (or multiplication in frequency domain) of the input and the impulse response (or frequency response, transfer) function of an equivalent LTI system consisting of the cascade of the original system and its *adjoint system* (simply put, the adjoint system of an LTI system is its time-reversed system). Note that interchange the order in the cascade of these two systems does not alter the (external) input-output "power" relationships.

For simplicity, only results for stationary input are given in Fig. 8.8. Results similar to Fig. 8.8 but more complex also hold for nonstationary input.

Figs. 8.7 and 8.8 hold for a system with an input applied at $t = -\infty$. As a consequence, the input and output of a linear time-invariant system are jointly wide-sense stationary if the input is wide-sense stationary and applied at $t = -\infty$. In fact, the following stronger results hold under the same conditions:

- The input and output are jointly strict-sense stationary.
- The output is ergodic if the input is ergodic.

If the input was not applied at $t = -\infty$, then the response of the linear time-invariant system is in general neither strict- nor wide-sense stationary because the existence of the nonzero initial condition certainly indicates that the choice of time origin is important.

As stated before, the response of a linear system consists of two parts: the response to input, known as *zero-state response*, and the response due to the fact that the system was not relaxed (i.e., the system has a nonzero initial condition), which is known as *zero-input response*. Most students are familiar only with the zero-state response. The convolution-based methods, including the time-domain analysis of Section 8.2, are applicable only to the zero-state response because the convolution relationship between the input and output holds only for the zero-state response. This is also the case for the methods based on Laplace, Fourier, or z-transforms unless the nonzero initial conditions are taken into account. The method based on the solution of a differential or difference can yield the total response by utilizing the initial condition. The state variable method will yield the total response.

8.5 Linear Systems with White Noise Input

The dependence of the response, along with its autocorrclation and power spectrum, of a linear time-invariant system to white noise $X(t)$ is given in Fig. 8.9 below. In general, the output is colored (autocorrelated), except for *memoryless systems*.

The *crosscorrelations* between the input and output are

$$R_{xy}(\tau) = R_x(\tau) * h(-\tau) = \int_{-\infty}^{\infty} h(-t)S_0\delta(\tau - t)dt = S_0h(-\tau) \qquad (8.23)$$

$$R_{yx}(\tau) = R_x(\tau) * h(\tau) = \int_{-\infty}^{\infty} h(t)S_0\delta(\tau - t)dt = S_0h(\tau) \qquad (8.24)$$

(8.24) also follows from (8.23) and the mirror image property: $R_{yx}(\tau) = R_{xy}(-\tau)$. The corresponding *power spectra* are given by

$$S_{xy}(\omega) \overset{(8.23)}{=} S_0H(\omega)^*, \qquad S_{yx}(\omega) \overset{(8.24)}{=} S_0H(\omega) \qquad (8.25)$$

The *autocorrelation* and *average power* of the output are given by

$$\begin{aligned} R_y(\tau) &= S_0\delta(\tau) * h(\tau) * h(-\tau) \\ &= S_0[h(\tau) * h(-\tau)] \qquad (8.26) \\ &= S_0\int_{-\infty}^{\infty} h(-t)h(\tau - t)dt = S_0\int_{-\infty}^{\infty} h(t)h(\tau + t)dt \end{aligned}$$

$$P_y = R_y(0) \overset{(8.26)}{=} S_0\int_{-\infty}^{\infty} |h(t)|^2dt \qquad (8.27)$$

$$\overset{(8.19)}{=} \frac{S_0}{2\pi}\int_{-\infty}^{\infty} |H(\omega)|^2d\omega \qquad (8.28)$$

In fact, the last two equations are also connected by the Parseval theorem.

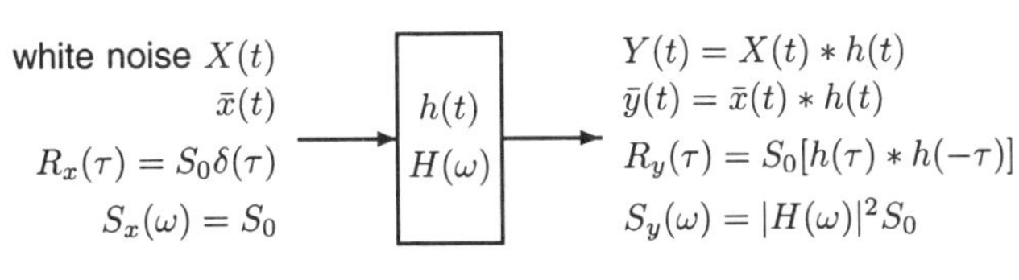

Figure 8.9: Input-output relationships of an LTI system with white noise input.

Example 8.10: Response of an Ideal Bandpass System to White Noise

An ideal ***bandpass system*** or ***bandpass filter*** with frequency response

$$H(\omega) = \begin{cases} H(\omega_c) & |\omega \pm \omega_c| < W/2 \\ 0 & \text{elsewhere} \end{cases} \tag{8.29}$$

has a ***bandwidth*** W, where ω_c is called its ***center frequency***. If it is excited by stationary white noise with zero mean and power spectrum $S_x(\omega) = S_0$, its output has the power spectrum

$$S_y(\omega) \overset{(8.17)}{=} |H(\omega)|^2 S_0 \overset{(8.29)}{=} \begin{cases} S_0|H(\omega_c)|^2 & |\omega \pm \omega_c| < W/2 \\ 0 & \text{elsewhere} \end{cases} \tag{8.30}$$

That is, the output is (bandpass) bandlimited white noise. Hence, its autocorrelation and power are given by

$$R_y(\tau) \overset{(7.25)}{=} \frac{S_0|H(\omega_c)|^2 W}{\pi} \frac{\sin(\tau W/2)}{\tau W/2} \cos(\omega_c \tau) \tag{8.31}$$

$$P_y \overset{(7.23)}{=} \frac{S_0|H(\omega_c)|^2 W}{\pi} \tag{8.32}$$

Its average output is always zero even if the input has nonzero mean $\bar{x}$: $E[Y(t)] = \bar{x}H(0) = 0$ since dc component cannot pass through an (ideal) bandpass system because $H(0) = 0$; that is, $\omega = 0$, the frequency of dc component, is not in its ***passband*** $(\omega_c - W/2, \omega_c + W/2)$.

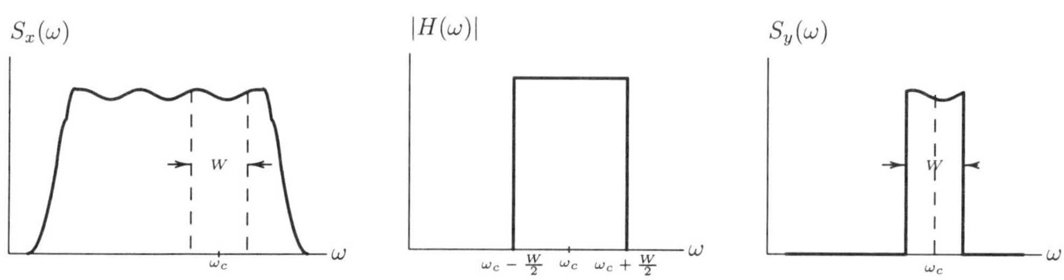

Figure 8.10: Input and output power spectra of an ideal bandpass system.

Example 8.11: Filtering — Noise Removal

Data file `e8_11x.dat` contains a noisy signal $X(t) = S(t) + N(t)$, where $S(t) = A\sin(\omega_0 \pi t + \theta)$ is the signal, with $\theta \sim \mathcal{U}(0, 2\pi)$, and $N(t)$ is white noise with power spectrum S_0. The bottom left plot in Fig. 8.11 depicts $X(t)$. The response of a bandpass filter with center frequency $\omega_c = \omega_0 = 800\pi$ and a 20Hz bandwidth to $X(t)$ can be obtained by P&R as follows:

1. Click "RPResponse" in the main window of P&R.
2. Click "Bandpass System." You will be prompted to enter a data file name.
3. Choose "data" subdirectory, enter file name "`e8_11x.dat`" and click "Ok."
4. Fill out the window "**Response Generator for Bandpass System**" as shown in Fig. 8.11 and click "Ok." Then the output $Y(t)$ is plotted on bottom right in Fig. 8.11.

Clearly, the noise in the output of the bandpass filter is reduced greatly.

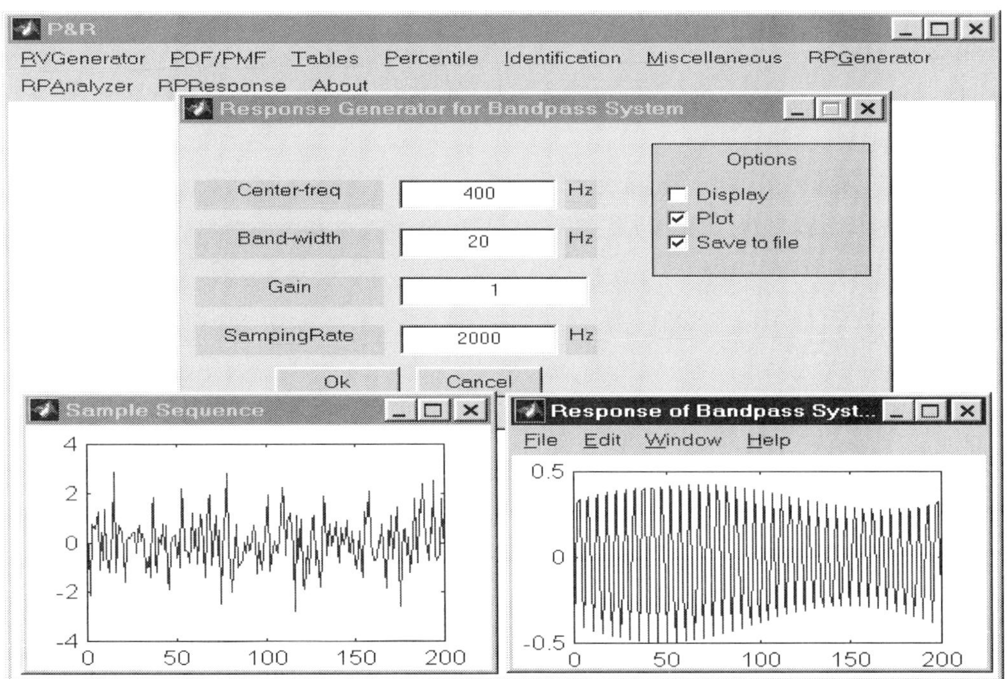

Figure 8.11: Noise reduction by a bandpass filter.

8.6 Equivalent Noise Bandwidth

Most signals have significant frequency components concentrated in a finite band, the width of which is called ***bandwidth*** of the signal. Various bandwidths can be defined in terms of how "significance" is interpreted. The bandwidth of a system can be defined as the width of the frequency range outside of which the frequency components will be attenuated to almost zero.

Example 8.12: Response of an Ideal Lowpass System to White Noise

An ideal ***lowpass system*** or ***lowpass filter*** with frequency response

$$H(\omega) = \begin{cases} H(0) & |\omega| < W \\ 0 & \text{elsewhere} \end{cases} \tag{8.33}$$

has a ***bandwidth*** W. Its response to stationary white noise with zero-mean and power spectrum $S_x(\omega) = S_0$ has power spectrum

$$S_y(\omega) \overset{(8.17)}{=} |H(\omega)|^2 S_0 \overset{(8.33)}{=} \begin{cases} |H(0)|^2 S_0 & |\omega| \leq W \\ 0 & \text{elsewhere} \end{cases}$$

and is thus (lowpass) bandlimited white noise. Hence, its autocorrelation and average power are given by (7.22)–(7.23) with S_0 replaced by $|H(0)|^2 S_0$. If the input has a nonzero mean $\bar{x}$, then its average output is $\bar{y} = H(0)\bar{x}$, which is nonzero if $\bar{x} \neq 0$.

Clearly, the output of a lowpass filter has significant frequency components only over $|\omega| < W$.

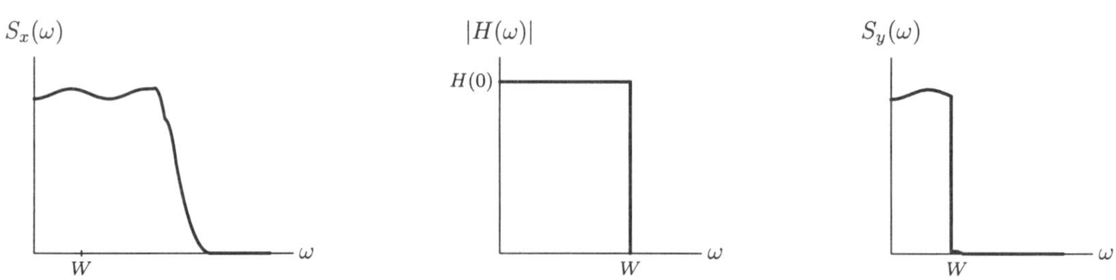

Figure 8.12: Input and output power spectra of an ideal lowpass system.

Various definitions of bandwidth are possible. For an ideal lowpass or bandpass system (or signal), their bandwidths under various definitions coincide. This should be the case since different definitions are based on different interpretations of what "significant" components really mean in signals and systems analysis. Owing to this, the differences among these bandwidths are small and may be neglected if the frequency response of the system is close to the ideal case.

No matter how the bandwidth of a system is defined, it should depend only on the parameters of the system itself, regardless of its input.

The correlation coefficient of the response of the ideal lowpass system (8.33) to *zero-mean* white noise is that of lowpass white noise, given by (7.26). The corresponding ***correlation time*** is

$$\tau_c \overset{(6.12)}{=} \int_0^\infty \rho_y(\tau)d\tau \overset{(7.26)}{=} \int_0^\infty \frac{\sin(W\tau)}{W\tau}d\tau = \frac{\pi}{2W}$$

The correlation coefficient of the response of the ideal bandpass system (8.29) to *zero-mean* white noise is that of bandpass white noise, given by (7.27). Thus, it can be seen that $\rho_y^{BP}(\tau)$ has two multiplicative components: the carrier $\cos(\omega_c\tau)$, which varies quickly, and the envelope $\frac{\sin(W\tau/2)}{W\tau/2}$, which varies slowly. The ***correlation time*** is often defined by the slow varying component alone

$$\tau_c \overset{(6.12)}{=} \int_0^\infty \rho_y(\tau)d\tau \overset{(7.27)}{=} \int_0^\infty \frac{\sin(W\tau/2)}{W\tau/2}\cos(\omega_c\tau)d\tau \approx \int_0^\infty \frac{\sin(W\tau/2)}{W\tau/2}d\tau = \frac{\pi}{W}$$

This is convenient and reasonable since autocorrelation (coupling) depends primarily on the magnitude (i.e., envelope) of the correlation coefficient.

Note that *the correlation time of the bandlimited white noise is inversely proportional to the bandwidth.* This makes sense because a smaller bandwidth implies that fewer frequency components can get through the system (filter) and thus those that do get through have a stronger correlation and thus a larger correlation time.

A bandpass system with a center frequency much higher than its bandwidth (i.e., $\omega_c \gg W$) is called a ***narrowband system***.

(8.32) can be used to measure the power spectrum of a random process $X(t)$ as follows. Apply this process as the input to an LTI system (filter) that has an adjustable center frequency ω_c and a very narrow bandwidth of W_n compared with that of $X(t)$. It then follows from (8.32) and (8.30) that

$$S_x(\omega_c) = \frac{S_y(\omega_c)}{|H(\omega_c)|^2} = \frac{\pi P_y(\omega_c)}{W_n|H(\omega_c)|^2} \tag{8.34}$$

Thus, the power spectrum at ω_c is obtained by measuring the average power $P_y(\omega_c)$ of the output when the filter is operational with the center frequency ω_c. This procedure can be repeated for different center frequencies ω_c.

Example 8.13: Generation of the Response of a Lowpass System by P&R

Suppose that the data record in file `e8_13x.dat` contains a sample function of a white noise input to a lowpass system with a cutoff frequency of 200 Hz and gain of 2. Then the response of the lowpass system can be obtained by P&R according to the following steps:

S1. Click "RPResponse" in the main window of P&R.
S2. Click "Lowpass System." You will be prompted to enter the name of a data file.
S3. Choose "data" subdirectory, enter the data file name "`e8_13x.dat`" and click "Ok."
S4. Fill out the window "**Response Generator for Lowpass System**" as shown in Fig. 8.13 and click "Ok." You will be prompted for a data file name since "Save to file" box is checked. Choose "data" subdirectory and enter the data file name "`e8_13y.dat`." Click "Ok." Then a MATLAB figure window will appear with a plot of the output $y(t)$.
S5. Follow the steps in Example 7.8 to get two power spectrum plots for the input and output contained in `e8_13x.dat` and `e8_13y.dat`, respectively. The two spectral plots correspond to the bottom left and right plots in Fig. 8.13.

Clearly, the input has a power spectrum almost constant (with small variations) over all frequencies while the power spectrum of the output is mostly limited to $f < 200$Hz.

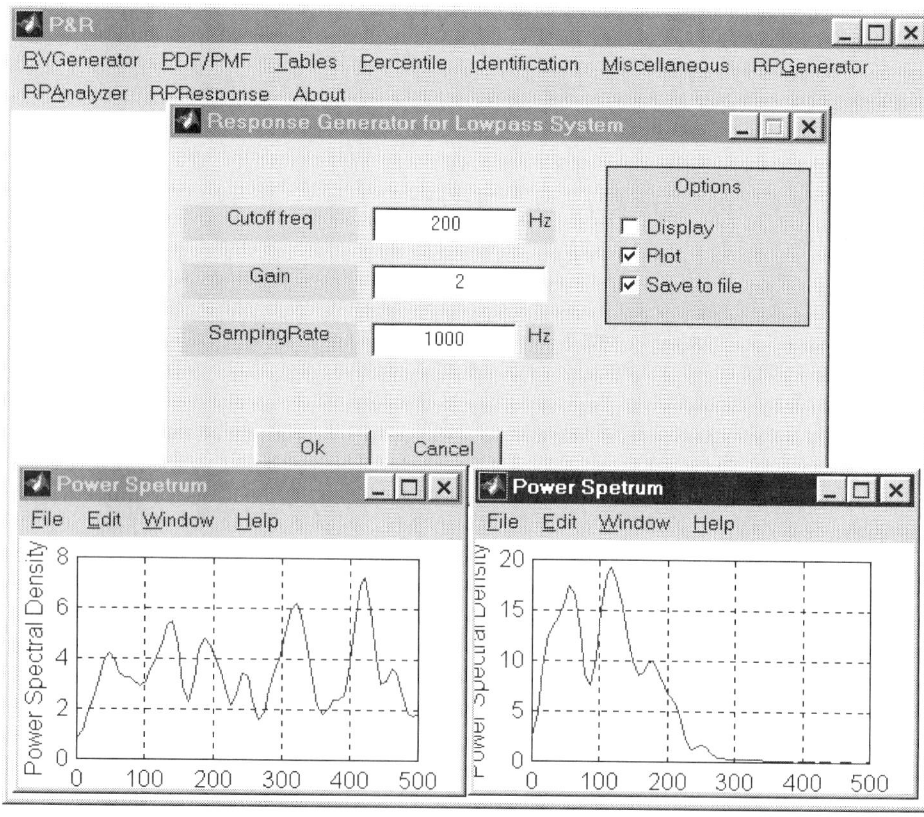

Figure 8.13: Generating Response of lowpass system by P&R.

Equivalent Noise Bandwidth

The previous three examples indicate that (lowpass or bandpass) bandlimited white noise is the response of an ideal (lowpass or bandpass) bandlimited system to white noise. Conversely, the response of a real-world system to white noise may be treated (accurately or crudely) as bandlimited white noise. As such, the bandwidth of this bandlimited white noise can be defined as that of the system. This is the concept of ***equivalent noise bandwidth***. It greatly simplifies signals and systems analysis when dealing with noise. Here the equivalence is in the sense that the actual system and the equivalent ideal bandlimited system have equal average output power and equal maximum gain when excited by white noise.

Similarly, the bandwidth of a colored noise process can be defined as that of the equivalent bandlimited white noise with both equal average power and equal maximum value of their power spectra.

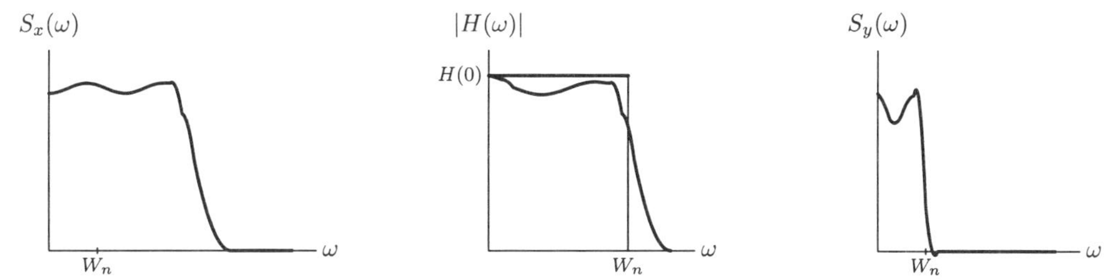

Figure 8.14: Input and output power spectra of a lowpass system.

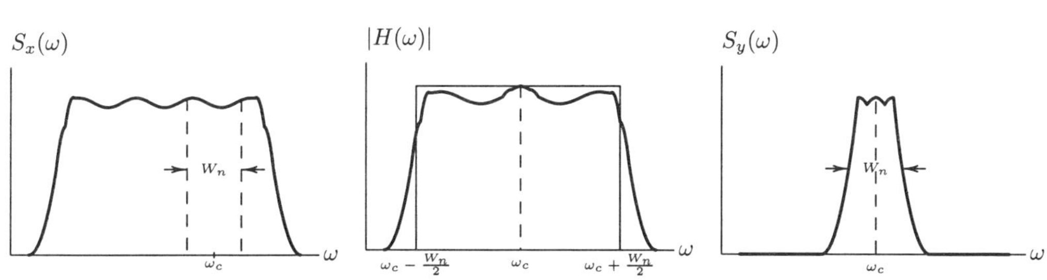

Figure 8.15: Input and output power spectra of a bandpass system.

Specifically, consider a bandlimited system with the frequency response $H(\omega)$ excited by white noise with $S_x(\omega) = S_0$. To simplify analysis, it is often desirable to replace (or approximate) its actual output $Y(t)$ by an "equivalent" bandlimited white noise process $N(t)$ with equal average power and intensity $S_n = S_0|H(\omega_n)|$, where ω_n is the peak location of $|H(\omega)|$. The average powers of the two processes

$$P_y = \frac{1}{2\pi}\int_{-\infty}^{\infty} S_y(\omega)d\omega = \frac{S_0}{2\pi}\int_{-\infty}^{\infty} |H(\omega)|^2 d\omega$$
$$P_n = \frac{S_n W_n}{\pi} = \frac{S_0|H(\omega_n)|^2 W_n}{\pi}$$

will be equal if the bandwidth W_n of $N(t)$ is chosen by

$$W_n = \frac{\int_{-\infty}^{\infty} |H(\omega)|^2 d\omega}{2|H(\omega_n)|^2} = \frac{\int_{0}^{\infty} |H(\omega)|^2 d\omega}{|H(\omega_n)|^2} \tag{8.35}$$

W_n so defined is known as the ***equivalent noise bandwidth*** of the system (the divider $|H(\omega_n)|^2$ is to ensure that two systems $H_1(\omega)$ and $H_2(\omega) = aH_1(\omega)$ have the same bandwidth). Thus, in a simplified analysis, the actual output can be replaced with the bandlimited white noise $N(t)$; or equivalently, the system is replaced with an ideal one defined by (8.29) or (8.33). Often, $\omega_n = 0$ for a lowpass system and $\omega_n = \omega_c$ for a bandpass systems, where ω_c is the center frequency of the passband. Then,

$$\text{bandpass system: } W_n = \frac{\int_{0}^{\infty} |H(\omega)|^2 d\omega}{|H(\omega_c)|^2} \tag{8.36}$$

$$\text{lowpass system: } W_n = \frac{\int_{0}^{\infty} |H(\omega)|^2 d\omega}{|H(0)|^2} = \frac{\pi\int_{-\infty}^{\infty} |h(t)|^2 dt}{\left|\int_{-\infty}^{\infty} h(t)dt\right|^2} \tag{8.37}$$

where the last equation above follows from the Parseval theorem and (8.6).

Since a colored noise process can be considered as the response of an LTI system to white noise input, its equivalent noise bandwidth may be defined by that of the system. Specifically, if it has $S_n(\omega) = |H(\omega)|^2 S_0$, then its equivalent (white) noise bandwidth is defined by (8.35).

Example 8.14: Signal-to-Noise Ratio of an RC Circuit

Assume the RC circuit of Example 8.2 has an input $X(t) = S(t) + N(t)$, where

- $N(t)$ is zero-mean white noise with $S_n(\omega) = S_0$
- $S(t) = A\sin(\omega_0 t + \phi)$ is a signal with nonrandom A and $\omega_0 < \frac{a}{2}\pi$, $a = \frac{1}{RC}$
- $\phi \sim \mathcal{U}(-\pi, \pi)$ is independent of $N(t)$

Since the system (RC circuit) has the power gain

$$|H(\omega)|^2 = H(\omega)H(\omega)^* = \frac{a}{j\omega + a}\frac{a}{-j\omega + a} = \frac{a^2}{\omega^2 + a^2}$$

it is lowpass and has the following equivalent noise bandwidth:

$$\begin{aligned} W_n &= \frac{\int_0^\infty |H(\omega)|^2 d\omega}{|H(0)|^2} = \int_0^\infty \frac{a^2}{\omega^2 + a^2} d\omega = \frac{\pi}{2} a \frac{1}{2\pi} \int_{-\infty}^\infty \frac{2a}{\omega^2 + a^2} e^{j\omega t} d\omega \Big|_{t=0} \\ &= \frac{1}{2} a\pi \mathcal{F}^{-1}\Big[\frac{2a}{\omega^2 + a^2}\Big]\Big|_{t=0} = \frac{1}{2} a\pi e^{-a|t|}\Big|_{t=0} = \frac{a}{2}\pi \end{aligned} \quad (8.38)$$

From Example 6.7, we have

$$\begin{aligned} R_s(\tau) &= \frac{1}{2} A^2 \cos(\omega_0 \tau) \\ S_s(\omega) &\overset{\text{Table 7.5}}{=} \frac{\pi}{2} A^2 [\delta(\omega - \omega_0) + \delta(\omega + \omega_0)] \end{aligned}$$

The response $S_o(t)$ to the sinusoidal component $S(t)$ has the power spectrum and power

$$S_{s_o}(\omega) = |H(\omega)|^2 S_s(\omega) = \frac{a^2 \pi A^2/2}{\omega^2 + a^2} [\delta(\omega - \omega_0) + \delta(\omega + \omega_0)]$$

$$P_{s_o} = \frac{1}{2\pi} \int_{-\infty}^\infty \frac{a^2 \pi A^2/2}{\omega^2 + a^2} [\delta(\omega - \omega_0) + \delta(\omega + \omega_0)] d\omega = \frac{a^2 A^2/2}{\omega_0^2 + a^2}$$

This average power also follows from the fact that for a pure sinusoidal input $X(t)$ with frequency ω_0, the input and output powers are related by

$$P_y = |H(\omega_0)|^2 P_x$$

As a special case, the output dc power and input dc power are related by

$$P_y^{dc} = |H(0)|^2 P_x^{dc}$$

The response $N_o(t)$ to the white noise component $N(t)$ has the power spectrum and power

$$S_{n_o}(\omega) = |H(\omega)|^2 S_n(\omega) = \frac{a^2 S_0}{\omega^2 + a^2}$$

$$P_{n_o} = \frac{1}{2\pi}\int_{-\infty}^{\infty} \frac{a^2 S_0}{\omega^2 + a^2} d\omega = \mathcal{F}^{-1}\Big[\frac{a^2 S_0}{\omega^2 + a^2}\Big]_{t=0} = \frac{a}{2} S_0 e^{-a|t|}\Big|_{t=0} = \frac{a}{2} S_0$$

This average power can also be obtained easily from the system's noise bandwidth using (8.32) and noting that $W_n = \frac{a}{2}\pi, \omega_c = 0$ for this example:

$$P_{n_o} = \frac{S_0 |H(\omega_c)|^2 W_n}{\pi} = \frac{a}{2} S_0$$

This demonstrates how easily the output power can be obtained using the equivalent noise bandwidth. Note that the *output power is proportional to the system's bandwidth.*

$S(t)$ and zero-mean $N(t)$ are orthogonal since they are independent and thus uncorrelated. Hence, the power spectrum of the output $Y(t)$ is the sum of their power spectra:

$$S_y(\omega) \overset{\text{Table 7.1}}{=} S_{s_o}(\omega) + S_{n_o}(\omega)$$

and the average power of the output is given by

$$P_y = P_{s_o} + P_{n_o} = \frac{a^2 A^2/2}{\omega_0^2 + a^2} + \frac{a}{2} S_0$$

The output ***signal-to-noise ratio*** (***SNR***) is given by

$$\frac{P_{s_o}}{P_{n_o}} = \frac{|H(\omega_0)|^2 P_s}{S_0 |H(0)|^2 W_n/\pi} = \frac{a A^2/S_0}{\omega_0^2 + a^2}$$

which is larger if the frequency ω_0 of the sinusoid is lower and/or the system's bandwidth is narrower. Its maximum is $A^2/(aS_0)$, which occurs when the desired signal is dc.

Note that the input SNR is 0 since the input noise is white noise, which has infinite power.

Example 8.15: Consider again Examples 8.4 and 8.5 with the RC circuit. From Example 8.4, the RC circuit has the following input-output relationship:

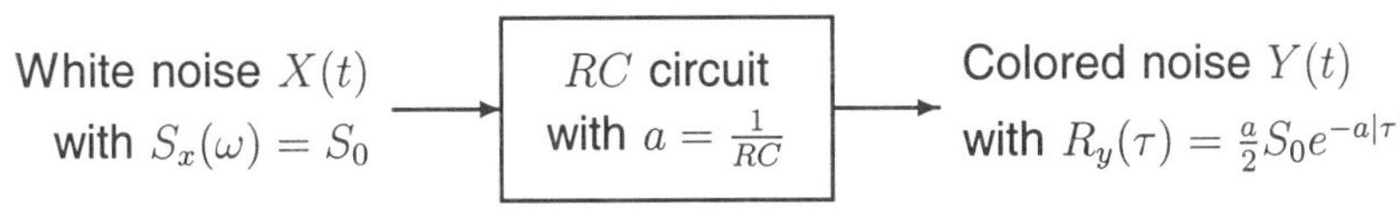

Its noise bandwidth is $\frac{a}{2}\pi$, from (8.38). Thus, $X_2(t)$ with $R_{x_2}(\tau) = \frac{b}{2}S_0 e^{-b|\tau|}$ can be treated as the response of an RC circuit, with $b = 1/(R_2C_2)$, to white noise input with $S(\omega) = S_0$. It has an equivalent bandwidth of $\frac{b}{2}\pi$. The autocorrelation of the response to the input $X_2(t)$, given by (8.16), can be rewritten as, for $S_2 = \frac{b}{2}S_0$,

$$\begin{aligned} R_{y_2}(\tau)\Big|_{S_2=\frac{b}{2}S_0} &= \frac{aS_2}{a^2-b^2}\left[ae^{-b|\tau|} - be^{-a|\tau|}\right]\Big|_{S_2=\frac{b}{2}S_0} \\ &= \frac{a}{2}S_0 e^{-a|\tau|}\frac{1}{1-(a/b)^2}\left[1 - \frac{a}{b}e^{-(b-a)|\tau|}\right] \end{aligned}$$

Thus, *for $b \gg a$, the response to the colored noise input is roughly the same as to white noise input*:

$$R_{y_2}(\tau)|_{S_2=\frac{b}{2}S_0} \approx \frac{a}{2}S_0 e^{-a|\tau|} = R_{y_1}(\tau)$$

or, from (8.21)–(8.22),

$$S_{y_2}(\omega)\Big|_{S_2=\frac{b}{2}S_0} = \frac{a^2}{\omega^2+a^2}\frac{b^2S_0}{\omega^2+b^2} = \frac{a^2S_0}{\omega^2+a^2}\frac{1}{1+(\omega/b)^2} \approx \frac{a^2S_0}{\omega^2+a^2} = S_{y_1}(\omega)$$

where the last approximate equation follows from the fact that almost all significant frequency components of the output are confined by the system's bandwidth $\frac{a}{2}\pi$; that is, $|\omega| \ll b$.

The implication of the above is quite significant for the simplification of analysis: *If the input is autocorrelated but with a large bandwidth ($\frac{b}{2}\pi$ for this example) compared with the system's bandwidth ($\frac{a}{2}\pi$ for this example), it may be replaced by (treated as) white noise* since the output will be approximately the same. As such, the *output power is proportional to the system's bandwidth.* This is a major reason why white noise model is so popular.

The most commonly used definition of the bandwidth of a system is the so-called ***half-power bandwidth***, also known as ***3-dB bandwidth***. It is defined as the width of the frequency band over which the power gain (i.e., power transfer function) $|H(\omega)|^2$ is larger than one half (3dB below) of the maximum gain; or equivalently, the magnitude is larger than $1/\sqrt{2}$ of its maximum. For the RC circuit of Example 8.4, since its power gain is $|H(\omega)|^2 = \frac{a^2}{\omega^2+a^2}$, which is lowpass and has a maximum gain of unity, its half-power frequency is $\omega = a$ and thus its half-power bandwidth is $W_{1/2} = a = \frac{1}{RC}$. It is seen that the equivalent noise bandwidth W_n of this system is $\pi/2$ times the half-power bandwidth. The equivalent noise bandwidth is also known as ***noise-equivalent bandwidth***.

An advantage of using the equivalent noise bandwidth (and any other bandwidth) is that it, along with the maximum or nominal power gain $|H(\omega_n)|^2$, provides an approximate description of a possibly very complex system. For such a system, an accurate analysis may be too complicated or too costly and quite often not worth it. Furthermore, relatively simple methods are available to measure the equivalent noise bandwidth and the maximum power gain of a system. Thus, equivalent noise bandwidths of actual systems provide a basis for the comparison of signal-to-noise ratio of the systems. However, this approximation is good only when the random input has a power spectrum that is nearly constant over a frequency band that is substantially greater than the system's bandwidth. Otherwise, great errors may be introduced by using the equivalent noise bandwidth.

8.7 Output of a Linear System as a Gaussian Process

We have studied in this chapter the first two moments (mean, mean-square value, autocorrelation, crosscorrelation) and power spectra of the output and input of a linear system. Two natural questions are the following. What are the marginal and joint distribution or density functions of the output? What is the joint distribution or density function between the input and output? Unfortunately, a general answer to these questions is yet to be found.

However, in the special case where the input to the linear system is a Gaussian process, the output is also a Gaussian process with the mean and autocorrelation given by (8.4) and (8.11) (or its nonstationary counterpart), respectively. This follows from the fact that every linear function of a set of jointly Gaussian RVs is a Gaussian RV (see (4.36)).

An important result related to equivalent noise bandwidth is the following.

If the input process is not Gaussian but is stationary and has a bandwidth much wider than that of the linear system, then the output random process is (more or less) approximately Gaussian no matter what distribution the input process has. This follows from the central limit theorem since the output is the sum of a large number of (nearly uncorrelated) RVs in this case. Note that bandwidth is inversely proportional to correlation time and that if system bandwidth is small, the number of terms in the summation is large.

In general, a random sequence $Y[n]$ that is the sampled version of a zero-mean colored Gaussian random process $Y(t)$ with autocorrelation $R(\tau)$ or power spectrum $S(\omega)$ can be generated as follows.

S1. Generate a white Gaussian random sequence $X[n]$ with unity variance using say MATLAB function `randn`.

S2. Apply $X[n]$ as input to an LTI system with frequency response $H(\omega)$ satisfying $|H(\omega)|^2 = S(\omega)$. Then, by (8.17), the output $Y[n]$ has the power spectrum $S(\omega)$.

Since the input to the linear system is Gaussian, the output $Y[n]$ is thus Gaussian with power spectrum $S(\omega)$.

Example 8.16: Generation of a Colored Gaussian Random Process

Use the method described above to generate a random sequence $Y[n]$ that is the sampled version of a zero-mean colored Gaussian process $Y(t)$ with autocorrelation $R(\tau) = ae^{-b|\tau|}$, $a, b > 0$.

From Table 7.5, the power spectrum of $Y(t)$ is given by

$$S(\omega) = \frac{2ab}{\omega^2 + b^2}$$

Note that this power spectrum can be decomposed as

$$S(\omega) = \frac{\sqrt{2ab}}{j\omega + b} \frac{\sqrt{2ab}}{-j\omega + b} = H(\omega)H(\omega)^* = |H(\omega)|^2$$

It is thus clear by (8.17) that the response of the LTI system

$$H(\omega) = \frac{\sqrt{2ab}}{j\omega + b}$$

to a white Gaussian process input $X(t)$ with power spectrum $S_x(\omega) = 1$ is then a Gaussian process with $R(\tau) = ae^{-b|\tau|}$. Note that the LTI system $H(\omega) = \frac{\sqrt{2ab}}{-j\omega+b}$ is neither stable nor causal.

The state-variable equation for the causal and stable LTI system $H(\omega) = \frac{\sqrt{2ab}}{j\omega+b}$ is

$$\dot{Y}(t) = -bY(t) + \sqrt{2ab}X(t)$$

with $X(t)$ as input and $Y(t)$ as output. Sampling (without holding) the above equation yields the following difference equation:

$$Y[n+1] = e^{-bT}Y[n] + \sqrt{a(1 - e^{-2bT})}X[n]$$

where T is the sampling interval. Thus, the required random sequence $Y[n]$ is generated using the above equation with an initial $Y[1]$ that is a zero-mean Gaussian RV with variance a and independent of $X[n]$.

We can use $Z[n] = Y[n] + \bar{z}[n]$ to generate a Gaussian sequence $Z[n]$ with the above autocorrelation but a nonzero mean $\bar{z}[n] = \bar{z}(nT)$.

For example, if $a = 5, b = 20, T = 0.1, \bar{z}(t) = 3.5$, then the following MATLAB routine generates the required Gaussian sequence of length 1000:

```
function y=rp_gauss_color(a,b,T,N)
x=randn(N-1,1);
y=zeros(N,1);
y(1)=sqrt(a)*randn(1,1);
c=exp(-b*T); c2=sqrt(a*(1-c*c));
for n=1:N-1
   y(n+1)=c*y(n)+c2*x(n)
end
return

a=5; b=20; T=0.1; N=1000; z_bar=3.5*ones(N,1);
y=rp_gauss_color(a,b,T,N);
z=y+z_bar;
```

8.8 Summary and Requirements

The response of a linear system to random input is random. For a linear time-invariant system (relaxed at $t = -\infty$), if its input is stationary, then its output and input are jointly stationary. Specifically, the *autocorrelation of the output* is related to that of the input by

$$R_y(\tau) = R_x(\tau) * h(\tau) * h(-\tau)$$

the *average output* is equal to the convolution of the average input with the impulse response and for a stationary input it is simply

$$\bar{y} = \bar{x}H(0)$$

where $H(0)$ is the dc gain of the system; the *crosscorrelations of the output and input* are given by $R_{xy}(\tau) = R_x(\tau) * h(-\tau)$ and $R_{yx}(\tau) = R_x(-\tau) * h(\tau)$.

In frequency domain, the *power spectrum of the output* and the *cross-power spectra of the output and input* are related to that of the input by

$$S_y(\omega) = |H(\omega)|^2 S_x(\omega)$$

and $S_{xy}(\omega) = S_x(\omega)H(\omega)^*$, $S_{yx}(\omega) = S_x(\omega)H(\omega)$.

In general, if the power spectrum of the input is rational, frequency-domain analysis is easier; if the autocorrelation of the input has a finite nonzero duration, however, the time-domain analysis may be simpler.

The analysis of a linear system with white noise input is much simpler than with colored noise input.

A *lowpass* (or *bandpass*) white noise process is the response of an ideal lowpass (or bandpass) system to white noise. Most autocorrelated (i.e., colored) noise can be approximated by the response to white noise input of a lowpass (or bandpass) system with the same (equivalent noise) bandwidth as that of the input.

The *equivalent noise bandwidth* of a linear system is the (approximate) bandwidth of its response to white noise. It is given by

$$W_n = \frac{\int_0^\infty |H(\omega)|^2 d\omega}{|H(\omega_n)|^2}$$

where ω_n is the central frequency of the passband for a bandpass system or zero for a lowpass system. To simplify system or signal analysis, a real-world linear system can be replaced approximately by an ideal bandlimited system with the same equivalent noise bandwidth and maximum gain, and the autocorrelated random input can be replaced by white noise if the input has a much wider bandwidth than the equivalent noise bandwidth of the system.

The response of a linear system to Gaussian process input is a Gaussian process.

Basic Requirements

- Know how to calculate the power spectra, correlation functions, and average output of the response of a linear time-invariant system to random input, in particular to white noise input.
- Know roughly when the frequency domain is preferred to the time domain and when it is not for system and signal analysis.
- Know how to calculate the equivalent noise bandwidth of a linear system.
- Master the concepts of bandlimited, lowpass, bandpass, narrowband, and passband. Understand the similarity of these concepts as applied to a system and to a signal (or a process).

The emphasis of the chapter is the relationships of power spectra, means and autocorrelations between the input and output.

8.9 Additional Examples

8.17 *Finite-time integrator.* Consider a finite-time integrator with impulse response

$$h(t) = \begin{cases} 1/T & 0 < t < T \\ 0 & \text{elsewhere} \end{cases} \tag{8.39}$$

Its input is a zero-mean white noise with $S(\omega) = S_0$.

(a) Show that the output is in some sense the average of its input (no matter what it is).
(b) Find the mean, autocorrelation, and power spectrum of the output.
(c) How large should T be if it is desired to have an output power of $10S_0$?
(d) Determine the equivalent noise bandwidth of the integrator.

Solution:

(a)

$$
\begin{aligned}
y(t) &= h(t) * x(t) \\
&= \int_{-\infty}^{\infty} h(\tau)x(t-\tau)d\tau \\
&= \int_0^{T} \frac{1}{T}x(t-\tau)d\tau \\
&\overset{\lambda = t-\tau}{=} -\int_t^{t-T} \frac{1}{T}x(\lambda)d\lambda \\
&= \frac{1}{T}\int_{t-T}^{t} x(\lambda)d\lambda \\
&= \text{average of the (immediate past) input over time period } T
\end{aligned}
$$

(b)

$$E[Y(t)] = E\left[\frac{1}{T}\int_{t-T}^{t} x(\lambda)d\lambda\right] = \frac{1}{T}\int_{t-T}^{t} E[X(\lambda)]d\lambda = 0$$

$X(t)$ is white noise input $\Rightarrow$ use (8.26) for $R_y(\tau)$:

$$R_y(\tau) = S_0 \int_{-\infty}^{\infty} h(t)x(t+\tau)dt$$

Determine the limits as follows:

$$
\left.\begin{array}{ll}
0 < t < T & 0 < t < T \\
0 < \tau + t < T \Rightarrow & -\tau < t < T - \tau
\end{array}\right\} \Rightarrow \max(0, -\tau) < t < \min(T, T-\tau)
$$

$$
\Rightarrow \begin{cases} 0 < t < T - \tau & 0 < \tau < T \\ -\tau < t < T & -T < \tau \leq 0 \end{cases}
$$

Thus:

$$
\begin{aligned}
R_y(\tau) &= S_0 \int_{-\infty}^{\infty} h(t)x(t+\tau)dt \\
&= \begin{cases} S_0 \int_0^{T-\tau} \frac{1}{T}\frac{1}{T}dt & 0 \leq \tau < T \\ S_0 \int_{-\tau}^{T} \frac{1}{T}\frac{1}{T}dt & -T < \tau < 0 \\ 0 & \text{elsewhere} \end{cases} \\
&= \begin{cases} \frac{S_0}{T^2}(T-\tau) & 0 \leq \tau < T \\ \frac{S_0}{T^2}(T+\tau) & -T < \tau < 0 \\ 0 & \text{elsewhere} \end{cases} \\
&= \begin{cases} \frac{S_0}{T^2}(T-|\tau|) & 0 \leq |\tau| < T \\ 0 & \text{elsewhere} \end{cases}
\end{aligned}
$$

In fact, we need only to work out $R_y(\tau) = \frac{S_0}{T^2}(T - |\tau|)$ for $0 \leq \tau < T$. Then, $R_y(\tau)$ for $\tau \leq 0$ can be obtained by symmetry $R(-\tau) = R(\tau)$.
From Example 7.1 with $A = 1/T$, we have

$$H(\omega) = ATe^{-j\omega T/2}\frac{\sin(\omega T/2)}{\omega T/2}\bigg|_{A=1/T} = \frac{\sin(\omega T/2)}{\omega T/2}e^{-j\omega T/2}$$

Then

$$S_y(\omega) = |H(\omega)|^2 S_x(\omega) = \frac{\sin^2(\omega T/2)}{(\omega T/2)^2}S_0 = \frac{2S_0}{\omega^2 T^2}[1 - \cos(\omega T)]$$

Alternatively:

$$\begin{aligned}
S_y(\omega) &= \mathcal{F}[R_y(\tau)] = \int_{-\infty}^{\infty} R_y(\tau)e^{-j\omega\tau}d\tau \\
&= \frac{S_0}{T^2}\int_{-T}^{T}(T - |\tau|)e^{-j\omega\tau}d\tau \\
&= \frac{S_0}{T^2}\left[\int_{-T}^{0}(T + \tau)e^{-j\omega\tau}d\tau + \int_{0}^{T}(T - \tau)e^{-j\omega\tau}d\tau\right] \\
&= \frac{S_0}{T^2}\left[T\int_{-T}^{T}e^{-j\omega\tau}d\tau + \int_{-T}^{0}\tau e^{-j\omega\tau}d\tau - \int_{0}^{T}\tau e^{-j\omega\tau}d\tau\right] \\
&= \frac{S_0}{T^2}\left[\frac{T}{-j\omega}e^{-j\omega\tau}\bigg|_{-T}^{T} + \frac{T}{-j\omega}\tau e^{-j\omega}\bigg|_{-T}^{0} - \frac{1}{-j\omega}\int_{-T}^{0}e^{-j\omega\tau}d\tau\right. \\
&\quad \left. - \left(\frac{1}{-j\omega}\tau e^{-j\omega}\bigg|_{0}^{T} - \frac{1}{-j\omega}\int_{0}^{T}e^{-j\omega\tau}d\tau\right)\right] \\
&= \frac{S_0}{-j\omega T^2}\left[\frac{T}{-j\omega}e^{-j\omega\tau}\bigg|_{-T}^{T} - \frac{T}{-j\omega}e^{-j\omega\tau}\bigg|_{-T}^{0} - \left(Te^{-j\omega T} - \frac{T}{-j\omega}e^{-j\omega\tau}\bigg|_{0}^{T}\right)\right] \\
&= \frac{1}{(-j\omega)^2}\frac{S_0}{T^2}\left[-1 + e^{j\omega T} + e^{-j\omega T} - 1\right] \\
&= \frac{2}{\omega^2}\frac{S_0}{T^2}[1 - \cos(\omega T)]
\end{aligned}$$

which agrees with the above.

(c) $P_y = R_y(0) = \frac{S_0}{T^2}[T - 0] = \frac{S_0}{T} = 10S_0 \Longrightarrow T = 1/10$.

(d)

$$\begin{aligned}
H(0) &= \lim_{\omega \to 0}\frac{\sin(\omega T/2)}{\omega T/2} = 1 \neq 0 \quad \Longrightarrow \quad \text{lowpass} \\
W_n &= \frac{1}{2|H(0)|^2}\int_{-\infty}^{\infty}|H(\omega)|^2 d\omega \\
&= \frac{2\pi}{2S_0}\underbrace{\frac{1}{2\pi}\int_{-\infty}^{\infty}|H(\omega)|^2 S_0 d\omega}_{R_y(0)}
\end{aligned}$$

$$
\begin{aligned}
&= \frac{\pi}{S_0} R_y(0) = \frac{\pi}{S_0} \frac{S_0}{T} \\
&= \frac{\pi}{T}
\end{aligned}
$$

Alternatively:

$$
W_n = \frac{1}{2} \int_{-\infty}^{\infty} |H(\omega)|^2 d\omega = \frac{1}{2} \int_{-\infty}^{\infty} \left[\frac{\sin(\omega T/2)}{\omega T/2} \right]^2 d\omega = \frac{2}{T} \int_{-\infty}^{\infty} \frac{\sin^2 x}{x} dx = \frac{2}{T} \frac{\pi}{2} = \frac{\pi}{T}
$$

Although the solution to the problem is not simple, it would be much more complicated were a frequency-domain technique is used for this problem.

8.18 *Output autocorrelation of a discrete-time system.* A linear system is described by the following difference equation

$$
y_n = x_n - x_{n-1}
$$

Find the autocorrelation $R_y(n, n-k)$ of y_n given that the input to the system is zero-mean stationary white noise with unity variance.

Solution:

$$
\begin{aligned}
R_y(n, n-k) &= E[y_n y_{n-k}] \\
&= E[(x_n - x_{n-1})(x_{n-k} - x_{n-1-k})] \\
&= E[x_n x_{n-k} - x_n x_{n-1-k} - x_{n-1} x_{n-k} + x_{n-1} x_{n-1-k}] \\
&= R_x(k) - R_x(k+1) - R_x(k-1) + R_x(k) \\
&= \begin{cases} 2 & k = 0 \\ -1 & k = \pm 1 \\ 0 & \text{elsewhere} \end{cases}
\end{aligned}
$$

Since it depends only on the time difference k, it is wide-sense stationary.

8.19 *Output signal-to-noise ratio.* A noisy signal $X(t) = S(t) + N(t)$ passes through a linear RL circuit with a frequency response $H(\omega) = \frac{R/L}{j\omega + R/L}$, where the zero-mean noise $N(t)$ with autocorrelation function $R_n(\tau) = K\delta(\tau)$ is independent of $S(t)$. Suppose that the power spectrum of the output $Y(t)$ is

$$
S_y(\omega) = \frac{R^2}{(\omega L)^2 + R^2} \left\{ \frac{\pi}{2} [\delta(\omega + \omega_0) + \delta(\omega - \omega_0)] + K \right\}
$$

(a) Find the average power of the input noiseless signal and the input signal-to-noise ratio.

(b) Find the power of the output noiseless signal, the power of the output noise power, and the output signal-to-noise ratio. Is the response of the linear circuit considered to a pure noiseless sine wave still a sine wave? What is the relation between the power of the response and that of the input?

Solution:

(a) The input power spectrum is given by

$$\begin{aligned} S_x(\omega) &= \frac{S_y(\omega)}{|H(\omega)|^2} \\ &= \frac{\frac{R^2}{(\omega L)^2+R^2}\left\{\frac{\pi}{2}[\delta(\omega+\omega_0)+\delta(\omega-\omega_0)]+K\right\}}{\frac{R^2}{(\omega L)^2+R^2}} \\ &= \frac{\pi}{2}[\delta(\omega+\omega_0)+\delta(\omega-\omega_0)]+K \end{aligned}$$

Since $S(t)$ and $N(t)$ are independent and $N(t)$ has zero mean, they are orthogonal, and thus

$$S_x(\omega) \overset{\text{Table 7.1}}{=} S_s(\omega) + S_n(\omega)$$

Hence, the noiseless input signal has the power spectrum

$$S_s(\omega) = S_x(\omega) - S_n(\omega) = \frac{\pi}{2}[\delta(\omega+\omega_0)+\delta(\omega-\omega_0)]$$

It has average power

$$\begin{aligned} P_s &= \mathcal{F}^{-1}[S_s(\omega)]_{\tau=0} \\ &= \mathcal{F}^{-1}\left\{\frac{\pi}{2}[\delta(\omega+\omega_0)+\delta(\omega-\omega_0)]\right\}_{\tau=0} \\ &= \left[\frac{1}{2}\cos(\omega_0\tau)\right]_{\tau=0} \\ &= \frac{1}{2} \end{aligned}$$

Since the input noise is a white noise process, which has infinite power, the input signal-to-noise ratio is zero.

(b) From part (a), the output noiseless signal has the power spectrum corresponding to the first term of $S_y(\omega)$ and thus the power

$$\begin{aligned} P_{s_o} &= \frac{1}{2\pi}\int_{-\infty}^{\infty}\frac{R^2}{(\omega L)^2+R^2}\frac{\pi}{2}[\delta(\omega+\omega_0)+\delta(\omega-\omega_0)]d\omega \\ &= \frac{1}{2\pi}\frac{\pi}{2}\left[\frac{R^2}{(\omega L)^2+R^2}\bigg|_{\omega=-\omega_0} + \frac{R^2}{(\omega L)^2+R^2}\bigg|_{\omega=\omega_0}\right] \\ &= \frac{1}{2}\frac{R^2}{(\omega_0 L)^2+R^2} \end{aligned}$$

The output noise power is

$$P_{n_o} = \frac{1}{2\pi}\int_{-\infty}^{\infty}\frac{R^2}{(\omega L)^2+R^2}K d\omega$$

$$= KR/(2L)\mathcal{F}^{-1}\left[\frac{2R/L}{\omega^2 + (R/L)^2}\right]_{\tau=0}$$
$$= KR/(2L)\left[e^{R|\tau|/L}\right]_{\tau=0}$$
$$= KR/(2L)$$

The output signal-to-noise ratio (SNR) is thus

$$\text{SNR} = \frac{1}{2}\frac{R^2}{[(\omega_0 L)^2 + R^2]KR/(2L)} = \frac{RL}{K[(\omega_0 L)^2 + R^2]}$$

8.10 Problems

8.1 *Differentiator with random input.* Let $h(t)$ be the impulse response of a linear differentiator such that its output $x(t) = \frac{dx(t)}{dt}$ is the derivative of its input $x(t)$. Let the random input be $X(t) = 5\cos(100\pi t + \phi) + 10$, where $\phi \sim \mathcal{U}(0, 2\pi)$.

(a) Are $X(t)$ and $Y(t)$ jointly wide-sense stationary?
(b) Find the crosscorrelation $R_{xy}(t+\tau, t)$ and the autocorrelation $R_y(t+\tau, t)$.
(c) Find the variances $\sigma_x^2(t)$ and $\sigma_y^2(t)$.
(d) Find the smallest value of τ in magnitude for which the output is uncorrelated with the input.

8.2 *Relation between output autocorrelation and input-output crosscorrelation.* Show that (8.10) is true.

8.3 *Response of linear system to white noise.* Consider a causal linear time-invariant system with an impulse response

$$h(t) = (1+t)[u(t) - u(t-2)] + 3tu(t-4)u(t)$$

with a white noise input of mean 10 and non-dc power spectrum $100V^2$/Hz.

(a) Is the system causal?
(b) Find the expected value of the output $E[Y(t)]$.
(c) Find the mean-square value and variance of the output.

8.4 *Time constant determination.* For the RC circuit of Example 8.4, if the input is zero-mean white noise with intensity $S_0 = 10^{-6}$V^2/Hz, find the smallest time constant RC (in seconds) such that the output has a variance (intensity) not larger than 100V^2.

8.5 *Response of a bandpass filter to white noise.* Zero-mean white noise of power spectral density $25V^2$/Hz enters a bandpass filter, operating at a center frequency $f_c = 2$Hz, with the transfer function $H(s) = \dfrac{s}{s^2 + 4s + 3}$. Find

(a) the power spectrum of the output
(b) the average power of the output

8.6 *Response of linear system to colored noise.* A linear time-invariant system with an impulse response $h(t) = 2e^{-2t}u(t)$ has a zero-mean noise input with a power spectrum $S(s) = \dfrac{1}{(1-s)(1+s)}$.

(a) Is the system causal?
(b) Find the expected value of the output $E[Y(t)]$.
(c) Find the power spectrum and average power of the output.

8.7 *Autocorrelation and power spectrum of derivative process.* Consider a differentiator such that its output $Y(t) = \frac{dX(t)}{dt}$ is the derivative of its input $X(t)$. Use the relationships of the autocorrelations and power spectra between its input and output to show (6.23) and (7.18).

8.8 *Response of cascade RC circuit to white noise.* Zero-mean white noise with $S_x(\omega) = S_0$ is the input to the LTI lowpass filter (system) consisting of two stages of the RC circuits as shown in Fig. 8.16, where $R_1C_1 \neq R_2C_2$. Assume that the loading effect can be neglected; that is, assume that the output of stage 1 is identical with or without stage 2. Find

(a) the mean and autocorrelation of the output
(b) the power spectrum and power of the output

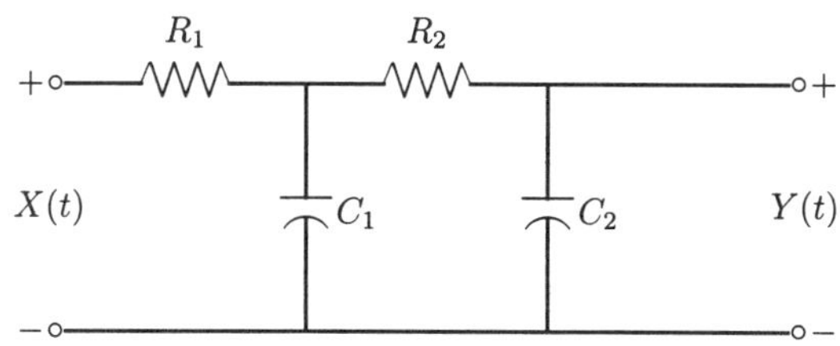

Figure 8.16: A cascade of two simple RC circuits with different parameters.

8.9 *Response of cascade RC circuit to white noise.* Zero-mean white noise with $S_x(\omega) = S_0$ is the input to the LTI lowpass filter (system) consisting of two identical stages of the RC circuits of Fig. 8.17. Assume that the loading effect can be neglected; that is, assume that the output of stage 1 is identical with or without stage 2. Find

(a) the mean and autocorrelation of the output
(b) the power spectrum and power of the output

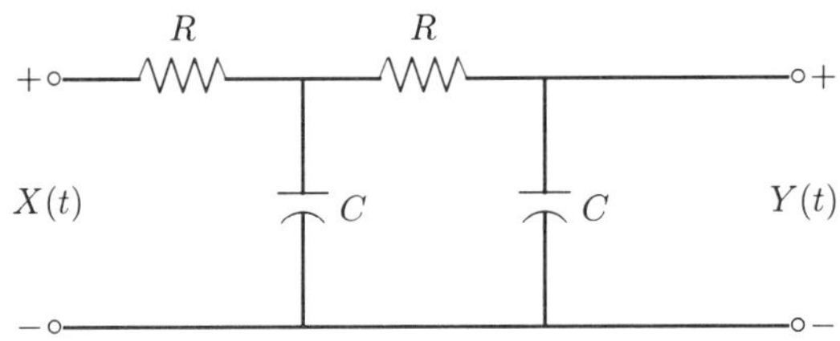

Figure 8.17: A cascade of two identical simple RC circuits.

8.10 *Autocorrelation and power spectrum of second derivative.* Given a stationary random process $X(t)$ and its autocorrelation function $R_x(\tau)$ and power spectrum $S_x(\omega)$, derive the autocorrelation and power spectrum of $Y(t) = \dfrac{d^2X(t)}{dt^2}$.

8.11 *State of linear system driven by white noise.* A linear system is described by the following state-space equation

$$\frac{d}{dt}X(t) = -aX(t) + N(t)$$

Suppose that this system is driven by zero-mean white noise $N(t)$ with power spectrum S_0. Find

(a) the frequency response $H(\omega)$ and impulse response $h(t)$ between the noise $N(t)$ and state $X(t)$
(b) the autocorrelation and power spectrum of the state $X(t)$
(c) the crosscorrelation and power spectrum between the state $X(t)$ and the noise $N(t)$

8.12 *Response of RLC circuit to white noise.* Zero-mean white noise with $S_x(\omega) = S_0$ is the input to the RLC circuit shown in Fig. 8.18. Find the mean, autocorrelation and power spectrum of the output, assuming that the loading effect can be neglected.

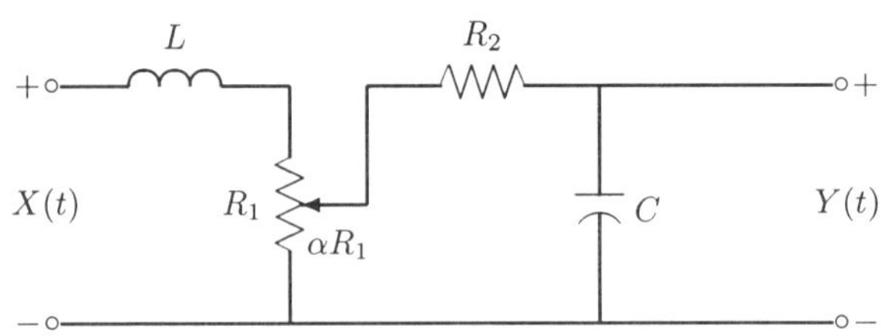

Figure 8.18: An RLC circuit.

8.13 *Response of ideal bandpass system to signal plus noise.* Consider the ideal bandpass system of Example 8.10 with input $X(t) = 2\cos(\omega_0 t+\phi)+N(t)$, where ω_0 is nonrandom;

$\phi \sim \mathcal{U}(0, 2\pi)$; $N(t)$ is zero-mean white noise with power spectrum S_0; ϕ and $N(t)$ are independent. Find

(a) the average output
(b) the average power of the output
(c) the ratio of output average power of the sinusoidal component to the noise component

8.14 *Finite-time integrator with colored noise input.* A finite-time integrator with the following impulse response has a random process $X(t)$ as its input, whose autocorrelation function is plotted in Fig. 8.19.

(a) Find the mean, autocorrelation, and power spectrum of the output.
(b) Explain how $X(t)$ can be generated using white noise generator. Hint: Use results from the above problem.
(c) Determine the bandwidth of the integrator.

Figure 8.19: Impulse response and input autocorrelation of a finite-time integrator.

8.15 *Response of two-stage finite-time integrator to white noise.* Two identical finite-time integrators are connected in cascade, each with impulse response given by (8.39). Its input is white noise with $S(\omega) = S_0/2$. Find the mean, autocorrelation, and power spectrum of the output.

8.16 *Response of RL circuit to random input.* A random process (voltage) $X(t)$ is applied to the RL circuit of Fig. 8.20.

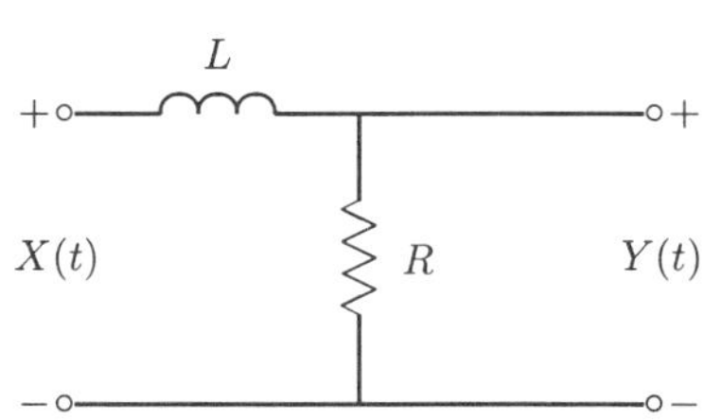

Figure 8.20: An RL circuit.

(a) Show that the frequency response function of the RL circuit is given by

$$H(\omega) = \frac{Y(\omega)}{X(\omega)} = \frac{a}{a + j\omega}$$

where $a = R/L$.

(b) Assume that the power spectrum of $X(t)$ is $S_x(\omega) = S_0$. What are the autocorrelation function $R_x(\tau)$ and average power P_x? Determine the power spectrum and the average power of $Y(t)$. Is $Y(t)$ white?

(c) Consider now the same RL circuit with $L = 1\ H$ and $R = 1\ \Omega$. The input voltage $X(t)$ is now a random process $F(t)$ corrupted by an independent additive white noise $N(t)$. $F(t)$ and $N(t)$ are jointly wide-sense stationary. The power spectral densities of $F(t)$ and $N(t)$ are, respectively,

$$S_f(\omega) = \begin{cases} 1 & |\omega| < 1 \\ 0 & \text{elsewhere} \end{cases} \qquad S_n(\omega) = 0.1$$

Determine the average SNR of the output process $Y(t)$.

8.17 *Response of RL circuit to exponentially correlated noise.* For the RL circuit of Fig. 8.20, if the input $X(t)$ is a colored noise process with $R_x(\tau) = e^{-c|\tau|}$, $c \neq R/L$, find the power spectrum and autocorrelation of the output $Y(t)$.

8.18 *LTI system with white noise input.* An LTI system with an impulse response $h(t) = 2e^{-4t}u(t)$ has a zero-mean white noise input with a power spectrum $S(\omega) = S_0$. Find

(a) the frequency response of this system
(b) the average output
(c) the power spectrum of the output
(d) the autocorrelation and the average power of the output
(e) the equivalent noise bandwidth of this system

8.19 *LTI system with white noise input.* An LTI system $h(t)$ has a zero-mean white noise input $X(t)$ with power spectrum $S(\omega) = S_0$. Find the crosscorrelation and the cross-power spectrum of the input and the output.

8.20 *Linear system with white noise input.* A linear system with impulse response function $h(t) = e^{-2t}u(t)$ is excited by zero-mean white noise with power spectrum $S_x(\omega) = 10\text{V}^2/\text{Hz}$.

(a) What is the autocorrelation of the input $X(t)$?
(b) Find the frequency response function $H(\omega)$ of the system and the power spectrum $S_y(\omega)$ of the output. Is the output white?
(c) Find the autocorrelation and average power of the output.
(d) Find the average power of the non-dc components of the output.

(e) Find the equivalent noise bandwidth of the system.

8.21 *Simplification by noise bandwidth.* An ideal lowpass linear system with bandwidth $W = \pi/24$ and $H(0) = 2$ is excited by colored noise $X(t)$ with power spectrum $S_x(\omega) = \dfrac{3}{(\omega^2+9)(\omega^2+1)}$. Find

(a) the equivalent noise bandwidth of the input
(b) the exact power spectrum $S_y(\omega)$ of the output
(c) the exact average power of the output
(d) the approximate average power of the output using equivalent noise bandwidth
(e) approximate average power of the output if the bandwidth of the ideal lowpass system is doubled

8.22 *Output power calculation using noise bandwidth.* A linear system with impulse response function $h(t) = e^{-3t}u(t)$ is excited by zero-mean white noise with power spectrum $S_x(\omega) = \dfrac{1}{(\omega^2+4)(\omega^2+1)} \text{V}^2/\text{Hz}$.

(a) Find the frequency response function $H(\omega)$ of the system and the power spectrum $S_y(\omega)$ of the output.
(b) Find the equivalent noise bandwidth of the input.
(c) Find the equivalent noise bandwidth of the system.
(d) Find the equivalent noise bandwidth of the output.

8.23 *Output power calculation using noise bandwidth.* A linear system with impulse response function $h(t) = e^{-t}u(t)$ is excited by zero-mean white noise with power spectrum $S_x(\omega) = \dfrac{40}{\omega^2+20^2}$. Find

(a) the frequency response function $H(\omega)$ of the system and the power spectrum $S_y(\omega)$ of the output
(b) the equivalent noise bandwidth of the input
(c) the equivalent noise bandwidth of the system
(d) approximate average power of the output using equivalent noise bandwidth

8.24 *Output distribution.* The RC circuit of Example 8.2 has an input that is a zero-mean stationary Gaussian white noise process with power spectrum $S(\omega) = S_0$. Find the marginal PDF of the output process at an arbitrary time. Find the joint PDF of the output at t_1 and $t_1 + \tau$.

8.25 *Distribution of LTI system output.* The RC circuit of Example 8.2 has a zero-mean stationary white noise input with power spectrum $S(\omega) = 20\pi$. Is the output a Gaussian process? Justify your answer. Give the marginal PDF of the output process at an arbitrary time if you can.

8.26 *Output and signal correlation.* An LTI system $h(t)$ has the response $Y(t)$ to the input $X(t) = S(t) + N(t)$, where $S(t)$ is a signal and $N(t)$ is noise. Show that

$$R_{ys}(\tau) = R_{xs}(\tau) * h(\tau)$$
$$R_{sy}(\tau) = R_{sx}(\tau) * h(-\tau)$$

8.27 *Puzzling questions.* Raise two technical questions of your own that you are puzzled by and that are closely related to the material covered in this chapter. These questions should be important, not too simple, and not too involved.

8.11 Computer Exercises

8.1 *System response and stability.* Consider two systems with the following transfer functions, respectively:

$$H_1(s) = \frac{2s+5}{s^2+30s+200}$$
$$H_2(s) = \frac{5s-6}{s^2+3s}$$

A random input $X(t)$ is applied to the above two systems, respectively.

(a) Are $H_1(s)$ and $H_2(s)$ stable?
(b) The data file `m8_1x1.dat` contains a discrete-time (with 1000 Hz sampling rate) sample function of $X(t)$. Use the companion software P&R to find the responses $y_1(t)$ and $y_2(t)$ of the systems to this input. Do $y_1(t)$ and $y_2(t)$ grow unbounded?
(c) The data file `m8_1x2.dat` contains another sample function of $X(t)$. Repeat (b) with this input.
(d) Discuss your results.

8.2 *Determination of dc gain.* The data files `m8_2x.dat` and `m8_2y.dat` contain the data records of an ergodic random input and its ergodic response of a linear time-invariant system, respectively. Write your own MATLAB routine to determine the dc gain of the system. Use the companion software P&R to verify your result.

8.3 *Verification of crosscorrelation.* The input to the RC circuit of Fig. 8.2 is a zero-mean Gaussian noise $X(t)$ with autocorrelation $R(\tau) = 2e^{-40|\tau|}$, where $R = 1000\Omega$ and $C = 10\mu F$.

(a) Use the companion software P&R to generate a 256-point sample function of $X(t)$. Use a sampling rate $f_s = 2000$ Hz. Save it as `m8_3x.dat`.
(b) Use P&R to generate a 256-point sample function of the output $Y(t)$. Use a sampling rate $f_s = 2000$ Hz. Save it as `m8_3y.dat`.

(c) Use P&R to plot the autocorrelations $R_x(i/f_s)$ of $X(t)$ and $R_y(i/f_s)$ of $Y(t)$ for $i = 0, 1, \ldots, 20$.
(d) Use P&R to plot the crosscorrelation $R_{yx}(\tau)$ of $Y(t)$ and $X(t)$ for τ_i, $i = 0, 1, \ldots, 20$. Note that $R_{yx}(\tau)$ is neither even nor odd symmetric. So you need to use the mirror image property to get $R_{yx}(\tau)$ for $\tau < 0$ from $R_{xy}(\tau)$.
(e) Find the theoretical $R_{yx}(\tau)$ and compare it with $R_{yx}(\tau)$ of (d).

8.4 *RC circuit as a lowpass filter.* The input to the RC circuit of Fig. 8.2 is zero-mean Gaussian white noise $X(t)$ with power spectrum 2, where $R = 1000\Omega$ and $C = 1\mu F$. Use the companion software P&R to do the following.

(a) Generate a 1024-point sample function of $X(t)$. Use a sampling rate of 2000 Hz.
(b) Plot the power spectrum of $X(t)$.
(c) Generate a 1024-point sample function of the output $Y(t)$. Use a sampling rate of 2000 Hz.
(d) Plot the power spectrum of the output $Y(t)$.
(e) Compare the power spectra of $X(t)$ and $Y(t)$ and comment on your results.

8.5 *Analysis of a lowpass filter.* A lowpass filter has a cutoff frequency of 100 Hz and unity gain. The input to the filter is a zero-mean Gaussian white noise $X(t)$ with power spectrum 4. Use the companion software P&R to do the following.

(a) Generate a 1024-point sample function of $X(t)$. Use a sampling rate $f_s = 2000$ Hz.
(b) Plot the power spectrum of $X(t)$.
(c) Plot the autocorrelation $R_x(i/f_s)$ of $X(t)$ for $i = 0, 1, \ldots, 70$.
(d) Plot the power spectrum of the output $Y(t)$.
(e) Plot the autocorrelation $R_y(i/f_s)$ of $Y(t)$ for $i = 0, 1, \ldots, 70$.
(f) Plot the crosscorrelation $R_{xy}(i/f_s)$ of $X(t)$ and $Y(t)$ for $i = 0, 1, \ldots, 70$.
(g) Plot $X(t)$ and $Y(t)$.
(h) Discuss your results.

8.6 *Noise reduction by a bandpass filter.* A bandpass filter has a center frequency of 500 Hz, a bandwidth of 200 Hz and a gain of 1. The input to the filter is $X(t) = S(t) + N(t)$, where $S(t) = 0.1 \sin(1000\pi t + \pi/6)$ and $N(t)$ is zero-mean Gaussian white noise with power spectrum 2. Use the companion software P&R to do the following.

(a) Generate a 1024-point sample function of $S(t)$ and $N(t)$ with a sampling rate of 2000 Hz. Save them as m8_6s.dat and m8_6n.dat, respectively. Use the following MATLAB routine to calculate $X(t)$ from $S(t)$ and $N(t)$ saved before:

```
cd data
load m8_6s.dat
load m8_6n.dat
x=m8_6s+m8_6n;
save m8_6x.dat x -ascii
```

```
cd ..
```

$X(t)$ is then saved to `m8_6x.dat`.

(b) Plot the power spectrum of $X(t)$.
(c) Plot the power spectrum of the output $Y(t)$.
(d) Compare the power spectra of $X(t)$ and $Y(t)$ and comment on your results.

8.7 *Bandpass filter design.* A bandpass filter has an adjustable center frequency and a bandwidth of 20Hz and a gain of 1. A data record of length 256 (with 4000 Hz sampling rate) of a noisy signal $X(t)$ is given in the data file `m8_7.dat`. The signal $X(t) = S(t) + N(t)$ is corrupted by zero-mean white noise $N(t)$, where $S(t)$ is a sinusoid with an unknown frequency. Use the companion software P&R to design the filter (i.e., to select the center frequency in Hz) such that the output of the filter has the maximum signal to noise ratio. You should set the segment length to the data record length in the power spectrum estimator. You may use a bandpass of a larger bandwidth to locate roughly the sinusoid first and then to fine tune the 20Hz filter's center frequency.

8.8 *Determination of noise bandwidth.* A data record of a signal $X(t)$ is given in the data file `m8_8.dat` with 2000 Hz sampling rate. Use the companion software P&R to determine its equivalent noise bandwidth in Hz.

8.9 *Generation of colored Gaussian process.* Following Example 8.16, write a MATLAB routine to generate a 1024-point colored Gaussian random sequence with autocorrelation $R(\tau) = 3.4e^{-15|\tau|}$ with sampling rate of 15Hz. Use the companion software P&R to plot the autocorrelation and power spectrum of the sequence generated. Discuss your results.

8.12 Self-Test Problems

8.1 Answer the following questions briefly.

(a) If a linear system has a zero dc gain, what is its average response to a stationary input?
(b) What kind of process is the response of a lowpass system to white noise? Is it white?
(c) Does the equivalent noise bandwidth of a system depend on the input?
(d) What can you say about the input power spectrum if the power spectrum of the output of a linear system is a pair of delta functions located at $\omega = \pm\omega_0$?
(e) What remains unchanged when replacing an (approximately) bandpass system with an ideal one with the same equivalent noise bandwidth?
(f) Will an ideal bandlimited system always improve the signal-to-noise ratio if the frequencies of the input signal are all within the passband of the system?
(g) Is the average response of a linear time-invariant system equal to the response of the system to the average input even if the input is a nonstationary process?

8.2 A noisy signal $X(t) = \cos(\omega_0 t + \theta) + N(t)$ passes through a linear circuit with a frequency response function $H(\omega) = \frac{R/L}{j\omega + R/L}$, where $N(t)$ is a zero-mean noise process with autocorrelation function $R_n(\tau) = K\delta(\tau)$; ω_0 is a known constant; θ is a random variable uniformly distributed over $(\pi, 2\pi)$; and $N(t)$ and θ are independent.

(a) Find the mean $\bar{x}(t)$ and autocorrelation $R_x(t+\tau, t)$ of $X(t)$.
(b) Is $X(t)$ wide-sense stationary? Is $N(t)$ wide-sense stationary or white? Justify your answer.
(c) Find the mean $\bar{y}(t)$ and autocorrelation $R_y(t+\tau, t)$ of the output $Y(t)$ of the linear circuit. Is $Y(t)$ wide-sense stationary?
(d) What is the power spectrum of $N(t)$. Find the power spectrum of $X(t)$. Find the power spectrum of $Y(t)$.

8.3 Consider a differentiator whose response $Y(t)$ to a random input $X(t)$ is

$$Y(t) = \frac{d}{dt}X(t)$$

(a) For this system, show that convolution of the input with the impulse response amounts to differentiation of the input.
(b) Find the average response of this system to the random input $X(t)$ in terms of the mean of $X(t)$.
(c) Find the average response of this system to the random input $X(t) = 2\cos(\omega_0 t + \phi)$, where ω_0 is nonrandom and $\phi \sim \mathcal{U}(\pi, 2\pi)$.
(d) For a system consisting of the cascade of n such differentiators, show that its average response to $X(t)$ is simply $\frac{d^n}{dt^n}E[X(t)]$.

8.13 Solutions to Self-Test Problems

8.1 (a) By (8.5), it is equal to zero no matter what the stationary input is.
(b) The response is lowpass white noise process. It is colored but with a flat power spectral density over a finite frequency band.
(c) It is independent of the actual input, although it is defined assuming the input is white noise. It is one of the inherent parameters of the system.
(d) By (8.17), the input power spectrum must also be a pair of delta functions located at $\omega = \pm\omega_0$ and the intensity is that of the output divided by $|H(\omega_0)|^2$. This indicates that the input is a sinusoid.
(e) Two things remain unchanged: the maximum gain and the average power of the system's response to white noise.
(f) Yes, because the signal part of the input will pass the system without attenuation while some frequency components (i.e., those that are outside of the system's passband) of the noise part of the input will not pass the system. As a result, the signal-to-noise

ratio is higher at the output than at the input because the output has the same signal power and a smaller noise power than the input.

(g) Yes, (8.4) is valid for nonstationary as well as stationary processes.

8.2 (a) The mean $\bar{x}(t)$ is

$$\begin{aligned}\bar{x}(t) &= E[X(t)] = E[\cos(\omega_0 t + \theta) + N(t)] \\ &= E[\cos(\omega_0 t + \theta)] + E[N(t)] \\ &= \int_{\pi}^{2\pi} \cos(\omega_0 t + \theta)\frac{1}{2\pi - \pi} d\theta + 0 \\ &= \frac{\sin(\omega_0 t + 2\pi) - \sin(\omega_0 t + \pi)}{\pi} \\ &= \frac{2}{\pi}\sin(\omega_0 t)\end{aligned}$$

and the autocorrelation $R_x(t+\tau, t)$ is

$$\begin{aligned}R_x(t+\tau,t) &= E[X(t+\tau)X(t)] \\ &= E\{[\cos(\omega_0 t + \omega_0\tau + \theta) + N(t+\tau)][\cos(\omega_0 t + \theta) + N(t)]\} \\ &= E[\cos(\omega_0 t + \theta)\cos(\omega_0 t + \omega_0 \tau + \theta)] + E[\cos(\omega_0 t + \theta)N(t+\tau)] \\ &\quad + E[N(t)\cos(\omega_0 t + \omega_0\tau + \theta)] + E[N(t+\tau)N(t)] \\ &= \frac{1}{2}E[\cos(\omega_0\tau) + \cos(2\omega_0 t + \omega_0\tau + 2\theta)] + E[\cos(\omega_0 t + \theta)]E[N(t+\tau)] \\ &\quad + E[N(t)]E[\cos(\omega_0 t + \omega_0\tau + \theta)] + R_n(\tau) \\ &= \frac{1}{2}[\cos(\omega_0\tau) + 0] + 0 + 0 + R_n(\tau) \\ &= \frac{1}{2}\cos(\omega_0\tau) + R_n(\tau) \\ &= R_x(\tau) \qquad \text{(depends only on } \tau)\end{aligned}$$

(b) Since $\bar{x}(t)$ is not constant, $X(t)$ is not wide-sense stationary. Since $\bar{N}(t) = 0 =$ constant and $R_n(t+\tau,t) = R_n(\tau)$, $N(t)$ is wide-sense stationary. Since $R_n(\tau) = K\delta(\tau)$, $N(t)$ is white [$N(t_1)$ and $N(t_2)$ are uncorrelated for every $t_1 \neq t_2$].

(c) Let $a = R/L$. Since

$$h(t) = \mathcal{F}^{-1}\left[\frac{a}{j\omega + a}\right] \overset{\text{Table 7.5}}{=} ae^{-at}u(t)$$

the mean $\bar{y}(t)$ is

$$\begin{aligned}\bar{y}(t) &= \bar{x}(t) * h(t) \\ &= \int_{-\infty}^{\infty} h(t-\tau)\bar{x}(\tau)d\tau\end{aligned}$$

$$= \int_{-\infty}^{t} ae^{-a(t-\tau)} \frac{2}{\pi} \sin(\omega_0 \tau) d\tau$$
$$= \frac{2a}{\pi} e^{-at} \frac{e^{a\tau}[a \sin(\omega_0 \tau) - \omega_0 \cos(\omega_0 \tau)]}{a^2 + \omega_0^2}\bigg|_{-\infty}^{t}$$
$$= \frac{2a}{\pi(a^2 + \omega_0^2)} [a \sin(\omega_0 t) - \omega_0 \cos(\omega_0 t)]$$

As a result, $Y(t)$ is not wide-sense stationary since its mean is not time invariant. Since $S_y(\omega)$ is given as below in part (d), the autocorrelation $R_y(\tau)$ is, compared with the Fourier transform pair of $R_x(\tau)$ and $S_x(\omega)$,

$$\begin{aligned} R_y(\tau) &= \mathcal{F}^{-1}\left\{\frac{1}{2}\frac{a^2}{\omega_0^2 + a^2}\pi[\delta(\omega+\omega_0) + \delta(\omega-\omega_0)] + \frac{a^2 K}{\omega^2 + a^2}\right\} \\ &= \frac{1}{2}\frac{a^2}{\omega_0^2 + a^2}\mathcal{F}^{-1}\{\pi[\delta(\omega+\omega_0) + \delta(\omega-\omega_0)]\} + \frac{aK}{2}\mathcal{F}^{-1}\left[\frac{2a}{\omega^2 + a^2}\right] \\ &\overset{\text{Table 7.5}}{=} \frac{1}{2}\frac{a^2}{\omega_0^2 + a^2}\cos(\omega_0 \tau) + \frac{Ka}{2}e^{-a|\tau|} \end{aligned}$$

(d) The power spectrum of $N(t)$ is

$$S_n(\omega) = \mathcal{F}[R_n(\tau)] = \mathcal{F}[K\delta(\tau)] = K$$

The power spectrum of $X(t)$ is

$$\begin{aligned} S_x(\omega) &= \mathcal{F}[R_x(\tau)] = \mathcal{F}[\frac{1}{2}\cos(\omega_0 \tau) + R_n(\tau)] \\ &= \mathcal{F}[\frac{1}{2}\cos(\omega_0 \tau)] + \mathcal{F}[R_n(\tau)] \\ &= \frac{1}{2}\mathcal{F}[\cos(\omega_0 \tau)] + S_n(\omega) \\ &\overset{\text{Table 7.5}}{=} \frac{\pi}{2}[\delta(\omega+\omega_0) + \delta(\omega-\omega_0)] + K \end{aligned}$$

The power spectrum of $Y(t)$ is

$$\begin{aligned} S_y(\omega) &= S_x(\omega)|H(\omega)|^2 \\ &= \frac{a^2}{\omega^2 + a^2}\left\{\frac{\pi}{2}[\delta(\omega+\omega_0) + \delta(\omega-\omega_0)] + K\right\} \\ &= \frac{1}{2}\frac{a^2}{\omega_0^2 + a^2}\pi[\delta(\omega+\omega_0) + \delta(\omega-\omega_0)] + \frac{a^2 K}{\omega^2 + a^2} \end{aligned}$$

8.3 (a)

$$\left.\begin{aligned} Y(t) &= h(t) * X(t) \\ Y(t) &= \tfrac{d}{dt}X(t) \end{aligned}\right\} \quad \Longrightarrow \quad h(t) * X(t) = \frac{d}{dt}X(t)$$

(b)

$$\text{average response} = E[Y(t)] \overset{(8.4)}{=} E[X(t)] * h(t) = \frac{d}{dt}E[X(t)]$$

(c)

$$\begin{aligned}
E[X(t)] &= \int_{\pi}^{2\pi} 2\cos(\omega_0 t + \phi)\frac{1}{2\pi - \pi}d\phi \\
&= \frac{2}{\pi}\sin(\omega_0 t + \phi)\Big|_{\pi}^{2\pi} \\
&= \frac{2}{\pi}[\sin(\omega_0 t + 2\pi) - \sin(\omega_0 t + \pi)] \\
&= \frac{4}{\pi}\sin(\omega_0 t) \\
\text{average response} &= \frac{d}{dt}E[X(t)] = \frac{4\omega_0}{\pi}\cos(\omega_0 t)
\end{aligned}$$

(d) Consider n cascade stages of such linear differentiators. Clearly

$$\begin{aligned}
\bar{y}(t) = \bar{x}_n(t) &= \frac{d}{dt}[\bar{x}_{n-1}(t)] \\
&= \frac{d}{dt}\left\{\frac{d}{dt}[\bar{x}_{n-1}(t)]\right\} \\
&= \frac{d^2}{dt^2}[\bar{x}_{n-2}(t)] \\
&= \cdots = \frac{d^n}{dt^n}[\bar{x}_0(t)] \\
&= \frac{d^n}{dt^n}\bar{x}(t)
\end{aligned}$$

That is,

$$\text{average response} = E[Y(t)] = \frac{d^n}{dt^n}E[X(t)]$$

9

OPTIMAL LINEAR SYSTEMS

A problem well put is a problem half solved.

An Axiom

In Chapter 8, we *analyzed* linear systems with random input. In this chapter, we *synthesize* linear systems that are optimal in some sense for some problems of practical importance.

Main Topics

- Signal-to-Noise Ratio
- The Matched Filter
- The Wiener Filter

The matched filter and Wiener filter are widely used in engineering, such as communications and signal processing.

9.1 Introduction

An ***optimal linear system*** is a linear system that is optimal in some sense. The optimal systems considered here have a random (noisy) input $X(t) = S(t)+N(t)$ that is the sum of a (deterministic or random) signal $S(t)$ and noise $N(t)$, as depicted below. Due to linearity, the output is the sum of the corresponding responses: $Y(t) = S_o(t) + N_o(t)$.

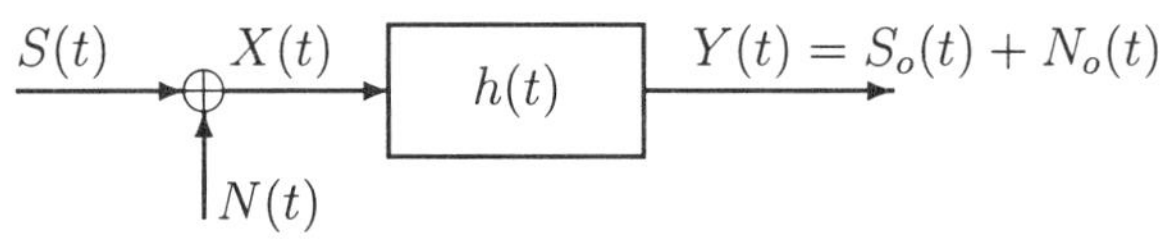

Figure 9.1: Input and output of an optimal system.

The task of such an optimal system is to provide an output that removes the noise (or enhances the signal) *as much as possible*.

How to quantify (or measure) the amount of noise removal (or signal enhancement) is application dependent, though general guidelines exist. Correspondingly, several popular ***optimality criteria*** are available, including the following.

- Maximization of *output* ***signal-to-noise ratio*** (SNR):

  The ratio of output signal power to output noise power at some specified time is maximized. This is typically used in e.g., digital communication and radar systems when the noiseless input signal is deterministic, such as to *detect* the presence/absence of a deterministic signal of a known shape or its occurrence time.
- Minimization of the ***mean-square error*** (MSE):

  The mean-square value of the difference between the system's output and the desirable input (noise-free) signal is minimized. This is typically used in e.g. speech signal processing when the noiseless input signal is random but with known characteristics, such as to *estimate* the value (form) of a signal that is known to be present.

  Use of *mean (average) error* is not a good choice because positive and negative errors will more or less cancel out each other. The mean-square error, rather than mean absolute error, is chosen because the former is much easier to handle mathematically than the latter.

9.2 Signal-to-Noise Ratio

In general, ***signal-to-noise ratio*** (***SNR***) for a noisy signal is defined as

$$\boxed{\text{SNR} = \frac{\text{(average) signal power}}{\text{(average) noise power}}}$$

It measures how "noisy" a signal is. Specifically, consider a signal $s(t)$ corrupted by noise $N(t)$ with average power P_n:

$$X(t) = s(t) + N(t)$$

- If $s(t)$ is deterministic, then SNR at $t = t_0$ is given by

$$\text{SNR}|_{t=t_0} = \frac{\text{signal power at } t = t_0}{\text{average noise power}} = \frac{|s(t_0)|^2}{P_n}$$

  The ***average SNR*** is

$$\text{SNR} = \frac{P_s}{P_n} = \frac{\langle |s(t)|^2 \rangle}{P_n}$$

  where P_s is the average power of $s(t)$ and $\langle |s(t)|^2 \rangle$ is the ***time average*** of $|s(t)|^2$:

$$\langle |s(t)|^2 \rangle = \lim_{T\to\infty} \frac{1}{2T} \int_{-T}^{T} |s(t)|^2 dt = \lim_{T\to\infty} \frac{1}{2T} \frac{1}{2\pi} \int_{-\infty}^{\infty} |S(\omega)|^2 d\omega$$

  where $S(\omega) = \mathcal{F}[s(t)]$ and the last equation above follows from the *Parseval theorem* of Fourier transforms. For example, if $s(t) = A\cos(\omega_0 t + \theta)$, then its average power is given by

$$\begin{aligned} P_s = \langle |s(t)|^2 \rangle &= \lim_{T\to\infty} \frac{1}{2T} \int_{-T}^{T} s^2(t) dt \\ &= \lim_{T\to\infty} \frac{1}{2T} \int_{-T}^{T} \frac{1}{2} A^2 [1 + \cos 2(\omega_0 t + \theta)] dt \\ &= \frac{1}{2} A^2 + \lim_{T\to\infty} \frac{1}{2T} \frac{\sin 2(\omega_0 T + \theta) - \sin 2(-\omega_0 T + \theta)}{2\omega_0} = \frac{1}{2} A^2 \end{aligned}$$

- If $s(t)$ is random, then, the ***average SNR*** is given by

$$\boxed{\text{SNR} = \frac{P_s}{P_n} \overset{\text{stationary } s(t)}{=} = \frac{R_s(0)}{P_n}}$$

Note that SNR in dB is $10 \log \text{SNR}$ not $20 \log \text{SNR}$ unless SNR is defined as amplitude (not power) ratio.

Example 9.1: Maximization of Output SNR

The input $X(t)$ to the lowpass RC circuit of Example 8.2 is a deterministic signal $s(t) = A\cos\omega_0 t$ corrupted by additive zero-mean white noise with $S_n(\omega) = S_0$. Determine R and C such that the steady-state SNR at the output is maximized.

The steady-state response to the input $s(t)$ is

$$s_o^{ss}(t) = |H(\omega_0)|A\cos[\omega_0 t + \angle H(\omega_0)] \overset{\text{Example 8.2}}{=} \frac{Aa}{\sqrt{\omega_0^2 + a^2}}\cos[\omega_0 t - \tan^{-1}(\omega_0/a)]$$

The output steady-state signal power and noise power and the steady-state output SNR are given by Example 8.14 as

$$P_{s_o} = \frac{A^2a^2}{2(\omega_0^2 + a^2)}, \qquad P_{n_o} = \frac{a}{2}S_0, \qquad \text{SNR} = \frac{A^2a}{S_0(\omega_0^2 + a^2)}$$

To find the maximum SNR, we take derivative and set it to zero:

$$\frac{d}{da}(\text{SNR}) = \frac{A^2}{S_0}\cdot\frac{\omega_0^2 + a^2 - a\cdot 2a}{(\omega_0^2 + a^2)^2} = \frac{A^2(\omega_0^2 - a^2)}{S_0(\omega_0^2 + a^2)^2} = 0$$

since $(\frac{u}{v})' = \frac{u'v - v'u}{v^2}$, which leads to

$$a^2 = \omega_0^2 \qquad \text{or} \qquad a = \frac{1}{RC} = \omega_0$$

This gives the maximum steady-state SNR:

$$\text{SNR}_{\max} = \left.\frac{A^2a}{S_0(\omega_0^2 + a^2)}\right|_{a=\omega_0} = \frac{A^2}{2\omega_0 S_0}$$

which can be verified to be the maximum by plotting SNR vs. a, or by checking

$$\frac{d^2}{da^2}\text{SNR} = \frac{A^2}{S_0}\frac{-2a(\omega_0^2 + a^2)^2 - (\omega_0^2 - a^2)\cdot 2(\omega_0^2 + a^2)2a}{(\omega_0^2 + a^2)^4} = \frac{A^2}{S_0}\frac{a^2 - 3\omega_0^2}{(\omega_0^2 + a^2)^4} < 0$$

in the neighborhood of $a = \omega_0$ because the point with zero first derivative and negative second derivative is a maximum.

This example illustrates how to ***optimize performance by parameter adjustment***, given the system structure. In this example, only one parameter (i.e., a) is involved. It would be more complicated if more than one parameter is involved.

9.3 The Matched Filter

A ***matched filter*** *is a system that maximizes the output SNR*. Consider the system of Fig. 9.2, where $s(t)$ is a deterministic signal and $N(t)$ is wide-sense stationary white noise with $S_n(\omega) = S_0$. We wish to find the matched filter (i.e., to determine $H(\omega)$ or $h(t)$ that maximizes the output SNR) for $s(t)$ at $t = t_0$.

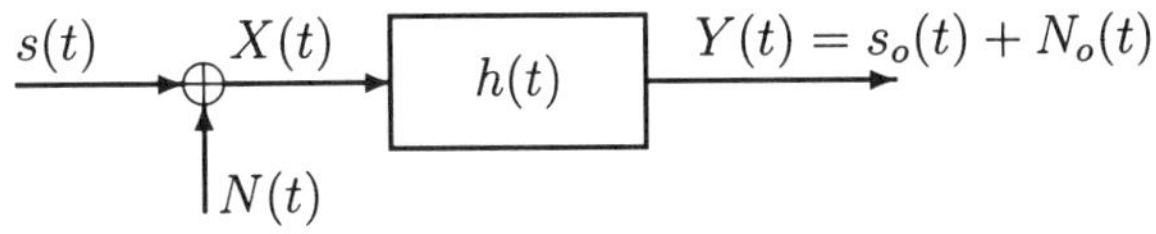

Figure 9.2: Input and output of a matched filter.

The output SNR at $t = t_0$ is given by

$$\text{SNR}|_{t=t_0} = \frac{|s_o(t_0)|^2}{P_{n_o}} = \frac{|h(t_0) * s(t_0)|^2}{(1/2\pi)\int_{-\infty}^{\infty} S_n(\omega)|H(\omega)|^2 d\omega} = \frac{\left|\int_{-\infty}^{\infty} h(t)s(t_0-t)dt\right|^2}{(S_0/2\pi)\int_{-\infty}^{\infty}|H(\omega)|^2 d\omega}$$

$$= \frac{\left|\int_{-\infty}^{\infty} h(t)s(t_0-t)dt\right|^2}{S_0\int_{-\infty}^{\infty}|h(t)|^2 dt} \qquad \text{(by Parseval theorem)}$$

$$\le \frac{\int_{-\infty}^{\infty}|h(t)|^2 dt \int_{-\infty}^{\infty}|s(t_0-t)|^2 dt}{S_0\int_{-\infty}^{\infty}|h(t)|^2 dt} = \frac{\int_{-\infty}^{\infty}|s(t_0-t)|^2 dt}{S_0}$$

The inequality above follows from the powerful ***Schwarz inequality***:

$$\left|\int_a^b f(t)g(t)dt\right|^2 \le \int_a^b |f(t)|^2 dt \int_a^b |g(t)|^2 dt \qquad \forall f(t), g(t) \tag{9.1}$$

where the equality holds iff (if and only if) $f(t) = Kg^*(t)$ on $[a, b]$ with K an arbitrary real-valued constant. Thus, the ***maximum SNR*** at $t = t_0$ is given by

$$\text{SNR}_{\text{max}} = \frac{1}{S_0}\int_{-\infty}^{\infty}|s(t_0-t)|^2 dt = \frac{1}{S_0}\int_{-\infty}^{\infty}|s(\tau)|^2 d\tau = \frac{\text{total energy in } s(t)}{\text{power spectrum of } N(t)} \tag{9.2}$$

which turns out to be irrelevant of t_0 and is achieved iff the equality holds, that is, the optimal $h(t)$ (i.e., ***matched filter for white noise***) is given by

$$\boxed{h(t) = Ks^*(t_0 - t)} \tag{9.3}$$

The Matched Filter for Colored Noise

The matched filter for the case of colored noise can be derived in frequency domain. Let $S_n(\omega)$ be the power spectrum of $N(t)$ and $S(\omega) = \mathcal{F}[s(t)]$. The SNR at $t = t_0$ is

$$\text{SNR}|_{t=t_0} = \frac{|s_o(t_0)|^2}{R_{n_o}(0)} = \frac{|\mathcal{F}^{-1}[S_o(\omega)]_{t=t_0}|^2}{\mathcal{F}^{-1}[S_{n_o}(\omega)]_{\tau=0}} = \frac{\left|\frac{1}{2\pi}\int_{-\infty}^{\infty} S(\omega)H(\omega)e^{j\omega t_0}d\omega\right|^2}{\frac{1}{2\pi}\int_{-\infty}^{\infty} S_{n_o}d\omega}$$

$$\leq \frac{\int_{-\infty}^{\infty} S_n(\omega)|H(\omega)|^2 d\omega \frac{1}{(2\pi)^2}\int_{-\infty}^{\infty}\frac{|S(\omega)|^2}{S_n(\omega)}d\omega}{\frac{1}{2\pi}\int_{-\infty}^{\infty} S_n(\omega)|H(\omega)|^2 d\omega} = \frac{1}{2\pi}\int_{-\infty}^{\infty}\frac{|S(\omega)|^2}{S_n(\omega)}d\omega$$

The inequality above follows from letting

$$F(\omega) = \sqrt{S_n(\omega)}H(\omega) \qquad G(\omega) = \frac{S(\omega)e^{j\omega t_0}}{2\pi\sqrt{S_n(\omega)}}$$

and the following ***Schwarz inequality*** (which is identical to (9.1) with t replaced by ω):

$$\left|\int_a^b F(\omega)G(\omega)d\omega\right|^2 \leq \int_a^b |F(\omega)|^2 d\omega \int_a^b |G(\omega)|^2 d\omega \qquad \forall F(\omega), G(\omega)$$

where the equality holds iff $F(\omega) = cG^*(\omega)$ on $[a, b]$ with c an arbitrary real-valued constant. Consequently, the ***maximum SNR*** at $t = t_0$ is given by

$$\text{SNR}_{\text{max}} = \frac{1}{2\pi}\int_{-\infty}^{\infty}\frac{|S(\omega)|^2}{S_n(\omega)}d\omega \tag{9.4}$$

which turns out to be irrelevant of t_0 and is achieved if and only if $F(\omega) = cG^*(\omega)$:

$$\sqrt{S_n(\omega)}H(\omega) = cG^*(\omega) = c\frac{S^*(\omega)e^{-j\omega t_0}}{2\pi\sqrt{S_n(\omega)}}$$

Thus, the ***matched filter for colored noise*** is given by

$$H(\omega) = \frac{c}{2\pi}\frac{S^*(\omega)}{S_n(\omega)}e^{-j\omega t_0} = \underbrace{1/\sqrt{S_n(\omega)}}_{\text{whitening filter}} \cdot \underbrace{C\,\overbrace{S^*(\omega)/\sqrt{S_n(\omega)}}^{\text{filtered signal}}\,e^{-j\omega t_0}}_{\text{matched filter for white noise}} \tag{9.5}$$

and by inverse Fourier transform, its impulse response satisfies

$$h(\tau) * R_n(\tau) = Cs^*(t_0 - \tau) \tag{9.6}$$

If $N(t)$ is white noise, then (9.6) reduces to (9.3), (9.4) reduces to (9.2) by Parseval theorem, and (9.5) becomes

$$H(\omega) = \frac{c}{2\pi S_0}S^*(\omega)e^{-j\omega t_0} = KS^*(\omega)e^{-j\omega t_0}$$

which agrees with (9.3), by the shifting and conjugation properties of Fourier transforms.

Remarks:

$h(t)$ is *matched* to the signal $s(t)$ at time $t = t_0$, hence the name. It is the input signal waveform run backwards in time and delayed by t_0.

The matched filter requires perfect knowledge of the noiseless input signal $s(t)$. For different time t_0, it is in general different since $h(t) = Ks^*(t_0 - t)$. Thus, even with perfect knowledge of $s(t)$, a system with fixed parameters cannot match the signal for different times.

The matched filter has a (nonzero) arbitrary real-valued constant K, meaning that it is optimal no matter what amplitude gain the filter has. This makes sense since the input signal and noise are scaled by the same gain and thus has no effect on output SNR.

The time t_0 at which the output SNR is maximized enters into the system only as a pure time delay. Its value is usually upon design and may be selected to make the matched filter causal.

The above matched filter is in general noncausal (physically unrealizable). If a system that is optimal of all causal systems is of interest, then $h(t) = 0, \forall t < 0$, implies that the lower limit $-\infty$ of all the time integrals above should be replaced by 0. As a result, the ***causal matched filter*** is given by

$$\boxed{h^{\text{causal}}(t) = Ks(t_0 - t)u(t)} \qquad u(t) = \text{unit step} \tag{9.7}$$

and the corresponding ***maximum SNR*** at $t = t_0$ is given by

$$\text{SNR} = \frac{\int_0^\infty |s(t_0 - t)|^2 dt}{S_0} = \frac{\int_{-\infty}^{t_0} |s(\tau)|^2 d\tau}{S_0} = \frac{\text{energy in } s(t) \text{ over } (-\infty, t_0]}{\text{power spectrum of } N(t)} \tag{9.8}$$

This SNR increases with t_0 and is in general smaller than the one given by (9.2) for the noncausal filter. Thus, if $s(t)$ vanishes after some time T, best SNR is achieved by choosing $t_0 = T$.

(9.5) implies that the *matched filter for colored noise* is a cascade of a ***whitening filter*** with $H(\omega) = \frac{1}{\sqrt{S_n(\omega)}}$ that converts colored noise input with $S_n(\omega)$ into white noise output and the *matched filter for white noise*.

The matched filter is optimal also under several other optimality criteria, such as the likelihood ratio and the smallest error probability. It is, however, not good for recovering (reconstructing) the input signal.

Example 9.2: A Causal Matched Filter

The *matched filter* for the signal

$$s(t) = Ae^{-\alpha t}u(t) \qquad \alpha > 0$$

at $t = t_0$ has the impulse response:

$$h(t) = Ks^*(t_0 - t) = KAe^{-\alpha t_0}e^{\alpha t}u(t_0 - t) = ce^{\alpha t}u(t_0 - t)$$

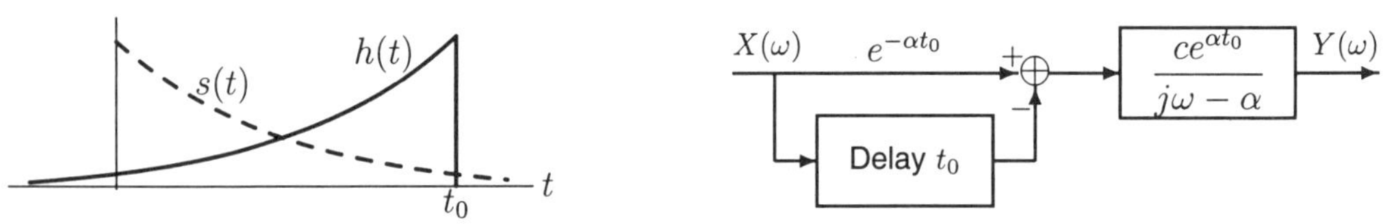

Figure 9.3: Input signal, impulse response, and block diagram of the matched filter.

Clearly, $h(t) \neq 0$, for some $t < 0$, meaning that the optimal (matched) filter is noncausal (physically unrealizable). However, a good realizable approximation to this filter is to force $h(t) = 0$ for all $t < 0$:

$$h(t) = ce^{\alpha t}u(t_0 - t)u(t) \Longrightarrow H(\omega) = \int_0^{t_0} ce^{\alpha t}e^{-j\omega t}dt = \frac{ce^{\alpha t_0}}{j\omega - \alpha}\left[e^{-\alpha t_0} - e^{-j\omega t_0}\right]$$

It is identical to the one given by the causal *matched filter* (9.7).

The maximum SNRs at $t = t_0$ for the causal and noncausal filters are

$$\text{SNR}_{\max}^{\text{causal}} \overset{(9.8)}{=} \frac{1}{S_0}\int_{-\infty}^{t_0} s^2(t)dt = \frac{1}{S_0}\int_0^{t_0} A^2e^{-2\alpha t}d\tau = \frac{A^2}{2\alpha S_0}[1 - e^{-2\alpha t_0}]$$

$$\text{SNR}_{\max} \overset{(9.2)}{=} \frac{1}{S_0}\int_{-\infty}^{\infty} s^2(t)dt = \frac{1}{S_0}\int_0^{\infty} A^2e^{-2\alpha t}d\tau = \frac{A^2}{2\alpha S_0}$$

Clearly, the realizable (suboptimal) matched filter gives an SNR (slightly) smaller than the maximum SNR. If a large t_0 is chosen (such a system is usually more costly), their difference is negligible.

This demonstrates how an "unpractical" theory can be useful:

optimal but unrealizable (theory) $\Longrightarrow$ realizable but suboptimal (practice)

9.4 The Wiener Filter

Consider the following system where the noiseless signal $S(t)$ is a random proccss with known characteristics.

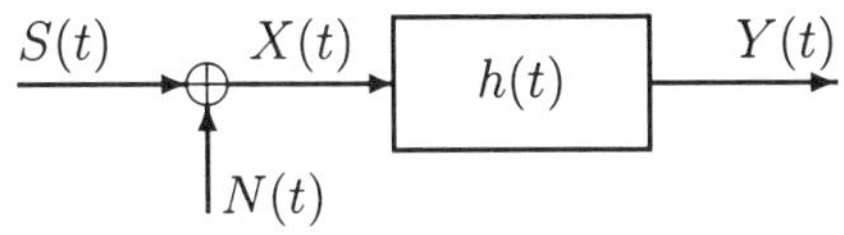

Figure 9.4: Input and output of a Wiener filter.

*The **Wiener filter** is an optimal system $h(t)$ whose output $Y(t)$ is an estimate of the signal $S(t+t_0)$ with a minimum **mean-square error**:*

$$\mathcal{E}(t_0) = E\{|S(t+t_0) - Y(t)|^2\} = E\{|S(t+t_0) - [X(t) * h(t)]|^2\}$$

where t_0 is an arbitrary time shift, which could be positive, zero, or negative.

As shown on the next page, this optimal system (***Wiener filter***) $h(t)$ is exactly the solution of the following ***Wiener-Hopf equation***:

$$R_{sx}(\tau + t_0) = h(\tau) * R_x(\tau) = \int_{-\infty}^{\infty} h(t) R_x(\tau - t) dt \tag{9.9}$$

where R_{sx} is the crosscorrelation between $S(t)$ and the noisy input $X(t)$.

(9.9) can be solved easily by Fourier transform to yield the (noncausal) ***Wiener filter** in frequency domain* [see (9.13)]:

$$\boxed{H(\omega) = \frac{S_{sx}(\omega)}{S_x(\omega)} e^{j\omega t_0}} \tag{9.10}$$

If the signal $S(t)$ and noise $N(t)$ are orthogonal, then

$$S_{sx}(\omega) \overset{(7.39)}{=} S_s(\omega) \qquad S_x(\omega) \overset{(7.38)}{=} S_s(\omega) + S_n(\omega)$$

and thus (9.10) becomes

$$H(\omega) = \frac{S_s(\omega)}{S_s(\omega) + S_n(\omega)} e^{j\omega t_0} = \frac{1}{1 + S_n(\omega)/S_s(\omega)} e^{j\omega t_0}$$

This makes good sense: the optimal filter suppresses the frequency band that contains a large (small) amount of noise (signal) energy, and has a pure time advance (if $t_0 > 0$) or delay (if $t_0 < 0$).

The Wiener-Hopf equation (9.9) can be derived as follows. Let $\mathcal{E}_h$ and $\mathcal{E}_{h+\epsilon g}$ be the mean-square error of the LTI systems $h(t)$ and $h(t) + \epsilon g(t)$, respectively, where ϵ is an arbitrary real-valued constant. Then, since

$$\begin{aligned} R_{sy}(t_0) &= E[S(t+t_0)Y^*(t)] = E[S(t+t_0)\{X(t) * h(t)\}^*] \\ &= E\left[S(t+t_0)\int_{-\infty}^{\infty} h(\tau)^* X^*(t-\tau)d\tau\right] = \int_{-\infty}^{\infty} h(\tau)^* E[S(t+t_0)X^*(t-\tau)]d\tau \\ &= \int_{-\infty}^{\infty} h(\tau)^* R_{sx}(t_0+\tau)d\tau \overset{(8.7)}{=} R_{sx}(t_0) * h^*(-t_0) \end{aligned} \tag{9.11}$$

we have

$$\begin{aligned} \mathcal{E}_h &= E\{|S(t+t_0)|^2 + |Y(t)|^2 - 2\text{Re}[S(t+t_0)Y^*(t)]\} = R_s(0) + R_y(0) - 2\text{Re}[R_{sy}(t_0)] \\ &= R_s(0) + R_x(0) * h(0) * h^*(0) - 2\text{Re}[R_{sx}(t_0) * h^*(-t_0)] \\ \mathcal{E}_{h+\epsilon g} &= R_s(0) + R_x(0) * [h(0) + \epsilon g(0)] * [h(0) + \epsilon g(0)]^* - 2\text{Re}[R_{sx}(t_0) * [h(-t_0) + \epsilon g(-t_0)]^*] \\ &= R_s(0) + R_x(0) * h(0) * h^*(0) - 2\text{Re}[R_{sx}(t_0) * h^*(-t_0)] \\ &\quad - 2\epsilon\text{Re}[R_{sx}(t_0) * g^*(-t_0)] + 2\epsilon\text{Re}[R_x(0) * h(0) * g^*(0)] + \epsilon^2 R_x(0) * g(0) * g^*(0) \\ &= \mathcal{E}_h + \epsilon^2 \underbrace{R_x(0) * g(0) * g^*(0)}_{=R_z(0)} + 2\epsilon\text{Re}[R_x(0) * h(0) * g^*(0) - R_{sx}(t_0) * g^*(-t_0)] \end{aligned}$$

where Re[·] stands for real part and $Z(t)$ is the response of the system $g(t)$ to input $X(t)$. If $h(t)$ is optimal, then, for every ϵ, $g(t)$,

$$\mathcal{E}_{h+\epsilon g} - \mathcal{E}_h = \underbrace{\epsilon^2 R_z(0)}_{\geq 0} + \underbrace{2\epsilon\text{Re}[R_x(0) * h(0) * g^*(0) - R_{sx}(t_0) * g^*(-t_0)]}_{\text{if } \neq 0, \text{ could choose } \epsilon \text{ so that } \mathcal{E}_{h+\epsilon g} < \mathcal{E}_h} \geq 0$$

and thus a necessary and sufficient condition for $h(t)$ to be optimal is, for every $g(t)$,

$$\begin{aligned} 0 &= \text{Re}[R_x(0) * h(0) * g^*(0) - R_{sx}(t_0) * g^*(-t_0)] \\ &= \text{Re}\left[\int_{-\infty}^{\infty} g^*(\tau)\int_{-\infty}^{\infty} h(t)R_x(\tau - t)dtd\tau - \int_{-\infty}^{\infty} g^*(\tau)R_{sx}(t_0+\tau)d\tau\right] \\ &= \text{Re}\left[\int_{-\infty}^{\infty} g^*(\tau)\left[\int_{-\infty}^{\infty} h(t)R_x(\tau - t)dt - R_{sx}(t_0+\tau)\right]d\tau\right] \end{aligned}$$

Since this has to be true for every $g(t)$, the above condition is equivalent to

$$\text{Re}[R_{sx}(t_0+\tau)] = \text{Re}\left[\int_{-\infty}^{\infty} h(t)R_x(\tau - t)dt\right] = \text{Re}[h(\tau) * R_x(\tau)] \tag{9.12}$$

which reduces to the ***Wiener-Hopf equation*** (9.9) for real-valued $X(t)$ and $h(t)$.

(9.10) follows from Fourier transforming (9.9) and its shifting and convolution properties:

$$\begin{array}{ccccc} R_{sx}(\tau+t_0) & = & h(\tau) & * & R_x(\tau) \\ \updownarrow & & \updownarrow & & \updownarrow \\ S_{sx}(\omega)e^{j\omega t_0} & = & H(\omega) & \cdot & S_x(\omega) \end{array} \tag{9.13}$$

The Wiener filter (9.10) is noncausal because the solution of (9.9) does not guarantee $h(t) = 0$ for $t < 0$. The causal ***Wiener filter*** is the solution of the following causal ***Wiener-Hopf equation***:

$$R_{sx}(\tau + t_0) = \int_0^\infty h(t) R_x(\tau - t) dt \tag{9.14}$$

which has lower limit 0 instead of $-\infty$. (9.14) is, however, considerably more difficult to solve than (9.9). Simple and general solution is available only for input with rational power spectrum, which is based on ***spectral factorization*** and ***whitening*** techniques.

In general, the Wiener filter is a cascade of a whitening filter and a Wiener filter for white input, as depicted below.

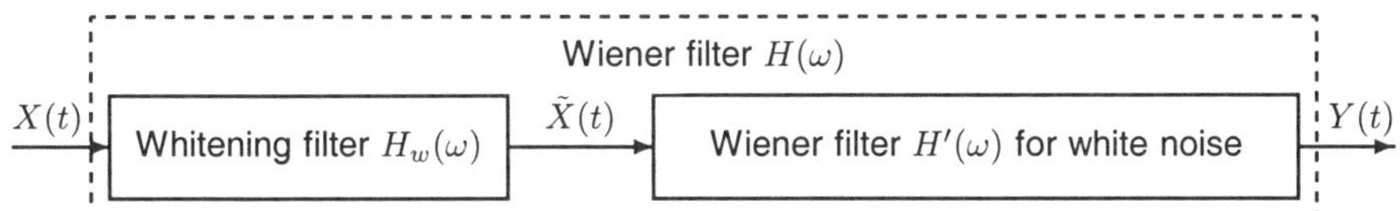

Figure 9.5: Constituents of the Wiener filter.

If the input $X(t)$ has a *rational power spectrum* $S_x(\omega)$, it can always be factorized as

$$S_x(\omega) = G(\omega) G^*(\omega) = |G(\omega)|^2 \tag{9.15}$$

where $G(s)$ has all its zeros and poles in the left-half s-plane.

It can be shown that the system $H_w(\omega) = 1/G(\omega)$

- is causal (and stable) and thus can be actually implemented
- is a ***whitening filter***: it converts *colored* input $X(t)$ into *white* output $\tilde{X}(t)$ since

$$S_{\tilde{x}}(\omega) = S_x(\omega)|H_w(\omega)|^2 = |G(\omega)|^2/|G(\omega)|^2 = 1 \quad \text{(constant)}$$

- preserves information: its output $\tilde{X}(t)$ is called ***innovations*** (i.e., new information process) that contains new and only new information about its input $X(t)$ and is thus more convenient to use. In other words, what the whitening filter does is the complete removal of the autocorrelation in $X(t)$.

If $h(t)$ is restricted to be causal (physically realizable), then following the same procedure as that led to (9.12), we have

$$\text{Re}[R_{sx}(t_0+\tau)] = \text{Re}\left[\int_0^\infty h(t)R_x(\tau-t)dt\right]$$

which reduces to the ***Wiener-Hopf equation*** (9.14) for real-valued $X(t)$ and $h(t)$. (9.14) and (9.9) also follow immediately from the so-called *orthogonality principle*, an important result in linear estimation that is not covered here.

For a rational power spectrum

$$S(\omega) = a^2\frac{(\omega-z_1)\cdots(\omega-z_N)}{(\omega-p_1)\cdots(\omega-p_M)} \qquad z_n \neq p_m$$

spectral factorization (9.15) can always be done due to the following properties:

- $S(\omega)$ is real-valued implies that $S(\omega) = S^*(\omega)$ and thus a^2 must be real and all complex zeros and poles must be in conjugate pairs.
- $S(\omega)$ is integrable over $-\infty < \omega < \infty$ (this yields the average power, except for the cases with white noise) implies that it cannot have real-valued poles and the spectrum must be proper (i.e., $N < M$).
- $S(\omega)$ is nonnegative implies that all real-valued zeros must be in even multiplicity.

Hence, we have

$$S(\omega) = \underbrace{a\frac{(\omega-z_1)\cdots(\omega-z_{N/2})}{(\omega-p_1)\cdots(\omega-p_{M/2})}}_{G(\omega)}\underbrace{a\frac{(\omega-z_1^*)\cdots(\omega-z_{N/2}^*)}{(\omega-p_1^*)\cdots(\omega-p_{M/2}^*)}}_{G^*(\omega)} \qquad N < M$$

where z_n have nonnegative imaginary parts and p_m have positive imaginary parts. Thus

$$\begin{aligned}G(s) &= G(\omega)|_{\omega=-js} \\ &= aj^{(M-N)/2}\frac{(s-jz_1)\cdots(s-jz_{N/2})}{(s-jp_1)\cdots(s-jp_{M/2})} \\ &= aj^{(M-N)/2}\frac{(s+\text{Im}[z_1]-j\text{Re}[z_1])\cdots(s+\text{Im}[z_{N/2}]-j\text{Re}[z_{N/2}])}{(s+\text{Im}[p_1]-j\text{Re}[p_1])\cdots(s+\text{Im}[p_{M/2}]-j\text{Re}[p_{M/2}])}\end{aligned}$$

has all its poles and zeros in the left-half s-plane. Such a system is said to have ***minimum phase***. Thus, $G(\omega)$ is clearly *stable*. It can also be shown to be *causal*.

The use of the whitening filter here is an excellent example of how important and useful the "unpractical" white noise model is.

For $t_0 = 0$ and white $\tilde{X}(t)$ with $R_{\tilde{x}}(\tau) = \delta(\tau)$, (9.14) has the solution (i.e., the Wiener filter)

$$h'(t) = R_{s\tilde{x}}(t)u(t) \qquad u(t) = \text{unit step} \tag{9.16}$$

The corresponding frequency response is given by

$$H'(\omega) = \int_0^{\infty} R_{s\tilde{x}}(\tau)d\tau = \int_0^{\infty} \mathcal{F}^{-1}\Big[\frac{S_{sx}(\omega)}{G^*(\omega)}\Big]e^{-j\omega\tau}d\tau = \Big[\frac{S_{sx}(\omega)}{G^*(\omega)}\Big]^+ \tag{9.17}$$

where as shown by (9.20), $\mathcal{F}[R_{s\tilde{x}}(\tau)] = \frac{S_{sx}(\omega)}{G^*(\omega)}$; $[F(\cdot)]^+$ stands for the causal (and stable) part of $F(\cdot)$ (i.e., the part with all zeros and poles in the left-half s-plane), defined by

$$F(\omega) = \mathcal{F}\{\mathcal{F}^{-1}[F(\omega)]\} = \underbrace{\int_{-\infty}^{0} \mathcal{F}^{-1}[F(\omega)]e^{-j\omega t}dt}_{[F(\omega)]^-} + \underbrace{\int_0^{\infty} \mathcal{F}^{-1}[F(\omega)]e^{-j\omega t}dt}_{[F(\omega)]^+}$$

$[F(\omega)]^-$ and $[F(\omega)]^+$ can be obtained from $F(\omega)$ by partial fraction expansion.

Consequently, the (causal) ***Wiener filter*** is given by

$$\boxed{H(\omega) = H_w(\omega)H'(\omega) = \frac{1}{G(\omega)}\Big[\frac{S_{sx}(\omega)}{G^*(\omega)}\Big]^+} \tag{9.18}$$

The corresponding minimum ***mean-square error*** for $t_0 = 0$ is given by

$$\begin{aligned}
\mathcal{E}_{\min} &= E\{[S(t) - Y(t)]^2\} = R_s(0) + R_y(0) - 2R_{sy}(0) \overset{(9.22)}{=} R_s(0) - R_y(0) \\
&= R_s(0) - \frac{1}{2\pi}\int_{-\infty}^{\infty} |H(\omega)|^2 S_x(\omega)d\omega \overset{(9.15)}{=} R_s(0) - \frac{1}{2\pi}\int_{-\infty}^{\infty} |H(\omega)G(\omega)|^2 d\omega \\
&= R_s(0) - \frac{1}{2\pi}\int_{-\infty}^{\infty} |H'(\omega)|^2 d\omega \qquad \text{[by Fig. 9.5 and (9.15)]} \\
&= R_s(0) - \int_{-\infty}^{\infty} |h'(t)|^2 dt \qquad \text{(by Parseval theorem)} \\
&= \begin{cases} R_s(0) - \int_0^{\infty} |R_{s\tilde{x}}(\tau)|^2 d\tau & \text{for causal Wiener filter} \\ R_s(0) - \int_{-\infty}^{\infty} |R_{s\tilde{x}}(\tau)|^2 d\tau & \text{for noncausal Wiener filter} \end{cases}
\end{aligned} \tag{9.19}$$

where the last equation above follows from the fact that the causal Wiener filter has its $h'(t)$ given by (9.16) while the noncausal Wiener filter has

$$h'(t) = R_{s\tilde{x}}(t)$$

Note that the causal filter has a larger MSE than the noncausal filter.

The solution of (9.14) for the white input $\tilde{X}(t)$, which has autocorrelation $R_{\tilde{x}}(\tau) = \delta(\tau)$ since $S_{\tilde{x}}(\omega) = 1$, is given by

$$h'(t) = R_{s\tilde{x}}(t + t_0)u(t)$$

where $u(t)$ is the unit step function. Similarly to (9.11), we have (replacing y with $\tilde{x}$) $R_{s\tilde{x}}(\tau) = R_{sx}(\tau) * h_w^*(-\tau)$, which leads to, by convolution and conjugate properties of Fourier transforms,

$$S_{s\tilde{x}}(\omega) = S_{sx}(\omega)H_w^*(\omega)$$

Hence, we have

$$\mathcal{F}[R_{s\tilde{x}}(\tau + t_0)] = S_{s\tilde{x}}(\omega)e^{j\omega t_0} = S_{sx}(\omega)H_w^*(\omega)e^{j\omega t_0} = \frac{S_{sx}(\omega)}{G^*(\omega)}e^{j\omega t_0} \tag{9.20}$$

Then, the causal Wiener filter for white input $\tilde{X}(t)$ is given by

$$\begin{aligned} H'(\omega) &= \mathcal{F}[h'(t)] = \int_0^\infty R_{s\tilde{x}}(\tau + t_0)e^{-j\omega\tau}d\tau \\ &= \int_0^\infty \mathcal{F}^{-1}\left[\frac{S_{sx}(\omega)}{G^*(\omega)}e^{j\omega t_0}\right]e^{-j\omega\tau}d\tau = \left[\frac{S_{sx}(\omega)}{G^*(\omega)}e^{j\omega t_0}\right]^+ \end{aligned}$$

which reduces to (9.17) for $t_0 = 0$. Consequently, the ***causal Wiener filter*** for random input with a rational power spectrum is given by

$$H(\omega) = \frac{1}{G(\omega)}\int_0^\infty \mathcal{F}^{-1}\left[\frac{S_{sx}(\omega)}{G^*(\omega)}e^{j\omega t_0}\right]e^{-j\omega t}dt = \frac{1}{G(\omega)}\left[\frac{S_{sx}(\omega)}{G^*(\omega)}e^{j\omega t_0}\right]^+$$

which clearly reduces to (9.18) for $t_0 = 0$.

Note that the noncausal Wiener filter (9.10) can also be written in a form similar to the above:

$$H(\omega) = \frac{S_{sx}(\omega)}{S_x(\omega)}e^{j\omega t_0} = \frac{1}{G(\omega)}\left[\frac{S_{sx}(\omega)}{G^*(\omega)}e^{j\omega t_0}\right]$$

The above methodology based on ***spectral factorization*** and ***whitening*** techniques is general. It also works for input $X(t)$ with a power spectrum that is not rational. The spectral factorization is very simple for a rational function but is quite difficult for a general function, which calls for a more sophisticated function theory. Since techniques for approximating arbitrary power spectra by rational functions are available, it is often sufficient in practice to consider only rational functions.

Note that (9.11) implies that

$$R_{sy}(\tau) = R_{sx}(\tau) * h^*(-\tau) \quad \Longrightarrow \quad S_{sy}(\omega) = S_{sx}(\omega)H^*(\omega)$$

Furthermore, for the (causal or noncausal) Wiener filter, we have

$$\text{(9.9) or (9.14)} \Longrightarrow R_{sx}(\tau) = R_x(\tau) * h(\tau) \Longrightarrow S_{sx}(\omega) = H(\omega)S_x(\omega) \tag{9.21}$$

$$\text{(9.11) and (9.21)} \Longrightarrow R_{sy}(\tau) = R_y(\tau) \Longrightarrow S_{sy}(\omega) = S_Y(\omega) \tag{9.22}$$

Example 9.3: Causal and Noncausal Wiener Filters for White Noise

Consider a noisy input $X(t) = S(t) + N(t)$, where $N(t)$ is zero-mean white noise with $S_n(\omega) = 1$, independent of the signal $S(t)$ with power spectrum $S_s(\omega) = \frac{3}{1+\omega^2}$.

$S(t)$ and zero-mean $N(t)$ are orthogonal since they are independent. Thus

$$S_x(\omega) \overset{(7.38)}{=} S_s(\omega) + S_n(\omega) = \frac{4+\omega^2}{1+\omega^2} = \underbrace{\frac{2+j\omega}{1+j\omega}}_{G(\omega)}\underbrace{\frac{2-j\omega}{1-j\omega}}_{G^*(\omega)} = G(\omega)G^*(\omega)$$

$$S_{sx}(\omega) \overset{(7.39)}{=} S_s(\omega)$$

Hence, we have

$$\frac{S_{sx}(\omega)}{G^*(\omega)} = \frac{3}{1+\omega^2}\frac{1-j\omega}{2-j\omega} = \frac{3}{(1+j\omega)(2-j\omega)} = \frac{1}{1+j\omega} + \frac{1}{2-j\omega}$$

Its causal (and stable) part is $\frac{1}{1+j\omega}$ (poles and zeros are in the left-half s-plane). Thus, the (non)causal *Wiener filters* are given in frequency domain by, for $t_0 = 0$,

$$H^{\text{causal}}(\omega) \overset{(9.18)}{=} \frac{1}{G(\omega)}\left[\frac{S_{sx}(\omega)}{G^*(\omega)}\right]^+ = \frac{1+j\omega}{2+j\omega}\frac{1}{1+j\omega} = \frac{1}{2+j\omega}$$

$$H(\omega) \overset{(9.10)}{=} \frac{S_{sx}(\omega)}{S_x(\omega)} = \frac{1}{G(\omega)}\frac{S_{sx}(\omega)}{G^*(\omega)} = \frac{3}{1+\omega^2}\frac{1+\omega^2}{4+\omega^2} = \frac{3}{4+\omega^2}$$

or in time domain by (from inverse Fourier transforms)

$$h^{\text{causal}}(t) = e^{-2t}u(t), \qquad h(t) = \frac{3}{4}e^{-2|t|}$$

The corresponding *minimum mean-square errors* are given by (9.19) as

$$\begin{aligned}
\mathcal{E}_{\min}^{\text{causal}} &= \frac{1}{2\pi}\int_{-\infty}^{\infty}[S_s(\omega) - |H^{\text{causal}}(\omega)|^2 S_x(\omega)]d\omega \\
&= \mathcal{F}^{-1}\left[\frac{3}{1+\omega^2} - \frac{1}{4+\omega^2}\frac{4+\omega^2}{1+\omega^2}\right]_{\tau=0} = \mathcal{F}^{-1}\left[\frac{2}{\omega^2+1}\right]_{\tau=0} = \left[e^{-|\tau|}\right]_{\tau=0} = 1
\end{aligned}$$

$$\begin{aligned}
\mathcal{E}_{\min} &= \frac{1}{2\pi}\int_{-\infty}^{\infty}[S_s(\omega) - |H(\omega)|^2 S_x(\omega)]d\omega \\
&= \mathcal{F}^{-1}\left[\frac{3}{1+\omega^2} - \frac{9}{(4+\omega^2)^2}\frac{4+\omega^2}{1+\omega^2}\right]_{\tau=0} = \mathcal{F}^{-1}\left[\frac{3}{\omega^2+4}\right]_{\tau=0} = \left[\frac{3}{4}e^{-2|\tau|}\right]_{\tau=0} = \frac{3}{4}
\end{aligned}$$

The Wiener Filter for an Arbitrary Desired Response

In fact, the Wiener filter can be used to minimize the mean-square error between the output and a desired output, which is not necessarily the time-shifted version of the input signal $S(t+t_0)$. Let $D(t)$ be the desired output. Then the mean-square error between the actual output $Y(t)$ and $D(t)$ is:

$$\mathcal{E} = E\{|D(t) - Y(t)|^2\} = E\{|D(t) - [X(t) * h(t)]|^2\}$$

The ***Wiener-Hopf equation*** for this problem is

$$R_{dx}(\tau) = h(\tau) * R_x(\tau)$$

where $R_{dx}(\tau)$ is the crosscorrelation between $D(t)$ and the noisy input $X(t)$. The corresponding ***Wiener filters*** are

$$\text{noncausal: } H(\omega) = \frac{S_{dx}(\omega)}{S_x(\omega)}$$

$$\text{causal: } H(\omega) = \frac{1}{G(\omega)}\left[\frac{S_{dx}(\omega)}{G^*(\omega)}\right]^+$$

which have the following minimum mean-square errors:

$$\text{noncausal: } \mathcal{E}_{\min} = R_d(0) - \int_{-\infty}^{\infty} R_{dx}(\tau)h(\tau)d\tau$$

$$\text{causal: } \mathcal{E}_{\min} = R_d(0) - \int_{0}^{\infty} R_{dx}(\tau)h(\tau)d\tau$$

9.5 Summary and Requirements

An *optimal linear system* is a linear system that is optimal under some optimality criterion. Two such optimal systems are studied in this chapter: the matched filter and the Wiener filter.

The *matched filter* maximizes the output *signal-to-noise ratio* (SNR), defined by

$$\text{SNR} = \frac{\text{(average) signal power}}{\text{(average) noise power}}$$

For the case of a deterministic signal $s(t)$ corrupted by white noise, the matched filter has the form

$$h(t) = Ks^*(t_0 - t)$$

where t_0 is the time at which the output SNR is to be maximized. Such a filter is in general noncausal. Its causal version is

$$h(t) = Ks^*(t_0 - t)u(t)$$

It has an SNR proportional to the total signal energy through time t_0, smaller than the signal's total energy, which corresponds to the SNR of the noncausal matched filter.

The *Wiener filter* minimizes the mean-square error between the output and the signal part of the input. The key to the Wiener filter is the *Wiener-Hopf equation*:

$$R_{sx}(\tau + t_0) = h(\tau) * R_x(\tau)$$

which relates the crosscorrelation $R_{sx}(\tau + t_0)$ (between input signal $S(t)$ and the noisy input $X(t)$) with the autocorrelation $R_x(\tau)$ of $X(t)$ through the system $h(t)$. Solving it using Fourier transform yields the (noncausal) *Wiener filter*

$$H(\omega) = \frac{S_{sx}(\omega)}{S_x(\omega)} e^{j\omega t_0}$$

where $S_{sx}(\omega)$ and $S_x(\omega)$ are the power spectra corresponding to $R_{sx}(\tau)$ and $R_x(\tau)$. The above filter is noncausal. If only causal systems $h(t) = 0, t < 0$ are considered, the corresponding Wiener-Hopf equation is in general not easy to solve. The general causal *Wiener filter* for colored noise is the cascade of a whitening filter $H_w(\omega) = \frac{1}{G(\omega)}$ and the Wiener filter $H'(\omega)$ for white noise:

$$H(\omega) = H_w(\omega)H'(\omega) = \frac{1}{G(\omega)} \left[\frac{S_{sx}(\omega)}{G^*(\omega)} e^{j\omega t_0} \right]^+$$

where $G(\omega)$ is such that $S_x(\omega) = G(\omega)G^*(\omega)$ and $[F(\omega)]^+$ is the causal and stable part of $F(\omega)$, which can be obtained easily by spectral factorization and partial fraction expansion if $S_x(\omega)$ is rational.

Basic Requirements

- Understand the concept of an optimal system and the underlying optimality criteria for the matched filter and the Wiener filter.
- Master the concept of signal-to-noise ratio.
- Know how to determine the causal and noncausal matched filters for white noise case and the corresponding maximum SNRs.
- Know how to obtain the causal and noncausal Wiener filters for input with rational power spectrum and the corresponding minimum mean-square errors.

9.6 Additional Examples

9.4 *Optimal linear differentiator in the presence of noise.* An LTI system is intended to be a differentiator; that is, its output is the differentiation of its input. If the input is noisy, however, the output could be extremely noisy. It is desired that the output is approximately the differentiation of the signal part $S(t)$ of the input $X(t) = S(t) + N(t)$, where $S(t)$ and $N(t)$ are uncorrelated and have zero mean and power spectra $S_s(\omega)$ and $_n(\omega)$, respectively. Find the optimal linear system that minimizes the mean-square error between its output $Y(t)$ and $\frac{d}{dt}S(t)$.

Solution: The optimal linear system that minimizes the mean-square error between its output $Y(t)$ and $\frac{d}{dt}S(t)$ is the Wiener filter with the desired output $D(t) = \frac{d}{dt}S(t)$. Since

$$R_{\dot{s}x}(\tau) \overset{(6.36)}{=} \frac{d}{d\tau}R_{sx}(\tau)$$

implies

$$S_{\dot{s}x}(\omega) = j\omega S_{sx}(\omega)$$

by (9.23), the optimal differentiator has the frequency response

$$H(\omega) = \frac{S_{dx}(\omega)}{S_x(\omega)} = \frac{S_{\dot{s}x}(\omega)}{S_x(\omega)} = \frac{j\omega S_{sx}(\omega)}{S_x(\omega)}$$

9.7 Problems

9.1 *SNR optimalization by parameter adjustment.* The RL circuit of Example 8.8 has a random input $X(t) = \sin(\omega_0 + \phi)$ corrupted by zero-mean white noise with $S_n(\omega) = S_0$, where $\phi \sim \mathcal{U}(0, 2\pi)$ is independent of $N(t)$.

(a) Find the average steady-state power of the response to the random signal.
(b) Determine the values of R/L and α such that the average output SNR is maximized.

9.2 *Matched filter for white noise.* Consider a deterministic signal

$$s(t) = \begin{cases} 10 & 0 < t < 20 \\ 0 & \text{elsewhere} \end{cases}$$

corrupted by white noise with $S(\omega) = 2$.

(a) Find the causal matched filter for $s(t)$ at an arbitrary but fixed time t_0. Find the output SNR of this filter.
(b) Find the smallest time t_0 at which this causal matched filter yields largest output SNR. Find its output SNR at this time.
(c) Find the response of the matched filter obtained in (b) to deterministic signal $s(t)$. Is it good to use the matched filter to recover $s(t)$ from the noisy input? Why?

9.3 *Signal recovery performance of matched filter.* For Example 9.2, find the response $s_o(t)$ of the causal matched filter with $t_0 = 5$ to $s(t) = 10e^{-t}u(t)$. Compare $s_o(t)$ with $s(t)$. Is it good to use the matched filter to recover $s(t)$ from the noisy input?

9.4 *Matched filter for triangular signal.* Consider the deterministic signal

$$s(t) = \begin{cases} \frac{T-|t|}{T} & |t| < T \\ 0 & \text{elsewhere} \end{cases}$$

corrupted by white noise $N(t)$ with power spectrum $S_n(\omega) = S_0$.

(a) Plot $s(t)$.
(b) Find the noncausal matched filter, matched to $s(t)$ at $t_0 = T/2$, and the corresponding output SNR.
(c) Find the causal matched filter, matched to $s(t)$ at $t_0 = T/2$, and the corresponding output SNR.
(d) Is it good to use the (causal or noncausal) matched filter to recover the input signal from the output? Justify your answer. If yes, how large is the error between the output and the input signal? If no, suggest an alternative filter for this job (signal recovery).

9.5 *Wiener filter.* The sum $X(t)$ of a signal $S(t)$ and noise $N(t)$ is a white random process with $S_x(\omega) = 3$, where $S(t)$ and $N(t)$ are orthogonal. The cross-power spectrum between $S(t)$ and $X(t)$ is $S_{sx}(\omega) = \frac{4}{\omega^2 + 1}$. Find the causal and noncausal Wiener filters (for $t_0 = 0$) in both frequency and time domains. Find the corresponding mean-square errors.

9.6 *Wiener filter.* We would like to design a causal Wiener filter $H(\omega)$ for an input $X(t) = S(t) + N(t)$, where the signal $S(t)$ and noise $N(t)$ are independent, with power spectra $S_s(\omega) = \frac{1}{\omega^2 + 1}$ and $S_n(\omega) = \frac{\omega^2}{\omega^2 + 1}$, respectively.

(a) Determine $H(\omega)$ for the Wiener filter and the mean-square error between $S(t)$ and the output $Y(t)$.
(b) Determine $H(\omega)$ for the optimal realizable Wiener filter and the mean-square error between $S(t)$ and the output $Y(t)$.

9.7 *Wiener filter.* Consider a noisy input $X(t) = 3S(t) + N(t)$, where $N(t)$ is zero-mean white noise with $S_n(\omega) = 1$, independent of the signal $S(t)$ with power spectrum $S_s(\omega) = \frac{1}{2 + \omega^2}$.

(a) Find the causal and noncausal Wiener filters in both time and frequency domains that minimize $E\{[S(t) - Y(t)]^2\}$.
(b) Find the minimum mean-square error $E\{[S(t) - Y(t)]^2\}$.

9.8 *From noncausal to causal Wiener filter.* The noncausal Wiener filter for a random input $X(t)$ with power spectrum $S_x(\omega) = 4$ is given by

$$h(t) = 2e^{-3|t|}$$

Find the causal Wiener filter in time domain for this input.

9.9 *From noncausal to causal Wiener filter.* The noncausal Wiener filter for a random input $X(t)$ with power spectrum $S_x(\omega) = \frac{1}{\omega^2 + 1}$ is given by

$$h(t) = 2e^{-3|t|}$$

Find the causal Wiener filter in time domain for this input.

9.10 *Causal Wiener filter for signal recovery.* A random signal $S(t)$ with $S_s(\omega) = \dfrac{1}{\omega^2+1}$ is corrupted by colored noise with $S_n(\omega) = \dfrac{a^2 S_0}{\omega^2+a^2}$ that is independent of $S(t)$. Find the causal Wiener filter that reconstructs the signal with a minimum $E\{[S(t)-Y(t)]^2\}$, where $Y(t)$ is the output of the filter.

9.11 *Signal recovery by Wiener filter.* A random signal $X(t)$ satisfies the following equation

$$\frac{d}{dt}X(t) = -X(t) + W(t) \tag{9.23}$$

where $W(t)$ is white with $S_w(\omega) = 3$. It is observed (measured) as $Z(t)$ in the presence of white noise $N(t)$ with spectrum $S_n(\omega) = 1$ and independent of $X(t)$:

$$Z(t) = X(t) + N(t) \tag{9.24}$$

(a) Find the autocorrelation and power spectrum of $X(t)$.
(b) Find the noncausal Wiener filter to recover $X(t)$ from $Z(t)$.
(c) Design the causal Wiener filter to recover $X(t)$ from $Z(t)$.
(d) Find and compare the mean-square errors of the two Wiener filters.

9.12 *Wiener filter.* Given a random process $X(t) = S(t)+N(t)$ where the random signal $S(t)$ and noise $N(t)$ are orthogonal and have zero mean and autocorrelations $R_s(\tau) = e^{-|\tau|}$ and $R_s(\tau) = e^{-2|\tau|}$, find the causal Wiener filter.

9.13 *Most probable value.* Suppose that a system's output $Y(t)$ is the response to the input $X(t)$ corrupted by additive noise $N(t)$ that is independent of $X(t)$, where $X(t)$ for any given time is a random variable $X(t) \sim \mathcal{U}(0,6)$ and the noise has the marginal PDF

$$f_n(x) = 0.5\delta(x-2) + 0.5\delta(x-14)$$

What is the most probable value of X if we observe that $Y = 8$?

9.14 *Optimal bandwidth.* We would like to design an ideal lowpass filter with the following frequency response

$$H(\omega) = \begin{cases} 1 & \leq W \\ 0, & \text{elsewhere} \end{cases}$$

for an input $X(t) = S(t)+N(t)$, where signal $S(t)$ and noise $N(t)$ are independent, with power spectra given below for the positive ω:

$$S_s(\omega) = \begin{cases} 1 & 0 \leq \omega \leq 20 \\ 0.5 & 40 \leq \omega \leq 60 \\ 0 & \text{elsewhere} \end{cases}$$

$$S_n(\omega) = \begin{cases} 1 & 10 \leq \omega \leq 50 \\ 0 & \text{elsewhere} \end{cases}$$

Determine the optimal bandwidth W such that the output SNR is maximized.

9.15 *Puzzling questions.* Raise two technical questions of your own that you are puzzled by and that are closely related to the material covered in this chapter. These questions should be important, not too simple, and not too involved.

9.8 Computer Exercises

9.1 *Matched filter.* A random process $X(t)$ is the sum of a deterministic signal $s(t) = \sin(120\pi t)$ and zero-mean Gaussian white noise $N(t)$ with intensity 2.

(a) Find the matched filter for time $t_0 = 2$ for this problem theoretically.
(b) Use the companion software P&R to generate a 2400-point discrete-time (with sampling rate of 600 Hz) sample function of $X(t)$.
(c) Write a MATLAB routine to generate the response of the above matched filter to the input $X(t)$.
(d) Calculate output SNR of the matched filter.
(e) Compare the output and input waveforms of the matched filter and comment on their difference. Can we have another LTI system that yields a smaller difference between the input and output? Can we have another LTI system that yields a larger output SNR?

9.2 *Wiener filter.* A random process $X(t)$ is the sum of a deterministic signal $S(t) = \sin(120\pi t + \theta)$ and zero-mean Gaussian white noise $N(t)$ with intensity 2, where $\theta \sim \mathcal{U}(0, 2\pi)$, θ and $N(t)$ are independent.

(a) Find the causal and noncausal Wiener filters for time $t_0 = 0$ for this problem theoretically.
(b) Use the companion software P&R to generate a 2400-point discrete-time (with sampling rate of 600 Hz) sample function of $X(t)$.
(c) Write a MATLAB routine to generate the response of the noncausal Wiener filter to the input $X(t)$.
(d) Calculate output SNR of the Wiener filters.
(e) Compare the output and input waveforms for the Wiener filters, respectively. Comment on their differences. Can we have another LTI system that yields a smaller difference between the input and output? Can we have another LTI system that yields a larger output SNR?

9.9 Self-Test Problems

9.1 Answer the following questions briefly.

(a) What is the optimality criteria for the matched filter and the Wiener filter?
(b) Is the matched filter for white noise case unique (i.e., are two matched filters for the same signal at a given time with the same white noise always identical)?
(c) Given an arbitrary deterministic signal corrupted by white noise, can we have an LTI system that is the matched filter for different time?
(d) Is it good to use the matched filter to recover the signal in the input?
(e) Which one has a smaller mean-square error, the noncausal or causal Wiener filter?
(f) Can a nonoptimal noncausal system have smaller mean-square error than the causal Wiener filter?
(g) Can a causal system have smaller mean-square error than the noncausal Wiener filter?
(h) What is the relation between the Wiener filter for colored noise and for white noise?

9.2 Find the causal Wiener filter for a random process $X(t)$ that is the sum of a random signal $S(t)$ with autocorrelation $R_s(\tau) = \frac{1}{2}e^{-|\tau|}$ and zero-mean noise $N(t)$ with autocorrelation $R_n(\tau) = \frac{1}{4}e^{-2|\tau|}$, where $S(t)$ and $N(t)$ are uncorrelated.

9.10 Solutions to Self-Test Problems

9.1 (a) The matched filter maximizes the output SNR while the Wiener filter minimizes the mean-square error between the output and the signal.
(b) No. The matched filters for white noise case are not unique. They can be different by a proportionality factor.
(c) No. The impulse response function of the matched filter depends on the time instant at which the match is sought and thus the matched filter in general cannot be time-invariant.
(d) No. The matched filter maximizes the output SNR but its output may differ substantially from its input signal and is thus not suitable for signal recovery.
(e) The noncausal Wiener filter in general has a smaller mean-square error than the causal one.
(f) Yes. The causal Wiener filter is optimal (have smallest mean-square error) only of all causal systems.
(g) No. The noncausal Wiener filter is optimal (have smallest mean-square error) of all (causal and noncausal) systems and thus a causal system cannot have smaller mean-square error than the noncausal Wiener filter.
(h) The Wiener filter for colored noise is the cascade of a whitening filter and the Wiener filter for white noise.

9.2 Since $S(t)$ and zero-mean $N(t)$ are uncorrelated, they are orthogonal. Thus

$$\begin{aligned} S_x(\omega) &\overset{(7.38)}{=} S_s(\omega) + S_n(\omega) \\ &= \frac{5+2\omega^2}{(1+\omega^2)(4+\omega^2)} \\ &= \underbrace{\frac{\sqrt{2}(\sqrt{2.5}+j\omega)}{(1+j\omega)(2+j\omega)}}_{G(\omega)} \underbrace{\frac{\sqrt{2}(\sqrt{2.5}-j\omega)}{(1-j\omega)(2-j\omega)}}_{G(\omega)} \\ &- G(\omega)G^*(\omega) \\ S_{sx}(\omega) &\overset{(7.39)}{=} S_s(\omega) = \frac{1}{1+\omega^2} \end{aligned}$$

Hence, we have

$$\begin{aligned} \frac{S_{sx}(\omega)}{G^*(\omega)} &= \frac{1}{1+j\omega}\frac{1}{1-j\omega}\frac{(1-j\omega)(2-j\omega)}{\sqrt{2}(\sqrt{2.5}-j\omega)} \\ &= \frac{2-j\omega}{\sqrt{2}(\sqrt{2.5}-j\omega)(1+j\omega)} \\ &= \frac{0.822}{1+j\omega} + \frac{0.115}{\sqrt{2.5}-j\omega} \end{aligned}$$

Its causal (and stable) part (poles and zeros are in the left-half s-plane) is

$$\left[\frac{S_{sx}(\omega)}{G^*(\omega)}\right]^+ = \frac{0.822}{1+j\omega}$$

Thus, the causal Wiener filter is given in frequency domain by

$$\begin{aligned} H(\omega) &\overset{(9.18)}{=} \frac{1}{G(\omega)}\left[\frac{S_{sx}(\omega)}{G^*(\omega)}\right]^+ \\ &= \frac{(1+j\omega)(2+j\omega)}{\sqrt{2}(\sqrt{2.5}+j\omega)}\frac{0.822}{1+j\omega} \\ &= 0.582\frac{2+j\omega}{\sqrt{2.5}+j\omega} \end{aligned}$$

A

COMPANION SOFTWARE P&R

The companion software package is written in MATLAB and runs in a Windows environment. It is entirely menu-driven and is intended to be extremely user-friendly. It does not require the knowledge of any programming language, including MATLAB. However, a basic knowledge of MATLAB helps the reader run P&R more effectively and smoothly. Many of the computer exercises included in this book would require too much computer programming effort without this companion software.

Fig. A.1 shows the main window of P&R, along with some of its menus.

Figure A.1: The main window of P&R.

P&R can be activated by double clicking on the P&R icon. It can also be activated by typing `main` at the MATLAB prompt when the current working directory is the one that houses P&R (i.e., where `main.m` of P&R is located).

P&R is capable of the following tasks:

- generating random numbers with a variety of commonly used one- and two-dimensional distributions
- plotting probability density (or mass) functions of a variety of commonly used one-dimensional distributions
- looking up tables of probability distributions and percentiles of a variety of commonly used one-dimensional distributions
- plotting histograms, empirical PDFs, and scatter diagrams of data
- identifying distributions of data
- generating random processes, including Gaussian processes, random sinusoids, and random pulse trains
- estimating means, correlation functions, and power spectra of random processes
- generating responses of linear systems
- solving all computer exercises (instructor's version only)
- performing other miscellaneous tasks

Generation of Random Numbers. P&R can be used to generate random numbers having the following distributions

- one-dimensional binomial, chi-square, exponential, Gaussian, log-normal, Poisson, Rayleigh, uniform, and Weibull distributions
- two-dimensional correlated as well as uncorrelated jointly Gaussian distributions

The procedure for generating random numbers is simple, as demonstrated in Example 3.33 for a chi-square distribution and in Example 4.16 for a jointly Gaussian distribution.

Plotting PDF/PMF Curves. The procedure for plotting the probability density (or mass) functions of binomial, chi-square, exponential, Gaussian, log-normal, Poisson, Rayleigh, uniform, and Weibull distributions by P&R is illustrated in Example 3.16 for a continuous distribution and in Example 3.17 for a discrete distribution. Knowledge of the PDF/PMF curves of various distributions is useful in, e.g., identifying the distribution of data.

Looking Up Distribution Tables. Example 3.12 illustrates the procedure for finding the probabilities (i.e., CDF) or tail probabilities of binomial, chi-square, exponential, Gaussian, log-normal, Poisson, Rayleigh, uniform, and Weibull distributions using P&R. It is because of this capability of P&R that no probability tables are provided in this book.

Determination of Percentiles. No percentile tables are provided in this book because P&R provides a simple means to determine the percentile points of binomial, chi-square, exponential, Gaussian, log-normal, Poisson, Rayleigh, uniform, and Weibull distributions. This can be done according to the following simple steps (using the chi-square distribution as an example):

S1. Click "Percentile" in the main window of P&R.

S2. Click "ChiSquare." The "**Percentiles of Chi-Square Distribution**" window will appear.

S3. Enter "2" for D.O.F. (degrees of freedom) and '0.99" for Percent, as shown in Fig. A.2. Click "Ok." The 99% percentile point of the (standard) chi-square distribution with two degrees of freedom will then appear in the grey box as $x = 9.2103$.

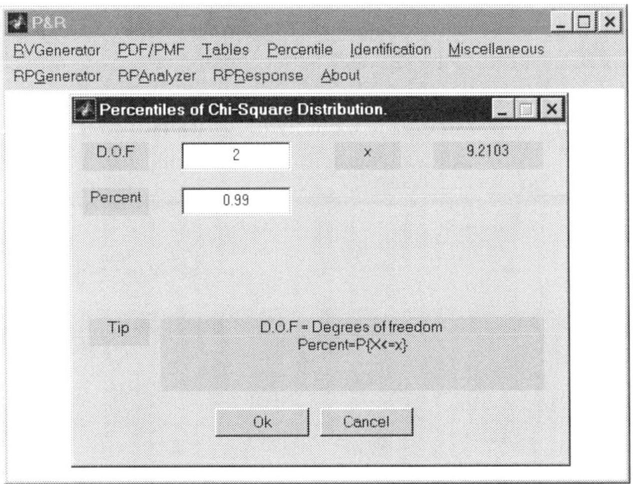

Figure A.2: Determination of chi-square percentiles by P&R.

Plotting Histograms. P&R can be used to plot the histograms of data, as illustrated in Example 3.34. The data should be arranged in columns. One histogram will be generated from each column. The corresponding histograms will be overlaid if two or more columns of data are present.

Distribution Identification. The "Identification" menu in P&R can be used to identify the one-dimensional distribution of a given set of data. The following example assumes that the data are contained in the file `e5_4.dat` in the subdirectory of P&R.

S1. Plot the histogram of the data as described above and illustrated in Example 3.34. The histogram is given in Fig. A.3.

S2. Choose a candidate distribution with a theoretical PDF/PMF curve resembling the histogram in shape.[1] Suppose that a chi-square distribution is chosen by an inspection of the histogram and Fig. 3.12.

S3. Click "Identification" in the main window of P&R. Point to "Estimation and Validation" and "1-D Distribution." Then choose (click) the candidate distribution identified above (i.e., "ChiSquare" for this example).

[1] Knowledge of the shapes of various PDF/PMF curves is clearly beneficial here. Such knowledge can be gained by plotting the PDF/PMF curves of various distributions using P&R as described above.

S4. Choose the correct subdirectory and the data file that contains the data (i.e., `e5_4.dat` under `data` subdirectory in this example). The "**Parameter Estimation**" window for the candidate distribution will appear with the estimated parameters of the distribution from the data, that is, the sample mean, sample variance, estimated σ and degrees of freedom for this example with the chi-square distribution.

S5. Click "Validate." Two curves will appear, one being the empirical PDF (i.e., the approximate PDF obtained from the data, which is the histogram with the correct scale as a PDF) and the other the theoretical PDF using the estimated parameters.

S6. If the two curves are sufficiently close, the candidate distribution with the estimated parameters may thus be identified as an approximate theoretical distribution of the data. Otherwise, repeat Steps 2 through 5 for another candidate distribution. For the chi-square distribution in the current example, the two curves are not close enough, as shown in Fig. A.3, and thus we may try a Weibull distribution. This then becomes exactly Example 5.4.

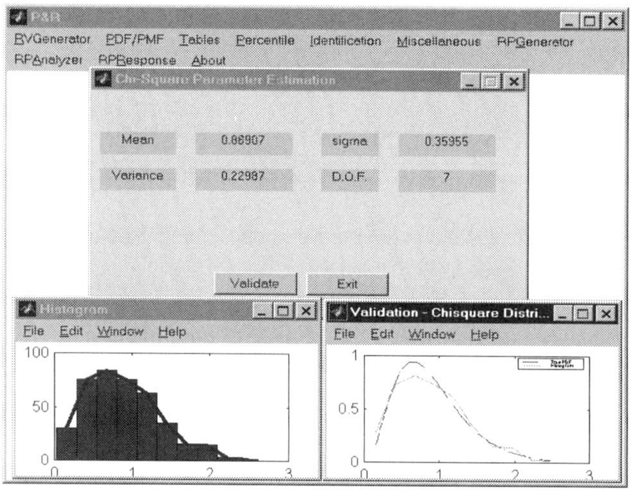

Figure A.3: Identification of the distribution of data by P&R.

Plotting Empirical PDFs. A plot of the empirical PDF of a set of data can clearly be obtained by following Steps 3 through 5 in the above "Distribution Identification" task by using any continuous distribution as the candidate.

Calculation of Sample Statistics. P&R can be used to calculate the sample statistics of data, including the sample mean and sample variance for one-dimensional distributions and additionally, the sample crosscorrelation, sample crosscovariance, and sample correlation coefficient for two-dimensional distributions. This is demonstrated in Example 4.8.

Plotting Scatter Diagrams. The scatter diagram of a set of two-dimensional data can easily be plotted using P&R, as illustrated in Example 4.8. The data should be arranged in two columns.

Generation of Gaussian Processes. It is a simple task to generate a white or correlated Gaussian random process with an exponential autocorrelation using P&R. This is illustrated in Example 6.18. The generated random process is in a column format.

Generation of Random Pulse Trains. Example 6.3 demonstrates the simple procedure for generating a pulse train with time-varying or time-invariant random amplitude and/or pulse width. The generated random process is in a column format.

Generation of Random Sinusoids. Sinusoids with random amplitude, phase, and/or frequency can be generated easily by P&R, as illustrated in Example 6.8. The generated random process is in a column format.

Time-Domain Analysis. Examples of time-domain analysis of random processes by P&R can be found in Examples 6.13 through 6.15 and in Example 6.17. The analysis includes the estimation/computation of means, variances, autocorrelations, autocovariances, correlation coefficients, crosscorrelations, crosscovariances, and crosscorrelation coefficients. The random process to be analyzed should be in a column format.

Frequency-Domain Analysis. P&R can be used to estimate the power spectra of random processes. This is demonstrated in Examples 7.8 through 7.10. The random process to be analyzed should be in a column format.

Generation of System Responses. P&R includes a response generator for LTI systems, including lowpass, bandpass, bandstop, highpass, and rational systems. This is demonstrated in Examples 8.9, 8.11 and 8.13. The input random or deterministic process should be in a column format.

Miscellaneous Tasks. P&R can also be used to perform some other tasks, including conducting Bernoulli trials (Example 2.25), calculating its relevant probabilities (Example 2.26), and performing linear regression (Example 5.14).

B

TIPS FOR USING MATLAB

MATLAB is a technical computing environment which is particularly convenient for mathematical operations involving vectors, matrices, and/or arrays. MATLAB also possesses good plotting capabilities for the visualization of data.

Although the companion software P&R runs under MATLAB, the use of P&R does not require the knowledge of MATLAB. However, a good knowledge of MATLAB does help the reader make better use of P&R mainly in two aspects: entering mathematical expressions into P&R and utilizing the results obtained by P&R.

Table B.1 presents a sample of MATLAB rules useful for entering a mathematical expression into P&R.

Table B.1: A sample of useful MATLAB rules for operations.

Operation	*Formula*	*MATLAB*	*Example*
Multiplication	ab	`a*b`	`2.5*4.6`
Power	a^b	`a^b`	`4.3^4`
Square root	$\sqrt{a}$	`sqrt(a)`	`sqrt(4.3)`
Absolute value	$\|a\|$	`abs(a)`	`abs(-4.3)`
Exponential function	e^{-x}	`exp(-x)`	`exp(-4.3)`
Sine function	$\sin x$	`sin(x)`	`sin(pi/3)`
	π	`pi`	
	1.5×10^{-8}	`1.5e-8`	
Element-by-element multiplication	$[a_i b_i]$ or $[a_{ij} b_{ij}]$	`a.*b`	`a=[1, 2]', b=[3, 4]'` `a.*b =[3, 8]'`
Element-by-element power	$[a_i^3]$ or $[a_{ij}^3]$	`a.^3`	`a=[2, 3]'` `a.^3=[8, 27]'`

Examples of MATLAB commands that are useful for utilizing the results of P&R are listed in Table B.2.

Table B.2: Examples of useful MATLAB commands.

MATLAB command	*Explanation*
`cd data`	Change working directory to subdirectory `data`
`cd ..`	Change working directory to parent directory
`hist(z,20)`	Generate histogram(s) of data contained in `z` using 20 bins
`load e4_1.dat`	Load data from file `e4_1.dat` to a matrix named `e4_1.dat`
`save e4_1.dat z -ascii`	Save variable `z` as ascii data into file `e4_1.dat`
`x=rand(100,1)`	`x` is a vector of 100 independent $\mathcal{U}(0,1)$ random numbers
`x=randn(100,1)`	`x` is a vector of 100 independent $\mathcal{N}(0,1)$ random numbers
`x=2*rand(100,1)-1`	`x` is the vector with each element equal to twice the corresponding random number minus 1
`y=mean(x)`	`y` is the average value of the elements of `x`, i.e., `y` is the sample mean of the data contained in `x`
`y=var(x)`	`y` is the sample variance of the data contained in `x`
`y=sqrt(x)`	Equivalent to `y=x.^(1/2)`, meaning $y_i = x_i^{1/2}$ or $y_{ij} = x_{ij}^{1/2}$ elements of `y` are the square of the corresponding elements of `x`
`Y=sqrtm(X)`	`Y` is the square-root matrix of `X`, i.e., $X = YY'$
`y=(x-a).^2`	Each element of `y` equals the square of the corresponding element of `x` minus `a`
`z=[x, y]`	`z` is the matrix formed by columns of `x` and `y`

The following two documents are good for getting started with MATLAB: *User's Guide* of the *Student Edition of MATLAB*, and *Getting Started with MATLAB*, which is distributed along with MATLAB. For more information, consult the *MATLAB User's Guide*.

Some tips for writing and running a MATLAB program are:

- Take advantage of the help capabilities whenever using the software. MATLAB provides two good online help mechanisms: either from the "Help" menu in the MATLAB window or by typing "help" followed by a word/command/function in question at the MATLAB prompt.
- Run your programs as well as P&R from a special directory on your hard disk. It takes much less time running a program using files on a hard disk than on a floppy disk.
- Start each program file with one or more comment lines explaining what the file is for. This will save you a lot of time later when trying to figure out what the file is for. A comment line is preceded by a % character.
- Before you start to make major modifications in a file that works, save the file under another name.
- Save frequently what you have written when you are writing a program.
- Use matrix operations directly as much as possible and avoid `do`, `for` and/or `while` loops if possible. This often reduces the computational time in MATLAB to a surprising degree.
- Avoid a program with too many lines. Break it down using smaller subroutines.

Index

Page numbers are underlined when they represent the main source of information.